Engineering Materials

For further volumes:
http://www.springer.com/series/4288

Vladislav Boronenkov · Michael Zinigrad
Leopold Leontiev · Edward Pastukhov
Mikhail Shalimov · Sergey Shanchurov

Phase Interaction in the Metal–Oxide Melts–Gas System

The Modeling of Structure, Properties and Processes

Springer

Prof. Vladislav Boronenkov[†]

Prof. Michael Zinigrad
Materials Research Center
Rector of Ariel University Center of Samaria, Ariel
Israel
e-mail: rector@ariel.ac.il

Prof. Leopold Leontiev
Member of Presidium of Russian Academy of Science
Institute of Metallurgy
Ural Branch of RAS
Ekaterinburg
Russia

Prof. Edward Pastukhov
Physical Chemistry of Metallurgical Melts Laboratory
Institute of Metallurgy
Ural Branch of RAS
Russian Academy of Science
Ekaterinburg
Russia

Prof. Mikhail Shalimov
Welding Technology Department
Ural Federal University
Ekaterinburg
Russia

Prof. Sergey Shanchurov
Civil Engineering Management Department
Ural State University of Economics
Ekaterinburg
Russia

[†] Prof. Boronenkov died in 2001.

This book has already been published in Russian by Ural Branch of the Russian Academy of Science Press 2010.
The English edition was edited by Professor Michael Zinigrad.

ISSN 1612-1317 e-ISSN 1868-1212
ISBN 978-3-642-27099-4 ISBN 978-3-642-22377-8 (eBook)
DOI 10.1007/978-3-642-22377-8
Springer Heidelberg Dordrecht London New York

Softcover reprint of the hardcover 1st edition 2012

Cover design: deblik, Berlin

Printed on acid-free paper

Springer is part of Springer Science+Business Media (www.springer.com)

Contents

Abbreviations

a	Anodic
abs	Absorption
ad	Additive, aerodynamic drag
add	Added
ads	Adsorbed
akerm	Akermanite
AP	Aluminum powder
Ast	Austenite-forming
atm	Atmospheric
av	Average
B	Boltzmann, Bohr
b	Boiling, binder
bcc	Body-centered cubic
bm	Base metal
br	Bridge
c	Cathodic
CAD	Computer-aided design
calc	Calculation
CESC	Centrifugal electroslag casting
CFSS	Carbon ferromanganese smelting slag
coat	Coating
col	Column
con	Constriction
cond	Condensation
cr	Critical
cryst	Crystallization
D	Debye
d	Dissociation
dd	Dipole–dipole
deox	Deoxidization
des	Desorption

diff	Different
dis	Dissolution
disch	Discharge
dq	Dipole–quadrupole
dr	Droplet
e	Electrode, electrostatic
EAF	Electric arc furnace
EAS	Electric arc surfacing
EBR	Electron-beam remelting
eff	Effective
eq	Equilibrium, equivalent
ESR	Electroslag remelting
ESS	Electroslag surfacing
EST	Electroslag technology
ESW	Electroslag welding
exc	Excess
exch	Exchange
exp	Experiment, experimental
f	Formation, front
fcc	Face-centered cubic
fl	Flux
fr	Front
Frt	Ferrite-forming
fs	Focal spot
g	Gas
geom	Geometric
hcp	Hexagonal close-packed
in	Incoming
init	Initial
inl	Inleakage
l	Liquid
lim	Limiting
liq	Liquidus
m	Melting
Me	Metal
mix	Mixture
mod	Modifier
MSR	Microslag refining
NI	Nonmetallic inclusion
oct	Octahedral
out	Outgoing
ox	Oxide
p	Particle
PAR	Plasma-arc remelting
pen	Penetration

po	Pool
polym	Polymerization
qu	Quasieutectic
r	Repulsive
RDF	Radial distribution function
reg	Regular
rel	Relative
rem	Removal
res	Residual
s	Solid
sat	Saturated
sk	Skin
sol	Solidus
solid	Slidification
st	Statistical
sur	Surfacing
tetr	Tetrahedral
vap	Vaporization
VAR	Vacuum arc remelting
VCS	Vanadium converter slag
w	Weld, welding

Chapter 1
Introduction

The current trends in the evolution of science and technology are characterized by the development, adoption, and widespread use of diverse models that are based both on the traditional experimental approach and on the use of the latest information technologies. It is already impossible to name a field of human activity in which modeling methods would not be used to some extent. Modeling (in the broad sense) is the main method for investigating complex systems in all areas of knowledge and is also used to obtain scientifically substantiated estimates of their characteristics for adopting solutions in various areas of engineering activity.

A similar approach is undoubtedly also implemented in the study of high-temperature physicochemical processes. Investigations of the interaction processes and properties of the individual components in a metal–slag–gas system are important primarily on the theoretical level for evaluating thermodynamic characteristics and revealing the kinetic laws and mechanisms of both specific chemical reactions and steps in them. At the same time, the results of such investigations are also important in the applied sciences. Obtaining metals and alloys with a required chemical composition and specified levels of service characteristics is related to the occurrence of complex physicochemical processes. The greatest contribution to shaping the products of a melting process is made by the chemical reactions that occur on the various phase boundaries in the metal–slag–gas system. A comprehensive study of their laws is needed to improve existing technological processes, to develop new processes, and to introduce them into practice. However, the search for an efficient technology based on ordinary laboratory and industrial experiments requires large time, financial, and labor resources. The development of methods for predicting and optimizing manufacturing processes based on computational experiments is an economically justified endeavor of current interest.

Prediction of the chemical composition of a metal and a slag, as well as their properties, based on mathematical modeling and simulation of the phase interaction not only opens up additional routes for improving the economic parameters of technological processes, but also allows their optimization in the design stage.

M. Zinigrad et al., *Phase Interaction in the Metal–Oxide Melts–Gas System*,
Engineering Materials, DOI: 10.1007/978-3-642-22377-8_1,

This book presents descriptions of mathematical models that allow prediction of the composition of the phases in various technological processes that are developed on the basis of a comprehensive physicochemical analysis of the reactions. The use of these methods of analysis requires knowledge of the corresponding thermodynamic and kinetic parameters for both the individual electrode processes and the reactions as a whole. For this reason, special attention is focused in this book on studying the thermodynamics and kinetics of the individual steps in complex pyrometallurgical processes involving boron, carbon, sulfur, tungsten, phosphorus, and other elements, and the influence of various factors on them is analyzed.

At the present time, the important conditions for improving, optimizing, and intensifying metallurgical and welding technologies include not only the continuation of the theoretical and experimental research in various fields of science, but also the development of modern methods for simulating processes and technologies. The possibility of simulation (including real-time simulation) appeared relatively recently, but with the rapid development of computer technology, the results of the research in this area have taken on real practical significance.

This is primarily due to the possibility of optimizing processes and technologies already in the design stage, while the conventional empirical methods for developing new technologies do not provide an optimal variant of the solution at all in the overwhelming majority of cases. The results of the simulation of technologies for obtaining metals and alloys, welding processes, and processes for depositing coatings allow the purposeful and cost-effective solution of the problem of obtaining materials with assigned composition, structure, and properties. The solution of the so-called “inverse” problem, i.e., the selection of the composition of the starting materials that ensure that the required result will be obtained is totally impossible by other methods.

This monograph presents an analysis and summarization of the result of theoretical and experimental work carried out during the period from 1965 to 2004 in the Ural State Technical University (Prof. V. Boronenkov, Prof. M. Zinigrad, Prof. M. Shalimov, and Prof. S. Shanchurov), in the Institute of Metallurgy of the Ural Branch of the Russian Academy of Sciences (Prof. L. Leont’ev and E. Pastukhov), and from 1994 in the Ariel University Center of Samaria, Israel (Prof. M. Zinigrad).

Chapter 2 examines general aspects of the development of models of technical objects or processes based on the systems approach. It is devoted to the “technology” of constructing models and to the description of the contents of the stages of this process.

Chapter 3 presents information about the current state of the theory of the constitution and properties of metal and slag melts and demonstrates various approaches to describing and calculating the structural characteristics of melts. The structure of several specific metal and oxide melts is examined.

Chapter 4 is devoted to the simulation of high-temperature processes based on thermodynamic and kinetic methods of analysis. Methods developed by the authors for predicting the equilibrium concentrations and reaction rates in a metal–slag–gas system are examined in detail. The influence of various parameters on the conditions for carrying out the reactions is analyzed. Calculated and experimental data are compared.

Chapter 5 presents mathematical models of several processes for producing and refining steel, which comprise a closed technological chain from the smelting to the casting of the metal.

Chapter 6 is devoted to the development of mathematical models of high-temperature physicochemical processes that occur during welding, surfacing, and the deposition of coatings. The physical models of the interactions and the assumptions and constraints adopted are examined in detail.

Chapter 7 examines methods for predicting the structure, phase composition, and properties of the metal that crystallizes under specific welding and surfacing conditions. Their application to several specific welding and surfacing variants is demonstrated. Data from experiments and calculations are compared.

The last chapter demonstrates the use of the models developed for perfecting high-temperature technologies for smelting and treating steels.

Chapter 2
Systems Analysis: The Strategy of Modeling

Modeling is one of the most widely used methods for studying various systems. It provides a way to reveal qualitative and quantitative interrelationships between various parameters of the object of investigation.

Modeling begins with defining a subject of investigation, i.e., a system of concepts that reflects the characteristics of the object that are significant for modeling. The systems approach is generally used in the analysis and synthesis of large systems. Systems analysis is the strategy for studying complex systems that include objects and processes of diverse physical nature, whose functioning is described by a large amount of information. The following principal steps underlie systems analysis [1–3]:

- clear formulation of the goal of the investigation;
- statement of the problem in terms of achieving the goal and defining an efficiency criterion for solving the problem;
- development of a detailed research plan with indication of the principal steps and directions for solving the problem;
- systematic advancement along the entire set of interrelated steps of the research plan;
- organization of successive approximations and repeated cycles in the individual steps;
- application of the principle of a descending analysis hierarchy and an ascending synthesis hierarchy in solving the component particular problems.

A basic method of investigation used in systems analysis is mathematical modeling, and a basic principle is the decomposition of a complex system into simpler subsystems, i.e., blocks. Accordingly, models are constructed by the building block principle: the overall model is subdivided into blocks, to which a comparatively simple mathematical description can be given [1–3]. Application of the decomposition technique to the analysis and mathematical modeling of large

M. Zinigrad et al., *Phase Interaction in the Metal–Oxide Melts–Gas System*,
Engineering Materials, DOI: 10.1007/978-3-642-22377-8_2,

technical systems, which are distinguished by a high degree of complexity and a multilevel structure, is especially effective [4–8].

Systems analysis is based on fundamental principles, whose fulfillment ensures the successful modeling and simulation of complex systems of diverse physical nature and information structure [9, 10]:

- the principle of a final goal—the absolute priority of the final goal;
- the principle of unity—combined consideration of the system as a whole and as a collection of interrelated parts;
- the principle of connectivity—considering any part of the system together with its connections to other parts and the surrounding environment;
- the principle of modularity—demarcating modules (blocks) in the system and consideration of the system as the complete collection of these modules;
- the principle of hierarchy—establishing the subordination of the modules;
- the principle of evolution—taking into account the variability of the system and its ability to evolve and store information;
- the principle of decentralization—combining the methods of centralization and decentralization in the solutions employed;
- the principle of indefiniteness—taking into account the indefiniteness and randomness in the system.

All of these principles are used in nearly any application of the systems approach. They have a very high degree of generality, but the existence of conditions under which any of the principles will be insignificant is possible in each particular case. For example, there may be no hierarchy in a system, which can be considered completely defined, etc. The practical implementation of these principles in the modeling of specific complex systems ensures the effectiveness of the models developed and optimization of the processes in the system. The use of these principles will be illustrated below in the modeling of objects and processes.

2.1 Methodology of Modeling

In systems theory a model is an approximate description of a system that reflects its specified properties, which are generally fundamental properties with respect to the modeling goal. The term model refers to a material or mentally represented object, which replaces the original object during a study, but maintains some typical features of it that are important for the particular investigation. A well constructed model is generally more accessible, informative, and convenient for the investigator than the real object. Investigations performed on models enable investigators to do the following:

- to understand how a specific object is organized and what its structure, fundamental properties, laws of evolution and self-evolution, and interactions with the surrounding environment are;

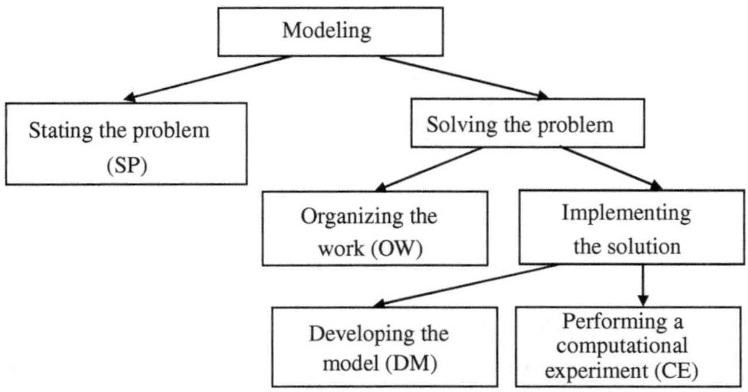

Fig. 2.1 Block diagram of the modeling process

Fig. 2.2 Iterative nature of the modeling process

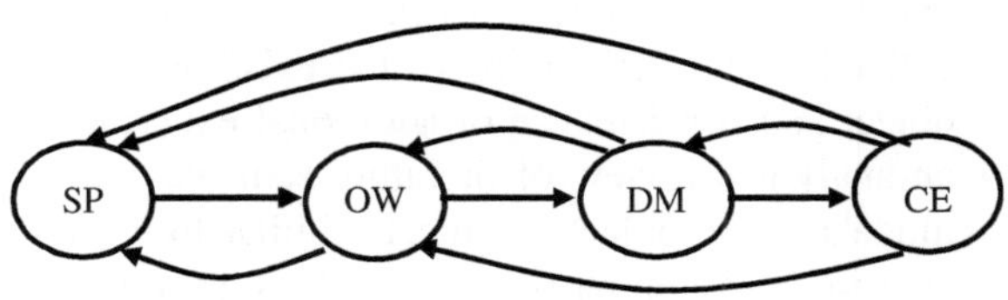

- to learn to control an object or process and to determine the optimal operating conditions for assigned goals and criteria;
- to predict direct and indirect consequences of the realization of assigned modes and forms of action on an object.

The methodology for creating models of complex systems can be represented in the form of a block diagram that reflects the sequence of the main stages of the modeling process (Fig. 2.1).

It should be noted that the modeling process has an iterative character, as a result of which the model that most faithfully describes the system under study is created. It is manifested in the continuous correction and modification of the model, which provide a way to obtain its optimal variant (Fig. 2.2).

We will consider each of the stages in the development of a model that are shown in Fig. 2.2 in greater detail.

2.2 Stating the Modeling Problem

The main goal of this stage is preliminary investigation of the modeling object and formulation of a meaningful (descriptive) statement of the modeling problem, which is generally not final and can be refined and made more specific during the development of the model. If the modeling object is a technological process or a

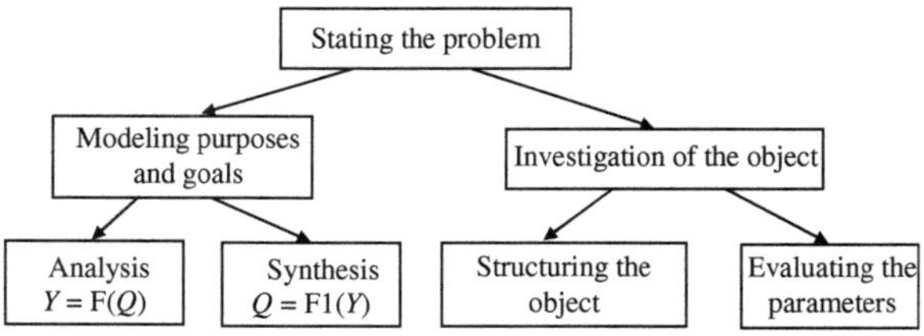

Fig. 2.3 Block diagram of the stage of stating the modeling problem

technical object, the meaningful statement of the modeling problem is very often called the technical statement of the problem. This stage includes the following operations:

- thorough inspection of the modeling object itself for the purpose of revealing the main factors and mechanisms that determine its behavior, as well as the parameters that allow description of the object being modeled;
- collection and verification of the existing experimental data for analogs of the object and performance of additional experiments when necessary;
- an analytical review of literature sources, analysis and comparison of existing models of the object or objects similar to it with one another;
- analysis and summarization of all the accumulated material.

We represent this stage of the modeling process in the form of a block diagram (Fig. 2.3). In the general case three principal modeling problems of complex systems can be distinguished: research, design, and control. From the theoretical point of view, research problems can involve obtaining new scientific knowledge, and from the applied point of view, they can involve optimization of technological processes. Designing refers to the creation of artificial man-made objects that have desirable properties. The control of processes or objects is possible, in particular, in the form of regulation and optimization. Depending on the problems posed, modeling goals that are uniquely related to data for performing research, design, or control are formulated.

Analysis and synthesis can be included among the basic goals of modeling.

Analysis is the calculation of the output parameters of a process from assigned values of input parameters. In other words, in analysis problems an evaluation is made according to the "What will happen if …?" principle.

Synthesis is the determination of the values of the input parameters of a process for which assigned values of the output parameters are achieved. In synthesis problems the question "What is needed in order for …?" is resolved. In this case the problem of determining the best operating conditions of a system can be solved, i.e., the modeling goal is optimization of the process.

A comparison of the problems and goals of modeling enables us to state that there is a correspondence between them. In particular, analysis is employed in research problems, synthesis is employed in design problems, and analysis and synthesis are employed together in control problems (Fig. 2.4). Modeling problems can be described in the form of target functions:

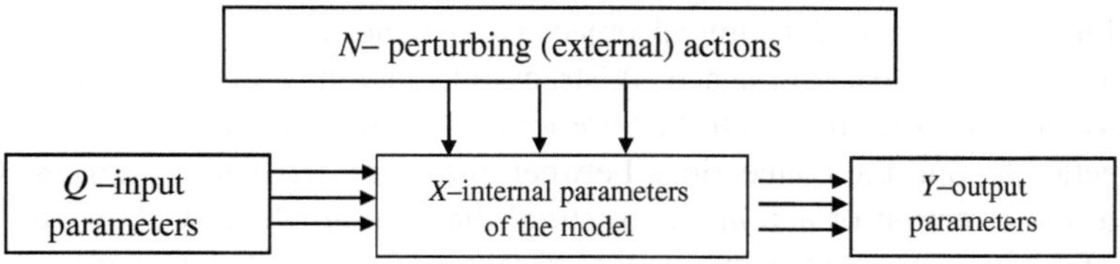

Fig. 2.4 Diagram of the purpose of modeling problems

- research: $Y = \mathrm{F}(Q, N, X)$;
- control: $Q = \mathrm{F1}(Y, N, X)$;
- design: $X = \mathrm{F2}(Q, Y, N)$;
- design with optimization of the parameters: $Q, X = \mathrm{F3}(Y, N)$.

When the properties of the object and resources available to the investigator are taken into account, the goals can be refined and corrected.

To achieve practical modeling goals, a need to consider the object under investigation not as something whole, but as a collection of separate interrelated elements often arises. In some cases this is specified by the complexity of the object of investigation itself, and in other cases it is due to the absence of necessary information about the object or appropriate mathematical machinery. The process of dividing a system into elements with indication of the connections between them is called structuring or decomposition. The division process can have a material, functional, algorithmic, or other basis. As a result of decomposition, the original system breaks down into subsystems, and the problem breaks down into subproblems that have known solutions or can be solved using approved methods. However, it is important not just to divide the whole into separate elements, but also to combine them in such a manner that they would again form a single whole. Therefore, analysis and synthesis methods are widely employed in this stage of the work. Analysis methods are used to divide the object under consideration into separate elements and study their properties. During decomposition, at least two criteria, namely completeness and simplicity, must be taken into account. Of course, some compromise between completeness and simplicity should be made during decomposition. This can be achieved if only elements that are significant with respect to the modeling goal are included in the structure of the system. The problem can be simplified by diminishing the inputs and outputs of each element in comparison to the original state. Accordingly, separate properties of the system as a whole and/or properties of its separate elements can be lost during analysis. Synthesis supplements analysis, since the correctness of the decomposition process can be evaluated only after all the elements have been added to the system. The synthesis process provides a way to join the elements to one another and to establish the relationship between them, i.e., to obtain a structural diagram of the object of investigation. The stable ordering of the system's elements and connections in space is called its structure. For example, a technological process is characterized by a definite sequence of technological operations in time, and a technical object is characterized by a definite arrangement of nodes and parts in space. A structural diagram represents the elements (blocks) and the order in which they are joined. Each element (block) is depicted by a rectangle, and the connections between them are depicted by arrows, which indicate the direction of action. A structural diagram graphically shows the existence of phenomena that occur consecutively and in parallel and closed loops caused by the presence of feedback in the system. The structure of the interaction of the phenomena allows well-founded selection of the ones that it would be expedient to include in the model. For example, phenomena that form a consecutive chain of connections between input and output parameters must be included in the model. The parallel phenomena whose contribution to the overall effect is small may be disregarded. Reverse connections can be disregarded if their strength is insignificant. Thus, rough quantitative estimates of the contributions of phenomena to the final result of the process are useful, and often necessary, for selecting the significant phenomena. The required accuracies of the modeling result and of the original data play an important role in selecting the phenomena that are taken into account in a model of a process. Obviously, a phenomenon can be disregarded only in cases where its contribution to the final result is smaller than the required modeling accuracy or smaller than the error caused by the inaccuracy of the original data.

In the general case parameters which describe the state and behavior of a modeling object can be divided into the following groups (see Fig. 2.4):

- the collection of input (controllable) actions on the object (Q);
- the collection of actions of the external environment (uncontrollable) (N);
- the collection of internal (intrinsic) parameters of the object (X);
- the collection of output characteristics (Y).

A description of each parameter and variable is given in the following form:

- definition and brief characterization;
- notation symbol and unit of measure;
- range of variation and accuracy of assignment;
- place where it is used in the model.

The number of parameters of all types in models is finite. Each of the parameters can have a different "mathematical" nature: it can be a constant or a function, a scalar or a vector, a crisp or fuzzy set, etc. Sometimes parameters can be assigned in the form of a qualitative characteristic. In the general case the following types of factors and interactions between them can be identified within a model [11, 12]:

- variables—factors that take arbitrary values, on which the behavior of the system or its state depends;
- parameters—factors that are taken into account in the model and can be set at fixed levels or according to assigned dependences;
- functional relations—equations that describe the interdependence between variables and parameters when the system functions;

- constraints—established limits of variation of parameter values or conditions that ensure a normal regime for the occurrence of processes in the system;
- target functions—parameters for optimization of the operation of the system, which must take extremum values to achieve the most efficient operating conditions of the system with consideration of the possible constraints, which depend both on the properties of the system and on its interaction with other systems and the environment.

To successfully construct a model, all of its parameters and variables must be considered, and their influence on the functioning process of the system as a whole must be evaluated. Isolating the most important factors that significantly influence the occurrence of the processes being modeled is important. The absence of such factors in a model can be the cause of incorrect solutions that are adopted on the basis of modeling [5, 13]. In addition, in this case the models are generally unfaithful [14, 15]. It is useful to divide all the factors which can be included in a model of the system under investigation into the following groups [16, 17]:

- essential factors, without whose consideration modeling is impossible;
- principal factors, which play an important role in modeling, but the system does not lose its basic qualities and features in their absence;
- additional factors, whose role in the process or object being modeled is small. They can be called secondary because their variation has practically no influence on the behavior of the system under investigation.

The comparative significance of the factors for any complex process can be evaluated from a quantitative comparison of the criteria obtained by the methods of similarity theory [18, 19].

2.3 Organizing the Work

In this stage of the work, specific approaches to creating a model of the system under investigation are selected on the basis of an analysis of the functional possibilities of different models. The sequence of operations can be described by the following block diagram (Fig. 2.5).

The principal types of material modeling are physical and analog. Both types of material modeling are based on the properties of geometric or physical similarity. For example, two geometric figures are similar if the ratios between all the corresponding lengths and angles are identical. If the similarity factor, i.e., the scale, is known, simple multiplication of the dimensions of one figure by the scale gives the dimensions of another geometric figure that is similar to it. Two phenomena are physically similar if specified characteristics of one can be used to obtain the characteristics of the other by a simple mathematical conversion, which is similar to a conversion from one system of units of measure to another. Similarity theory deals with the study of the conditions for similarity between phenomena. There is

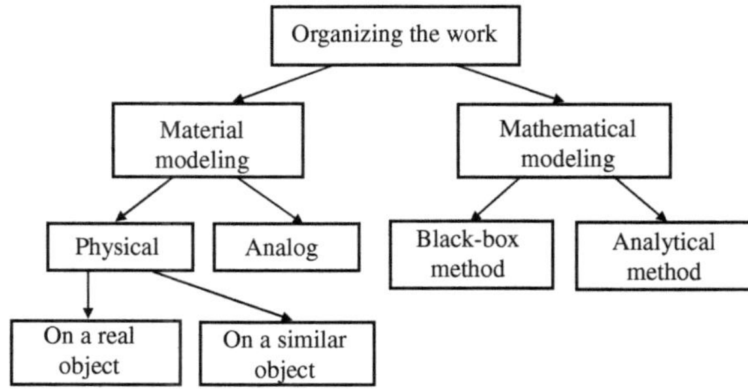

Fig. 2.5 Block diagram of the stage of organizing the work

presently a whole system of dimensionless criteria sets that were obtained by the methods of dimensional analysis [18–20]. They can be used to analyze the features of numerous processes that occur in technological systems. This refers mainly to heat and mass transfer processes. The most important and widely used similarity criteria are listed in Table 2.1.

The possibility of investigating different characteristics on a real object as a whole or on a part of it is also utilized in material modeling and simulation.

The performance of investigations on a real object is called full-scale modeling. The experimental results are also treated using similarity theory. When the object functions in accordance with the stated goal, the laws governing the course of a real process can be revealed. Types of full-scale experiments, such as production and system tests, have a high degree of reliability. A full-scale experiment in the form of material modeling is characterized by the extensive use of automation tools, the employment of diverse information processing tools, and the possibility of human intervention during its performance.

A physical model is a system, apparatus, or device that reproduces a full-scale object on some scale with maintenance of the physical (dynamic) similarity between the processes in the model and in the real system. Physical modeling is a form of modeling, in which an enlarged or reduced material analog is matched to the real object. The analog permits investigation, generally under laboratory conditions, followed by transference of the properties of the processes and phenomena studied from the model to the real object using similarity theory. Physical modeling generally provides a way to obtain reliable results for fairly simple systems.

Analog modeling is based on finding a matching mathematical description for qualitatively different objects [5, 21–24]. Analog modeling is a form of modeling that is based on the similarity of processes and phenomena of different physical nature that can be described in a similar manner by mathematical relations and by logical and structural diagrams (Table 2.2).

Models of the physical and analog types are material reflections of the real object and are closely related to it by their geometric, physical, and strength characteristics. In fact, the process of investigating models of these types reduces

Table 2.1 Principal similarity criteria from dimensional analysis that are used in the investigation of heat and mass transfer processes [18]

Name	Symbol	Formula	Physical meaning
Fourier number	Fo	$\mathrm{Fo} = \frac{at_0}{L_2}$	Ratio between the rate of variation of the conditions in the surrounding medium and the rate of restructuring of the temperature field in a body
Biot number	Bi	$\mathrm{Bi} = \frac{aL}{\lambda}$	Ratio of the temperature drop in a body to the thermal head between the medium and the body
Reynolds number	Re	$\mathrm{Re} = \frac{VL}{v}$	Ratio of the inertial force to the viscous force
Froude number	Fr	$\mathrm{Fr} = \frac{V^2}{gL}$	Ratio of the inertial force to the gravitational force
Euler number	Eu	$\mathrm{Eu} = \frac{\Delta p}{\rho V^2}$	Ratio of the pressure drop to the dynamic head
Péclet number	Pe	$\mathrm{Pe} = \frac{VL}{a}$	Ratio of the convective heat transfer rate to the conductive heat transfer rate
Prandtl number	Pr	$\mathrm{Pr} = \frac{v}{a}$	Ratio of the temperature change in a medium upon passage of a heat flux to the velocity change in the medium upon passage of a momentum flux
Nusselt number	Nu	$\mathrm{Nu} = \frac{\alpha L}{\lambda}$	Ratio of the convective heat transfer rate across a surface to the conductive heat transfer rate across the surface
Mach number	M	$\mathrm{M} = \frac{V}{a}$	Ratio of the flow velocity of a gas to the speed of sound in it.
Bond number	Bo	$\mathrm{Bo} = \frac{\rho g L^2}{\sigma}$	Ratio between the capillary and gravitational forces acting in a system
Weber number	We	$\mathrm{We} = \frac{\sigma}{g(\rho_1 - \rho_2)L^2}$	Measure of the pressure created by interfacial tension forces upon contact between phases
Marangoni number	Ma	$\mathrm{Ma} = \frac{L\Delta\sigma}{\mu a}$	Ratio between the thermocapillary forces and the viscous forces

Note Here L is the characteristic dimension; t is the characteristic time; V is the characteristic velocity; g is the acceleration of gravity; ρ and σ are the density and surface tension of the liquid phase; α, γ, and v are the thermal diffusivity, thermal conductivity, and kinematic viscosity of the material, respectively

Table 2.2 Similarity between fields of different nature in continuous media [1]

Physical process	Process parameter	Coefficient	Relationship between quantities
Heat conduction	Temperature (T)	Thermal conductivity (λ)	Fourier's law (heat flux density) $q = -\lambda$ grad T
Diffusion	Concentration (C)	Diffusion (D)	Fick's law (flux density of matter) $q = -D$ grad C
Direct-current electric field	Potential (U)	Conductivity (G)	Ohm's law (current density) $j = -G$ grad U
Strain (elastic)	Displacement (x)	Elastic modulus (E)	Hooke's law (normal mechanical stress) $\sigma = E$ grad x
Viscous flow of a liquid or gas	Velocity of layers (V)	Dynamic viscosity (μ)	Newton's law (shear stress) $\tau = -\mu$ grad V

to performing a series of full-scale experiments, in which a physical or analog model is used instead of the real object.

From the methodological point of view, two principal approaches to the mathematical modeling of complex systems can be singled out: the black-box method and the analytical method [25–27].

The black-box method is used in cases where the internal structure of the system is unknown or is not of interest to the investigator. The structure of the object of investigation is ignored if its state is characterized only by input parameters, output parameters, and perturbing actions and there is no information about the internal structure of the object. A mathematical model is constructed by establishing a relationship between the input and output parameters through an investigation of the response of the object of investigation to external actions. The methods of experiment planning, as well as variance, regression, and correlation analysis, are widely used for this purpose. The main merits of the black-box method include its simplicity, the highly developed level of its mathematical machinery, and the ensured quality of the modeling result. The shortcomings of this method are associated mainly with the small amount of information that can be derived from the models obtained and the impossibility of evaluating the true causes of the phenomena occurring in the system studied.

Under the analytical approach a model is constructed on the basis of a study of the internal structure of the phenomena occurring in the system. In this case the input and output parameters of the model are related to one another using fundamental physical and physicochemical laws that take into account the internal structure of the modeling object. The mathematical description of the process obtained has a high information content, a broad area of application, and universality. The use of such a model allows understanding the course of the process, and it can easily be combined with optimization procedures. The analytical approach greatly reduces the volume of experiments and improves the efficiency of modeling.

2.4 Developing the Model

This stage is the most critical, because the process studied is simplified specifically here in accordance with the requirements for accuracy, completeness, and information content that are imposed on the models. In the general case there can be several models of a process, which are intended for different goals and which describe the process from different standpoints. The requirements imposed on these models will also be different. The structure of this stage can be represented by the following diagram (Fig. 2.6).

The importance of this stage of the modeling process is confirmed by the need for constant monitoring of its results, which are obtained when each of the procedures presented is performed. This is functionally ensured by iterations during development of the model (Fig. 2.7).

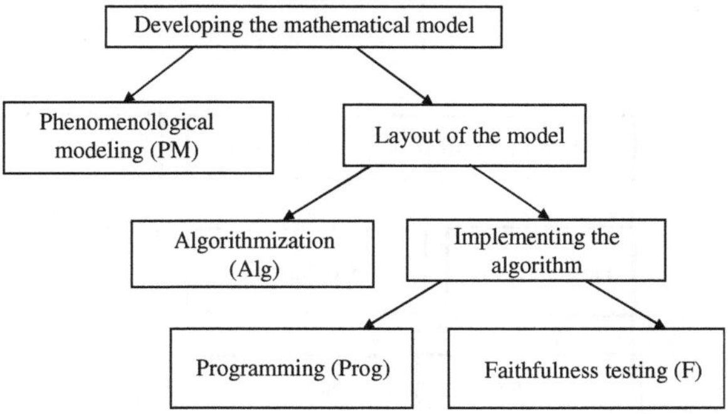

Fig. 2.6 Structural diagram of the stage of developing the mathematical model

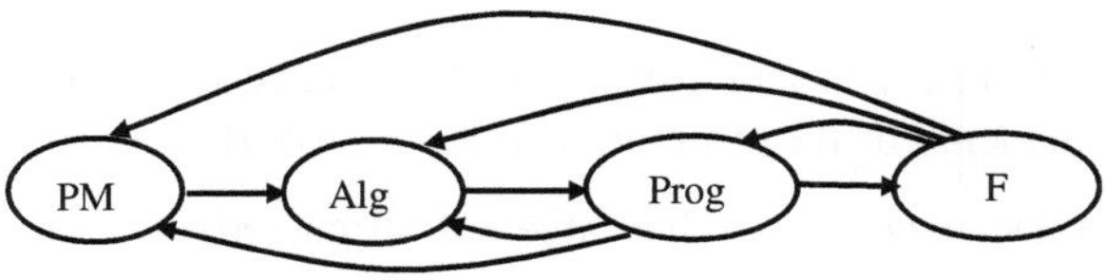

Fig. 2.7 Functional diagram of the stage of developing the mathematical model

We will consider each of the stages singled out on the functional diagram (Fig. 2.7) in detail.

2.4.1 Phenomenological Modeling

Phenomenological modeling is carried out on the basis of the existing information about the modeling object and the requirements placed on the model being created. A phenomenological model usually contains information about the nature and parameters of elementary phenomena in the system under investigation, about the type and extent of the interaction between them, and about the place and significance of each elementary phenomenon in the overall functioning process of the system [28]. Among all the modeling stages, creation of the phenomenological model has been formalized to the smallest extent. However, a highly general approach to the implementation of this stage of the modeling process can be pointed out (Fig. 2.8).

Phenomenological modeling is based on the results of the preceding stages, in which preliminary structuring of the system was performed. When models of real objects and phenomena are constructed, it is generally necessary to deal with a lack of sufficient information. For any object investigated the initial state, the actions, and the distribution of properties associated with them are generally known with some degree of uncertainty. This is due to the limited nature of the number of parameters used in the model, the specific accuracy of the experimental data, and the existence of several factors that are difficult to take into account.

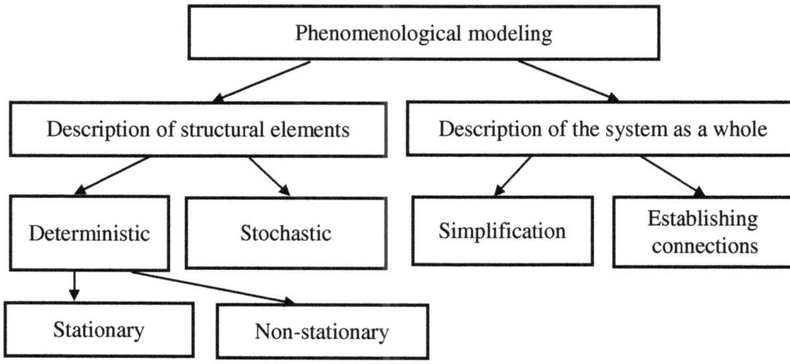

Fig. 2.8 Structural diagram of the phenomenological modeling stage

During construction of a model, the uncertainty of the parameters and individual elements of the structure can be described most simply in the following manner.

- Deterministic—The values of all or certain parameters of the model are specified by deterministic quantities, and a specific integer, a real number or complex number, or an appropriate function corresponds to each parameter. This situation corresponds to complete certainty of the structural units of the system.
- Stochastic—The values of all or certain parameters of the model are set by random quantities, which are assigned by probability densities. Cases of normal (Gaussian) and exponential distributions of the random quantities have been investigated most thoroughly. In certain cases the probability density is determined by treating a restricted experimental sampling of data, which can have an effect on the modeling results.

When the analytical approach is used, the mathematical model generally reduces to a system of differential, integrodifferential, or integral equations of various types. The solution of a given system can uniquely specify the state of the object being modeled and its output parameters at a given time. This uniqueness of the solution also characterizes the deterministic nature of the model. At the same time, random deviations of the state parameters, inputs, and outputs are inherent to any real process. However, these deviations are still insignificant and can be disregarded, and the model itself can be considered deterministic.

The stochastic character of a model is associated with the presence of factors that are not monitored and are simultaneously significant both at the inlet of the system and within it. It is these factors that introduce uncertainty into the modeling results.

The division of parameters into stationary and non-stationary parameters is used for models when one of the independent arguments is time or there is a characteristic that determines the direction of the process or the sequence of its steps.

Stationary models are generally used to describe various flows (liquid, gas, heat) in the case of constancy of the conditions at the flow inlet and outlet. Processes for which the state of the object at each fixed point in space does not vary with the passage of time are called stationary. For such processes time can be eliminated from the list of independent variables. If time or its analog must be employed as one of the significant independent variables of a model, the model is called non-stationary. Such models are usually considerably more complex than stationary models and require large time expenditures for computer realization. In stationary models the input and output parameters, as well as the state parameters of the system, can vary with time, but the functional relationship between them always remains unchanged. The response of a non-stationary system depends both on the current time and on the time of application of the input action. In the case of a shift of the input signal with time without any change in its shape, not only are the output parameters shifted with time, but their shape is also altered. Static models are a special case of stationary models. Their distinguishing feature is the absence of a dependence of all the parameters of the system on the time.

In this stage the level of detail of the phenomenological model at which the model is represented in the form of separate, relatively self-contained subsystems is determined. Naturally, the results obtained during structuring of the modeling object must be taken into account here. The necessary connections between the subsystems identified are revealed. Then each of the subsystems is analyzed separately. Special attention is focused on the conditions for terminating the decomposition of the model. In the general case two such conditions can be singled out.

- The subsystems identified can be described by fundamental laws of physics, chemistry, or other disciplines or by conservation laws.
- Further decomposition is impossible because the necessary data or information is not available.

Thus, in the general case the mathematical model of a system is a multilevel formation consisting of interacting elements that are combined to form the subsystems of the different levels. The result is the creation of a hierarchical sequence of models, each of which reflects the behavior of the system both on a different level of detail and in different physicochemical processes that occur in the system (thermal, hydrodynamic, mass transfer, and other processes). Each subsystem (element) identified has its own characteristic interactions with the other subsystems (elements) and with the external environment [16, 17, 25, 26]. The mathematical model of the system must consist of mathematical models of the separate elements and mathematical models of the interactions between the elements and the external environment. Of course, each element of the mathematical model can have mathematical description with a different level of detail. It is important that the input and output parameters of all the elements of the model would be mutually consistent to ensure that a closed system of equations would be obtained for the mathematical model of the process as a whole. As has already been noted, it is not possible to obtain a description of all the elements and their interactions solely on the basis of fundamental laws. Therefore, for the practical completion of this stage at some level of detail, it is necessary to employ empirical relations and to use definite simplifications. Among the most frequently used methods for simplifying a system, the following can be singled out.

- Stationary (steady-state) processes are analyzed instead of non-stationary processes.
- Random parameters that have a small spread of values are assigned using their mean deterministic values.
- Certain parameters take constant values during the entire modeling process.
- Conditions under which the parameters take specified values are assigned.
- A type of distribution specified by empirical data regarding the behavior of the system under different conditions is assigned for stochastic variables.

Final establishment of the connections between the elements of the system implements the structuring of the phenomenological model of the system under investigation.

In conclusion, we present a recommended sequence of steps for phenomenological modeling:

- analyze the technical and economic reasons behind the statement of the problem;
- establish quantitative criteria for achieving the goal;
- establish and analyze cause-and-effect relationships between the principal physicochemical phenomena and build a structural model of the process;
- analyze the significance of the phenomena;
- define the modeling goal.

2.4.2 Algorithmization

In the case under consideration, algorithmization refers to defining a sequence of mathematical and logical operations that must be performed to obtain a result. This stage of the modeling process can be represented in the form of the following structural diagram (Fig. 2.9).

A diagram is a convenient form for representing the logical structure of models of functioning processes of systems. The logical diagram of a modeling algorithm indicates the time-ordered sequence of logical operations that are associated with the solution of a modeling problem. In the general case three methods for constructing algorithms are distinguished: the verbal, tabular, and graphical (block diagrams and graph diagrams) methods. Generalized and detailed diagrams of modeling algorithms are devised in the various stages of modeling [29, 30].

A generalized (consolidated) diagram of a modeling algorithm specifies the general order of the operations for modeling a system without any refining details. It shows what must be done in each successive step of modeling.

A detailed diagram of a modeling algorithm contains refinements that are lacking in the generalized diagram. A detailed diagram shows what must be done in each successive step of modeling and how it should be done.

The set of symbols defined in ISO 5807:1985 (State Standard 19.701-90) "Information processing—Documentation symbols and conventions for data,

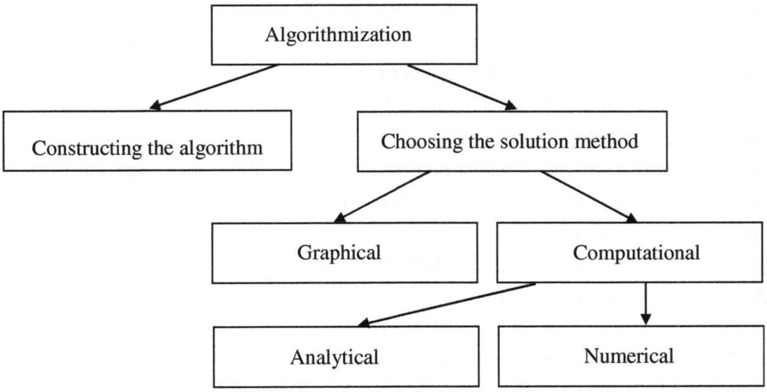

Fig. 2.9 Structural diagram of the algorithmization stage

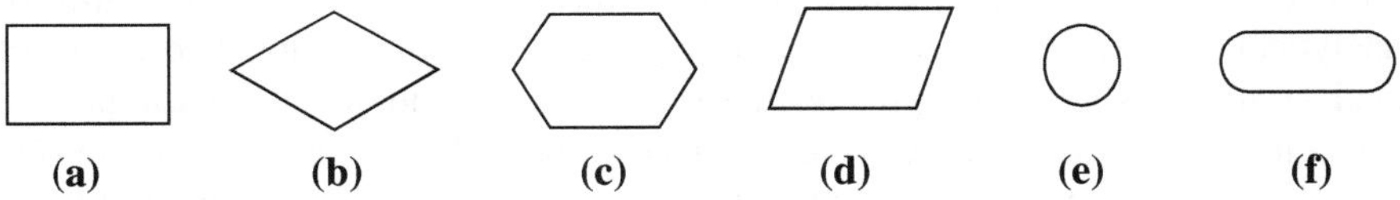

Fig. 2.10 Symbols used in modeling algorithms

program and system flowcharts, program network charts and system resources charts" are used to draw these diagrams. Some of the most commonly used symbols in the practice of modeling are shown in Fig. 2.10.

They include basic (*a*), specific (*b–d*), and special (*e* and *f*) symbols:

- *a*—Process—This symbol represents any kind of data processing function.
- *b*—Decision—This symbol represents a decision or a function that has one input and several alternative outputs, only one of which can be activated after calculation of the conditions specified within this symbol.
- *c*—Preparation—This symbol represents modification of a command or a group of commands for the purpose of influencing a certain subsequent function.
- *d*—Manual operation—This symbol represents any process performed by a human.
- *e*—Connector—This symbol represents an exit from one part of the flowchart and entry to another part of the same flowchart and contains the same unique identification.
- *f*—Terminator—This symbol represents an exit to the external environment or an entry from the external environment, or the start or end of the algorithm flowchart.

To use the mathematical models developed, the dependence of certain parameters of the modeling object that were unknown *a priori* and satisfy a certain system of equations must be found. Thus, the search for a solution to a problem reduces to finding certain dependences of the quantities sought on the initial parameters of the model. All the methods for solving the corresponding problems that make up the "core" of mathematical models can be subdivided into graphical, analytical, and numerical methods.

Graphical methods are used least frequently. They give satisfactory results when experimental data are treated and relatively simple equations are solved. Graphical methods can also be used for differentiation and integration.

A method for investigating models is classified as an analytical method if it provides a way to obtain output parameters in the form of analytical expressions, i.e., expressions in which a set of arithmetic operations is used. Algebraic expressions that employ a finite number of arithmetic operations, operations that involve raising a number to an integer power, and root extraction are a special case of analytical expressions. Analytical methods for investigating models are more effective, because they provide a way to study the properties of the modeling object with little computational resources by employing conventional, well developed mathematical methods for the analysis of analytic functions. It is significant that the application of analytical methods is possible without using a computer. In addition, knowledge of analytical expressions for the parameters sought makes it possible to investigate the fundamental properties of the object and its qualitative behavior and to devise new hypotheses regarding its internal structure. It should be noted that the possibilities of analytical methods depend on the level of development of the corresponding branches of mathematics. The currently existing mathematical models provide a way to obtain analytical solutions only for relatively simple mathematical models in a narrow range of parameter values.

In most cases numerical methods, which provide a way to obtain only approximate values of the parameters sought, are used to investigate models. The replacement of real relationships by discrete relationships is common in numerical methods. It is generally achieved by switching from a function of a continuous argument to functions of a discrete argument. The solution of the discrete problem obtained is taken as an approximate solution of the original mathematical problem. When numerical methods are used, a computational algorithm that provides a solution of the discrete problem after a finite number of steps must be developed. Algorithms that require a smaller number of operations to achieve the same accuracy are called *economic* or *efficient*. The *accuracy* of an algorithm refers to the possibility of solving the original problem with an assigned error after a finite number of operations. The magnitude of the error can vary from operation to operation. If the error increases without bound during the calculations, the algorithm is called *unstable* or *divergent*. Otherwise, the algorithm is called *stable* or *convergent*. Three principal components of the resultant error can be distinguished.

- The *unavoidable error* is associated with the inaccurate assignment of the original data of the problem.
- The *method error* is associated with an improper transition to the discrete analog of the original problem.

- The *rounding error* is associated with the finite number of digits in the numbers represented in a computer.

Numerical methods are applicable only to proper mathematical problems. This places a significant restriction on the use of these methods in mathematical modeling. A problem is called proper if it has a unique and stable solution for any values of the original data.

The following numerical methods are used most widely to solve differential equations:

- the finite-difference method (FDM);
- the finite-element method (FEM);
- iterative methods.

2.4.3 Programming

The next stage is the development of computer programs that implement each of the mathematical models of the process on the basis of a program flowchart. A program flowchart represents an interpretation of the logical diagram of a modeling algorithm based on a specific algorithmic language. The program flowchart reflects the logic of the computer implementation of the model using specific modeling software. A set of standard symbols is also used to represent the program flowchart. Before the programming is performed, the program flowchart is verified. For this purpose, each operation represented in the program flowchart and the corresponding operations in the logical diagram of the model are tested. Sometimes control experiments are performed for this purpose. The process of developing reliable and efficient software is no less complicated than the development of the preceding stages in the creation of a mathematical model. Successful resolution of this question is possible only after the modern algorithmic languages, programming technologies, and available software have been thoroughly mastered and knowledge of the technological possibilities and features of the computer implementations of the methods of computational mathematics and experience in solving similar problems have been gained. Modern programming is an independent science with its own fundamental principles, approaches, and methods. Therefore, specialists in this area, i.e., programmers, are enlisted to implement this stage of the work. For this reason, we will dwell only on the general approaches to the implementation of a mathematical model in the form of a computer program.

Most programs that implement mathematical models consist of three parts:

- a preprocessor for preparing and testing of the original data;
- a processor for solving the problem and implementing the computational experiment;
- a postprocessor for displaying the results obtained.

These three component parts can be implemented in the form of a single program only for relatively simple cases. To create modern models of the behavior of liquids, gases, and solids, each of the parts indicated can include an entire set of programs. Preprocessors and postprocessors have special value in modern computer-assisted design (CAD) systems, and they significantly shorten the time needed to retrieve data and evaluate modeling results. A program is more reliable and is created in less time when standard software elements, i.e., packages of applied programs and databases, are used to the maximum extent.

2.4.4 Testing the Faithfulness of a Model

The faithfulness of a mathematical model usually refers to the degree of correspondence of the results obtained using it to the experimental data or a test problem, as well as its ability to satisfy the goals for which it was constructed. Before proceeding to test the faithfulness of a model, convincing evidence of the correct combined functioning of all of its algorithms and programs is needed, i.e., debugging and testing of the programs written are performed. Syntactic, semantic, algorithmic, and other errors in the programs are thereby revealed and eliminated, and the speed and reliability of the computational algorithm chosen are evaluated.

Testing the faithfulness of a model has two goals:

- to ascertain the validity of the collection of hypotheses that were formulated in the preceding stages of the work;
- to ascertain that the accuracy of the results obtained corresponds to the specified accuracy.

The answer to the question regarding the accuracy of modeling depends on the requirements placed on the model and on its purpose. The accuracy of the experimental results or the special features of the statements of the test problems must be taken into account here. For models that are intended for performing estimated calculations, an accuracy of 10–15% is considered satisfactory. For models used in controlling and monitoring systems, the required accuracy must be no less than 1–2%.

The qualitative agreement and quantitative agreement of comparison results are generally distinguished. In a qualitative comparison only the agreement of certain characteristic features in the distribution of the parameters investigated is required. In fact, in a qualitative comparison the agreement of the form of the distribution function of the parameters is evaluated to determine whether it is a decreasing or increasing function, whether it has one or several extrema, etc. A quantitative comparison is made only after there is satisfactory qualitative agreement.

In this stage several aspects of the faithfulness of the model, viz., its consistency, stability, and realism, are tested [6, 27].

The first aspect of faithfulness is the consistency of the model with common sense. First of all, it is necessary to determine the consistency of the model with

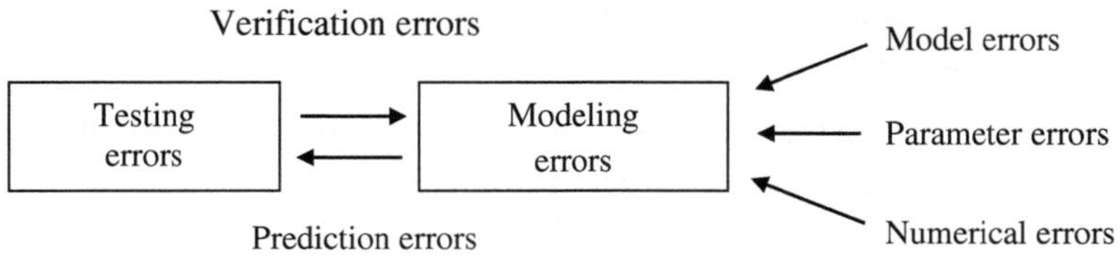

Fig. 2.11 Structure of the errors in the computer simulation and testing of the faithfulness of a model

fundamental physical laws and its compatibility with the action of the most important laws of mechanics and thermodynamics. The model must not contradict the general conceptions regarding the process that were derived from its experimental study. If simulation results confirm the consistency of the model in all the situations indicated, it can be stated that there is a possibility to compare different design solutions using it.

To test for stability (robustness), the sensitivity of the output parameters of the model to small changes in the input parameters within the range of applicability of the model is analyzed. The ability of the model to adjust itself to the influence of these deviations on simulation results is evaluated.

The third aspect of testing for faithfulness is evaluation of the realism of the model, i.e., the correspondence of simulation results with the required accuracy to special cases, for which there are actual data. Testing for realism requires setting up and performing real experiments.

When problems associated with the faithfulness of a model arise, the correction process should begin with a systematic analysis of all possible causes of a disparity between the simulation and experimental results. First, the model should be investigated, and its degree of faithfulness with different values of the variable parameters should be evaluated. If the model is unfaithful in a range of parameters that is interesting to the investigator, an attempt should be made to refine the values of the constants and initial parameters of the model. If this does not produce positive results, the system of hypotheses adopted must be modified. This solution actually means going back to the initial stages of the model development process. Accordingly, it can involve not only profound changes in the mathematical statement of the problem, but also changes in the methods used to solve it, complete revision of the software, and a new faithfulness testing cycle. Therefore, a solution involving modification of the system of hypotheses adopted should be considered from all points of view and adopted only in cases where all other possibilities for improving the faithfulness of the model have been exhausted.

The principal types of errors associated with evaluation of the faithfulness of a model are classified in the following manner (Fig. 2.11) [16, 17]:

- errors in verification of the model using results of control experiments;
- modeling errors associated with the effect of factors that are not taken into account in the model;

- testing errors associated with evaluation of the faithfulness of the model by comparing calculated and experimental data.

The modeling errors have the most complex structure and are subdivided into model errors, parameter errors, and numerical errors.

The end result of all of the stages of the model development process described is a computer model, which can be used to obtain required results from evaluations of characteristics of the process studied (the analysis problem) or to perform a search for the optimal structures, algorithms, and parameters of the system (the synthesis problem).

2.5 Performing a Computational Experiment

The rapid advancement of computer technology has not only helped to speed up various types of calculations, but has also opened new, broader possibilities for the mathematical modeling of complex processes. One of the most important techniques for applying mathematical models in a modern scientific investigation is mathematical (computational) experimentation. It is used both during the preliminary analysis of the systems being modeled (in identifying the parameters of the model and in testing for faithfulness) and during the synthesis of design solutions. A computational experiment is an organized set of investigations, in which devices and processes are studied on the basis of mathematical models using a computer, their behavior under different conditions is simulated, and the optimal parameters and regimes of functioning systems or systems being designed are found. The need to use computational experimentation as a research method is due to the fact that solving modern, extremely complex scientific technical problems by traditional methods has become difficult and, in some cases, impossible.

A computational experiment has many features in common with a real (full-scale) experiment [16, 17].

First, the repeated triggering of the performance of calculations according to a mathematical model on a computer is equivalent to the repetition of trials in a full-scale experiment. The performance of each individual calculation is similar to the performance of a physical experiment: the investigator "switches on" the equations and then follows what happens in the system (as in a real trial).

Second, in a real experiment measurements are performed with some error, and in a computational experiment the solution obtained is generally not exact (especially when numerical methods are used), has a discrete character, and provides only an approximate description of the behavior of the object or process being studied. This is due, in particular, to the fact that many initial parameters of a model are often known only approximately, as are the initial and boundary conditions. All these factors have an effect on the final result of the calculations to a certain extent.

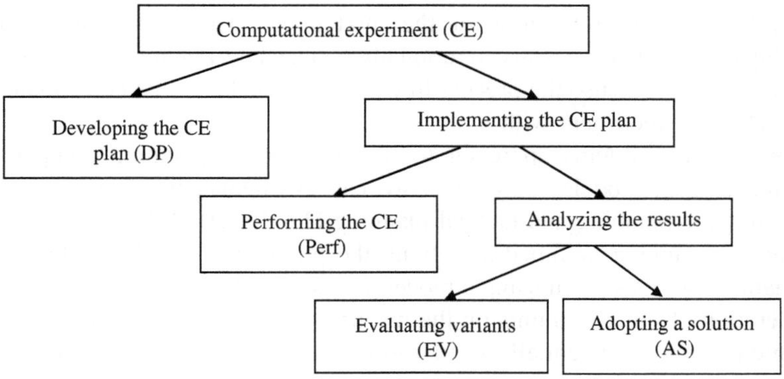

Fig. 2.12 Structural diagram of the computational experiment stage

While a computational experiment has properties in common with a full-scale experiment, it also has some merits and advantages that are inherent only to it. We note the most important among them [16, 17, 27, 31].

- Full-scale experiments often require considerable resources, unique equipment, and the energy and efforts of a large number of specialists. A computational experiment is distinguished by the fact that it is performed on mathematical models of objects, rather than on the objects themselves. The same model can be used to study processes of different physical nature.
- In a full-scale experiment each trial must be set up from the start, while in a computational experiment it is sufficient to modify coefficients of the model and the initial or boundary conditions. Any combination of parameters in the equations of a model can be assigned, even over a fairly broad range.
- A computational experiment can be more easily controlled. This is especially important when the dimensions of the region where the process takes place are small and its duration is short.
- In the case of a dependence of the system being studied on a large number of parameters, consideration of the influence of each of them separately is possible. This is impossible for a full-scale experiment.

The sequence of operations in each stage of the work can be represented in the form of a structural diagram (Fig. 2.12).

It must be noted that the calculation procedure itself should be thought out so that the maximum amount of information would be obtained with the smallest expenditures of resources. In other words, a computational experiment must be planned. The plan specifies the number and order of performance of the calculations on the computer, as well as the techniques for storing and statistically treating the simulation results. The particular problems that can be solved when computational experiments are planned include decreasing the expenditures of machine time on simulations, increasing the accuracy and reliability of the simulation results, testing the faithfulness of the model, etc. Thus, when mathematical modeling is used, it is necessary to rationally plan and design not only the model of the system itself, but also the process in which it is used, i.e., the performance of a computational experiment with it.

Applying the systems approach to the problem of planning computational experiments with models of systems, we can distinguish two planning components: strategic planning and tactical planning [11, 12, 29, 30].

Strategic planning is aimed at solving the problem of obtaining the required information about a system using a model that is implemented on a computer with consideration of the constraints on the resources available to the experimentalist. Strategic planning is essentially similar to the external designing performed when the system is created, except that here the process of modeling the system is the object.

Tactical planning refers to determining the procedure for carrying out each series of tests of the mathematical model that are specified by the experiment plan. For tactical planning there is also an analogy to internal designing of the system, in which the process of working with the model is regarded as the object. The more complex is the mathematical model, the more important is the stage of tactical planning of the computational experiment that is performed immediately before the simulation.

The end result of the computational experiment consists of data expressed in an exact quantitative form from a simulation performed in exact accordance with the plan developed. These data form the basis for the next stage of the work: their comparison with theoretical predictions and data from full-scale experiments. If a significant disparity is observed between the results of the computational experiment and the empirical data, the model is corrected both in the direction of making it more complex and in the direction of simplifying it, and the computational experiment "cycle" is repeated on the improved foundation. This stage of the realization of the computational experiment is continued as long as data that more accurately describe the system under study are obtained as a result of refinements of the mathematical model and the calculation algorithm. The last version of the data can be used to interpret the results and adopt a solution as applies to the modeling goals. Thus, the end result of the computational experiment consists of detailed and specific practical recommendations that are expressed in an exact quantitative form and achieve the specified goal.

Of course, all the procedures in this stage of modeling have an iterative character, which can be expressed in the form of a diagram (Fig. 2.13).

A more general end result of the performance of the computational experiment and analysis of its results is a fully developed mathematical model that is faithful to the assigned conditions and is intended for studying, predicting, and optimizing complex multi-parameter processes.

However, it cannot be assumed that computational experimentation is an ideal method of investigation that has no shortcomings. There are several problems associated with the realization of a computational experiment when complex applied problems are modeled.

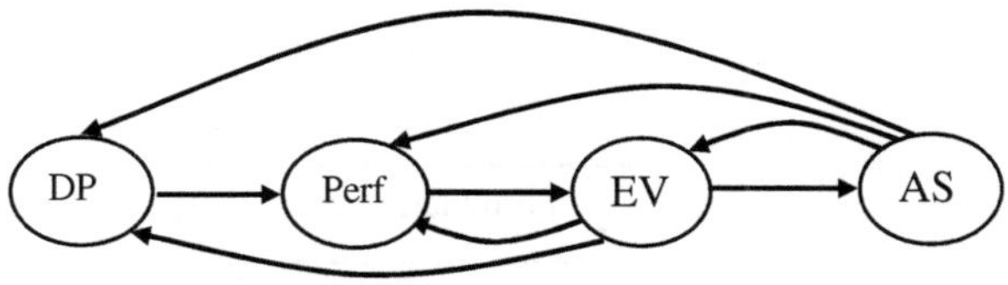

Fig. 2.13 Functional diagram of the computational experiment stage

1. When the number of independent factors included in the mathematical model is large, the investigator may fail to detect fundamental laws and may encounter problems in isolating the main factors that influence the system under study.
2. A computational experiment is limited in the same sense as a full-scale experiment, i.e., it yields discrete information for a certain particular combination of parameters.
3. Since numerical methods are used to solve the system of equations that appear in the mathematical model, a situation in which a loss of stability of the calculations due to a flaw in the algorithm can be interpreted as a loss of stability of the real physical system is possible.
4. It is difficult to evaluate the error due to the replacement of continuous functions by discrete functions and of differential equations by finite-difference expressions. In addition, the possibility of the accumulation of errors due to errors in the rounding of numbers during their processing in the computer must be taken into account.
5. The accuracy of an approximate solution of the problem obtained using classical difference schemes depends on the numerical method chosen, its parameters, and the parameters of the system under investigation.

Therefore, it may be stated that a computational experiment can never replace either a physical experiment or a theoretical analysis of the object or process under study. Only a reasonable combination of all three methods of investigation is a necessary condition for success in solving many problems [32–34].

In conclusion, we present the modeling process in the form of a generalized structural diagram (Fig. 2.14).

2.6 Conclusion

The current trends in the development of high-temperature processes are characterized by the creation of information technologies and their incorporation into scientific research, as well as into technical and technological developments. The technical basis of information technologies is the current, highly efficient computer technology, the methodological basis is systems analysis, and the solution method is mathematical modeling and computational experimentation.

The development of computer hardware and of general and special software gives investigators a powerful tool for solving scientific–technical and engineering problems. The use of "artificial intelligence" enables them to devise and

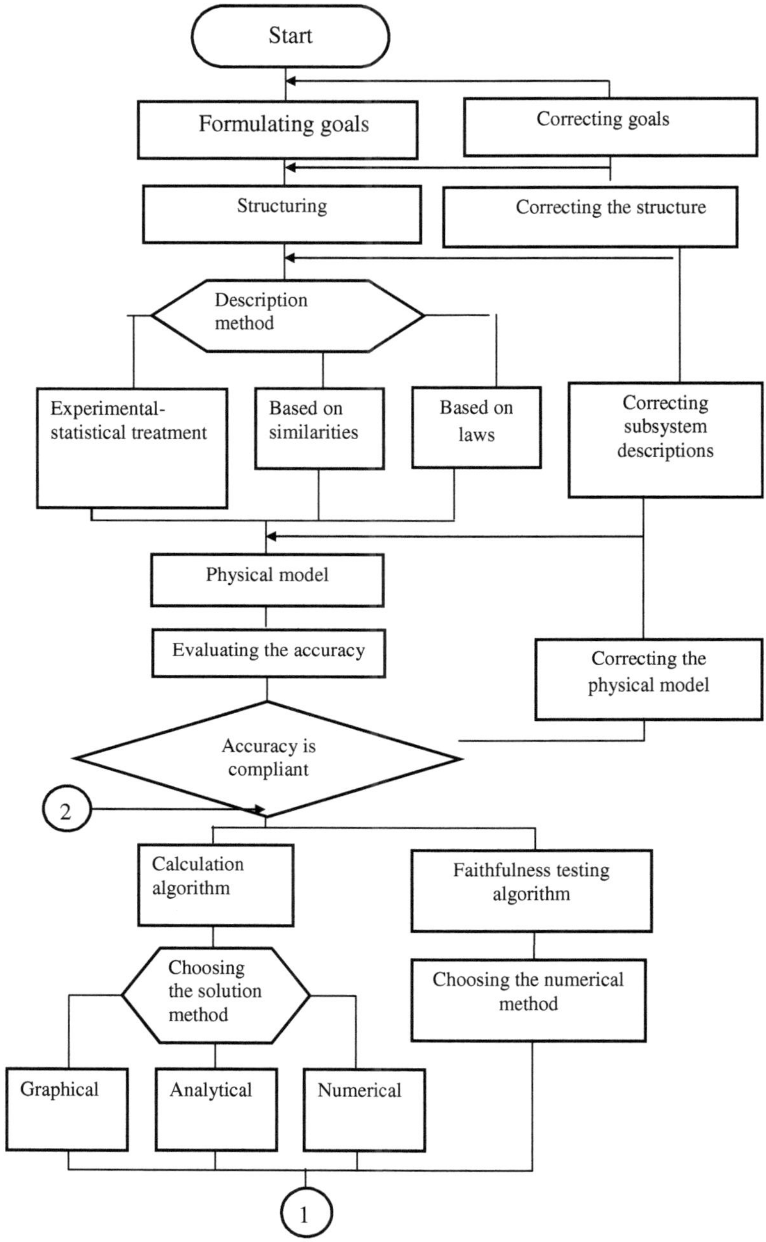

Fig. 2.14 Block diagram of the modeling process

implement complex mathematical models that are simultaneously accurate and meaningful. However, the availability of modern computer equipment and an ability to utilize its continually expanding possibilities are necessary, but not

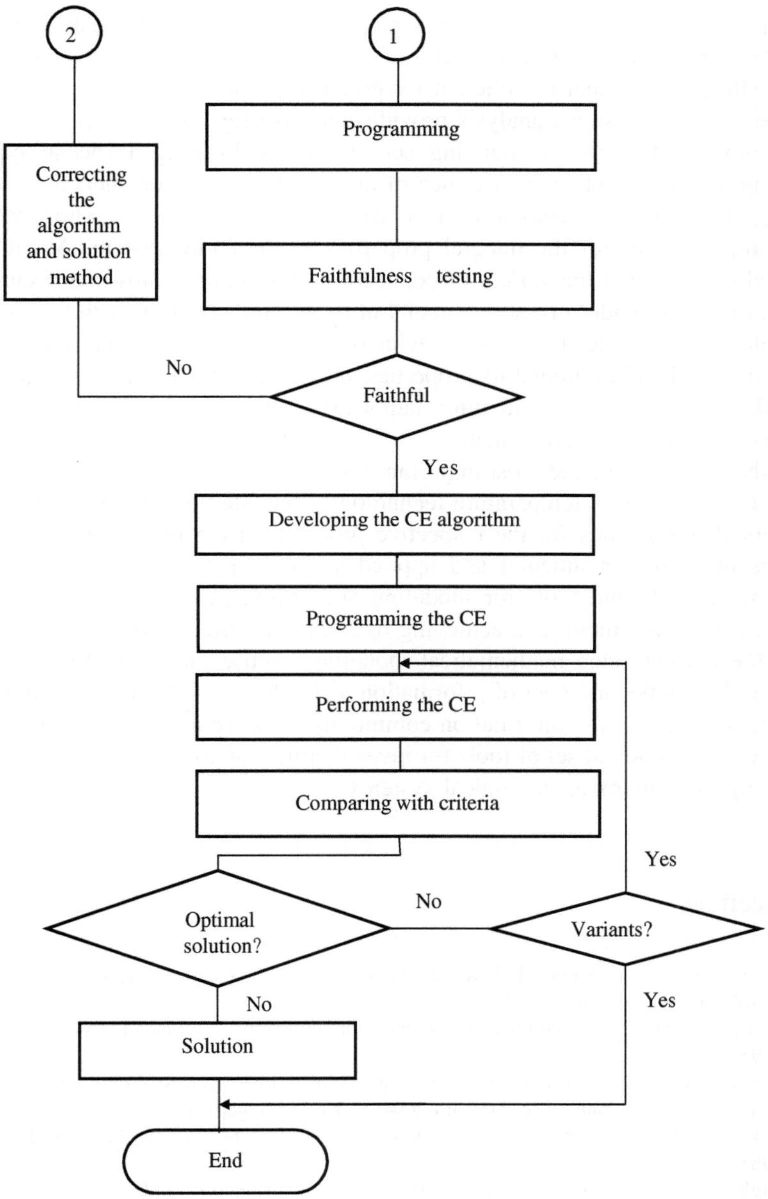

Fig. 2.14 Continued

sufficient conditions for creating effective information technologies. One of the most important conditions for creating faithful mathematical models for complex high-temperature processes is an understanding of the essence of the interactions that occur in multicomponent systems and the laws describing them. Therefore, the investigator's training and professional knowledge in a specific area are fundamental conditions for the development of complete mathematical models that can be used to predict the behavior of objects or the occurrence of processes in them with accuracy that is sufficient for practical purposes.

The theory of systems analysis provides the investigator with a powerful tool for comprehending the surrounding activity by establishing a special type of thinking. Comprehending the essence of an object, process, or phenomenon, the investigator treats it as consisting of a large number of elements, whose interrelationship provides for the integral properties of the entire system. Attention is focused on revealing the variety of connections and relationships that occur both within the object under investigation and in its interrelationships with the external environment. To understand the behavior of the object, process, or phenomenon being modeled and to predict its properties, the investigator must reveal the forms in which matter, energy, and information are transferred from one element to another and ways to control them.

In the current stage the most important conditions for further improvement and intensification of high-temperature technologies include both the development of theoretical foundations for the respective processes using the latest advances in various fields of fundamental and applied sciences and the creation of highly efficient methods and tools for modeling and simulating high-temperature processes, monitoring them, and achieving optimal real-time control.

In the current stage mathematical modeling, whose methods and tools have become the intellectual core of information technologies, is being "incorporated" into the structure of the information community. Modern modeling methods have become a fundamental set of tools for investigating and comprehending processes occurring in complex technological systems.

References

1. Kafarov, V.V.: Cybernetic Methods in Chemistry and Chemical Engineering (in Russian), p. 448. Khimia, Moscow (1985)
2. Kheir, N.A.: Systems Modeling and Computer Simulation, p. 760. Marcel Dekker, New York (1995)
3. Systems Modeling and Simulation: Theory and Applications. In: Post-Proceedings of the 3rd Asian Simulation Conference AsiaSim 2004, p. 733, Springer (2004)
4. Diakoptics, Kron G.: The Piecewise Solution of Large Scale Systems. McDonald, London (1963)
5. Zeigler, B.P., Kim, T.G., Praehofer, H.: Theory of Modeling and Simulation, p. 510. Academic Press, San Diego (2000)
6. Samarskii, A., Mikhailov, A.: Principles of Mathematical Modeling, p. 360. Taylor and Francis, London (2002)
7. Sarkisyan, S.A., Akhundov, V.M., Minaev, É.S.: Large Technical Systems. Analysis and Prognosis of Development (in Russian). p. 350. Nauka, Moscow (1977)
8. Williams, C.P.: Quantum Computing and Quantum Communications, p. 480. Springer, Berlin (1999)
9. Gubanov, V.A., Zakharov, V.V., Kovalenko, A.N.: Introduction to Systems Analysis (in Russian), p. 232. Leningrad State University Press, Leningrad (1988)

10. Cha, P.D., Rosenberg, J.J., Dym, C.L.: Fundamentals of Modeling and Analyzing Engineering Systems, p. 488. Cambridge University Press, Cambridge (2003)
11. Shannon, R.E.: Systems Simulation The Art and Science. Prentice–Hall publishing, Englewood Cliffs (1975)
12. Peterson, G., Ashenden, P., Teegarden, D.: The System Designer's Guide to VHDL-AMS, vol. 139, p. 880. Marcel Dekker, New York (2002)
13. Surguchev, G.D.: Mathematical Modeling of Steel Smelting Processes (in Russian), p. 224. Metallurgiya, Moscow (1978)
14. Hartmann, K., Lezki, E., Schäfer, W. (eds.): Statistische Versuchsplanung und Auswertung in der Stoffwirtschaft. VEB Deutscher Verlag fur Grundstoffindustrie, Leipzig (1974)
15. Lorenzen, T.J.: Design of Experiments, vol. 5, p. 432. Marcel Dekker, New York (1993)
16. Berezovskii, B.M.: Mathematical Models of Arc Welding, Vol. 1: Mathematical Modeling and Information Technologies, Models of the Weld Pool and Joint Formation (in Russian), p. 585. South Ural State University Press, Chelyabinsk (2002)
17. Arisawa, H., Kambayashi, Y.: Conceptual Modeling for New Information Systems Technologies, p. 517. Springer, Berlin (2002)
18. Gukhman, A.A.: Introduction to Similarity Theory (in Russian), p. 296. Vysshaya Shkola, Moscow, (1973)
19. Szirtes, T.: Applied Dimensional Analysis and Modeling, p. 790. McGraw–Hill, New York, (1997)
20. Huntley, H.E.: Dimensional Analysis. Dover, New York (1999)
21. Norenkov, I.P.: Principles of Computer-Assisted Design (in Russian), p. 360. Moscow State Technical University Press, Moscow (2000)
22. Michael, C., Ismail, M.: Statistical Modeling for Computer-Aided Design of MOS VLSI Circuits, p. 212. Kluwer, Boston (1993)
23. Ashikhmin, V.N., et al.: Introduction to Mathematical Modeling (in Russian), p. 336. Dover, Moscow, (2000)
24. Advani, S.G., Sohn, H.M.V., Sozer, E.M.: Process Modeling in Composites Manufacturing, p. 456. Marcel Dekker, New York (2002)
25. Spirin, N.A., Shvydkii, V.S., Lobanov, V.I., Lavrov, V.V.: Introduction to the Systems Analysis of Thermophysical Processes in Metallurgy (in Russian), p. 205. Ural State Technical University Press, Ekaterinburg (1999)
26. Galenko, P.K., Zhuravlev, V.A.: Physics of Dendrites: Computational Experiments, p. 192. World Scientific publishing, Singapore, (1994)
27. Anishchenko, L.M., Lavrenyuk, S.Yu.: Mathematical Principles for Designing High-Temperature Technological Processes (in Russian), p. 80. Nauka, Moscow, (1986)
28. Westmoreland, P.R., Kollman, P.A., Chaka, A.M., Cummings, P.T., Morokuma, K., Neurock, M., Stechel, E.B., Vashishta, P.: Applying Molecular and Materials Modeling, p. 587. Kluwer, Dordrecht (2002)
29. Sovetov, B.Ya., Yakovlev, S.A.: Modeling of Systems (in Russian), p. 343. Vysshaya Shkola, Moscow (2001)
30. Anderson, V.L., McLean, R.A.: Design of Experiments A Realistic Approach, p. 440. Marcel Dekker, New York (1974)
31. Borowko, M.: Computational Methods in Surface and Colloid Science, p. 984. Marcel Dekker, New York (2000)
32. Zarubin, V.S.: Mathematical Modeling in Technology (in Russian), p. 496. Moscow State Technical University Press, Moscow (2001)
33. Chapman, W.L., Bahill, A.T., Wymore, A.W.: Engineering Modeling and Design, p. 363. CRC Press, Boca Raton (1992)
34. Samarskii, A.A., Mikhailov, A.P.: Mathematical Modeling: Ideas, Methods, and Examples (in Russian), p. 320. Nauka publishers, Moscow (1997)

Chapter 3
Constitution and Model Description of the Structural Characteristics of Metallurgical Melts

The available experimental material on the properties and constitution of various melts (metals, oxides, sulfides, salts, and molecular liquids) has confirmed their complex microheterogeneous constitution and has shown that the structural parameters of pure and multicomponent solutions vary with the nature, temperature, and composition of the substance. The idea that a liquid phase is interrelated to the structure and properties of the solid phase, which was advanced back in the 1930s by Frenkel, has been confirmed. Quasicrystalline models, which reflect the unity of the real forces of interatomic interaction on the two sides of the melting range, have had great success in describing liquids.

The application of the terms bcc- and fcc-like packing to melts does not mean that there are polymicrocrystals in them. It only points to a tendency of a group of atoms in the condensed state to take on a specific type of ordering.

An analysis of the results of structural investigations of various liquids showed [1] that they have a definite short-range order, which is characteristic of the particular class of substances and which is close to the structure of the respective substances in the solid state in most cases. When metals are melted and their temperature is raised further or when the composition of multicomponent melts is altered, structural transformations associated with changes in the nature of the chemical bonds between atoms and their relative arrangement can occur.

3.1 Structural Characteristics of Melts

3.1.1 Radial Distribution Function and Structure Factor

The structure of substances is described quantitatively using the radial distribution function $\rho(R)$, which indicates the probability of finding any atom (molecule) at

M. Zinigrad et al., *Phase Interaction in the Metal–Oxide Melts–Gas System*,
Engineering Materials, DOI: 10.1007/978-3-642-22377-8_3,

the distance R from the atom taken as the center. The dimensionless probability distribution function $g(R)$ is related to $\rho(R)$ by the expression

$$g(R) = \rho(R)/\rho_0, \tag{3.1}$$

where ρ_0 is the average atomic density in the system.

In the case of simple liquids, $g(R)$ is identical to the pair correlation function, which actually appears in any thermodynamic equation describing a liquid phase [2–4]. The radial distribution function is calculated from the structure factor $a(s)$, which is measured in diffraction experiments:

$$a(s) = I(s)/Nf^2(s). \tag{3.2}$$

Here $I(s)$ is the intensity of the scattered radiation as a function of the wave vector $s = 4\pi \sin\theta/\lambda$ (θ is half of the scattering angle, and λ is the wavelength of the radiation); $f(s)$ is the atomic factor, which characterizes the intensity of the scattering by an isolated atom; N is the number of scattering centers. In the case of neutrons, the atomic factor does not depend on the scattering vector. The functions $G(R)$ and $a(s)$ are related to one another by the equations [5]

$$g(R) = 1 + \frac{1}{2\pi^2\rho_0 R}\int_0^\infty s[a(s) - 1]\sin(sR)\mathrm{d}s, \tag{3.3}$$

$$a(s) = 1 + \frac{4\pi\rho_0}{s}\int_0^\infty R[g(R) - 1]\sin(sR)\mathrm{d}R. \tag{3.4}$$

The quantity $h(R) = g(R) - 1$ is called the complete (total) correlation function. It consists of two components. The first term $C(R)$ (the direct correlation function) characterizes the pair interaction between the atoms, and the second term characterizes the interaction of the central atom with all the other atoms. The function $C(R)$ is defined by the equation

$$C(R) = 1 + \frac{1}{2\pi^2\rho_0 R}\int_0^\infty sC(s)\sin(sR)ds, \tag{3.5}$$

in which

$$C(s) = \frac{a(s) - 1}{a(s)}. \tag{3.6}$$

Unlike $a(s)$, the function $C(s)$ is concentrated in the region of small values of s. Thus, the experimental structure factor $a(s)$ can be used to find the functions $g(R)$ and $C(R)$, which can then be used to obtain quantitative information about the structure and to calculate the pair interaction potentials. It can be seen from Eqs. 3.3 and 3.5 that diffraction data in the range of wave vector values from 0 to infinity are needed to calculate $g(R)$ and $C(R)$. However, the experimental conditions for measuring intensity curves are restricted both in the direction of small scattering angles and in the direction of large scattering angles. The maximum values is $s = 4\pi/\lambda$. For hard Mo Kα radiation ($\lambda = 0.071$ nm), we obtain $s_{\text{max}} = 177\ \text{nm}^{-1}$. In practice the upper integration limit is usually 100–120 nm^{-1}. On the small-scattering-angle side the structure factor is extrapolated to its value at $s = 0$, which is determined from the well-known thermodynamic relation

$$a(0) = k_{\text{B}} T \rho_0 \kappa_T. \tag{3.7}$$

Here k_{B} is Boltzmann's constant; T is the temperature; and κ_T is the isothermal compressibility.

Several methods for extrapolating the structure factor into regions of large and small scattering vectors that are inaccessible to experiment have been proposed (see [6]). In some papers experimental $a(s)$ and $g(R)$ curves were extended into the region of large values of s and R using analytical expressions. The parameters contained in them were found from the condition that information about the short-range atomic order is embedded in the long-range maxima of the structure factor and from the condition that the course of the radial distribution function (RDF) at large distances determines the scattering curve at small values of s. The behavior $a(s)$ at small values of s was obtained by Fourier transformation of the RDF after it had been extended into the region of large distances.

Liquid metals have been divided into three groups according to the form of the structure factor [7]. The first group includes the alkali and noble metals, as well as Al, Pb, Fe, and Ni. The main peak of their structure factor is symmetric and undergoes changes as the temperature is increased. Near the melting point its height is $H = 2.5$, and the ratio of the position of the second maximum to the position of the first maximum is $s_2/s_1 = 1.86$. The second group includes Ga, Ge, Sn, Sb, Bi, etc., for which the first peak on the $a(s)$ curves has a secondary maximum on the large-scattering-angle side. The ratio is $s_2/s_1 = 1.95$, and the height is $H = 2.20$. The form of $a(s)$ exhibits practically no dependence on the superheating of the melt. The metals Mg, Zn, Cu, and Hg, whose structure factor has an asymmetric first peak, can be assigned to the last group. Here $a(s)$ varies only slightly with temperature, $H = 2.5$ as in the first group, but the ratio s_2/s_1 differs from 1.86.

However, to draw a reliable conclusion regarding the constitution of a liquid, the shape and position not only of the first peaks, but also of the long-range peaks of the structure factor and the RDF must be analyzed. An analysis shows [8] that the long-range RDF maxima also reflect features of the ordering in melts. Diffraction from the second and subsequent atomic spheres makes a considerable contribution to the decay rate and to the positions of the structure factor maxima. Popel', Manov, and Spiridonov [9] analyzed the radial distribution functions for 21 metals in the range from 0.2 to 1.1 nm. The RDFs were obtained from the structure factors in [7], which were discussed above. As a result, all the metals were divided into two groups. Elements that have one of the compact structures (fcc, hcp) or a bcc structure before melting were assigned to the first group. After a small amount of superheating above the melting point, the following values of the inverse ratios of the horizontal coordinates of the first and subsequent peaks are characteristic here: $R_2/R_1 = 1.79–1.92$, $R_3/R_1 = 2.64–2.77$. The second group includes metals which display a high percentage of covalent character in their bonds and a low coordination number in the crystalline state. In their melts $R_2/R_1 = 1.94–1.97$, and $R_3/R_1 = 2.82–2.95$.

3.1.2 Partial Structure Factors

Interpretation of the results of diffraction investigations of complex liquids containing atoms of different kinds is a more a more difficult problem, because only a weighted-mean atomic distribution function can be calculated from one experimental interference curve. In this case the coherent X-ray scattering intensity and the RDF are related to one another by the equations [10]

$$a(s) = 1 + \int_0^\infty 4\pi R^2 \sum_i \sum_j n_i k_i k_j (\rho_{ij} - \rho_0) \frac{\sin(sR)}{sR} dR, \tag{3.8}$$

$$\begin{aligned} G(R) &= 4\pi R^2 \sum_i \sum_j n_i k_i k_j \rho_{ij} \\ &= 4\pi R^2 \rho_0 \left(\sum_i n_i k_i \right)^2 + \frac{2R}{\pi} \int_0^\infty s[a(s) - 1] \sin(sR) ds, \end{aligned} \tag{3.9}$$

in which $a(s) = I(s)/\Sigma n_i f_i^2(s)$, n_i is the atomic fraction of the component, and k_i is its relative atomic factor averaged over the scattering angle:

$$k_i = \frac{f_i(s)}{\left[\sum n_i f_i^2(s)\right]^{1/2}}. \tag{3.10}$$

In a multicomponent system each species of atoms has its own characteristic nearest neighbors. Therefore, here a detailed description of the structure requires knowledge of the partial distribution functions of the atoms. To obtain them, as many independent diffraction experiments as there are types of coordination of the atoms in the melt must be performed.

A melt containing atoms of two kinds is characterized by the three radial distribution functions $\rho_{ij}(R)$, which indicate the probability of finding an atom of type i at a distance R from an atom of type j. To determine them using Eq. 3.3, the three corresponding partial structure factors $a_{ij}(s)$ must be known. According to the Faber–Ziman formalism, the latter are related to the total structure factor $a(s)$ by the equation [11]

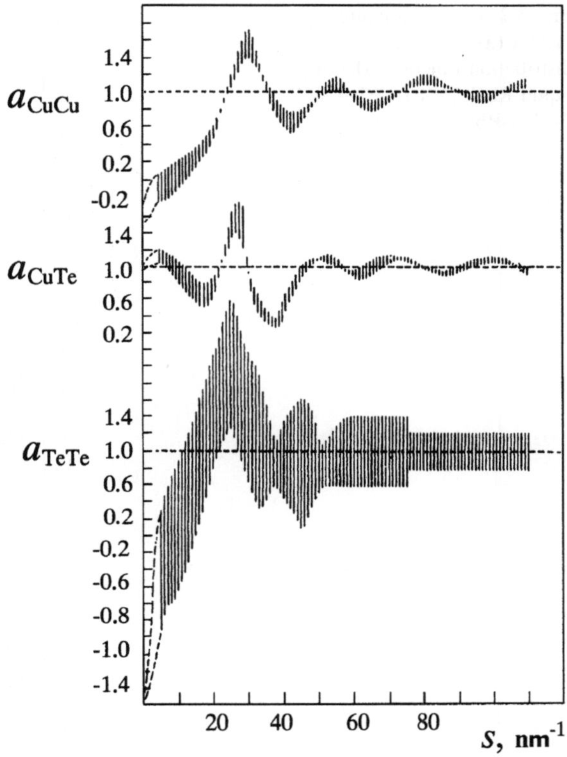

Fig. 3.1 Three partial structure factors of a Cu_2Te melt at 1,000°C [14]

$$a(s) = n_i^2 k_i^2 a_{ii}(s) + n_j^2 k_j^2 a_{jj}(s) + 2n_i n_j k_i k_j a_{ij}(s). \tag{3.11}$$

To determine the functions $a_{ij}(s)$ using expression (3.11), three experimental intensity curves which would be distinguished by the scattering power f_i of the components must be available. For this purpose, three independent diffraction experiments (X-ray diffraction analysis, neutron diffraction analysis, and electron diffraction analysis) can be carried out, or three samples of a specific composition that contain different isotopes of one of the elements in the alloy can be studied by neutron diffraction analysis. Investigations of this kind have been carried out on melts of Cu_6Sn_5 [12, 13], CuTe and Cu_2Te [14], and CuSe [15], whose scattering power was varied by using different isotopes of copper. The partial structure factors for Cu_2Te are presented with their determination error in Fig. 3.1. The structure factor a_{CuCu} is similar to that for pure copper, but a_{TeTe} differs significantly from the structure factor of pure tellurium. The partial structure factors of molten KCl and CsCl [16], NaCl [17], RbCl [18], AgCl [19], and CuCl [20] have been determined using chlorine isotopes. Typical plots of $a_{ij}(s)$ and $g_{ij}(R)$ for chlorides are presented in Fig. 3.2. Unlike the total radial distribution function, the plots of the function for a pair of oppositely charged atoms

Fig. 3.2 Partial structure factors (**a**) and radial distribution functions (**b**) of liquid RbCl [18]: *1*—Rb–Cl, *2*—Rb–Rb, *3*—Cl–Cl

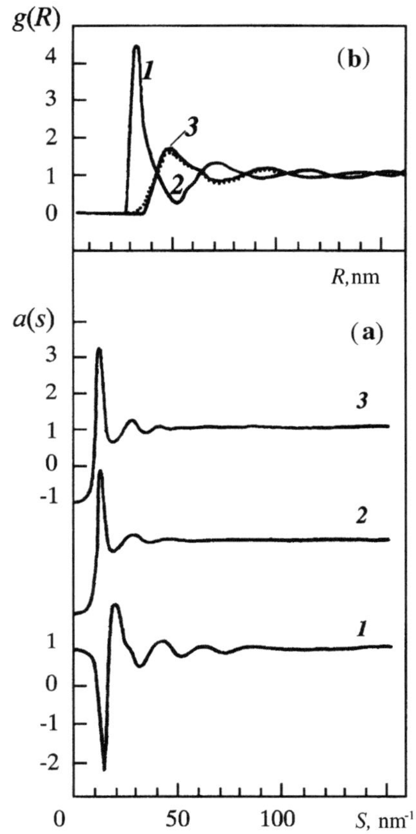

$g_{+-}(R)$ has a well isolated first maximum. The functions $g_{++}(R)$ and $g_{--}(R)$ differ only slightly from one another.

Another technique for finding three partial structure factors in a binary alloy without using isotopes was proposed in [21]. It is based on the anomalous scattering of X-rays by matter near the absorption edge, where the frequency of the radiation approaches the characteristic frequency of an electron in an atom. In the general case the atomic scattering factor is written in the following form:

$$f_i(s) = f_{i0}(s) + \Delta f_i, \tag{3.12}$$

where $f_{i0}(s)$ is the atomic scattering factor for radiation with a wavelength much smaller than the absorption edge λ_{edge} and Δf_i is a correction for the anomalous dispersion, which contains real ($\Delta f_i'$) and imaginary ($i\Delta f_i''$) parts. The dependence of the latter on the wavelength of the primary radiation is shown in Fig. 3.3. By studying the intensity of the scattering of X-radiation with different wavelengths, two of which lie close to the absorption edge, while one is far from it, three expressions of type (3.11) can be written, and the partial structure factors $a_{ij}(s)$ can

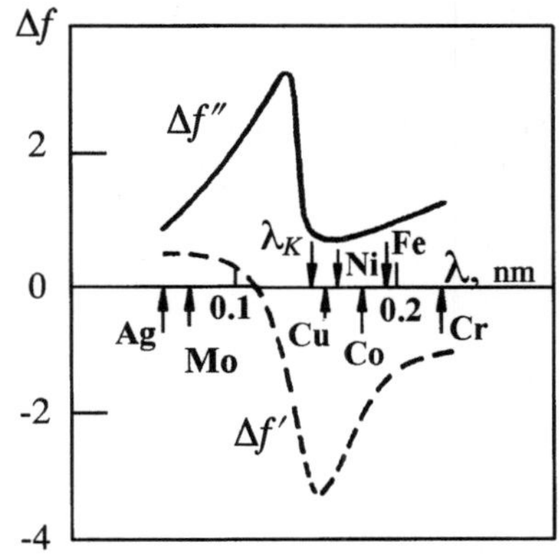

Fig. 3.3 Dependence of the corrections for anomalous dispersion for nickel atoms on the wavelength of the X-radiation [21]

be calculated from them. By combining Mo Kα, Cu Kα, and Co Kα radiation, $a_{Ni-Ni}(s)$, $a_{Si-Si}(s)$, and $a_{Ni-Si}(s)$ were calculated for Ni–Si melts using this method and were found to be practically independent of the composition (Fig. 3.4).

The problem of finding partial structure factors is simplified if it is assumed that they do not depend on the concentration of the components. Then, values of the total structure factors for three or more melt compositions that were found experimentally in one type of radiation can be used to solve Eq. 3.11. The partial functions $a_{ij}(s)$ in the Au–Sn [22], Mg–Zn and Al–Au [23, 24], Cu–Sn [25], Na–K and Al–Mg [26], Hg–In and Hg–Tl [27], and Tl–Te [28] systems have been calculated in this way. An analysis of the data reveals that for almost all the melts $a_{ii}(s)$ and $a_{jj}(s)$ coincide with the structure factors of the pure components. The relatively good convergence of the values of $a(s)$ obtained by the neutron method and calculated from the X-ray data for $a_{ij}(s)$ under the assumption that they are not dependent on the composition of the alloy should be noted in this context.

3.1.3 Model Description of Structural Characteristics

Among the numerous theories that have been devised to describe the constitution of real liquids, the theories based on the notion that the short-range order in the arrangement of the atoms and their thermal motion is similar in the liquid and solid states have had the greatest success (see [12, 29–36]). They include the quasi-crystalline, cybotactic, cluster, island, and other models based on the results of diffraction and physicochemical studies.

The quantitative methods for interpreting the constitution of liquids are generally based on an analysis of experimental interference curves and radial distribution functions and their comparison with structural characteristics calculated from quasicrystalline models [33, 37–44]. For model calculations the use of a correct "blurring" factor for the crystal structure is of great importance [30, 33, 45].

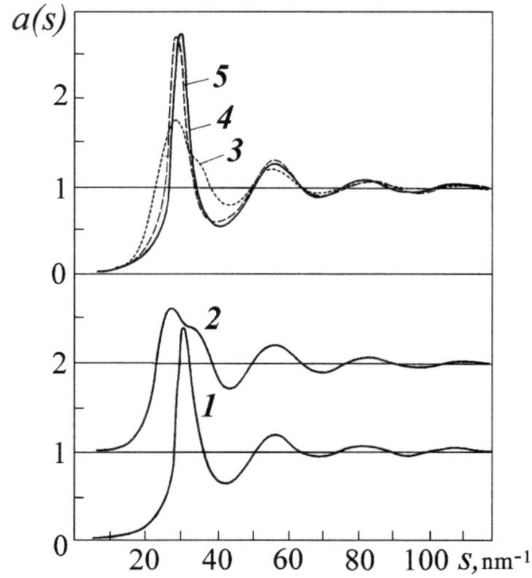

Fig. 3.4 Structure factors of pure liquid Ni (*1*) and Si (*2*) and Si–Si (*3*), Ni–Ni (*4*), and Ni–Si (*5*) partial structure factors in Ni–Si melts [21]

The blurring of interference curves and RDFs in liquids is caused by the following: (a) the increase in the amplitude of the thermal vibrations of the atoms relative to their equilibrium positions; (b) the statistical spread of the positions of the vibration centers of the atoms due to the fluctuation mechanism of particle motion, which results in a dependence of the relative displacement of any pair of atoms on the distance between them; (c) the loss of long-range order with sharp decreases in the dimensions of the ordered regions and the number of atoms in them.

The intensity of coherent X-ray scattering in a melt containing two kinds of particles can be calculated from the equation [46]

$$a(s) = 1 + \frac{1}{\sum n_i k_i^2} \times \left\{ \sum_{p=1}^{2} K_p \left[\sum_j K_{1j} \frac{\sin(sR)_{ij}}{sR_{ij}} \exp\left(-b_{ij}s^2\right) + \sum_j K_{2j} \frac{\sin(sR)_{ij}}{sR_{ij}} \exp\left(-b_{ij}s^2\right) \right] \right\} \quad (3.13)$$

if the arrangement of the atoms, the distances R_{ij} between them, and the values of $bij = \overline{u_{ij}^2}/2$ (see below) are assigned. In formula (3.13) the summation in the square brackets is carried out over the coordination spheres of the first and second components.

The mean square displacement of the atoms $\overline{u_i^2}$ for the ith coordination sphere in a crystal at high temperatures and in melts can be represented in the form [37]

$$\overline{u_i^2} = \overline{u_T^2} + \overline{u_i^2}(R), \quad (3.14)$$

where the first and second terms are the mean squares of the deviations associated with the thermal vibrations of the atoms and their diffusive displacement, respectively. The quantity $\overline{u_i^2} = \gamma_i \overline{u_\infty^2}$, where u_∞^2 characterizes the square of the displacement of a pair of atoms that are separated from one another by a large distance, and γ_i is the coupling constant, which is close to unity for the second coordination sphere. Representing the radial distribution function in the form of a sum of partial density functions approximated by exponential curves of the Gaussian type [29, 30, 34, 35, 47–49], for a monatomic liquid we can write

$$4\pi R^2 \rho(R) = \sum_i \frac{z_i}{\sqrt{2\pi\gamma_i \overline{u_\infty^2}}} \exp\left[-\frac{(R-R_i)^2}{2\gamma_i \overline{u_\infty^2}}\right]. \tag{3.15}$$

Here z_i is the number of atoms in the i-th coordination sphere; R_i is its radius; $\overline{u_i^2}$ describes the dispersion of the Gaussian curve. The use of this expression to find interference curves leads to the equation

$$a(s) = 1 + \sum_i z_i \frac{\sin(sR_i)}{sR_i} \exp\left(-\frac{\overline{u_i^2} s^2}{2}\right). \tag{3.16}$$

Figure 3.5 presents plots of the dependence of $\overline{u_i^2}$ on the number of the coordination sphere for crystalline and liquid gold and potassium [48]. Within the first three or four coordination spheres, the mean square deviation between atoms in the molten state increases rapidly, and then it varies slowly. For the shortest distance between atoms, the value of $\overline{u_1^2}(R)$ is close to 0; therefore, $\overline{u_1^2} \approx \overline{u_T^2}$, and in both states of aggregation the values of $\overline{u_i^2}$ at the melting point are approximately equal to one another. They can be calculated in the Debye approximation from the following equation, which is usually used [35, 50] to take into account the influence of the temperature on the diffraction of X-rays:

$$\overline{u_T^2} = \frac{6h^2 T}{4\pi^2 m k_B \theta_D^2} \left[\frac{1}{4}\frac{\theta_D}{T} + \Phi\left(\frac{\theta_D}{T}\right)\right]. \tag{3.17}$$

Here $\overline{u_T^2}/2$ is the mean square displacement of the atoms from their equilibrium positions; m is the mass of an atom; h and k_B are Planck's constant and Boltzmann's constant; θ_D is the Debye characteristic temperature. For $\theta_D/T \leq 1$, $\left[\frac{1}{4}\frac{\theta_D}{T} + \Phi\left(\frac{\theta_D}{T}\right)\right] \approx 1$. Then

$$\overline{u_T^2} = \frac{298T}{M\theta_D^2}, \tag{3.18}$$

The atomic weight M for complex substances is replaced by M_{lim}:

$$\frac{1}{M_{lim}} = \frac{n_1}{M_1} + \frac{n_2}{M_2}, \tag{3.19}$$

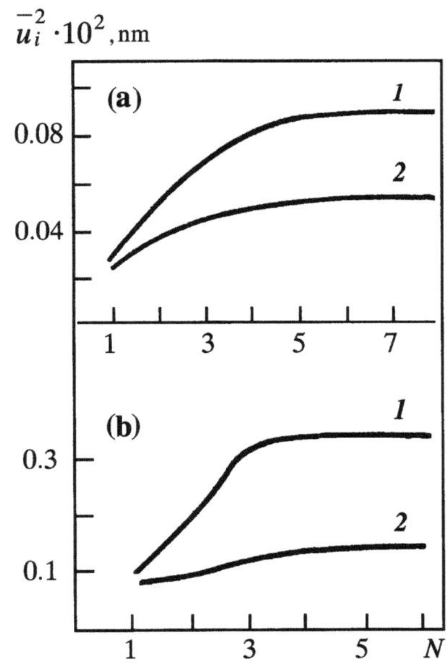

Fig. 3.5 Dependence of the mean square deviation of atoms on the number of the coordination sphere for gold (**a**) and potassium (**b**) in the liquid (*1*) and crystalline (*2*) states [48]

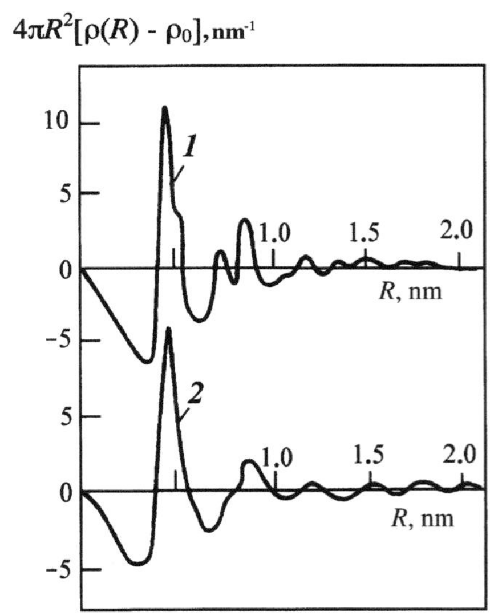

Fig. 3.6 Difference radial distribution curves of atoms in crystalline (*1*) and liquid (*2*) potassium [48]

where n_i and M_i are the atomic fractions and atomic weights of the components of the compound. The radial distribution curves of the atoms constructed according to Eq. 3.15 for a crystalline grain of specific dimensions using the dependence of the mean square deviation on the distance provide a picture that is identical to the RDF for the liquid state (see Fig. 3.6).

Taking into account the collective character of the thermal motion of the atoms in a liquid and averaging the molar volumes, Prins and Petersen [51] (see also [30])

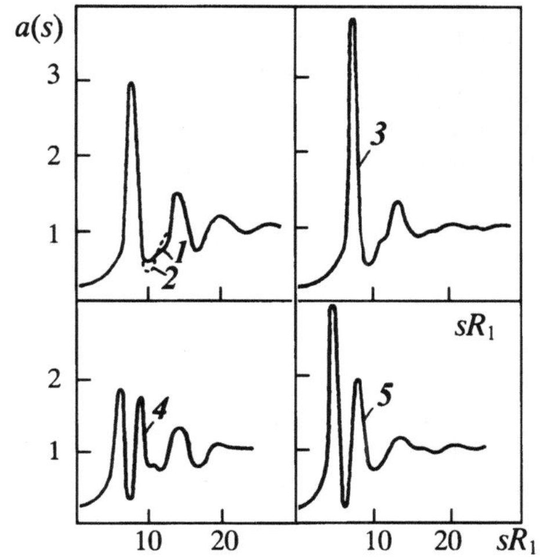

Fig. 3.7 Intensity curves calculated [51] for crystal structures: *1*—hexagonal close-packed; *2*—face-centered cubic; *3*—body-centered cubic; *4*—simple cubic; *5*—diamond structure

obtained the expression $b_{ij} = 0.0025R_i^2$ for the coefficient b_{ij}. The diffraction curves (Fig. 3.7), which they calculated from formula (3.16) with the exponential factor $\exp\left(-0.0025R_i^2s^2\right)$ for various structures of metals, satisfactorily describe the experimental data. Good agreement between experimental intensity curves and curves calculated according to the structure of solid phases has been obtained for several pure transition metals and their alloys [41], as well as for oxides [43]. In metallic melts the value of $\overline{u_1^2}$ for the shortest interatomic distance was found from the characteristic temperature of the substance investigated using Eq. 3.18, and the values for the subsequent coordination spheres were found using a semi-empirical formula from the condition that $\overline{u_i^2}$ is proportional to the square of the distance between the atoms. In the case of liquid oxides, a constant value of the mean square deviation of the atoms $\overline{u_T^2}$ for all the coordination spheres considered, which is determined only by their thermal vibrations and does not take into account the translational motion of the particles, was used.

The influence of the number of atoms in isolated ordered regions (microcrystalline grains) on the blurring of the diffraction pattern was examined in [38, 39, 52–54]. As the dimensions of the crystalline grains decreases, broadening of the interference lines occurs. The intensity curves calculated for randomly oriented crystalline grains measuring 2.0–3.0 nm are similar to the analogous curves for liquids. Even in the case of a crystal containing 729 atoms at 0 K, the interference lines are partially fused and are not completely resolved, and only 24 reflections out of the 35 for a large crystal with an fcc structure are present on the curve. The splitting of the main maximum on interference curve into two maxima that correspond to the (111) and (200) reflections for fcc packing begins when the number of atoms exceeds 100 [53]. At this number of atoms, a maximum corresponding to the (200) reflection, which is characteristic of the bcc structure at small scattering angles, is clearly distinguished [52].

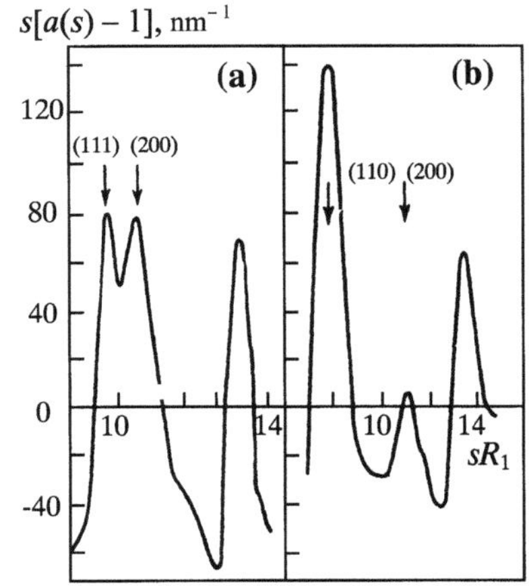

Fig. 3.8 Interference curve calculated [39] for the fcc (**a**) and bcc (**b**) packing of iron. The arrows indicate the positions of the interference lines in crystalline samples of γ- and α-iron

It was shown in [39] that if the blurring of the coordination spheres both as a consequence of the loss of the long-range order and as a result of the thermal vibrations of the atoms is taken into account, the first maximum on the intensity curve for the fcc structure of iron should bifurcate at a correlation radius that includes 15 coordination spheres (~ 1.0 nm). When 10 coordination spheres are taken into account, the result is one intensity peak located between the (111) and (200) reflections. An unbifurcated, narrower, and higher first maximum, as well as second and third maxima that are displaced toward larger scattering vectors (Fig. 3.8), are characteristic of the bcc packing of iron. In this context we note that considerable asymmetry of the first maximum of the structure factor, which corresponds to the (200) reflection for an fcc lattice, is observed for liquid aluminum [55–57]. Conversely, the diffraction curves of molten iron, nickel, and palladium [58] have a symmetric and narrower main peak and a strongly blurred, small subsequent maximum, which corresponds to (200) interference in the bcc structure.

It should be stressed that the height and width of the first peak depend on the size of the ordering regions, while the arrangement of the peaks at larger scattering angles is specified by the short-range order. The height of the first maximum on the intensity curve increases rapidly, and its width decreases, as the number of atoms in the cluster is increased from 13 to 100. They achieve constant values in ordered regions containing several thousand atoms [54]. Therefore, comparisons of experimental and model interference curves should not be confined to consideration of only the small-angle peaks, and the intensity curves must be analyzed and compared over the entire range of scattering angles. Another technique for interpreting diffraction data was described in [59]. Instead of calculation of the theoretical intensity curves from assigned values of z_i and R_i followed by comparison of these curves with experimental curves, determination of the coordination numbers from an experimental interference curve was proposed in that paper. For this purpose a system of equations, which has the following form for pure metals, is created:

$$
\begin{aligned}
a(s_1) - 1 &= \frac{\sin s_1 R_1}{s_1 R_1} z_1 + \frac{\sin s_1 R_2}{s_1 R_2} z_2 + \cdots + \frac{\sin s_1 R_i}{s_1 R_i} z_i, \\
a(s_2) - 1 &= \frac{\sin s_2 R_1}{s_2 R_1} z_1 + \frac{\sin s_2 R_2}{s_2 R_2} z_2 + \cdots + \frac{\sin s_2 R_i}{s_2 R_i} z_i, \\
&\cdots\cdots\cdots\cdots\cdots\cdots\cdots\cdots\cdots\cdots \\
a(s_i) - 1 &= \frac{\sin s_i R_1}{s_i R_1} z_1 + \frac{\sin s_i R_2}{s_i R_2} z_2 + \cdots + \frac{\sin s_i R_i}{s_i R_i} z_i.
\end{aligned}
\tag{3.20}
$$

Experimental values of the structure factor for an arbitrary value of s_i are inserted on the left-hand side. By assigning the type of structure (bcc, fcc, etc.) and thereby the law governing the variation of the interatomic distances R_i, specific values of the z_i are obtained. A comparison of these values with the corresponding coordination numbers in a crystal allows drawing conclusions regarding the short-range order in the liquid.

For binary and ternary systems a conclusion regarding the constitution of melts can also be drawn on the basis of a comparison of calculated and experimental values of the short-range order parameters: the position s_1 of the main diffraction maximum, the position R_1 of the first RDF maximum, and its area A_1. For an alloy containing two kinds of particles, from Eq. 3.9 we have

$$
A = \int_{R_1}^{R_2} 4\pi R^2 \sum_i \sum_j n_i k_i k_j \rho_{ij}(R) = \sum_i \sum_j n_i k_i k_j z_{ij} = n_1 k_1^2 z_{11} + n_2 k_2^2 z_{22} + n_1 k_1 k_2 z_{12} + n_2 k_1 k_2 z_{21}. \tag{3.21}
$$

Here z_{ij} is the number of atoms of type j around an atom of type i.

In the case of a statistical distribution of the atoms in a melt of the solid solution type, we can write

$$
A_{\text{st}} = z_{\text{st}} (n_1 k_1 + n_2 k_2)^2, \tag{3.22}
$$

where z_{st} is determined by the coordination numbers z_1 and z_2 of the pure liquid components:

$$
z_{\text{st}} = n_1 z_1 + n_2 z_2, \tag{3.23}
$$

Then

$$
R_{\text{st}} = \frac{n_1 z_1 R_1 + n_2 z_2 R_2}{z_{\text{st}}}. \tag{3.24}
$$

If the atoms of each kind form their own microregions (quasieutectic structures of atoms of the pure components), we can write

$$
A_{\text{qu}} = n_1 k_1^2 z_{11} + n_2 k_2^2 z_{22},, \tag{3.25}
$$

$$R_{\mathrm{qu}} = \frac{n_1 k_1^2 z_{11} R_1 + n_2 k_2^2 z_{22} R_2}{A_{\mathrm{qu}}}. \tag{3.26}$$

For neighboring atoms of different kinds, we have

$$A_{\mathrm{diff}} = n_1 k_1 k_2 z_{12} + n_2 k_1 k_2 z_{21}, \tag{3.27}$$

Since by virtue of the stoichiometry $n_1 z_{12} = n_2 z_{21}$, we have

$$A_{\mathrm{diff}} = 2 n_1 k_1 k_2 z_{12}. \tag{3.28}$$

Calculation formulas for more complex quasieutectic structures (a chemical compound and a pure element, a statistical distribution of the two components and a pure component, etc.) have been presented in [42]. However, it should be noted that the poor resolution of the main RDF maximum in most melts makes determination of the value of A difficult.

In several publications [8, 9, 60–63] experimental interference curves for binary melts and pure metals were described well by a model that includes two types of symmetry in the arrangement of the atoms: spherical symmetry with a high coordination number, which is characteristic of metallic or dispersion bonds, and chain symmetry with a low coordination number, which corresponds to a directed interaction. This model is based on a piecewise analysis of the experimental diffraction curve and consideration of not only the first, but also all the subsequent structural maxima of the radial distribution function. Conclusions, which are not always consistent with the quasicrystalline models, were drawn under such an approach. For metals with compact packing, zigzagged chains with $z_1 = 4$, which correspond to the arrangement of the atoms in the most densely populated crystallographic planes of premelting structures, and regions of spherical packing with $z_1 = 8$–10 were revealed in the solid state. In metals with a loose premelting structure, linear chains are maintained in the liquid. In binary alloys, regions with a statistically averaged arrangement of particles of a different kind can be added to these structural components. As the temperature is raised, the length and fraction of the chains in the melt decrease, and the number of atoms in the ordered regions with spherical packing increases. A comparison of the model and experimental structure factors showed that 75–85% of the atoms are found in chains.

The rigid-sphere model is widely used for theoretical calculations of the interference curves for one- and two-component melts (see [33]). Here the structure factor is a function of two parameters, viz., the packing density and the diameter of a rigid sphere, which can be determined from the height and position of the first maximum on the experimental plot of $a(s)$ [64]. This model describes the experimental data fairly well for liquid metals that have a dense crystalline structure. However, such agreement is not observed for metals with a loose structure in the solid state.

The information about the structure of melts that is usually provided by a diffraction experiment is averaged over the volume, angular coordinates, and correlations between the various kinds of atoms. Therefore, several techniques for constructing computer models from diffraction data are used to analyze the features of the short-range ordering. These techniques can be divided into groups. One group includes the traditional molecular dynamics and the Monte Carlo methods. They require exact knowledge of the interatomic potentials, whose determination is a complex problem that leads to ambiguous solutions. It is assumed that the potentials have a pair type of interaction. The results of calculations by these methods showed [65, 66] that the structure of liquid metals expressed in terms of the distribution function and the structure factor is determined mainly by the repulsive part of the potential and the position of the potential well and depends only slightly on the long-range part of the interaction potential.

The other group includes the reverse Monte Carlo method, which has recently been actively employed. This method does not require knowledge of the interatomic potentials and makes it possible to obtain three-dimensional models of the structure of substances in the unordered state that are quantitatively consistent with the experimental data. It uses a three-dimensional mesh with periodic boundary conditions and the mathematical machinery of the standard Monte Carlo method, but the quantity minimized here is not the potential energy, but the difference between the experimental structure factor $a_{\mathrm{exp}}(s_i)$ and the calculated structure factor $a_{\mathrm{calc}}(s_i)$:

$$\chi^2 = \sum_i \frac{\left[a_{\mathrm{calc}}(s_i) - a_{\mathrm{exp}}(s_i)\right]^2}{\sigma^2(s_i)}, \tag{3.29}$$

where χ is the minimization parameter and $\sigma(s_i)$ is the experimental error. The calculated structure factor $a_{\mathrm{calc}}(s_i)$ is found by Fourier transformation of the function $g(R)$ determined from the three-dimensional model obtained.

The reverse Monte Carlo method has been used to simulate the structure of molten iron and tin over a large range of temperatures [67]. The advantage of this method is manifested especially clearly when binary melts are investigated, because it provides a way to obtain partial distribution functions of the alloy components. The method has been used successfully [68] to find the structure factors $a_{ij}(s)$ of melts of the Fe–Sn system (Fig. 3.9).

3.1.4 Pair Interaction Potential

Liquid metals. To calculate the structure and properties of substances, the interparticle interaction law must be known. In simple liquids the interatomic potential depends only on the distance between the particles. Molten metals can be regarded as simple liquids. A metallic bond is not directed and is not saturated. Moreover, the ions of many metals have the electronic structure of a noble gas and have spherical symmetry.

A simple and commonly used potential in liquid-state physics is the 12-6 Lennard–Jones potential, in which the repulsive and attractive forces are inversely

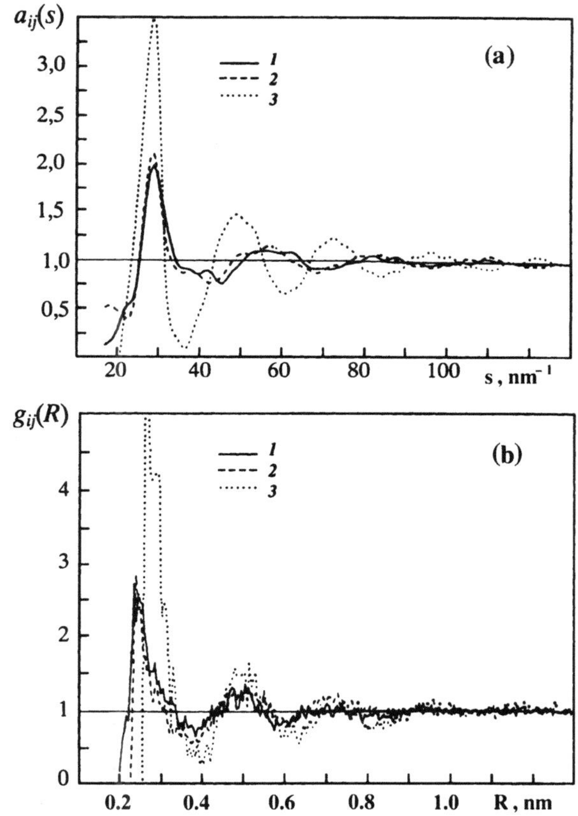

Fig. 3.9 Partial structure factors (**a**) and partial pair distribution functions of atoms (**b**) for an $Fe_{61}Sn_{29}$ melt at 1433 K: *1*—Fe–Fe; *2*—Fe–Sn; 3—Sn–Sn [67]

proportional to the distance raised to the power of 12 and 6, respectively. Consideration of the screening of positive metal ions by electrons leads to a long-range oscillatory potential $\varphi(R)$. The part of $\varphi(R)$ that is related to attraction is proportional tocos$(2k_fR)R^{-3}$, where k_f is radius of the Fermi sphere. In this case the total potential can be represented in the form [69, 70]

$$\varphi(R) = \frac{A}{R^{12}} - \frac{B\cos(2k_fR)}{R^3} \tag{3.30}$$

or by the more exact expression [1]

$$\varphi(R) = \frac{A}{R^{10}} - \frac{B}{R^6} + \frac{C}{R^3}\cos(2k_fR + a)\ \exp(-\Delta k_fR). \tag{3.31}$$

Here the first term describes the repulsion of the ions at short distances, the second term describes the short-range attraction caused by the correlation effects of the electrons in the ion shells, and the third term describes the long-range interaction through the conduction electrons. The coefficient $\Delta k_f \approx 0.05$ takes into account the blurring of the Fermi surface in liquid metals. The constants A, B, and C are expressed in terms of the effective diameter of the atom d_{eff}, the depth of the potential well, and the amplitude of the long-range oscillations, and the parameter a is chosen from the condition $\varphi(d_{\mathrm{eff}}) = 0$.

Oscillatory potential (3.30) has been used in the framework of the cell method to calculate several thermodynamic properties of Fe–Ni, Fe–Zr, and Ni–Si melts [69, 70].

Data on the interparticle interaction forces in liquid metals can be obtained from the results of diffraction studies. Three types of equations that relate the pair potential to structural characteristics of liquids have found extensive application [71, 72].

1. The Bogolyubov–Born–Green equation

$$\varphi(R) = -k_{\mathrm{B}}T \, \ln g(R) + \int_{0}^{\infty} \frac{d\varphi(t)}{dt} G(t, R)\, dt, \tag{3.32}$$

where

$$G(t, R) = \frac{\pi\rho_0}{R} g(t) \int_{-t}^{+t} (t^2 - q^2)(q + R)[g(|q + R|) - 1]\, dq,$$

$$-t \leq q \leq t; \quad 0 \leq t \leq R_{\max}.$$

2. The Percus–Yevick equation

$$\varphi(R) = k_{\mathrm{B}}T \ln\left[1 - \frac{C(R)}{g(R)}\right]. \tag{2.33}$$

3. The hyperchain equation

$$\varphi(R) = k_{\mathrm{B}}T[h(R) - C(R) - \ln g(R)]. \tag{3.34}$$

The total correlation function $h(R)$ and the direct correlation function $C(R)$ are defined by Eqs. 3.3 and 3.5. The principal approximations used to derive these equations were the Kirkwood superposition approximation, which expresses the three-particle correlation function in terms of pair distribution functions, and the pair interaction approximation.

Calculations of $\varphi(R)$ using expressions (3.32)–(3.34) were first performed in [73] for liquid alkali metals, Hg, Al, Pb, and Ar. An oscillatory potential was

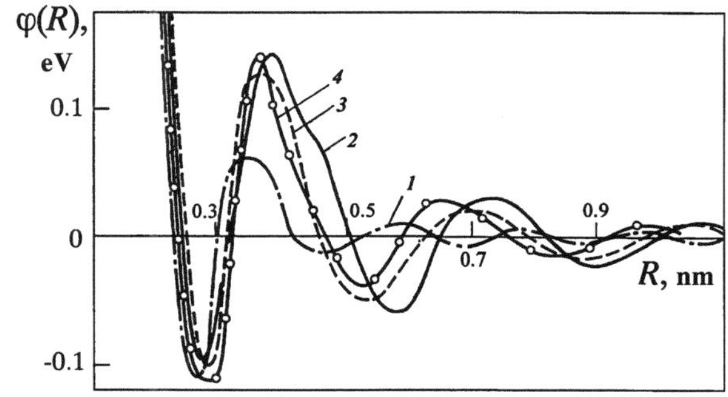

Fig. 3.10 Pair interaction potentials for liquid iron: *1*—calculation based on the Bogolyubov–Born–Green equation [75]; *2*, *4*—calculation based on the Percus–Yevick equation and X-ray structural data of different investigators [77]; *3*—calculation based on Eq. 3.30

obtained for the metals, while the potential for argon was described by a curve of the Lennard-Jones type. The results of subsequent calculations of $\varphi(R)$ for 25 elements were presented in [74–78].

Figure 3.10 presents the interparticle interaction potentials in molten Fe that were found by different investigators using Eqs. 3.30, 3.32, and 3.33. The potential functions for each element are close to one another with respect to the position and depth of the first minimum, but they differ with respect to the oscillation length and amplitude.

The reasons for the divergence of the data from different investigators and the conditions of applicability of the equations were examined in detail in review [6]. It was noted that the Percus–Yevick and hyperchain approximations give the best results when they are used for substances with short-range interaction forces and a low density, i.e., nonmetallic liquids. The principal error in the calculated values of the pair potential is introduced by the inexact determination of the structure factor in the region of small scattering vectors, where the Fourier transform of the direct correlation function $C(R)$ is localized. The influence of small-angle scattering on $\varphi(R)$ can be seen from the following data [6]. Figure 3.11 shows there types of interpolations of the experimental plot of $a(s)$ for gallium in the range of values of s from 0 to 12 nm^{-1}. In the first variant (curve 2) all the values of $a(s)$ are constant and equal to $a(0)$. In the second case a linear interpolation form (curve 3) was used, and in the third variant a quadratic interpolation form (curve 4) was used. Curve 1 is experimental. The pair potential was calculated using the hyperchain equation with allowance for the different course of the structure factor as $s \to 0$. The greatest deviation of the calculated values of $\varphi(R)$ from exact curve 1 is observed in the case in which quadratic interpolation was used.

Recently, the pair interaction potential in liquid metals was determined using the equation obtained in [79] on the basis of the pseudopotential method:

$$\varphi(R) = -\frac{z_e^2 e^2}{R} + \frac{V}{\pi^2} \int\limits_0^{\infty} F(s) \frac{\sin sR}{R} s^2 ds. \tag{3.35}$$

Here z_e is the valence; e is the change of an electron; V is the atomic volume. The first term gives the direct ion–ion interaction, and the second term takes into

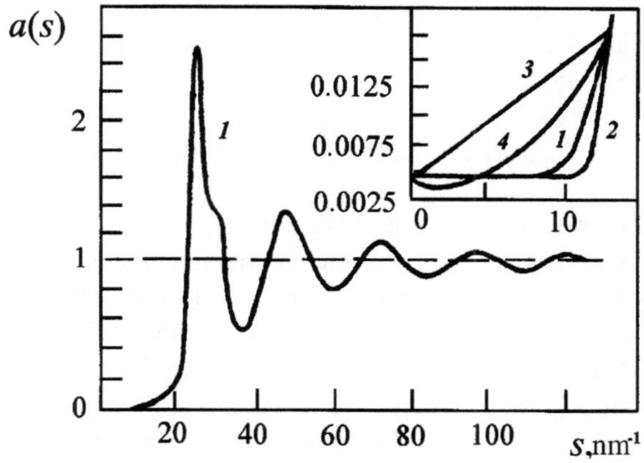

Fig. 3.11 Structure factor of liquid gallium: *1*—experiment; *2–4*—various methods for interpolating the $a(s)$ curve in the range $0 < s < 12\ \text{nm}^{-1}$ [6]

account the influence of the electrons on the ion–ion interaction. The characteristic function $F(s)$ is related to the pseudopotential form factor and the dielectric constant. The interatomic potentials in the liquid Al, Mg, and Zn, in Al–Mg and Al–Zn alloys [76, 78, 80], and in Cu [81] have been determined by this method.

The influence of the form of the pair potential on the structure factor and the radial distribution function has been investigated in several studies [78, 82]. For this purpose values of $\varphi(R)$ calculated by the methods described above were used in computer experiments to simulate the structure of molten Na, Al, Fe, and Ni. It was found that variation of the repulsive branch and the character of the oscillations of $\varphi(R)$ within reasonable limits does not have a strong effect on the form of the $a(s)$ and $g(R)$ curves and that the position and depth of the main potential minimum are decisive. The RDFs for iron and aluminum calculated by a molecular dynamics method with different potentials [78, 83] are close to the experimental $g(R)$ curves. However, it was noted in [84] that the long-range part of the pair potential has a significant influence on the determination of the properties of liquid metals using correlation functions. Calculations, for example, of the viscosity

$$\eta = \frac{2\pi}{15}\rho_0^2\left(\frac{m}{k_B T}\right)^{1/2}\int_0^\infty \frac{d\varphi(R)}{dR}g(R)R^4 dR \tag{3.36}$$

and the surface tension

$$\sigma = \frac{\pi}{8}\rho_0^2\int_0^\infty \frac{d\varphi(R)}{dR}g(R)R^4 dR \tag{3.37}$$

require knowledge of the exact form of the potential not only in the region of the main minimum, but also at large distances.

Ion melts. In calculations of the structural and thermodynamic characteristics of liquid simple salts, the following form of the pair potential is generally used (see [82, 85]):

$$\varphi_{ij}(R) = \xi_i \xi_j R^{-1} + b \cdot \exp\left[B\left(d_{ij} - R\right)\right] + C_{ij} R^{-6} + D_{ij} R^{-8}, \tag{3.38}$$

where ξ is the ionic charge and d_{ij} is the sum of the ionic radii. The first term describes the electrostatic interaction, the second term describes the repulsive energy, and the last two terms describe the dipole–dipole and dipole–quadrupole interactions. The constant b has the same value for all alkali metal chlorides, and B is a general parameter for three ion pairs in each salt. The energy parameters C_{ij} and D_{ij} are calculated from spectroscopic data, and the values of the constants in Eq. 3.38 for salts of Li and Cl, NaCl, and KCl were presented in [86]. The partial radial distribution functions and the interaction curves of pairs of ions of the same species and of different species were found by computer simulation methods [87]. The contributions of the various interactions to the total potential energy φ_{tot} were determined. The ratio between values of the electrostatic (φ_{e}), repulsive (φ_{r}), dipole–dipole (φ_{dd}), and dipole–quadrupole (φ_{dq}) energies is equal, for example, to 100:13:3.5:0.3.

3.2 Constitution of Metallic Melts

Numerous results of diffraction studies have shown [1, 30] that many of the liquid metals studied have a definite structure, which is close to the structure of the respective substance in the crystalline state, at the melting point and sometimes when the melt is superheated considerably. When several metals are melted and the temperature is raised further, as well as when other elements are introduced, structural changes associated with a change in the nature of the chemical bonds between the atoms and their relative arrangement can occur. Therefore, liquid metals can be divided into the following groups.

1. Metals which maintain the short-range order of the high-temperature crystalline phase upon melting. They include all metals which have been investigated by diffraction methods and have a bcc structure (the alkali metals Li, Na, K, Rb, and Cs, the alkaline earths Ca and Ba, and the 3*d* transition metals Ti, V, and Cr) or an fcc structure (Mg) before melting. It should be noted that about 40 elements have bcc packing of their atoms before melting; therefore, maintenance of this packing in the liquid state should be expected.
2. Elements in which a change in structure occurs upon melting or upon some superheating above the melting point. This group includes Ge Si, Te, and Se, which melt with a semiconductor–metal or semiconductor–semiconductor transition and become liquids with a single structure upon further heating, as well as metals that have either compact packing of the atoms (Al, Ag, Au, Cu,

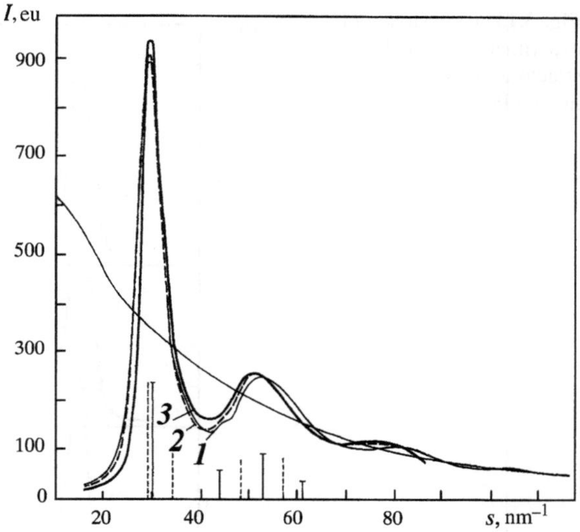

Fig. 3.12 Intensity curves for molten Fe: *1*—1,550; *2*—1,680; *3*—1,750°C. *Vertical lines*—positions of the interference in the bcc (*solid lines*) and hcp (*dashed lines*) structures

Fe, Co, Ni, Pd) or loose structures (Bi, Sn, Hg, In, Ga, Sb), which are due to the high percentage of covalent bonds, before melting.

3. Binary alloys. Here, as in the crystalline state, the structural transitions are associated with a change in the interatomic interaction under the influence of the second component and the increase in temperature.

3.2.1 Iron, Nickel, Cobalt

The early studies [39, 74, 88–93] of the constitution of liquid iron by diffraction methods referred mainly to the 1,550–1,620°C temperature range. The authors of most of these publications arrived at the conclusion that the body-centered cubic (bcc) packing characteristic of the crystal structure at high temperatures is maintained upon melting. The possibility of the simultaneous existence of clusters of atoms with bcc and fcc structures in molten iron was postulated in [91]. Diffraction experiments performed at higher temperatures (up to 1,750°C) revealed [41, 58, 94, 95] an abrupt change in the short-range parameters at approximately 1,650°C, which is evidence of the occurrence of structural transformations in liquid iron.

Experimental intensity curves (Fig. 3.12) and RDFs at 1,550–1,750°C [41, 58] show that the positions of the first and second maxima of the intensity curve correspond to the (110) and (200) reflections of δ-Fe, which has bcc packing of the atoms, at 1,550°C and to the (111) interference line of γ-Fe (the fcc structure) at 1,680 and 1,750°C [50].

Superheating iron to 130°C above the melting point causes not only displacement of all the maxima of the intensity curve toward smaller scattering angles

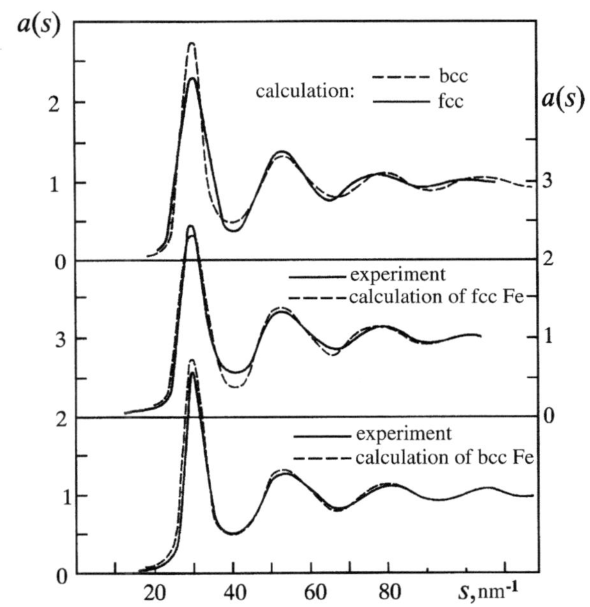

Fig. 3.13 Comparison of experimental and calculated structure factors for molten Fe

(which is especially noticeable at the third and fourth maxima), but also an abrupt decrease in the height and an increase in the half-height width Δs of the main peak, which becomes asymmetric on the large scattering angle side. The latter may be attributed to the fact that γ-Fe has a comparatively strong (200) interference line at $s = 34.2\ \mathrm{nm}^{-1}$.

Figure 3.13 compares the experimental and theoretical structure factors [1]. The latter were calculated for 8 coordination spheres in a bcc lattice ($R_1 = 0.258$ nm, $\overline{u_1^2} = 6 \times 10^{-4}\,\mathrm{nm}^2$) and for 10 coordination spheres in an fcc lattice ($R_1 = 0.262$ nm, $\overline{u_1^2} = 7 \times 10^{-4}\,\mathrm{nm}^2$). The experimental curve at 1,550°C practically coincides with the calculated curve for the body-centered structure, and the curves at 1,680 and 1,750°C are described well by the more compact packing of the atoms.

Additional information about the structure of molten ion is provided by an analysis of the radial distribution functions.

The first RDF maxima for Fe, Co, and Ni near the melting points (Fig. 3.14), as in the case of Ti, V, and Cr, have marked asymmetry, which, according to its position, corresponds to the second coordination sphere with bcc packing of the atoms.

When liquid iron is heated to ~1,650°C, the shortest interatomic distance increases abruptly by 0.005 nm (to 0.258–0.261 nm [58, 95]) and becomes close to the value $R_1 = 0.2604$ nm for γ-Fe with an fcc structure.

It is should be noted that the value of R_1 for all the liquid metals that have been investigated remains constant or increases slightly when the melt is heated. In this respect the increase in the interatomic distance in molten Fe with increasing temperature is anomalous.

Fig. 3.14 Radial distribution curves of atoms in molten cobalt (*1*), nickel (*2*), and iron (*3*) near their melting points and radial distribution of iron atoms at 1,750°C (*4*). Curves 2, 3, and 4 are shifted along the vertical axis by four, nine, and thirteen units [95]

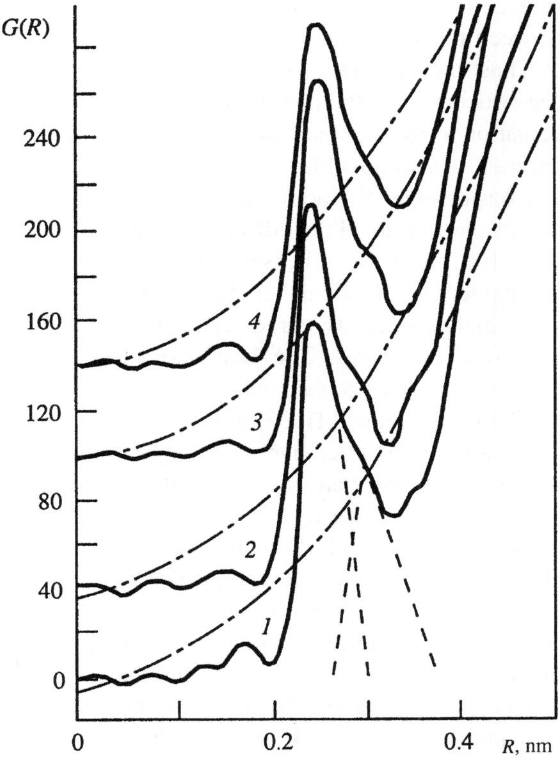

The variation of the short-range order parameters in Fe melts that was discovered in diffraction investigations is consistent with numerous indirect data from the investigation of the temperature dependence of structure-sensitive properties. For example, polytherms of the magnetic susceptibility [96–98], density [97, 99, 100], kinematic viscosity [101, 102], activation energy for viscous flow [101], and solubility of hydrogen [96] have discontinuities at ~1,650°C.

It should be noted that the hexagonal close packed ε-Fe phase, which combines the overlapping of the d^6 subshells in the bcc structure with the more compact atomic packing in the fcc lattice, appears in the crystalline state at high pressures [103]. This modification of Fe has also been discovered in Fe–N and Fe–C solutions [103].

Iron atoms have a $3d^6 4\,s^2$ outer electronic configuration. On the basis of the +2 valence that is often encountered in compounds, it is assumed [30, 103] that iron atoms are doubly ionized in the metallic state. The exchange interaction between the d^6 electronic subshells of Fe^{2+} ions leads to the bcc structure. The solid–liquid phase transition does not alter the electronic state of the system; therefore, molten ion has short-range order that corresponds to the high-temperature δ modification. The reorganization of the short-range order upon increases in the temperature of the melt can be caused by additional ionization of Fe^{2+} ions to Fe^{3+} (the higher valence form) [30] or excitation of d^6 electrons, upon which they are distributed between localized and delocalized states.

Thus, the short-range order of molten ion, which bears some similarity to the bcc structure near the melting point, undergoes a structural transformation with a transition to fcc-like packing at ~1,650°C. Coexistence of these two structures in the melt is also possible.

Considerable alteration of the short-range order parameters upon melting was discovered in all the diffraction studies of liquid nickel [39, 41, 58, 93, 95, 96, 104, 105]. Despite this, several investigators concluded that fcc packing is maintained in liquid nickel. However, it was already shown in [39] that the experimental intensity curves are not described by the more compact cubic and hexagonal structures. The best agreement with experiment was observed for models based on bcc packing.

The values found [41, 58, 95] for the short-range order parameters of liquid Ni, Co, and Pd, unlike Fe, diverge strongly from the characteristics of these elements in the crystalline state already at the melting point. The positions of the first maxima on the intensity curves do not correspond to the strongest interference lines of solid Ni and Pd, which have fcc lattices, and the shortest interatomic distance decreases by 0.009 nm. This value is comparable to the shortening of the radius of the atoms when their coordination changes from twelve to eight [106]. An increase in the temperature to 1,750°C smoothly broadens the peaks of the intensity curves and shifts them slightly toward smaller values of the scattering angles. The positions of the remaining maxima remain practically unchanged. The area under the first peak on the radial distribution curves of the atoms, which specifies the coordination number z_1, drops as the temperature is raised at the expense of the small asymmetric part. When its resolution is symmetric, the area remains close to 8 square units. The significant asymmetry of the first RDF maxima of liquid cobalt and nickel found in [95] allows their division into two peaks (Fig. 3.14), whose positions ($R_1 = 0.249$ nm, $R_2 = 0.298$ nm) and areas ($z_1 \approx 8$, $z_2 \approx 5$ atoms) are close to the parameters for bcc packing of the atoms. A comparison with the crystalline state provides evidence that the short-range order in Ni, Co, and Pd changes already upon melting. A comparison of the experimental intensity curves with the calculated curves for eight coordination spheres under the assumption of the bcc structure of Ni reveals good agreement between them (Fig. 3.15).

Cobalt (d^7s^2), nickel (d^8s^2), and palladium (d^{10} or d^9s^1) belong to subgroup VIII and crystallize in the more compact cubic lattice, because the Co^{2+}, Ni^{2+}, and Pd^{2+} ions have an electronic configuration with a spherically symmetric outer s subshell in the metallic state [103]. However, an $s \rightarrow d$ electronic transition is not ruled out in the excited state. The overlap and exchange interaction between the d subshells of neighboring ions possibly leads to body-centered cubic or tetragonal packing (8 + 2 + 4 coordination) of the atoms. Khrushchev [30] associates the possible alteration of the short-range order in the melt with reorganization of the electronic structure of the ions due to the detachment of the third valence electron and its transition into the "free" state.

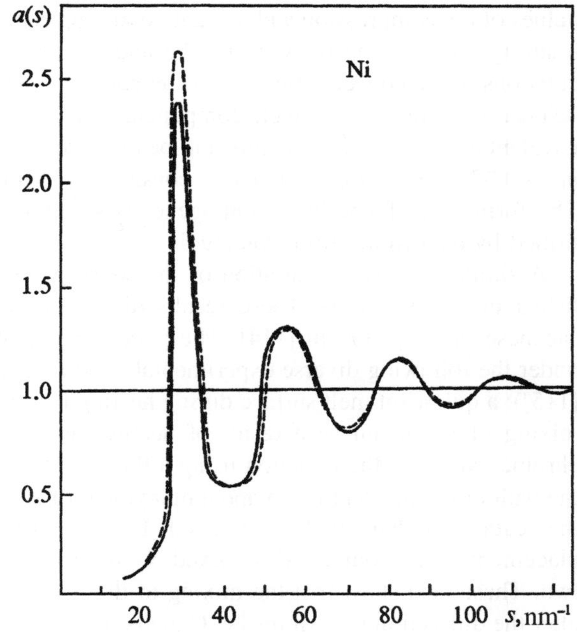

Fig. 3.15 Structure factor of molten Ni at 1,470°C: *Solid line*—experiment; *dashed line*—calculation for the bcc structure

Unlike Fe, in molten Ni the structure-sensitive properties vary monotonically with increasing temperature.

3.2.2 *Iron–Carbon*

The study of the constitution of iron–carbon melts has been the subject of numerous publications. However, the opinions regarding their structure are contradictory. For example, some investigators [96, 107] regard liquid pig iron as a colloidal solution, assuming that it contains equilibrium microclusters of carbon, whose number and size determine the physical properties of the solutions. Conversely, according to other investigators, the variation of the activity and partial energy parameters of carbon in Fe–C melts (see reviews [108–110]) indicate a comparatively strong interaction between atoms of different kinds. It is assumed that carbon dissolves in liquid iron by being inserted into octahedral interstitial spaces in the quasilattice. Then, the thermodynamic properties of a melt are described by the laws that are valid for austenite up to a carbon content C $\approx$ 3.5 wt.% (15 at.%). The number of octahedral cavities in the γ-iron lattice in which carbon atoms can be found amounts to $^1/_4$ of their total number (the Fe_4C structure). At larger carbon concentrations ($n_C > 0.15$) the Fe_3C structure is realized. Here the number of interstitial spaces for accommodating C atoms increases to $^1/_3$. In comparison to γ solid solutions, Fe–C melts form with larger values of the compression and a negative mixing energy (see [111]). This attests to a stronger interaction between the Fe and C atoms in the liquid state. The calculations of the concentration of current carriers in [112] indicate that some of the valence electrons of the melt components participate in the formation of Fe–C covalent bonds. The most significant localization of electronic states is observed at $n_C > 0.07$ (1.5 wt.%). It increases considerably with increasing carbon content. The formation of non-ideal homogeneous solutions in Fe–C melts is also confirmed by data from diffraction studies.

A similar pattern of variation of the short-range order parameters was established in [41, 91, 113]. These results differ somewhat from the results of the Japanese investigators in [114]. The good agreement between the data obtained under the following diverse experimental conditions is especially noteworthy (see [115]): a quiescent melt surface during heating by a resistance furnace and intense mixing of the metal as a result of heating by high-frequency currents; monochromatization of the radiation using a flat crystal of a monochromator placed on the path of the incident beam and employment of a system of differential filters on the scattered radiation; the use of soft Fe Kα and hard Mo Kα X-radiation; displacement of the counter with a fixed X-ray tube and synchronous rotation of the two. These data show that the mixing of the metal during the recording does not alter the diffraction pattern for Fe–C melts. In other words, the surface layer of the liquid is not enriched with one of the components.

The results of an analysis of the intensity curves (Fig. 3.16) and the radial distribution functions obtained with small degrees of overheating above the liquidus line and after relatively prolonged maintenance of the melt before the X-ray diffraction experiment were presented in [1, 41].

Since the relative scattering power of carbon atoms, which can be defined by the expression

$$K_C^2 = \left(\frac{f_C^2(s)}{n_C f_C^2(s) + n_{Fe} f_{Fe}^2(s)} \right)_{av},$$

in which n_C and $f_C(s)$ are the mole fraction and the atomic scattering factor of C, is approximately 50 times smaller than that of Fe atoms, the diffraction pattern for Fe–C melts will be determined mainly by the packing of the Fe atoms. Therefore, the changes in the interference curves should be associated with structural transitions in Fe.

Figure 3.17 presents experimental plots (curves 4–8) of the variation of various short-range order parameters as functions of the composition of Fe–C melts. The following three segments can be distinguished on the plots: 1) less than 0.5 wt.%, 2) from 0.5 to 3.5 wt.%, 3) from 3.5 to 6 wt.% C. Small (less than 0.5 wt.%) additions of C to Fe slightly increase the interatomic distance R_1 and shift the main maximum of the structure factor toward smaller scattering angles. The coordination number remains constant. Raising the carbon content to 3.5 wt.% does not alter the values of s_1 and R_1, but the number of nearest neighbors z_1 and the mean square deviation of the atoms u^2 calculated from the half-width of the coordination peak increases

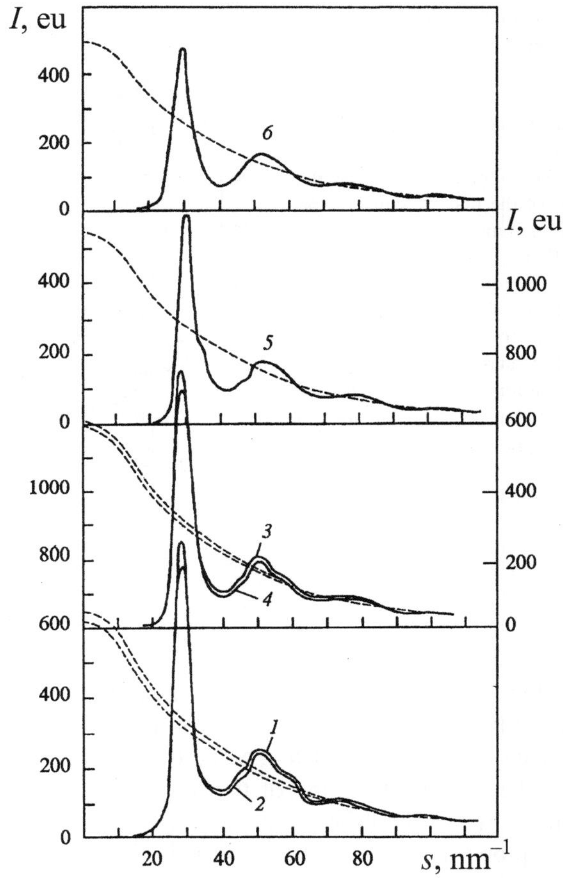

Fig. 3.16 Intensity curves for Fe–C melts containing carbon: *1*—0.5 wt.%, 1,550°C; *2*—1.5 wt.%, 1,450°C; *3*—2 wt.%, 1,400°C; *4*—3 wt.%, 1,300°C; *5*—4 wt.%, 1,200°C; *6*—6 wt.%, 1,650°C

sharply and reaches a maximum, at which $z \approx 12$, as in γ-Fe. An increase in the coordination number with constancy of the interatomic distance should decrease the molar volume of the liquid. This is also observed (see Fig. 3.18) on the density isotherms in the range of carbon concentrations from C $\approx$ 1 to 4 wt.%. An increase in the carbon content from 4 to 6 wt.% results in lowering of the value of z_1, an increase in R_1, and a sharp decrease in density.

The intensity curves for hypoeutectic alloys distinctly display the asymmetry of the second maximum (Fig. 3.16). This may be a consequence of the presence of two structures in the melt or the formation of ordered regions of large dimensions. The latter can cause decreases in the blurring and the width Δs of the diffraction maxima and in their partial separation. In fact, additions of up to 2 wt.% C to liquid Fe lower the value of Δs, i.e., increase the radii of the coherent scattering regions and, consequently, the coordination number. Below 3 wt.% C the first coordination peaks have a symmetric shape relative to R_1, which is characteristic

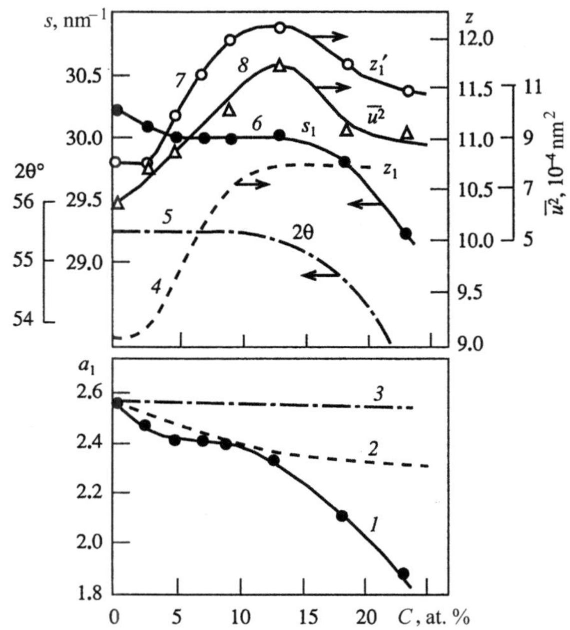

Fig. 3.17 Dependence of the short-range order parameters of Fe–C melts on the carbon content: *1–3*—experimental (*1*) and calculated dependences for the Fe_4C + α-Fe (*2*) and C_n + α-Fe (*3*) models of the height a_1 of the first structure factor maximum; *5, 6*—positions of 2θ [1] and s_1 of the first maximum of the intensity curves; *4, 7*—coordination numbers for symmetric (z_1) [113] and complete (z'_1) isolation of the first RDF maxima; *8*—mean square deviation $\overline{u^2}$ of the atoms

of the fcc structure. In pure Fe, which has bcc packing of the atoms, the first RDF maximum is asymmetric. Some asymmetry also appears on the large-interatomic-distance side in melts with 4.5 and 6 wt.% C.

The stronger lowering of the height of the main maximum on the intensity curves, which is observed at a smaller width than in the case of pure Fe and is not proportional to the Fe content, is noteworthy. It attests to the formation of interstitial solutions in Fe–C melts. In fact, when a colloidal solution is present, the height a_1 of the first maximum of the structure factor should remain nearly constant due to the low scattering power of C atoms and the maintenance of the short-range order between the Fe atoms (curve 5, Fig. 3.17). However, the experimental values of a_1 (curve 7, Fig. 3.17) vary similarly to the values of s_1, deviating strongly away from curve 3. In the range of carbon concentrations from C $\approx$ 1 to 3 wt.%, they coincide with curve 2, which was calculated under the assumption of an ordered interstitial solution of carbon in γ-Fe octahedral cavities with the formation of an Fe_4C structure. When C > 3.5 wt.%, curves 1 and 2 diverge strongly.

The data presented indicate that carbon (0.5–3.5 wt.%) stabilizes the structure of liquid γ-Fe. Its insertion into interstitial spaces with dimensions (0.106 nm) close to the diameter of C ions (0.11 nm) scarcely alters the interatomic distances in the melt and the positions of the main interference maxima (the values of R_1 and s_1 are constant). The small decreases in s_1 and a_1, the increase in R_1, and the constancy of z_1 at carbon concentrations in the range C < 0.5 wt.% are apparently due to the dissolution of C in the δ-Fe quasilattice [116]. For example, its insertion into octahedral cavities, which have small dimensions (0.0392 nm), causes strong tetragonal distortion of the bcc packing and, consequently, changes in the position and height of the main interference maximum of the intensity curves. The region for the existence of ordered regions with a bcc structure in Fe–C melts is broadened considerably compared with δ solid solutions. This fact was previously noted in [117] in the context of an investigation of the radiating power and density. According to the data from diffraction studies, the bcc–fcc structural transition in Fe–C alloys occurs above the liquidus line at a carbon concentration $C \approx 0.5$ wt.%, while the maximum solubility of carbon in solid δ-Fe at 1,493°C equals 0.1 wt.%.

The abrupt changes in the structural characteristics at carbon concentrations $C > 3.5$ wt.% indicate a new change in the short-range order in the liquid. This change can be the fcc(γ-Fe)-fcc(ε-Fe) transformation [116]. The latter structure has been discovered in pure iron at high pressures, as well as after the superfast quenching of Fe–C melts [118]. When hypoeutectic alloys (1.5–3.0 wt.% C) were cooled at a fast rate, an austenite phase appeared as the main component (80–90 wt.%). The quenching of high-carbon samples (4.3–4.7 wt.%) fixed the interstitial solid solution of C in the hexagonal close-packed modification of iron. The content of the ε phase reached values as high as 45–47 wt.%.

Figure 3.19 compares experimental interference curves and curves calculated using various models. For melts with a carbon content $C = 0.5$ to 3.0 wt.%, the diffraction pattern is described well by a quasieutectic structure, whose elements are ordered regions with the short-range order of γ-iron (Fe_4C) and δ-iron. At 6 wt.% C the experimental structure factor is more consistent with that for the Fe_3C model, i.e., an interstitial solution of carbon in the ε-Fe quasilattice. In this case the position of the first maximum of the calculated curve approximates the experimental value $s_1 = 0.293$ nm if the interatomic distances $R_{\text{Fe–C}} = 0.185$ to 0.215 nm and $R_{\text{Fe–Fe}} = 0.249$ to 0.268 nm, which exist in the structure of cementite (a deformed ε-Fe lattice), are used. In the latter case each carbon atom is surrounded by six carbon atoms located at the corners of a trigonal prism, and the Fe–C bonds have a directed covalent character. The increased strength of the covalent bond in melts with a carbon content $C > 3.5$ wt.% is evidenced by the lowering of the coordination number and the mean square displacement of the atoms.

The occurrence of structural transformations in Fe–C melts is also confirmed by the course of property–composition curves (Fig. 3.18). The carbon concentrations at which the change in the short-range order observed in diffraction experiments occurs correspond to abrupt changes in the physicochemical properties. Isotherms of the concentration dependences of Fe–C melts display minima of the kinetic viscosity, the activation energy for viscous flow, the magnetic susceptibility, and the adiabatic compressibility coefficient, a significant increase in the specific resistance, and extreme changes in the density and surface tension near 0.5 wt.% C [96, 119–123], i.e., in the region of the transition from δ-like packing of the atoms to a structure with the γ-Fe short-range order. A new sharp change in the course of the property–composition curves, which attests to reorganization of the short-range order in the melts, is observed at $C > 4.0$ wt.%. The decrease in the magnetic

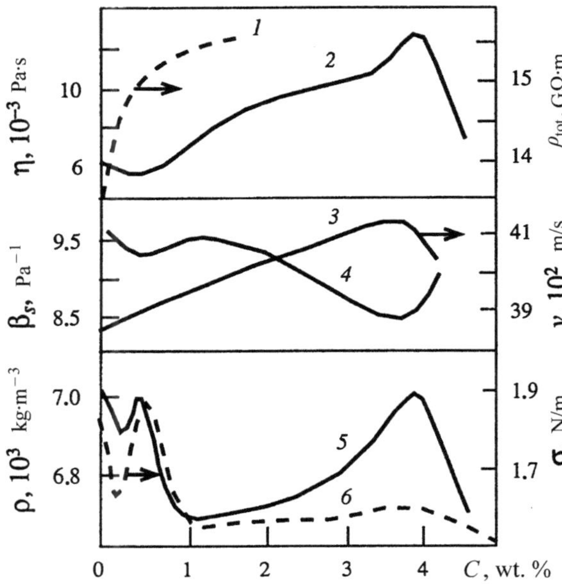

Fig. 3.18 Concentration dependence of the physicochemical properties of Fe–C melts at the liquidus temperature: *1*—specific resistance [123]; *2*—viscosity [120]; *3*, *4*—speed of sound and adiabatic compressibility [122]; *5*, *6*—density and surface tension [99]

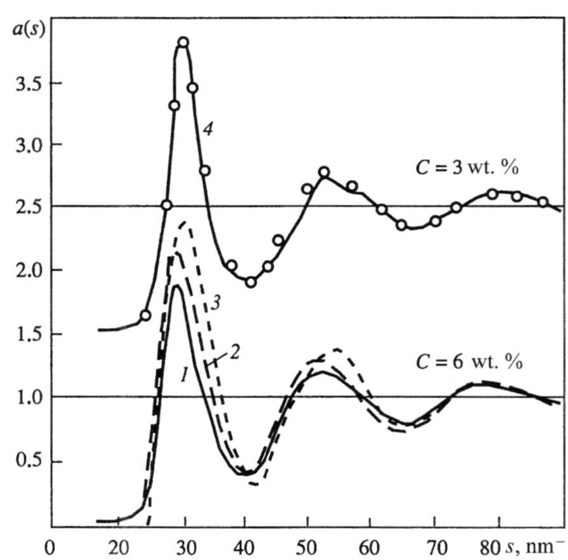

Fig. 3.19 Experimental (*1*) and calculated (*2*–*4*) structure factors for Fe–C melts: *2*—model of Fe_3C (ε-Fe); *3*—model of Fe_4C (γ-Fe); *4*—model of the Fe,C + α-Fe quasieutectic

moment with increasing carbon concentration may indicate enhancement of the covalent character of the bonds [121].

Thus, the results of diffraction and physicochemical studies lead to the conclusion that two types of structural transformations take place in iron–carbon melts: δ-Fe (bcc) → γ-Fe (fcc) at C ≈ 0.5 wt.% and γ-Fe → ε-Fe (fcc) at

C $\approx$ 4.0 wt.%. Low-carbon melts (up to 0.5 wt.%) have bcc packing of the atoms, as does pure iron at the melting point. In the range of carbon concentrations from 0.5 to 4.0 wt.%, a quasieutectic structure, in which the ordered regions are interstitial solutions of carbon in iron with bcc and fcc (the Fe_4C structure) packing of the iron, is observed. The percentage of γ–Fe increases with increasing carbon concentration. Solutions containing more than 4 wt.% C have clusters with short-range order of the γ-iron (Fe_4C) and ε-iron (Fe_3C) types.

It is noteworthy that the ambiguity of the theories regarding the state of carbon in Fe–C melts is attributable not only to difficulties of an experimental nature, but also to the fact that it depends on the holding time of the melt and the temperature. The influence of these factors on alloys that are close to the eutectic composition (4.3 wt.% C) is especially strong and is governed by the duration of the micro-dissolution of carbon in iron.

The effects of holding a sample in the liquid state and of the temperature on the structural characteristics of a melt containing 14.7 at.% C (3.7 wt.% C) were investigated on a diffractometer with a high resolving power and a high scattering angle reading accuracy in [124]. The results of the X-ray structural analysis are presented in Fig. 3.20. After the melt is held for more than 30 min, besides the main RDF maximum at $R_1 = 0.253$–0.263 nm, an additional peak appears at $R_2 = 0.30$–0.33 nm in the 1,250–1,450°C temperature range and becomes more distinct as the temperature is raised. A scattering intensity that is 2–2.5 times greater than for pure iron (curve 5) is observed on the structure factor curves in the region of small scattering vectors ($s = 19$–24 nm^{-1}). According to Rostovskaya and coworkers [124], these data attest to the microheterogeneous structure of Fe–C melts, which can be described by clusters of atoms of two types. The clusters of one type form on the basis of compact packing of the Fe atoms, and the positions of the main RDF maxima are associated with them. The other types of atomic formations, in which the average concentration of C atoms is significantly higher than in the clusters of the first type, cause the increased intensity on the $a(s)$ curves and the additional peak on the RDF, and when the sample is held for a short time after melting, they have packing of the cementite type, as does the crystal. In fact, in the Fe_3C crystal lattice the Fe–Fe distances are equal on the average to 0.254 nm, and the Fe–C distances are equal on the average to 0.204 nm, in agreement with the position of the first RDF maximum of the melt. The Debye reflections of cementite fall under the first diffraction maximum, broadening it and lowering its height.

When the holding time of the melt at 1,250°C is increased to 30 min (curves 1 and 2 in Fig. 3.20), as well as when the temperature is raised to 1,450°C (curves 3 and 4 in Fig. 3.20), the microclusters of the cementite type disintegrate, redistribution of the carbon atoms between the microclusters occurs, and the structural characteristics of the melt change. The structure of the melt is described by the model of a quasieutectic, whose elements are orderings of atoms of different kinds with the short-range order of γ-Fe and ε-Fe.

The slowness of the process of the formation of microclusters in liquid iron from atoms of different kinds with different carbon contents in them during cooling

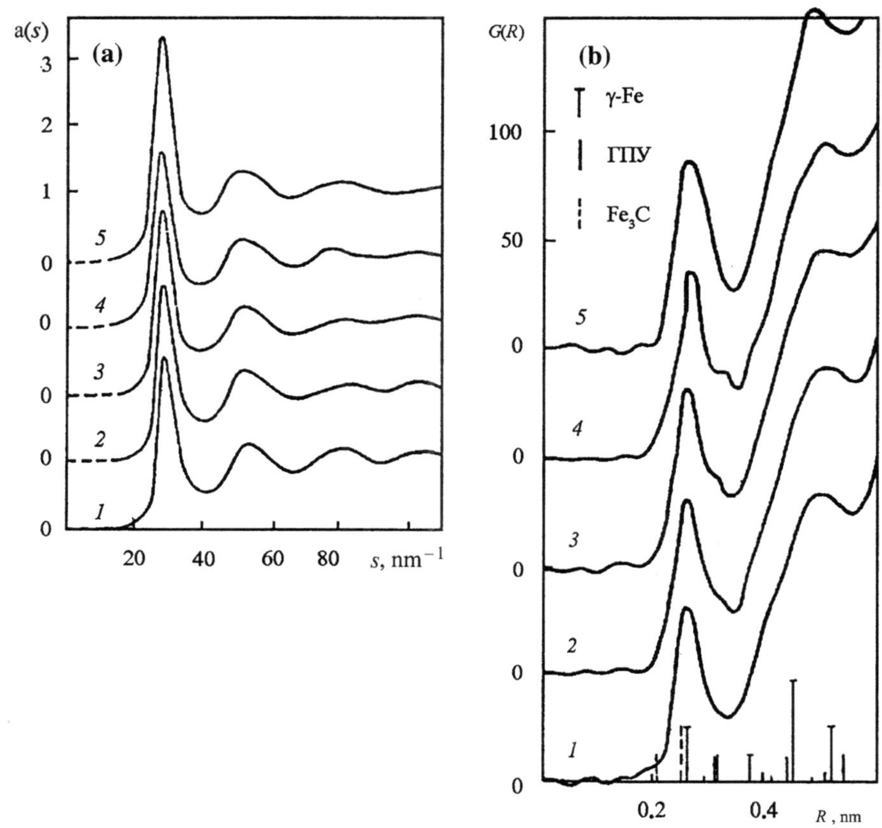

Fig. 3.20 Structure factors (**a**) and RDF (**b**) of $Fe_{86.3}C_{14.7}$ (3.7 wt.% C) [124]: *1*—at 1,250°C without holding in the liquid state; *2*—at 1,250°C with holding for 30 min; *3*—at 1,350°C; *4*—1,450°C; *5*—iron at 1,600°C (*vertical lines*—coordination spheres for different types of packing of the atoms)

from high temperatures is indicated by the hysteresis of the structural characteristics obtained during the heating and cooling of melts containing 3.2 and 3.9 wt.% C [125].

3.2.3 Iron–Boron

Alloys of this system are characterized by the fact that because of the weak scattering power of boron atoms and the relatively small boron concentration, changes in the diffraction pattern obtained as a result of X-ray diffraction analysis of melts should be associated with changes in the packing of the iron atoms under the influence of boron.

Fig. 3.21 Concentration dependence of the structural characteristics of iron–boron melts near the liquidus temperature [129]

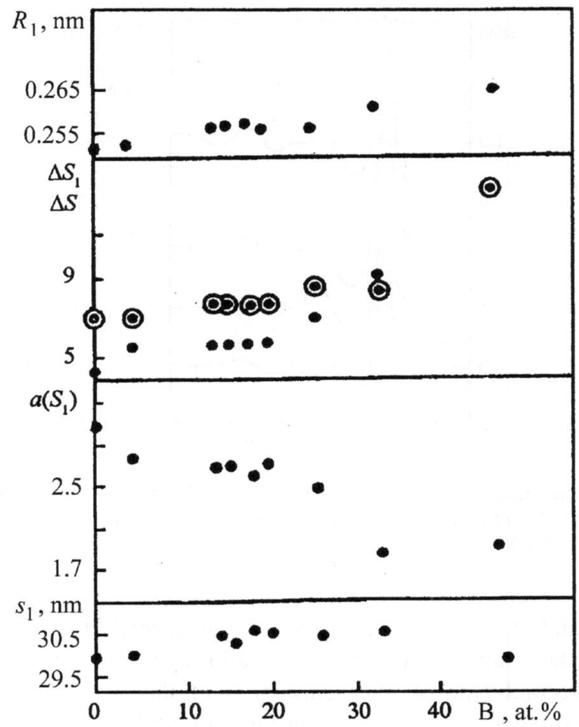

The temperature and concentration dependences of the structure of Fe–B melts were studied in [126–128]. Attention was focused on alloys with eutectic and hypereutectic compositions, including the stable intermetallic compounds Fe_2B and FeB and the metastable Fe_3B phase, which forms when an amorphous ribbon crystallizes. The constitution of hypoeutectic alloys (less than 17 at.% B), for which several peculiarities were discovered on isotherms of physical properties of the melts, was thoroughly investigated at 1,600°C in [129]. All the investigators arrived at the conclusion that Fe–B melts are microheterogeneous systems and that neither a statistical distribution of the atoms of different kinds nor a quasieutectic structure, which presumes the presence of microregions of iron and boron atoms, is realized in them. The structural components of these melts consist of two types of atomic microclusters with small and large (boride-like) concentrations of B atoms. In the hypereutectic alloys, the structural parameters vary sharply, beginning at the composition of the compound Fe_3B (Fig. 3.21). The quantitative ratio between the microclusters of the two types varies with increasing temperature. The structural characteristics of the melts, such as s_1, R_1, and z, are determined mainly by the structural component of the first type, which has short-range order that is similar to a solution of boron atoms in close-packed γ-Fe. Therefore, its contribution decreases with increasing boron content, as is conspicuously manifested in the values of $a(s_1)$

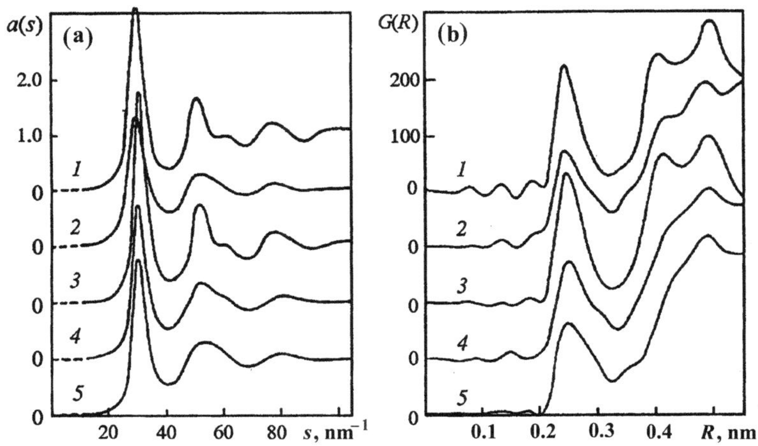

Fig. 3.22 Structure factors (**a**) and RDF (**b**) of the alloys $Fe_{96}B_4$ (*5*), $Fe_{80}B_{20}$ (*3*, *4*), and $Fe_{74}B_{26}$ (*1*, *2*) in the liquid (*2*, *4*, *5*) and amorphous (*1*, *3*) states at T, °C: *2*—1,300, *4*—1,250, *5*—1,450, *1*, *3*—20 [127]

and Δs and in the shape of the second diffraction peak and the first RDF maximum for hypereutectic alloys (Fig. 3.22). The short-range order of the second type of microclusters is of the type observed in Fe_3B or FeB with a strong interaction between the atoms of different kinds. The compound FeB is the most stable. It has a higher melting point (1,550°C) and a sharp maximum on the liquidus line. Scattering on these atomic formations occurs on the small-angle branch of the first interference maximum, increasing the intensity of the branch and broadening the maximum. The RDF displays a secondary peak at a distance of $\sim$0.3 nm and a weak hump on the left-hand branch of the main maximum at $\sim$0.2 nm, which are clearly attributable to the Fe–Fe and Fe–B interactions in the intermetallic compounds. In fact, the parameters of the rhombic FeB lattice are equal to 0.295–0.304 nm, and the Fe–B distance is slightly greater than 0.2 nm.

The constitution of the alloy $Fe_{82}B_{18}$, which is close to the eutectic composition, was studied in [128] in the 1,250–1,550°C temperature range in a regime of heating and cooling from a high temperature. At the liquidus temperature the structural characteristics of this and other near-eutectic alloys and of pure iron differ significantly. Compared with Fe, the first diffraction maximum of the melts is shifted toward larger values of s, its height $a(s)$ is decreased, its width Δs is increased, and a higher intensity is observed in the small-angle portion of the maximum. It follows from the RDF that the shortest interatomic distance R_1 is increased and there is noticeable asymmetry of the first maximum in the melts.

Increasing the temperature of the $Fe_{82}B_{18}$ melt by 200°C scarcely alters the structural parameters. Structural changes occur in the melt only when it is heated to 1,550°C. Then the position of the first diffraction maximum is shifted by

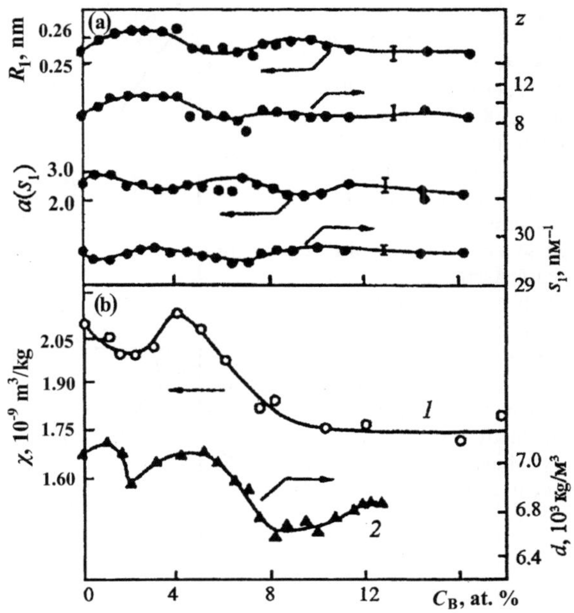

Fig. 3.23 Isotherms of structural characteristics of melts of the Fe–B system at 1,600°C (**a**) and their properties at 1,500°C (**b**) [129]: *1*—magnetic susceptibility; *2*—density

0.5 nm^{-1} toward smaller scattering angles, its height decreases, and the first interatomic distance increases by 0.005 nm. In the opinion of Zelinskaya and coworkers [128], a qualitative structural change, i.e., the formation of microclusters based on the hexagonal type of packing in ε-Fe, which appears in iron at high pressures and is characterized by a more symmetric relative arrangement of the Fe and B atoms than in, for example, Fe_2B, occurs in microregions with a high boron content at elevated melt temperatures.

Diffraction studies of the $Fe_{82}B_{18}$ melt performed in a heating and cooling regime revealed hysteresis of the structure. It was found that the structure factor curves and the RDFs of the melt obtained during heating to 1,550°C and subsequent cooling at a slow rate are not identical. During cooling, the structure of the high-temperature state is "inherited."

Figure 3.23 presents the concentration dependence of the structural characteristics and physical properties of hypoeutectic Fe–B melts at 1,600°C. The correlation established in [129] between the non-monotonic character of the variation of the structural characteristics and properties of the melts, such as the density and magnetic susceptibility, is of interest here. This finding confirms that the short-range order of a melt determines its properties.

3.2.4 *Iron–Silicon*

X-ray structural analysis of liquid iron silicides was carried out in [41, 44]. It was noted in these studies that the diffraction curves of the X-ray scattering depend

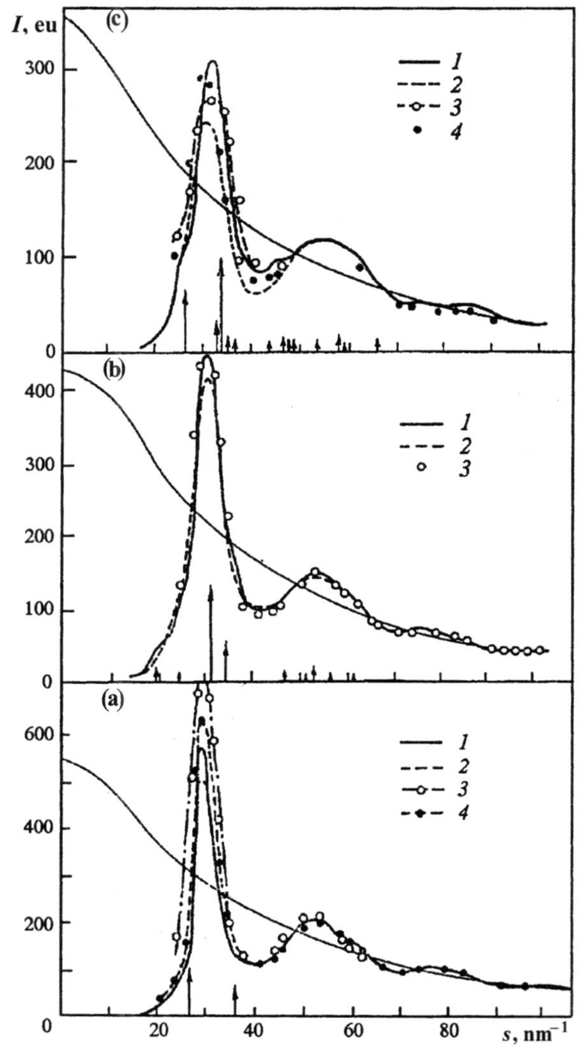

Fig. 3.24 Intensity curves for Fe_3Si (**a**), FeSi (**b**), and $FeSi_2$ (**c**) melts: *1*, *2*—experimental for 1,300 and 1,500 (**a**), 1,430 and 1,700 (**b**), and 1,240 and 1,700°C (**c**), respectively; *3*, *4*—calculation for the structures of the α'-Fe_3Si phase and the FeSi + δ-Fe quasieutectic (**a**), the compound ε-FeSi (**b**), and the compound α-$FeSi_2$ and the FeSi + Si eutectic (**c**), respectively. The vertical lines indicate the positions of the interference lines for crystalline phases of α'-FeSi, ε-FeSi, and α-$FeSi_2$

strongly on the composition of the Fe–Si melt and that a completely statistical distribution of the atoms and a quasieutectic based on the pure components are not realized in any of the solutions.

Figure 3.24 presents experimental intensity curves for the alloys Fe_3Si, FeSi, and $FeSi_2$ [41]. According to the diagram, the last two compounds exist at the liquidus temperatures. The compound FeSi is stronger, since it has a higher melting point and a sharp maximum.

The positions of the maxima of the interference curve for liquid Fe_3Si coincide with those for pure δ-Fe, but they diverge from the positions of the strong interference lines from the crystalline α'-Fe_3Si phase (Fig. 3.24a). For the liquid silicides FeSi (1,430 and 1,700°C) and $FeSi_2$ (1,240°C), the positions of at least two peaks on the intensity curves (Fig. 3.24b and c) correspond exactly to reflections from solid ε-FeSi and α-$FeSi_2$ phases. At 1,700°C the intensity curve for iron disilicide differs from the one at the melting point. The first maximum is considerably lower, is strongly broadened, and is displaced toward smaller scattering angle values. Raising the temperature to a point 450°C above the liquidus line results in reorganization of the short-range order in the arrangement of the atoms. Here clusters with short-range order of the type observed in the ε-FeSi phase along with a random distribution of Fe and Si become stable.

Intensity curves for the following models of the constitution of liquid alloys were calculated using Eq. 3.13.

1. A statistical distribution of the Si atoms based on the δ-Fe bcc lattice.
2. The quasieutectic based on pure Fe and Si.
3. The α'-Fe_3Si structure and the 0.5FeSi + 0.5Fe quasieutectic (for the Fe_3Si melt).
4. The FeSi structure based on α-Fe with $R_1 = 0.248$ nm, which has short-range order similar to the ε-FeSi phase.
5. The α-$FeSi_2$ structure and the quasieutectic consisting of 0.5(ε-FeSi) + a solution of Fe in Si (for the $FeSi_2$).

A comparison of the calculated and experimental curves shows that the first and second models are not realized in Fe–Si melts. Good agreement with experiment (see Fig. 3.24) is provided by the models of the structures of the crystalline mono- and disilicide, as well by the quasieutectics described in variants 3 and 5.

Let us compare the variation of the area A_1 under the first peak of the experimental radial distribution function with the melt composition with the corresponding values calculated from the equations for a statistical distribution of the Fe and Si atoms under the assumption of maintenance of the short-range order of the crystalline compounds in a melt.

In the structure of iron monosilicide [130–132] each metal atom is surrounded by seven Si atoms at an average distance of 0.24 nm and by six Fe atoms at a distance of 0.275 nm. Each Si atom, in turn, is surrounded by seven Fe atoms and six Si atoms at the same distances. Hence, the area of the first coordination peak equals

$$A_{\mathrm{FeSi}} = n_{\mathrm{Fe}} k_{\mathrm{Fe}} k_{\mathrm{Si}} z_{\mathrm{Si}}(\mathrm{Fe}) + n_{\mathrm{Fe}} k_{\mathrm{Fe}}^2 z_{\mathrm{Fe}}(\mathrm{Fe}) + n_{\mathrm{Si}} k_{\mathrm{Fe}} k_{\mathrm{Si}} z_{\mathrm{Fe}}(\mathrm{Si}) + n_{\mathrm{Si}} k_{\mathrm{Si}}^2 z_{\mathrm{Si}}(\mathrm{Si}) = 11.2.$$

Here $z_i(j)$ denotes the number of atoms of type i around an atoms of type j, which is taken as the central atom. The value of A_{FeSi} is greater than the experimental area $A_1 = 10$ square units. Assuming that 90% of the atoms in the melt form clusters with the structure of the ε-FeSi phase and that 10% of the iron and silicon atoms are distributed randomly, we obtain the value

$$A \cong n_{\mathrm{FeSi}} A_{\mathrm{FeSi}} = 10.2,$$

since in this case for a statistical distribution we have

$$A_{st} = (n_{Fe}k_{Fe} + n_{Si}k_{Si})^2[n_{Fe}z_{Fe}(Fe) + n_{Si}z_{Si}(Si)] \approx 0.$$

The maximum of the calculated coordination peak is found at the "weighted-mean" distance

$$R_1 = \frac{R_{Fe-Si}A_{Fe-Si} + R_{Fe-Fe}A_{Fe-Fe} + R_{Si-Si}A_{Si-Si}}{A_{FeSi}} = 0.257\,\text{nm},$$

which agrees well with experiment (0.255 nm).

In crystalline iron disilicide the atoms have the following closest environment [130–132]:

Fe–Si (8)	0.235 nm
Fe–Fe (4)	0.268 nm
Si–Fe (4)	0.235 nm
Si–Si (6)	0.236 (1), 0.268 (4), 0.277 (1) nm.

This coordination of the particles corresponds to $A = 10$ square units and the average value $R_1 = 0.250$ nm. If 20% of the atoms in the melt are distributed randomly, $A = 8.6$ square units, which is close to the experimental data. At 1,700°C for the quasieutectic consisting of complexes with the structure of the ε-FeSi phase (~50%) and with a random distribution of the atoms of different kinds, we have $A = 8.4$ square units, while $A_1 = 8.6$ square units.

The area calculated for the α'-Fe_3Si structure is equal to 13.1 square units. This is considerably greater than the experimental value $A = 10.2$ square units. Better agreement between the experimental and calculated values of the areas was obtained for Fe_3Si under the assumption of the quasieutectic structure for the 0.5(ε-FeSi) + 0.5Fe melt. In the latter case $A = 10.9$ square units. It should be noted that the structure of the lowest silicide of iron includes two types of non-equivalent bonds [130]: metal–metal and metal–silicon. Possibly, in the liquid state these bonds break, and the Fe–Fe + Fe–Si quasieutetic structure, rather than the short-range order of α'-Fe_3Si, is realized. Then, the Fe–Si clusters, which have a heterogeneous environment of atoms, can build up like the α'-Fe_3Si phase on the basis of an ordered substitutional solution in which iron atoms are replaced by silicon in the α-Fe structure.

Kita, Zeze, and Morita [133] subsequently performed a detailed X-ray structural analysis of the Fe–Si system at intervals of approximately 10 at.% Si. The dependence of the short-range order parameters on the Si content in comparison to our data is shown in Figs. 3.25 and 3.26. It was noted that the diffraction curves and the RDFs change sharply when the silicon content is greater than 40 at.%, while at smaller concentrations the structural characteristics of the melts differ only slightly from those for pure iron.

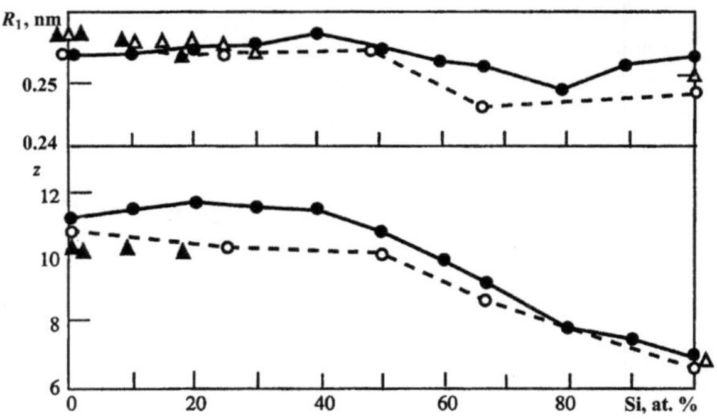

Fig. 3.25 Short-range order parameters of Fe–Si melts at the liquidus temperature: ○, Δ—our data; ●, ▲—[133]

3.3 Oxide Melts

The first X-ray diffraction investigations of the constitution of liquid systems containing oxygen ions were conducted on molten lead borates and alkali metal carbonates, sulfates, and nitrates (see [32, 134]). It was shown that the same complex anions (BO_3^{2-}, CO_3^{2-}, SO_4^{2-}, and NO_3^-) are observed in the liquid as in the crystal lattices.

The structure factors for different molten nitrates of univalent metals are similar in the region of large scattering vectors ($s > 100\ nm^{-1}$), which specifies coordination of the atoms within an NO_3^- complex, but differ in the small-angle region ($s < 100\ nm^{-1}$), which specifies coordination within M^+–M^+, M^+–NO_3^-, and NO_3^-–NO_3^- pairs. The interatomic distances within an NO_3^- complex do not depend on the size and charge of the metal ion and are close to those in the solid state.

In transition metal oxides the simultaneous presence of metal ions of different valence, whose concentrations are specified by the temperature and partial pressure of oxygen in the vacuum chamber, must be taken into account. The RDFs were calculated using published values of the relative densities of the following: $Fe_{1-x}O$ [135], V_2O_5 [136], SiO_2 [135], $Fe_{1-x}O$–SiO_2 [135, 137], $Fe_{1-x}O$–CaO [137], V_2O_5–CaO [138]. However, there are no density data for many oxide metals. In these cases the average atomic density ρ_0 was calculated from a plot of the radial distribution $\rho(R)$ in a range of values of R that are smaller than the atomic diameter [92], where the RDF should be equal to 0 by virtue of the mutual impenetrability of the particles. Then Eq. 3.9 is replaced by the relation

$$4\pi R^2 \rho_0 \left(\sum n_i k_i\right)^2 = -\frac{2R}{\pi} \int_0^\infty [a(s) - 1] s \sin(sR) ds = G(R), \qquad (3.39)$$

in which ρ_0 is determined from the linear dependence of $G(R)$ on R^2. However, because of the experimental errors in the determination of the interference curve and the normalizing factor for the conversion to electronic units, as well as the use of a finite integration limit in Eqs. 3.9 and 3.39, the radial distribution function

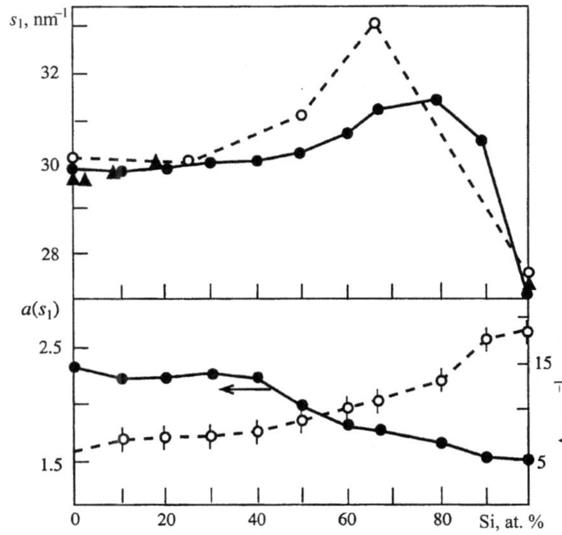

Fig. 3.26 Short-range order parameters of Fe–Si melts at the liquidus temperature: ○—our data; ●, ▲—[133]

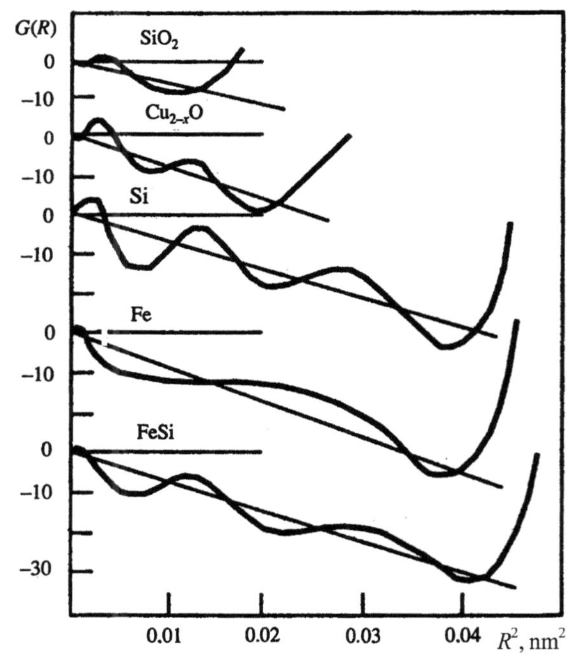

Fig. 3.27 Variation of the function $G(R)$ with the square of the distance at small values of R

always oscillates about 0, and the $G(R)$ curve oscillates about ρ_0. The behavior of the function $G(R)$ at small distances is shown in Fig. 3.27. The values of ρ_0 that were calculated from the slope of the straight lines and average the small oscillations of the $G(R)$ curve for a series of metal and oxide melts coincide to within 5% with the data obtained by other special methods:

		Fe	Ni	Si
		1,550°C	1,470°C	1,430°C
ρ_O, 10^3 nm^{-3}	Calculated from the RDF	0.073	0.080	0.058
	Experiment	0.076	0.0823	0.0548
		[139]	[139]	[139]
		FeSi	$Fe_{1-x}O$	SiO_2
		1,430°C	1,400°C	1,770°C
$(\Sigma n_i k_i)^2$		0.89	0.72	0.87
ρ_O, 10^3 nm^{-3}	Calculated from the RDF	0.070	0.080	0.062
	Experiment	0.0662	0.081	0.0631
		[108]	[135, 137]	[135]

Unlike the case of metal systems, alkali metal halides, and alkaline earth halides, the first maxima on the RDFs for liquid oxides are isolated, providing evidence of a considerable percentage of covalent bonding between the metal and oxygen atoms.

3.3.1 Ferrous Oxide $Fe_{1-x}O$

The structure of molten ferrous oxide has been investigated by X-ray structural analysis at 1,673–1,693 K for Fe_2O_3 contents equal to 9.5 mol% ($Fe_{0.88}O$) [140] and 5.3 mol% ($Fe_{0.95}O$) [141]. Table 3.1 presents the interatomic distance R_i and the coordination numbers z_i of the melts and compares them with those in crystalline wüstite and magnetite. It is sent that at relatively large concentrations of trivalent iron, the short-range order parameters of liquid $Fe_{0.88}O$ are close to the values in the structure of Fe_3O_4. Decreasing the Fe_2O_3 content to 5.3 mol% results in increases in $R_{Fe–O}$ and $z_{Fe–O}$.

The radial distribution function (Fig. 3.28a) has a completely isolated maximum at $R = 0.195$ nm and a series of subsequent peaks at $R \approx 0.28$, 0.32, and 0.38 nm. As follows from the data in Table 3.1, the values of R_1, R_2, and A_1 are close to the iron–oxygen and iron–iron interatomic distances and to the coordination of the particles in the structure of magnetite. However, it is difficult to attribute the decreases in the shortest interatomic distance and the coordination number observed upon melting solely to the appearance of Fe^{3+}–O^{2-} bonds in view of the small percentage of these bonds. Apparently, it is largely caused by the breaking of Fe–O bonds upon the crystal–liquid phase transition and the formation of comparatively small atomic clusters, in which the structure of the solid is maintained (Zarshitskii's island model; see [1]). The terminal ions of the complex, which have a smaller number of nearest neighbors and, as a result, have incompletely saturated bonds, intensify the interaction with particles within the cluster.

Table 3.1 Parameters of the short-range order in crystalline and molten iron oxides

Parameter	FeO		Fe_3O_4		$Fe_{0.95}O$	$Fe_{0.88}O$
	Crystal				Melt	
R_1 (nm)	Fe–O	0.215	$(Fe–O)_{oct}$	0.206	0.205	0.195
R_2 (nm)	Fe–Fe	0.305	$(Fe–O)_{tetr}$	0.187	–	–
	O–O	0.305	$(Fe–Fe)_{oct}$	0.291	0.315	0.28
	Fe–O	0.372	O–O	0.291	–	0.32
			$Fe_{oct}–Fe_{tetr}$	0.341	–	–
			$O–Fe_{tetr}$	0.341	–	–
z_1	6		$(Fe–O)_{oct}$	6	5.7	4.5
			$Fe_{tetr}–O$	4	–	–

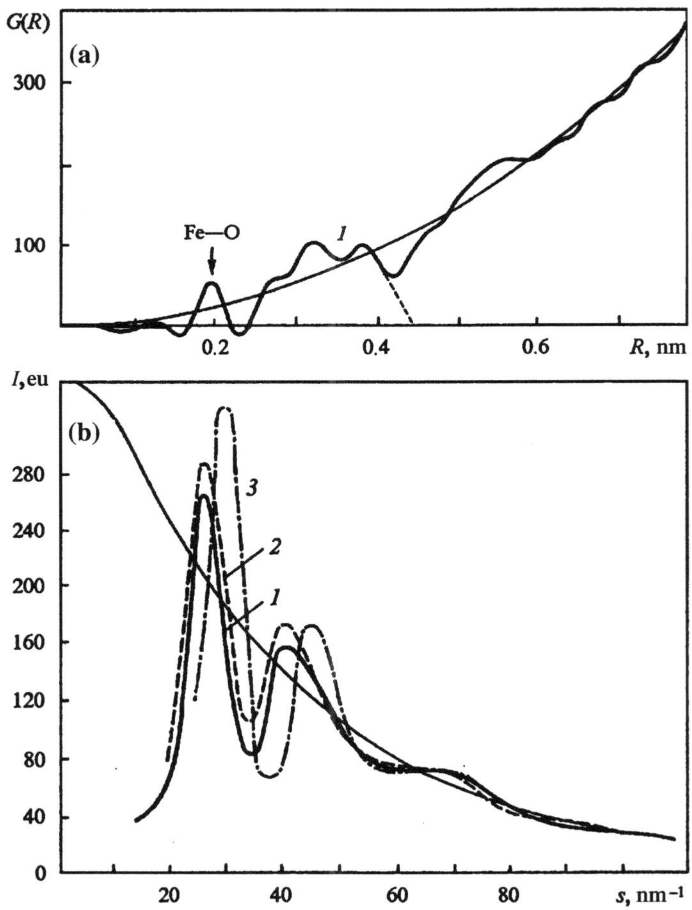

Fig. 3.28 Radial distribution function (**a**) and intensity curves (**b**) for molten $Fe_{0.88}O$: *1*—experiment at 1,400°C; *2*, *3*—calculation based on the structures of Fe_3O_4 and $Fe_{1-x}O$, respectively

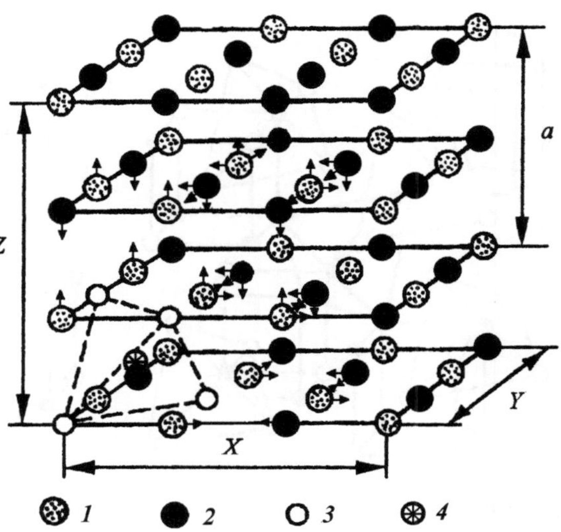

Fig. 3.29 One octant of a model complex of wüstite [143]: *1*—oxygen anions; *2*—iron octacations; *3*—octahedral vacancies; *4*—one of eight tetrahedral vacancies, which is occupied by an iron cation; *a*—edge of the wüstite unit cell with the NaCl structure

In nonstoichiometric solid ferrous oxide $Fe_{1-x}O$ with the NaCl structure, some of the octahedral points of the divalent iron sublattice are free. It has been shown (see [142–144]) that the trivalent iron atoms form complexes within which the Fe_3O_4 structure is realized in the crystal by occupying tetrahedral cavities of the NaCl lattice (Fig. 3.29). Under the action of defect associates, the oxygen ions in the complex do not form the ideal face-centered cubic packing observed in NaCl, but are displaced slightly toward the tetrahedral points (in Fig. 3.29 the directions of displacement of the ions are indicated by arrows). A calculation of the lengths of the bonds between the ions with allowance for their displacements showed that the shortest bond is observed between the tetracations and the four nearest oxygen ions. The length of three of the bonds is equal to 0.194 nm, and the length of the fourth bond is 0.187 nm. The distance from an octacation to its oxygen environment fluctuates from 0.202 to 0.231 nm. In pure magnetite these distances are equal to 0.185 and 0.206 nm, respectively. At elevated temperatures the divalent Fe atoms and half of the trivalent Fe atoms are distributed statistically among the octahedral sites, as is conveyed by the formula $Fe^{3+}_{tetr}(Fe^{2+}Fe^{3+})_{oct}O_4$.

To elucidate the constitution of an $Fe_{0.88}O$ melt, the intensity curves for two structural models were calculated.

Wüstite with vacancies in the sublattice of iron atoms. The Fe^{3+} and Fe^{2+} ions are located at octahedral points of the NaCl lattice. An isolated complex measuring ~1.1 nm, which contains 32 oxygen atoms and 28 iron atoms in accordance with the composition of liquid ferrous oxide, was chosen. The interatomic distances (nm) calculated for the value $R_1 = 0.200$ nm and the coordination numbers averaged over the Fe^{2+} and O^{2-} ions with such ordering are shown below. The latter deviate strongly from the corresponding values in an infinite FeO crystal (the values in parentheses):

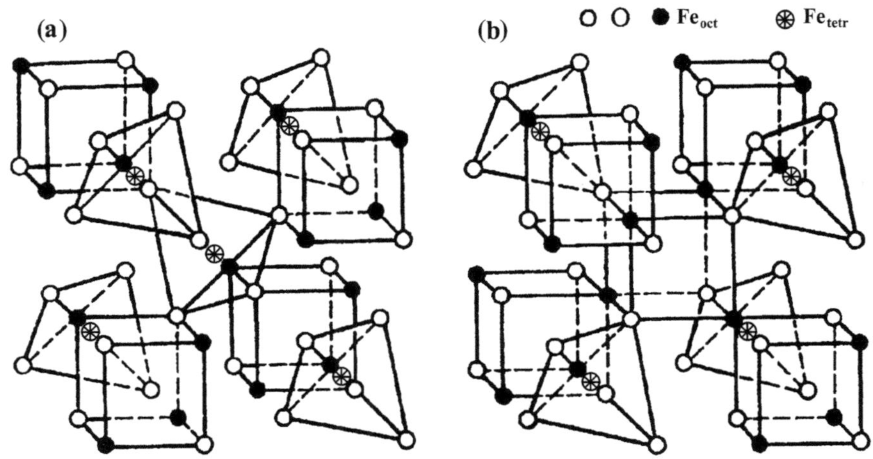

Fig. 3.30 Structure of model complexes of Fe_3O_4 consisting of five (**a**) and four (**b**) tetrahedrons

$Fe^{2+}(O^{2-})$	4.5 (6)	$O^{2-}(Fe^{2+})$	0.200
	6.8 (12)	$Fe^{2+}(O^{2-})$	0.282
	3.4 (8)	$O^{2-}(Fe^{2+})$	0.346
	3.0 (6)	$Fe^{2+}(O^{2-})$	0.400
	9.0 (24)	$O^{2-}(Fe^{2+})$	0.428
	6.8 (24)	$Fe^{2+}(O^{2-})$	0.490

A constant structure blurring factor, which was calculated for the characteristic temperature $\theta_D = 540$ K found in [143] from X-ray diffraction data for solid wüstite, was used for all the coordination spheres. The calculated interference curve is not consistent with experiment (Fig. 3.28b). The first and second maxima are displaced by 0.03–0.04 nm toward larger scattering angles. Decreasing the dimensions of the complex and the number of atoms in it by a factor of approximately 2 slightly lowers the intensity of the maxima, but does not alter their positions.

The Fe_3O_4 structure. Two groups of atoms containing 28 oxygen ions and 16 iron ions at octahedral points and five or four filled tetrahedra with Fe^{3+} ions at the center were singled out from the magnetite crystal lattice (Fig. 3.30). To make the concentrations of di- and trivalent iron in the complex correspond to the composition of the melt, vacant sites in the iron sublattice were filled with Fe^{2+} ions, and Fe^{3+} ions were simultaneously removed from tetrahedra. It was assumed that four octahedral sites can be filled for each tetracation eliminated. A complex containing 28 oxygen ions, 21.5 iron octacations, and 3.13 iron tetracations corresponds to $Fe_{0.88}O$. A comparison of the ion fractions of the components in a model complex of $Fe_{0.88}O$ with the composition of the melt investigated reveals that one-fourth of the cations of trivalent iron can be located in octahedral positions:

		O^{2-}	Fe_{oct} (Fe^{2+} + Fe^{3+})	Fe_{tetr} (Fe^{3+})
Number and mole fraction (values in parentheses) of ions in model complexes	$(Fe_3O_4)_{av}$	28 (0.58)	16 (0.33)	4.5 (0.09)
	$Fe_{0.88}O$	28 (0.53)	21.5 (0.41)	3.13 (0.06)
			Fe^{2+} ions	Fe^{3+} ions
Mole fraction of ions in the melt		0.53	0.39	0.08

The interatomic distances (nm) calculated for the ideal magnetite lattice using a value of the unit-cell parameter equal to 0.825 nm and the averaged coordination numbers (the values in parentheses) are presented below:

	Fe_{oct}	Fe_{tetr}	O
Fe_{oct}	0.291 (4.7)	0.341 (1.3)	0.206 (4.1)
	0.412 (1.6)	0.536 (1.2)	0.357 (3.3)
	0.505 (3.8)		0.461 (8.3)
Fe_{tetr}	0.341 (8.9)	0.357 (0.6)	0.185 (3.6)
	0.536 (8.9)	0.583 (1.6)	0.341 (6.6)
			0.449 (3.9)
			0.536 (4.3)
O	0.206 (3.1)	0.176 (0.5)	0.291 (7.0)
	0.357 (2.6)	0.341 (0.8)	0.412 (2.7)
	0.461 (6.8)	0.449 (0.5)	0.510 (7.1)
		0.536 (0.4)	

The diffraction curve calculated from these parameters accurately describes the experimental intensity curve with respect to both the amplitude and position of the maxima and minima (see Fig. 3.28b, curves 1 and 2).

The area under the isolated peak on the RDF (A_1) and the total area under the three subsequent maxima located in the range of values from $R = 0.24$ to 0.42 nm correspond to the values which were calculated from the coordination numbers. For example,

$$A_1 = k_{Fe}k_O[n_{Fe_{oct}}z_O(Fe_{oct}) + n_{Fe_{tetr}}z_O(Fe_{tetr}) + n_OZ_{Fe_{oct}}(O) + n_Oz_{Fe_{tetr}}(O)] = 1.88,$$

while in the experiment $A_1 = 1.9$ square units. The value $A_2 = 10.8$ square units is close to the experimental value of 10.3 square units.

3.3.2 The FeO–Fe_2O_3–SiO_2 System

We investigated the structure of SiO_2–FeO melts in [144, 145] at various temperatures. The data from X-ray structural analysis attest to the maintenance of orderings

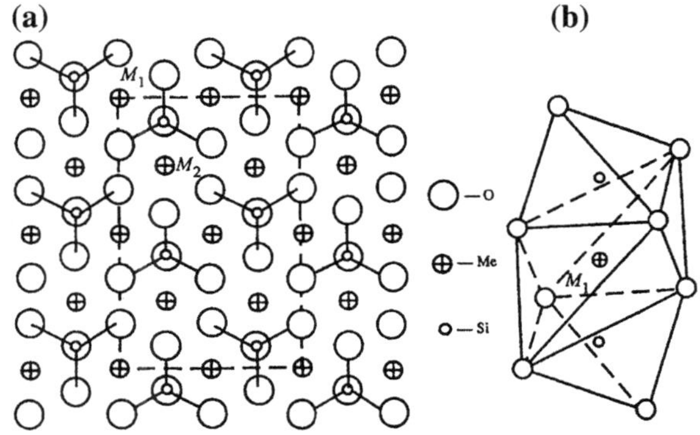

Fig. 3.31 Diagram of the crystal structure of an orthosilicate: **a**—projection of the structure onto the (010) plane; **b**—the orthosilicate fragment $(MeSi_2O_8)^{6-}$

based on the structure of the congruent compound Fe_2SiO_4 (fayalite) in the melts. It is characterized by two-layer compact packing of the oxygen atoms (Fig. 3.31). The FeO_6 octahedra form zigzagged chains joined by SiO_4 tetrahedra in such a manner that the latter each share three common edges with octahedra from another chain. Thus, each oxygen atom is bonded to three iron ions and one silicon ion.

The first RDF peaks for these melts at the melting points, which correspond to the shortest Fe–O and Si–O interionic distances, as in the case of liquid SiO_2, are completely isolated (Fig. 3.32). One exception is the melt that corresponds in composition to iron orthosilicate. The RDF for this melt displays an additional maximum at $R = 0.255$ nm, which is close to the Fe–Si (0.253 nm) and O–O (0.258, 0.261 nm) interatomic distances in the structure of fayalite (Fe_2SiO_4).

The subsequent peaks on the RDF for the alloys with 32, 38, and 47 mol% SiO_2 at $R = 0.32$ and 0.38 nm correspond to the Fe–Fe (0.315 and 0.326 nm) and Fe–O (0.379 nm) distances in an Fe_2SiO_4 crystal. For the alloys containing 12 and 27 mol% SiO_2, the second RDF maximum is broadened on the short-distance side (Fig. 3.32, curves 2 and 3) as a consequence of the appearance of Fe–Fe coordination with a distance equal to 0.29 nm, which occurs in molten iron oxides. The latter indicates the existence of microregions with the short-range order of nonstoichiometric wüstite in the melts rich in FeO.

An analysis of the intensity curves (Fig. 3.33) shows that for the compositions with a large SiO_2 content (32, 38, and 47 mol%) there is a clearly expressed maximum at $s = 29\ nm^{-1}$, which is also displayed by molten fayalite (curve 3). In addition, there is a peak at $s = 15–16\ nm^{-1}$, which corresponds in position to the main maximum on the interference curve of SiO_2. The melts containing 12 and 27 mol% SiO_2 have intensity curves that are similar to those for liquid FeO. This is evidence of the quasieutectic constitution of SiO_2–FeO melts. The elements of the quasieutectic are clusters with the fayalite, wüstite, and silicon dioxide structures. Figure 3.33 (curve 3) compares the experimental and model diffraction curves for the melt with 32 mol% SiO_2. The latter was calculated for a fragment of the fayalite structure, in which the iron ions occupy two inequivalent sites, namely, M_1 and M_2 sites (Fig. 3.31), and have different bond lengths with oxygen ions. The Fe–O distance at an M_2 site is greater than the distance at an M_1 site; therefore, when the melt

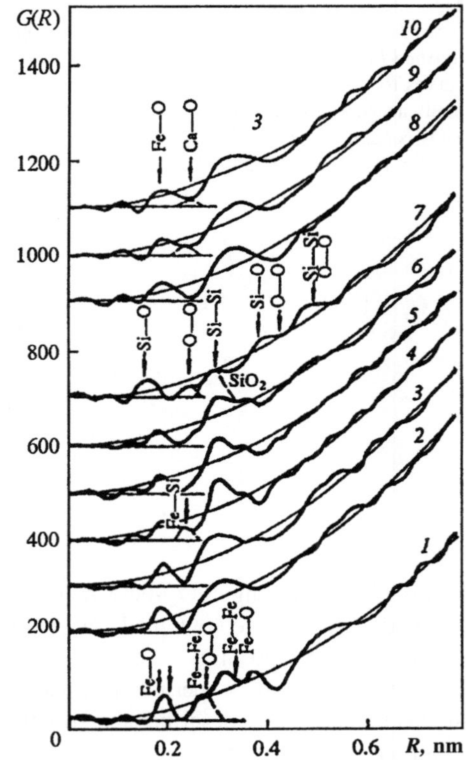

Fig. 3.32 Radial distribution functions for $FeO–Fe_2O_3–SiO_2$ (1–7) and $FeO–Fe_2O_3–CaO$ (8–10) melts at the liquidus temperature. $FeO + Fe_2O_3$ content, mole%: *1*—100; *2*—88; *3*—73; *4*—68; *5*—62; *6*—53; *7*—0; *8*—78; *9*—72; *10*—63

is melted and subsequently heated, disordering of the crystal structure presumably occurs at the weaker bonds of iron at an M_2 site with oxygen.

The following coordination numbers and shortest interionic distances, which were averaged over all the ions in the complex, were used for the calculation:

Fe		Si		O	
Fe–O (3.6)	0.217 nm	Si–O (4.0)	0.163 nm	O–Si (0.8)	0.163 nm
Fe–Si (0.8)	0.253	Si–Fe (2.0)	0.253	O–Fe (1.8)	0.217
Fe–Si (0.4)	0.296	Si–Fe (1.0)	0.296	O–O (2.4)	0.266
Fe–Fe (0.8)	0.315	Si–O (1.0)	0.300	O–O (0.8)	0.280
Fe–Fe (1.6)	0.326	Si–O (1.0)	0.350	O–Si (0.2)	0.300
Fe–Fe (1.6)	0.335	Si–O (1.0)	0.410	O–Fe (1.0)	0.335

The position and degree of blurring of the maxima on the model and experimental diffraction curves are in good agreement with one another. There is small amount of divergence in the values of the intensity at $s = 30–50\ nm^{-1}$, which can be corrected by taking into account the partial dissociation of fayalite in the melt

$$Fe_2SiO_4 \rightarrow 2FeO + SiO_2.$$

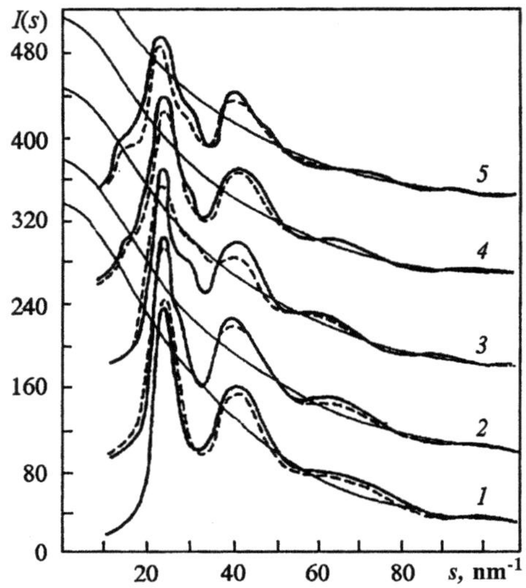

Fig. 3.33 Experimental (*solid lines*) and calculated (*dashed lines*) intensity curves for FeO–Fe_2O_3–SiO_2 melts near the liquidus line. SiO_2 content, mole%: *1*—12; *2*—27; *3*—32; *4*—38; *5*—47

For the remaining SiO_2–FeO melt compositions, the model intensity curves (dashed lines) were obtained by summation of the scattering curves for Fe_2SiO_4, FeO, and SiO_2.

Comparison with experiment reveals that the Fe_2SiO_4 + FeO + SiO_2 quasieutectic accurately describes the structure of the melts when the contents of the components in it correspond to 40–50% dissociation of fayalite. The Fe_2SiO_4 + FeO ($SiO_2 < 33\%$) and Fe_2SiO_4 + SiO_2($SiO_2 > 33\%$) eutectics give underestimated values of the intensity in the region of the main peak, and at high concentrations of silica, the asymmetry at s = 15–16 nm^{-1} is poorly displayed.

The areas under the first RDF peaks calculated for the model of a complex quasieutectic are close to the experimental values. The calculations were performed using the formula

$$A_1 = n_{FeO}A_{FeO}k^2_{FeO} + n_{SiO_2}A_{SiO_2}k^2_{SiO_2} + n_{Fe_2SiO_4}A_{Fe_2SiO_4}k^2_{Fe_2SiO_4},$$

where n_i, A_i and k_i denote the mole fractions, areas under the first RDF maxima, and the relative scattering powers of the melt components.

The similar constitution of liquid FeO–SiO_2 alloys is manifested in the position of the first RDF maximum, which is specified by the shortest Fe–O and Si–O distances. In nonstoichiometric ferrous oxide R_1 = 0.195 nm, and in silicon dioxide it equals 0.163 nm. Simple mixing of these oxides or the quasieutectic based on them should have given values of R_1 not exceeding 0.195 nm. The experimentally observed extremum value of R_1 (Fig. 3.34) is caused by the formation of clusters with the short-range order of the chemical compound Fe_2SiO_4, in which the Fe–O bond is longer than in $Fe_{1-x}O$ (in fayalite $R_{Fe(I)-O}$ = 0.213, 0.212, and 0.223 and $R_{Fe(II)-O}$ = 0.223, 0.211, 0.229, and 0.207 nm [146], which give R_{av} = 0.217 nm). An increase in the concentration of SiO_2 to 33% increases

Fig. 3.34 Influence of the SiO_2 concentration on the position of the first maximum of radial distribution curves

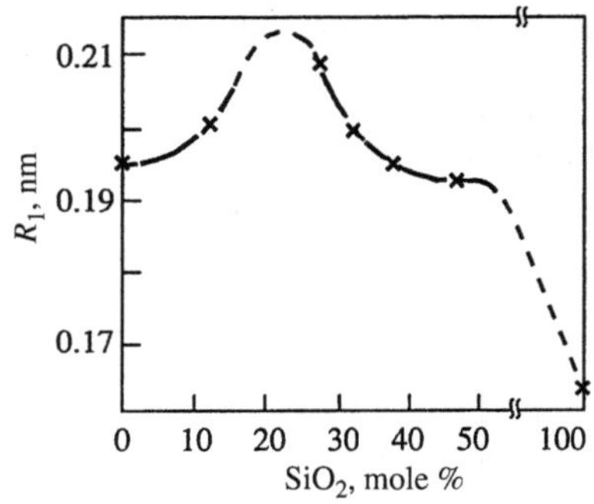

both the mole fraction of the Fe_2SiO_4 orderings, i.e., the number of long Fe–O bonds, and the fraction of short Si–O distances. This results in the appearance of a maximum on the plot of the dependence of R_1 on the composition at approximately 23% SiO_2. When the relative scattering power of the atoms ($k^2_{Fe} = 2.5$, $k^2_{Si} = 0.58$, $k^2_O = 0,15$) and the coordination numbers in the melt are taken into account, the calculated value of the average Me–O distance in iron orthosilicate is equal to 0.205 nm, which differs only slightly from the experimental value 0.200 nm. The latter value includes a definite contribution from relatively short Fe^{3+}–O^{2-} bonds. Apparently, the experimentally observed increase in R_1 with increasing temperature of the melts should be attributed to a decrease in the concentration of Fe_2O_3.

The variation of the parameters of the Si–O, Si–Si, Fe–O, and Fe–Si pair correlations with the composition of FeO–SiO_2 melts was examined on the basis of an RDF analysis in [147] (Fig. 3.35). The coordination number for the Si–O pairs does not depend on the content of iron oxides and is close to 4, confirming the existence of SiO_4 tetrahedra in the FeO–SiO_2 melts, as in liquid SiO_2. The Si–Si bond length, which determines the distance between the silicon–oxygen tetrahedra, decreases continuously with increasing SiO_2 concentration up to a composition close to fayalite Fe_2SiO_4, and remains constant thereafter. The latter value is comparable to the value for an Si–O–Si bond angle smaller than 180° and thus indicates polymerization of the SiO_4 tetrahedra and the formation of complex silicate anions. The Fe–O and Fe–Si correlations vary in a similar manner. However, it is difficult to attribute such a course of the variation of the Fe–O interatomic distance to the formation of associates with the fayalite structure in the melt. Apparently, when the content of iron oxides is increased above 67 mol%, i.e., when "free" clusters of atoms with the short-range order of $Fe_{1-x}O$ appear in the melt, the Fe–O bond length should decrease by virtue of the fact that it is considerably smaller in pure liquid iron oxides than in fayalite. The values of the Fe–Si bond lengths presented, which are considerably smaller than the Fe–O and Fe–Si distances in the fayalite structure, also seem questionable.

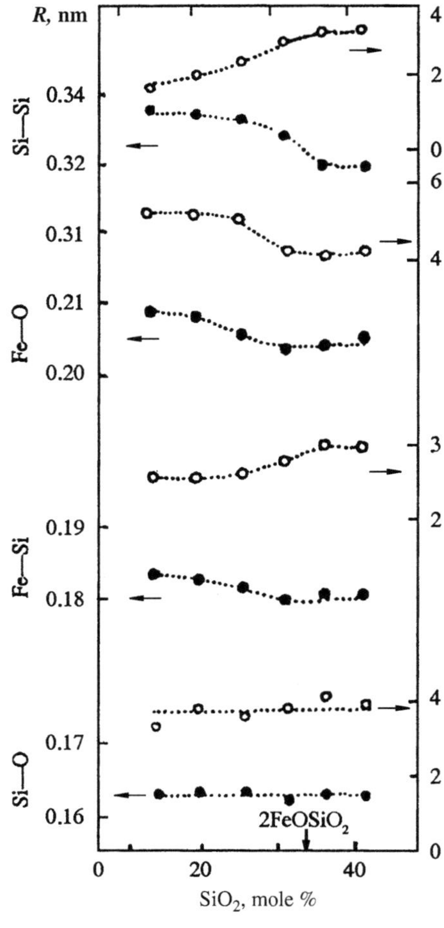

Fig. 3.35 Variation of the distances and coordination numbers of ion pairs in $FeO-SiO_2$ melts

3.3.3 The $FeO–Fe_2O_3–CaO$ System

The structure of two groups of iron calcium oxide melts with different fractions of trivalent iron in the range $Fe^{3+}/(Fe^{3+} + Fe^{2+}) = 0.15–0.30$ [1] and with a value greater than 0.9 [148] has been studied by X-ray structural analysis.

The radial distribution curves of the atoms for all the compositions investigated (Fig. 3.32, curves 8–10) exhibit a weekly expressed peak at $R \approx 0.24$ nm, as a result of which the first maximum, which is assigned to the shortest Fe–O coordination, is not isolated. Superheating above the liquidus line at 200°C results in clearer separation of the additional peak. Its position matches the sum of the ionic radii of Ca^{2+} (0.104 nm) and O^{2-} (0.136 nm). The shortest Fe–O distance in the melts containing CaO is decreased compared with liquid wüstite, apparently because of the increase in the relative fraction of trivalent iron at tetrahedral sites,

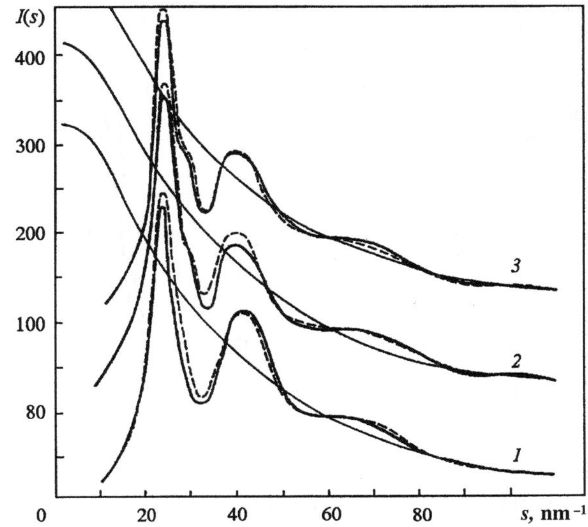

Fig. 3.36 Experimental (*solid lines*) and model (*dashed lines*) intensity curves for $FeO–Fe_2O_3–CaO$ melts at the liquidus temperature. CaO content, mole%: *1*—22, *2*—28, *3*—37

and the main maximum of the interference curves is shifted toward smaller scattering angles and essentially maintains its position both when the composition is varied and when the temperature is increased. The displacement of the maximum (23.5 nm^{-1} instead of the value of 25.0 nm^{-1} in liquid wüstite) is apparently caused by the formation of clusters of atoms in the melt with a structure like that of one of the calcium ferrites, since the strong interference lines for crystalline dicalcium ferrite, monocalcium ferrite, and hemicalcium ferrite lie in the range 23–24 nm^{-1} [149–151] (Fig. 3.36). The most stable compound is dicalcium ferrite. According to the diagram of state of the $CaO–Fe_2O_3$ system, only it melts without decomposing at a comparatively high temperature. The shortest Fe–O and Ca–O distances in melts containing calcium oxide are close to those in the compound $2CaO·Fe_2O_3$ [149] (nm):

	Octahedron		Tetrahedron		
Fe–O (6)	0.197	Fe–O (4)	0.187	Ca–O (7)	0.233–0.267
O–O (12)	0.269–0.279	O–O (5)	0.293–0.298	Ca–O (1)	0.282

A continuous series of solid solutions based on FeO has been discovered in an FeO–CaO system containing 37.5 mol% CaO, 53.3 mol% FeO, and 9.2 mol% Fe_2O_3 [152], whose composition scarcely differs from the composition of the melt investigated with the highest CaO content. It can, therefore, be assumed that in liquid $FeO–Fe_2O_3–CaO$ alloys there are atomic clusters that take the form of a solution of calcium oxide in wüstite along with complexes that have the structure of dicalcium ferrite.

Figure 3.37 presents interference curves calculated for the following models.

1. A solution of calcium oxide (37 mol%) in nonstoichiometric iron oxide (see Sect. 3.3.1, model 1). All the Fe^{2+}, Fe^{3+}, and Ca^{2+} ions are located at

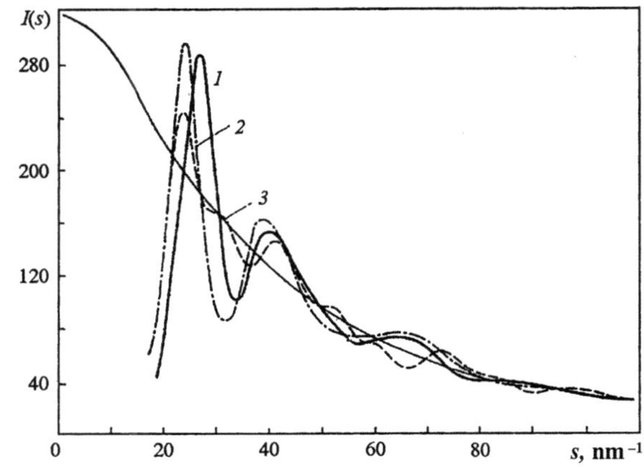

Fig. 3.37 Calculated intensity curves for $FeO–Fe_2O_3–CaO$ melts. The numbers on the curves are the numbers of the models described in the text

octahedral points, some of which are vacant. The model gives an intensity curve that differs strongly from the experimental diffraction curve (Fig. 3.37b, solid curve 3). The maxima are displaced 3 nm^{-1} toward larger scattering angles. The peak that is clearly recorded at $s \approx 28\ nm^{-1}$ in the experiment is absent.

2. A solution of calcium oxide (10 mol%) in nonstoichiometric iron oxide, whose short-range order is similar to the Fe_3O_4 structure (see Sect. 3.3.1, model 2). All the Fe^{2+} and Ca^{2+} ions and some of the Fe^{3+} cations are located at octahedral sites. Compared with $Fe_{1-x}O$, the concentration of trivalent iron ions at tetrahedral sites in the $Fe(Ca)O_{1-x}$ solution is increased from 6 to 10 mol%. This model leads to agreement between the positions of the first maximum of the interference curves. However, the secondary peak at $s \approx 28\ nm^{-1}$ is also absent.
3. The structure of dicalcium ferrite $2CaO{\cdot}Fe_2O_3$ with the following interatomic distances (nm) and coordination numbers (the values in parentheses) [149]:

	Fe	Ca	O		Fe	Ca	O
Fe	0.275 (4)	0.331 (6)	0.187 (1)	Ca	0.331 (6)	0.367 (2)	0.256 (8)
	0.367 (2)	0.616 (4)	0.197 (1)		0.616 (4)	0.390 (2)	0.330 (1)
	0.390 (2)	0.644 (2)			0.644 (2)	0.544 (2)	
	0.459 (4)			O	0.187 (1)	0.256 (2)	0.274 (12)
	0.536 (2)				0.197 (1)		0.296 (5)
							0.343 (1)

As in the experiment (see Fig. 3.37b, curve 3), here a maximum appears at $s \approx 28\ nm^{-1}$ on the model diffraction curve (see Fig. 3.37). However, the intensities of the first two peaks differ slightly from the experimental values.

4. A quasieutectic of complexes that represent a solution of CaO in FeO (model 2) and clusters of atoms with the short-range order of the compound $2CaO{\cdot}Fe_2O_3$

(model 3). This structure produces calculated intensity curves that are close to the experimental curves (Fig. 3.37b) both with respect to the positions of the maxima and with respect to their intensity. The concentration of the complexes in the melts has the following values:

Melt composition (mole%)			Concentration of complexes (%)	
CaO	FeO	Fe_2O_3	$2CaO{\cdot}Fe_2O_3$	$Fe(Ca)O_{1-x}$
22	72	6	20	80
28	65	7	30	70
37	53	10	50	50

The data presented are evidence that in $FeO–Fe_2O_3–CaO$ melts, along with the formations that have short-range order of the calcium ferrite type, there are clusters of atoms with a structure similar to nonstoichiometric iron oxide, where the Fe^{2+} ions at some of the octahedral sites are replaced by Ca^{2+} cations and the Fe^{3+} ions occupy octahedral and tetrahedral points. Raising the temperature of the melts by 250–300°C (as in the case of iron silicates) scarcely alters the diffraction curves and the RDFs. The presence of di- and trivalent iron in the structure of liquid wüstite in identical crystallographic positions, as well as the large concentration of $Fe(Ca)O_{1-x}$ clusters, ensures a high value of the electronic conductivity even when the calcium oxide content in the melt is comparatively large (see [153]). In liquid $FeO–SiO_2$ alloys, the mole fraction of the $Fe_{1-x}O$ clusters and the electronic component of the conductivity drop nearly to 0 when the composition corresponds to the compound $2FeO{\cdot}SiO_2$ (33 mol% SiO_2).

According to the data in [148], in $CaO–Fe_2O_3$ melts (33–60 mol% CaO) the shortest Fe–O distance and the first coordination number corresponding to this pair decrease with increasing CaO content up to 50 mol% from 0.203 to 0.196 nm and from 4.3 to 3.5 atoms, respectively, and then remain unchanged. This means that the iron ions are found predominantly at tetrahedral sites when the CaO content is >50 mol%. The parameters of the Ca–O pairs for all the compositions remain constant ($R_{\text{Ca–O}} = 0.237$ nm and $z_{\text{Ca–O}} \approx 6$) and indicate octahedral coordination of the atoms. The results obtained were attributed to the formation of local orderings that resemble the structures of monocalcium ferrite and dicalcium ferrite.

3.3.4 The SiO_2–CaO–MgO System

A melt containing 39, 40, and 21 mol% SiO_2, CaO, and MgO, respectively, has been studied [154, 155]. The composition falls in the region of akermanite $Ca_2MgSi_2O_7$. The melt contains an additional 1.2 mol% CaO and an additional 1.4 mol% MgO above the stoichiometric composition. Since the diffraction patterns obtained at 1,673, 1,773, and 1,873 K differ only slightly from one another,

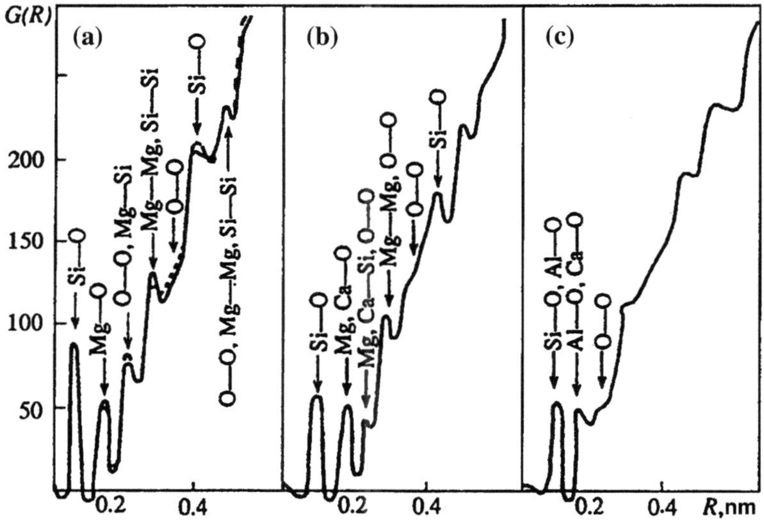

Fig. 3.38 RDF for silicate melts near the *liquidus line*: **a**—MgO–SiO_2 (*dashed curve*—for 2,073 K); **b**—MgO–CaO–SiO_2; **c**—Al_2O_3–CaO–SiO_2

the RDF and $a(s)$ curve for only +1,723 K are shown in Figs. 3.38 and 3.39. The short-range order of the melt could not be described by a structure inherited from akermanite. The position of the second RDF coordination peak in this case should be shifted markedly toward larger distances compared with experiment ($R_2^{exp} = 0.227$ nm, $R_2^{akerm} \geq 0.240$ nm). The average shortest distance between the calcium, magnesium, and oxygen ions, which is equal to 0.227 nm, can be obtained under the assumption that the dissociation products that appear when akermanite is melted are the simple orthosilicates Ca_2SiO_4 and Mg_2SiO_4:

$$2Ca_2MgSi_2O_7 \rightarrow 2Ca_2SiO_4 + Mg_2SiO_4 + SiO_2.$$

Recalculation of the original composition of the silicate for the composition of the orthosilicate gives the following, mole%: Ca_2SiO_4, 59; Mg_2SiO_4, 31; SiO_2, 10.

The coordination number of silicon with respect to oxygen found from the size of the area under the Si–O RDF maximum at $R_1 = 0.166$ nm is equal to 4.1, confirming the tetrahedral coordination of the silicon cation in the melt analyzed. The maximum height of the second RDF peak corresponds to the shortest Mg–O and Ca–O distances. The value $R_2 = 0.227$ nm falls between the values of $R_{Mg\text{–}O}$ and $R_{Ca\text{–}O}$ in MgO–SiO_2 (0.220 nm) and CaO–SiO_2 (0.240 nm) melts and was obtained under the assumption that all of the magnesium and only 18% of the original calcium are found in an octahedral oxygen environment.

The values of $z_{Ca\text{–}O}$ and $z_{Mg\text{–}O}$ are equal to 5.9 (in crystals their values are equal to 6). The total fraction of the Mg and Ca ions that are located at octahedral M_1 sites (Fig. 3.31) in the orthosilicates Mg_2SiO_4 and Ca_2SiO_4 amounts to 0.118, which is approximately equal to 50% of their total number in the melt, that is, 0.256.

The short-range order of the melt is represented by an element of the orthosilicate structure $[Me_nSi_{n+1}O_{4(n+1)}]^{2(n+2)-}$, where n is the number of Mg and Ca cations in a complex. It follows from a comparison of the RDFs of the silicate melts (Fig. 3.38) that the peaks (besides the first and second peaks) of the magnesium silicate melt are blurred more strongly and are displaced slightly toward larger values of R. As the amount of dissimilar structural components is increased, i.e., as the microheterogeneity of the melt is increased, the peaks of the corresponding RDF should be blurred more strongly. The experimental atomic density curve of the magnesium silicate melt is the superposition of two RDFs, one with short-range order based on magnesium orthosilicate (up to 62%) and one with short-range order based on microclusters of pure SiO_2 (up to 38%). Apparently, the microheterogeneity of the melt is less strongly expressed for calcium magnesium silicate because the differences between the structures of Mg_2SiO_4 and Ca_2SiO_4 are smaller than the differences between the structures of Mg_2SiO_4 and SiO_2. When large calcium cations are introduced into the magnesium silicate melt, the dimensions of the ordering regions in the latter increase slightly. By preferentially occupying M_2 sites (Fig. 3.31), calcium drives apart the $(SiO_4)^{4-}$ tetrahedra to large distances from one another and promotes the formation of large ortho complexes. The third and subsequent RDF maxima (Fig. 3.38) were identified with maxima which are superpositions of different Me–Me, Me–O, and O–O distances (Me denotes Si, Mg, and Ca) in the orthosilicates (Mg_2SiO_4 and Ca_2SiO_4). It was noted that the model of $(SiO_4)^{4-}$ tetrahedra that are regularly oriented relative to one another along with placement of the Ca and Mg cations within microclusters that correspond to the short-range order in olevines with random orientation of the calcium cations within these regions allows satisfactory interpretation of the experimental results. In this model calcium and magnesium are distributed statistically in the ordered microclusters.

3.3.5 The SiO_2–CaO–Al_2O_3 System

Information about the structure of a melt of the SiO_2–CaO–Al_2O_3 system containing 45, 36, and 19 mol% of each component, respectively, was presented in [154, 155]. When the experimental results were analyzed, the following was taken into account. The assignment of silicates according to grade to different groups is determined by the value of the ratio of the concentration of oxygen to the concentration of silicon. If this ratio is ≥ 4, the structure consists of isolated $(SiO_4)^{4-}$ groups. At the same time, application of this principle to alumosilicates is inadvisable because of the dualism of aluminum. Its ability to come into contact with four oxygen atoms increases the total percentage of cations that have a possibility to form tetrahedral groups. In addition, determination of the distribution of the Si and Al cations among the tetrahedral groups is complicated by the fact that their scattering powers are close to one another. Consequently, the pattern of the populating of the tetrahedra by these cations can be established only by

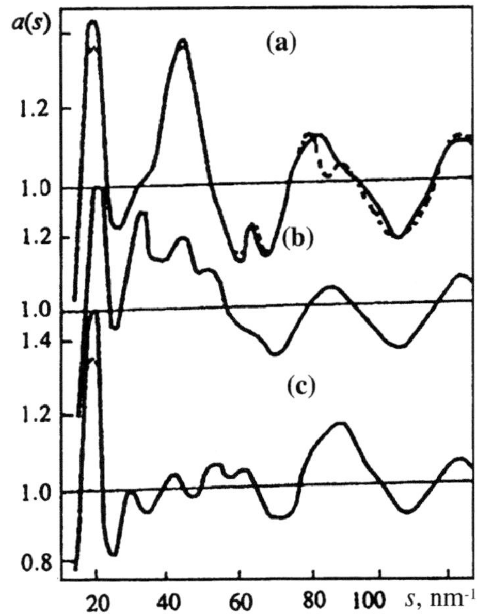

Fig. 3.39 Structure factor curves of silicate melts: **a**—MgO–SiO_2 (*dashed curve*—for 2,073 K); **b**—MgO–CaO–SiO_2; **c**—Al_2O_3–CaO–SiO_2 (*dashed curve*—for 293 K)

comparing their dimensions, since an $(AlO_4)^{5-}$ tetrahedron occupies a somewhat larger volume than does $(SiO_4)^{4-}$.

Diffraction patterns from the SiO_2–CaO–Al_2O_3 sample investigated were obtained in an argon atmosphere at 1,873 K and after solidification of the sample. Figures 3.38 and 3.39 present the RDFs and structure factors. A special feature of these diffraction patterns is the insignificant differences between them. They are distinguished only by the fact that the first $a(s)$ peak obtained at room temperature is higher than the corresponding peak at 1,873 K (Fig. 3.39). The first RDF maximum (Fig. 3.38) is located at $R = 0.169$ nm, which is slightly greater than the "average" Si–O distance (0.166 nm). This can be attributed to the entry of aluminum into four-coordination. Inasmuch as nearly all the Si is bonded to oxygen in SiO_4 groups, the fraction of Al cations that form $(AlO_4)^{5-}$ tetrahedra was determined from the value of the area under the first RDF maximum [154]:

$$\frac{A}{2k_{\mathrm{O}}(n_{\mathrm{Si}}k_{\mathrm{Si}} + n_{\mathrm{Al}}k_{\mathrm{Al}})} \cong 4,$$

where n_{Si} is the concentration of Si in the melt, n_{Al} is the fraction of Al cations that form tetrahedra, and k_{O}, k_{Si}, and k_{Al} are the scattering powers of the elements. It was found that $n_{\mathrm{Al}} \approx 0.04$. This quantity of aluminum amounts to about 60% of its total content in the melt analyzed and, in addition, to approximately 21% of the total contribution of Si and Al to the formation of tetrahedral groups.

With consideration of the fact that the scattering powers of Al and Si are close, the values of R_1 were calculated as the additive sum of $R_{\mathrm{Si-O}}$ and $R_{\mathrm{Al-O}}$ in first coordination using the values obtained for the concentrations of Al and Si in tetrahedra (0.21 and 0.79, respectively):

$$R_{Si-O,\,Al-O} = R_{Si-O} \cdot n_{Si} + R_{Al-O} \cdot n_{Al} = 0.169\ \text{nm}.$$

A special feature of the RDF of this melt is the fact that it decays at larger values of R (≈ 1.0 nm) than in the case of numerous other oxide melts (≈ 0.5 nm). The Al cations interact electrostatically with the anionic $(SiO_4)^{4-}$ and $(AlO_4)^{5-}$ groups more strongly than do the Ca cations. Therefore, the placement of Al^{3+} in six-coordination (in an M_1 site, Fig. 3.31) causes stronger bonding between these anions with the resultant formation of larger $\left[Al_nSi_{n+1}O_{4(n+1)}\right]^{(n+4)}$ clusters and $[(Al,Si)O_3]_n^{m+2}$ chains of the mullite type. Since the bonding between these groups is implemented mainly by Ca cations, it is not as strong as the bonding within the groups themselves, which, in the final analysis, leads to arbitrary relative orientation of the olivine $\left[Me_nSi_{n+1}O_{4(n+1)}\right]^{2(n+2)-}$ (Me = Si, Al) fragments and the mullite $[(Al,Si)O_3]_n^{m+2}$ fragments in the melt. The presence of such large formations in the melt, as well as the irregularity of the placement of the Al atoms in the mullite chains, account for the strong tendency of the alumosilicate investigated to form a glass.

3.3.6 The $MnO–TiO_2–SiO_2$ System

The constitution of $MnO–TiO_2–SiO_2$ ternary silicate systems containing 35–50 mol% MnO, 0–50 mol% TiO_2, and 0–50 mol% SiO_2 was studied by X-ray diffraction analysis in [156] at temperatures that were 50–200 K above the liquidus line. All the intensity curves of the melts of the $MnO–TiO_2$ binary system (except the sample containing 67 mol% TiO_2) exhibited a high sharp first maximum, which may be evidence of a high degree of ordering in the melt. The melt containing 67 mol% TiO_2 has a low, asymmetric, and slightly broadened maximum, which is usually observed in microheterogeneous systems.

The first RDF coordination maximum has a complex form. For the melts containing 30, 33, and 67 mol% TiO_2, a secondary peak is observed on the small-R side at $R \approx 0.190$ nm, while the main maximum is located near 0.220 nm. For the composition with ~ 50 mol% TiO_2, there is only one maximum, which is shifted toward shorter distances.

The coordination numbers were calculated under the assumption that contributions to the first complex maximum are made by coordinations of Ti–O and Mn–O atoms:

$$A = 2n_{Ti}k_{Ti}k_O z_{Ti(O)} + 2n_{Mn}k_{Mn}k_O z_{Mn(O)}.$$

It was also assumed that the coordination polyhedra for titanium and manganese are identical, i.e., $z_{Mn(O)} = z_{Ti(O)}$.

The analysis of the structure factors and RDFs provided a basis for advancing the following tentative conclusions [156]. (1) The melts whose compositions are close to $MnTiO_3$ consist of coordination polyhedra that are similar in shape and dimensions. Approximately six oxygen atoms are arranged around the metal atom. (2) The melts that are close in composition to Mn_2TiO_4 consist of polyhedra of a different kind, although $z_{Me(O)} \approx 6$. This is supported by the presence of the secondary maximum on the short-distance side, whose appearance may be caused by the presence of manganese ions with a valance greater than 2 and low-valence titantium ions. A similar picture was noted in crystal structures of the spinel type, which include Mn_2TiO_4. (3) The Mn_2TiO_4 and $MnTiO_3$ melts are structurally homogeneous. (4) When the TiO_2 content is increased (above the composition of $MnTiO_3$), a microheterogeneous constitution of the melt, in which there are microclusters with a structure of the $MnTiO_3$ type and microclusters with the short-range order of titanium oxides, is possible.

The results of investigations of the MnO–TiO_2 and MnO–SiO_2 binary systems were used to interpret the diffraction data for the MnO–TiO_2–SiO_2 melts in [156].

A distinguishing feature of the experimental interference curves of the multi-component melts is the fact that in the alloys in which TiO_2 replaces SiO_2, an increase in the height of the first maximum of the $I(s)$ curve, which is similar in shape and position to the corresponding maximum for an $MnTiO_3$ melt, is observed as the titanium oxide content is increased. For the melts in which the concentration of SiO_2 is 50 mol%, the shape and height of the first maximum are close to those observed for an $MnSiO_3$ melt. These data are evidence of the microheterogeneous constitution of the MnO–TiO_2–SiO_2 melts, in which the principal coexisting microclusters correspond to $MnTiO_3$ and the decomposition products of $MnSiO_3$ (Mn_2SiO_4 and SiO_2). The calculation was performed under the assumption that the coexisting clusters in a microheterogeneous melt make additive contributions to the structure factors with consideration of the expression

$$[a(S) - 1] = \sum_{i=0}^{n}\sum_{j=0}^{n}\frac{n_i n_j f_i f_j}{\langle f^2\rangle}[a_{ij}(S) - 1],$$

where $\langle f^2\rangle$ is the square of the average atomic scattering factor of the melt. The following relation was obtained for MnO–TiO_2–SiO_2 melts [1]:

$$[a(S) - 1]_{\text{melt}} = \frac{n_{\text{Si}}^2(\text{melt})}{n_{\text{Si}}^2(\text{MnSiO}_3)} \cdot \frac{\langle f^2\rangle(\text{MnSiO}_3)}{\langle f^2\rangle(\text{melt})}[a(S) - 1]_{(\text{MnSiO}_3)} + \frac{n_{\text{Ti}}^2(\text{melt})}{n_{\text{Ti}}^2(\text{MnTiO}_3)} \cdot \frac{\langle f^2\rangle(\text{MnTiO}_3)}{\langle f^2\rangle(\text{melt})}[a(S) - 1]_{(\text{MnTiO}_3)}.$$

The contribution

$$\frac{2n_{\text{Si}}n_{\text{Ti}}f_{\text{Si}}f_{\text{Ti}}}{\langle f^2\rangle}[a_{\text{Si-Ti}}(S) - 1],$$

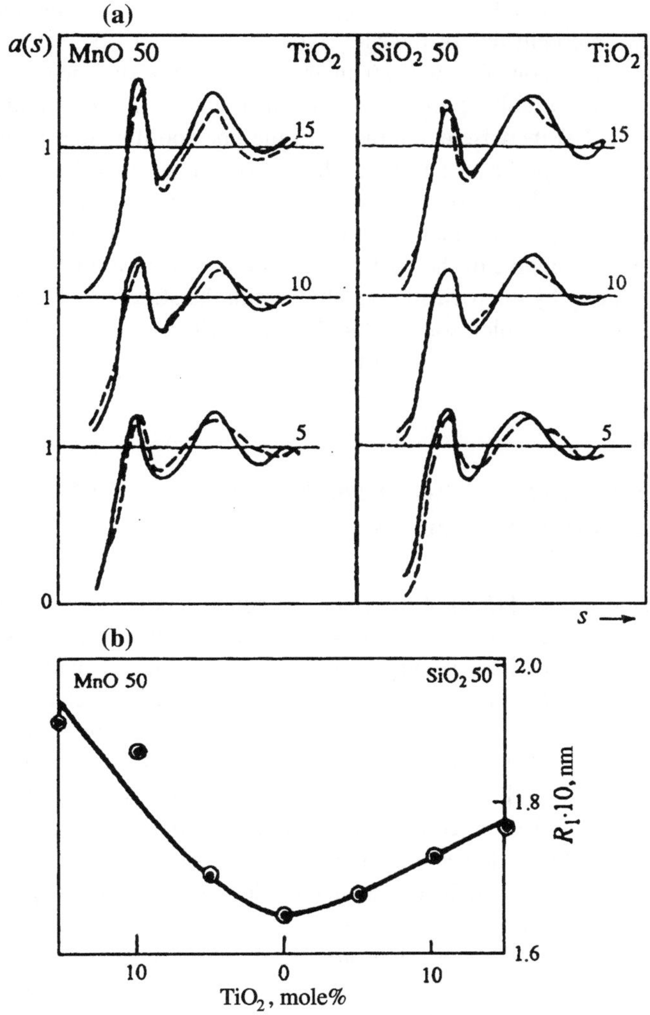

Fig. 3.40 Comparison of the experimental (*solid line*) and calculated (*dashed line*) structure factor for melts of the MnO–TiO_2–SiO_2 system (**a**) and dependence of the radius of the first coordination sphere R_1 on the composition (**b**)

which is small when the interaction on the microcluster boundaries is weak, was disregarded. For compositions with 50 mol% MnO, the experimental $a(s)$ curve is described satisfactorily (Fig. 3.40a) by a model, according to which the melts contain microclusters with the structures of a solid solution of $MnSiO_3$ in $MnTiO_3$ (up to 50 mol%) and the decomposition products of $MnSiO_3$ (Mn_2SiO_4 and SiO_2). The melts containing 50 mol% SiO_2 are satisfactorily described by the summation of $a(s)$ for liquid $MnTiO_3$ and $MnSiO_3$. In the former case microclusters with the Mn_2SiO_4 structure form in a larger amount than in the case of the ordinary decomposition of $MnSiO_3$ after melting, while in the latter case the melt is more enriched with microclusters of the SiO_2 type.

The same conclusions follow from an analysis of the RDF. The $G(R)$ curves for melts containing 50 mol% SiO_2 are similar in shape to the analogous curve for liquid $MnSiO_3$. The increase in the radius of the first coordination sphere for these compositions is insignificant and is attributed to a small contribution from Ti–O coordination (Fig. 3.40b). For compositions with 50 mol% MnO, the increase in the radius of the first coordination sphere with increasing TiO_2 content is more significant. Such a course of the RDF indicates that the constitution of this melt is determined mainly by microclusters of atoms with the structure of a solid solution based on $MnTiO_3$, in which some of the MnO_6 octahedra in the $MnTiO_3$ ilmenite motif are replaced by SiO_4 tetrahedra.

3.3.7 Constitution and Electrical Properties of Melts Containing Iron Oxides

Numerous studies (see [153]) of the electrical conductivity, thermoelectric power, transport numbers, and cathode and anode current efficiencies of melts containing transition metal oxides have provided evidence that electron conduction occurs in them. Semiconductor properties are displayed especially clearly when the dependence of electrical characteristics on the partial pressure of oxygen P_{O_2} is investigated. Since melting the oxides does not significantly alter the type of bonds between the atoms or their mutual short-range coordination, the laws governing the development of electron conduction in liquid oxides remain the same as in the solid state [157]. In the lower oxides of transition metals ($Fe_{1-x}O$, $Cu_{1-x}O$) the conductivity increases, and in the higher oxides (V_2O_{5-x}, TiO_{2-x}) it decreases with the oxygen excess.

FeO–Fe_2O_3 melts exhibit a high specific conductance κ (270 $\Omega^{-1}\cdot$ cm^{-1} at $P_{O_2} = 10^{-9}$ atm and 1,400°C), an increase in its value with the partial pressure of oxygen, and a positive temperature coefficient, which are characteristic of hole semiconductors. An addition of $\approx$30 mol% SiO_2 results in predominant ionic conductivity, disappearance of the dependence of κ on P_{O_2}, and a drop in the value κ to 2.7 $\Omega^{-1}\cdot$cm^{-1} ($P_{O_2} = 10^{-9}$ atm).

In the FeO–Fe_2O_3–CaO–SiO_2–Al_2O_3 system a dependence of κ on P_{O_2} appears when the content of iron oxides is greater than 20 mol%. In the range of partial pressures of oxygen from 10^{-10} to 10^{-8} atm, the conductance increases with increasing P_{O_2}. When $P_{O_2} = 10^{-8}$ to 1 atm, the value of κ drops sharply at first and then remains nearly unchanged. The decrease in κ is caused by the transition from *p*-type to *n*-type conduction due to the appearance of appreciable concentrations of trivalent iron in the melt.

In FeO–Fe_2O_3–CaO melts increasing the calcium oxide concentration from 0 to 33 mol% decreases the value of the conductance from 200 to 40 $\Omega^{-1}\cdot$ cm^{-1} and increases the anode current efficiency of iron from 2.5 to 17% [157]. Stated differently, a transition from predominantly electron conduction to a mixed type of conduction with an appreciable fraction assigned to the ionic component is also observed in this system. However, this transition is shifted toward higher (compared with the SiO_2 content) concentrations of calcium oxide.

Like iron silicate and iron–calcium melts, $FeO–Fe_2O_3–CaO–SiO_2$ melts also have a mixed type of conduction [158–160]. The electronic component appears when the percentage of SiO_2 is low and the content of the iron and calcium oxides is high. It increases with increasing values of the product of the mole fractions of the di- and trivalent iron ions [160]. The experimental data point to a disproportionally high fraction of the current that is transported by Ca^{2+} ions compared with the fraction transported by Fe^{2+}. This was attributed in [158] to the partial association of divalent iron cations with silicate anions.

Figure 3.41 shows the variation of the fraction of electron conduction with the composition for melts of the following systems: (1) $FeO–Fe_2O_3–SiO_2$, (2) $FeO–Fe_2O_3–CaO$; (3) $FeO–Fe_2O_3–CaO–SiO_2$; (4) $FeO–Fe_2O_3–CaO–SiO_2–Al_2O_3$. While electron conduction becomes predominant in systems 1 and 3 at $FeO + Fe_2O_3$ concentrations greater than 80 and 60 mol%, respectively, the predominance of electron conduction is observed in system 4 at a considerably lower concentration of iron oxides (about 20 mol%). This peculiarity can be attributed to the different effects of silicon, calcium, and aluminum cations on the polarization of oxygen anions and the concentration of trivalent iron in the melt.

As was previously pointed out (see Sect. 3.3.1), liquid nonstoichiometric iron oxide has the structure of Fe_3O_4. The electron conduction in magnetite includes the electron transfer from an Fe^{2+} ion to an Fe^{3+} cation at equivalent (octahedral) crystallographic sites [161]. The wave functions of the iron 3*d* electrons in the oxide do not overlap, and the electrons are found in localized levels [162]. Therefore, electron exchange between iron ions of different valence, which requires a small energy expenditure, apparently occurs through the 2*p* band of the oxygen ions [163] according to the scheme [164]

$$Fe^{3+}\left(3d^5\right) + O^{2-}\left(2p^6\right) = Fe^{2+}\left(3d^6\right) + O^{-}\left(2p^6\right)$$

and is maintained in multicomponent systems even at a small concentration of iron oxides, at which the distances between the iron ions are great and it is difficult to presume the occurrence of a direct (cation–cation) interaction.

Since Si^{4+} cations have a greater polarizing strength and more strongly deform the electron cloud of the O^{2-} anion than do Al^{3+} and Ca^{2+} ions, introducing them into an iron oxide melt strongly hinders the reaction that occurs with the formation of an electron–hole pair. As the X-ray structural investigations showed, the number of atomic clusters with the structure $Fe^{3+}_{tetr}(Fe^{3+},\ Fe^{2+})_{oct}O_4^{2-}$, in which electron exchange is possible, decreases sharply in compositions that are close to the chemical compound Fe_2SiO_4. In this range of SiO_2 concentrations, hole conduction nearly vanishes.

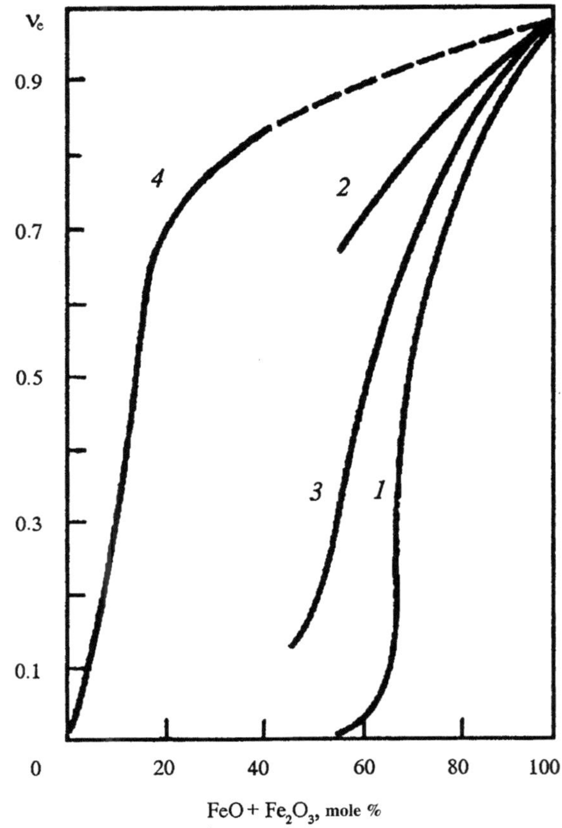

Fig. 3.41 Dependence of the fraction of the electronic conductivity on the content of iron oxides in melts: *1*—Fe_xO_y–SiO_2; *2*—Fe_xO_y–CaO [157]; *3*—Fe_xO_y–CaO–SiO_2 [158]; *4*—Fe_xO_y–CaO– SiO_2–Al_2O_3

In the iron–calcium melts a large part of the calcium oxide is bound to Fe_3O_4 in 2CaO·Fe_2O_3, and a small part is dissolved in $Fe_{1-x}O$. At the same time, Ca^{2+} cations statistically replace Fe^{2+} ions at octahedral sites, thereby diminishing electron exchange. However, while in the iron silicate system the concentration of Fe^{3+} ions decreases sharply with the addition of SiO_2, in the FeO–Fe_2O_3–CaO melts the stability of the trivalent iron ions increases together with CaO. In addition, at the same concentration of iron oxides, the mole fraction of $Fe_{1-x}O$ clusters with the magnetite structure is considerably greater in the FeO–Fe_2O_3–CaO melts than in the FeO–Fe_2O_3–SiO_2 system. In the FeO–Fe_2O_3–SiO_2 system the replacement of up to 20 mol% FeO by CaO results even in some increase in the fraction of the current transferred by electrons [158].

Alumina (Al_2O_3) exhibits amphoteric properties in oxide melts. The Al^{3+} cation can be found in both four-coordination and six-coordination with oxygen ions. An addition of Al_2O_3 to a melt has little influence on the ratio between the concentrations of the di- and trivalent iron ions. At the same time, electron conduction becomes predominant in the FeO–Fe_2O_3–CaO–SiO_2–Al_2O_3 system when the iron oxides amount to 20%. Apparently, the Al^{3+} ions, being located at tetrahedral points of the $Fe_{1-x}O$ complex, have no influence on the content of Fe^{2+} and Fe^{3+} ions at octahedral sites, which participate in electron exchange, but they lower the concentration of Ca^{2+} and Si^{4+} cations, which decrease the electronic conductivity.

3.3.8 Calculation of the Activities and Ionic Composition of Slag Melts Based on Polymer Theory

Many slag systems employed in metallurgical technologies contain relatively large amounts of silicon oxide and aluminum oxide, which can form complex ions under certain conditions. Several methods [165–168] that provide a way to describe the experimental data with satisfactory accuracy have been developed for calculating the activities of the components in oxide melts. However, only the theory of polymer melts provides a way to simultaneously calculate the ionic composition of an oxide melt [169, 170].

3.3.8.1 Calculation of the Activities and Ionic Composition of Multicomponent Silicate Melts

Methods that provide a way to calculate the ionic composition and activity of the components in Me^IO–$Me^{II}O$–SiO_2 ternary systems [171] and in multicomponent Me^IO–$Me^{II}O$–$\cdots$–Me^nO–SiO_2 oxide melts [172] on the basis of polymer theory have been developed. When the method described in [172] is used, all the cations in the melt are averaged energetically, i.e., in practice, the MeO–SiO_2 binary system with the following averaged polymerization constant k_p is considered:

$$\ln k_p = \frac{N_{Me^IO}}{\sum_{i=1}^{n} N_{Me^iO}} \ln k^I_{2,0} + \cdots + \frac{N_{Me^nO}}{\sum_{i=1}^{n} N_{Me^iO}} \ln k^n_{2,0}, \tag{3.40}$$

where $N_{Me^IO}, N_{Me^{II}O}, \ldots, N_{Me^nO}$ are the mole fractions of the oxides Me^IO, $Me^{II}O, \ldots, Me^nO$, respectively, and $k^I_{2,0}, k^{II}_{2,0}, \ldots, k^n_{2,0}$are the polymerization constants in the $Me^IO - SiO_2$, $Me^{II}O - SiO_2, \ldots, Me^nO - SiO_2$ binary systems, respectively.

In this section, we will develop a method for calculating the ionic composition and activities of the components in $Me^IO - Me^{II}O - \ldots - Me^hO - Me_2^{h+1}O - Me_2^{h+2}O - \ldots - Me_2^nO - SiO_2$ melts based on the approach proposed by Esin with allowance for the composition of the cation environment of the complex anions. In a binary melt the Me cations are neighbors of all the anions, but in an n-component system the $Si_iO_{3i+1-c}^{2(i+1-c)-}$ anions have a cation environment containing different amounts of Me^I, $Me^{II}, \ldots, Me^n$, which corresponds to ionic clusters of the type $Me^{I}_{i+1-c-m-\cdots-k} \times Me^{II}_{m} \ldots Me^{n}_{2k}Si_iO_{3i+1-c}$. We assume that $k^{I}_{2,0}$, $k^{II}_{2,0}$, …, $k^{n}_{2,0}$ are the equilibrium constants of the reactions

$$2Me^{I}_{2}SiO_4 = Me^{I}_{3}Si_2O_7 = Me^{I}O, \tag{3.41}$$

$$2Me^{II}_{2}SiO_4 = Me^{II}_{3}Si_2O_7 = Me^{II}O, \tag{3.42}$$

…………………………………………

$$2Me^{n}_{4}SiO_4 = Me^{n}_{6}Si_2O_7 = Me^{n}_{2}O. \tag{3.43}$$

We neglect the energy difference between the addition and self-closing of SiO_4^{4-} groups, i.e., we assume that the values of $k^{I}_{2,0}$, $k^{II}_{2,0}$, …, $k^{n}_{2,0}$ do not depend on i and c, as was done in [170, 171, 173, 174].

We note that as a consequence of the assumption that $k^{I}_{2,0}$, $k^{II}_{2,0}$, …, $k^{n}_{2,0}$ are not dependent on i and c for any polymerization reaction, including a reaction of the type

$$\begin{aligned}&(i-1+c+m+l+\cdots+k)Me^{I}_{3}Si_2O_7 + 2mMe^{II}SiO_4\\&+2lMe^{III}SiO_4+\cdots+2kMe^{n}_{4}SiO_7 = mMe^{II}_{3}Si_2O_7\\&+lMe^{III}_{3}Si_2O_7+\cdots+kMe^{n}_{6}Si_2O_7+Me^{I}_{i+1-c-m-l-\cdots-k}\\&\times Me^{II}_{m}Me^{III}_{l}\ldots Me^{n}_{2k}SiO_{3i+1-c}+(i-2-2c+2m\\&+2l+\cdots+2k)Me^{I}_{2}Si_iO_4,\end{aligned} \tag{3.44}$$

the standard change in the Gibbs free energy is proportional to the number of moles of MeO (n_{MeO}) that participate in the reaction. Since $n_{MeO} = 0$ for reaction (3.44), ΔG^0 is equal to 0 under the assumptions made, and $k_{(4)} = 1$.

Then, for the reaction

$$\begin{aligned}&iMe^{I}_{2}SiO_4 + mMe^{II}O + lMe^{III}O + \cdots + kMe^{n}_{2}O\\&= Me^{\,I}_{i+1-c-m-l-\cdots-k}Me^{II}_{m}Me^{III}_{m}\ldots Me^{n}_{2k}Si_iO_{4i+1-c}\\&+(i-1+c+m+l+\cdots+k)Me^{I}O,\end{aligned} \tag{3.45}$$

which is the sum of reactions (3.41)–(3.44), in the approximation of ideal ionic solutions the equilibrium constant equals

$$k_{i,c} = \frac{(k_{2,0})^{i+1+c+m+l+\cdots+k}}{(k^{II}_{2,0})^m(k^{III}_{2,0})^l\ldots(k^{n}_{2,0})^k} = \frac{(N_{i,c,m,\ldots k})(N^{I}_{O^{2-}})^{i-1+c+\cdots k}}{(N_{1,4})^i(N^{II}_{O^{2-}})^m\ldots(N^{n}_{O^{2-}})^k}, \tag{3.46}$$

where $N^{I}_{O^{2-}}$, $N^{II}_{O^{2-}}$, …, $N^{n}_{O^{2-}}$ are the ion fractions of "free oxygen" that is bonded to Me^{I}, Me^{II}, …, Me^{n} ions, respectively, $N^{I}_{1,4}$ is the ion fraction of the SiO_4^{4-} anion, whose cation environment consists only of Me^{I} cations. We find the equation describing the distribution of the complex anions from relation (3.46):

$$N_{i,c,m,\ldots,k} = \frac{N^{\mathrm{I}}_{\mathrm{O}^{2-}}}{k^{\mathrm{I}}_{2,0}} \left(\frac{k^{\mathrm{I}}_{2,0} N^{\mathrm{I}}_{1,4}}{N^{\mathrm{I}}_{\mathrm{O}^{2-}}}\right)^{i} \left(\frac{k^{\mathrm{I}}_{2,0}}{N^{\mathrm{I}}_{\mathrm{O}^{2-}}}\right)^{c} \left(\frac{k^{\mathrm{I}}_{2,0} N^{\mathrm{II}}_{\mathrm{O}^{2-}}}{k^{\mathrm{II}}_{2,0} N^{\mathrm{I}}_{\mathrm{O}^{2-}}}\right)^{m} \times \left(\frac{k^{\mathrm{I}}_{2,0} N^{\mathrm{III}}_{\mathrm{O}^{2-}}}{k^{\mathrm{III}}_{2,0} N^{\mathrm{I}}_{\mathrm{O}^{2-}}}\right)^{l} \times \cdots \times \left(\frac{k^{\mathrm{I}}_{2,0} N^{n}_{\mathrm{O}^{2-}}}{k^{n}_{2,0} N^{\mathrm{I}}_{\mathrm{O}^{2-}}}\right)^{k}. \tag{3.47}$$

As in [171], we introduce the notations $N^{\mathrm{I}}_{1,4} = U$, $k^{\mathrm{I}}_{2,0}/N^{n}_{\mathrm{O}^{2-}} = V$, $\left(k^{\mathrm{I}}_{2,0}, N^{\mathrm{II}}_{\mathrm{O}^{2-}}/k^{\mathrm{II}}_{2,0}, N^{\mathrm{I}}_{\mathrm{O}^{2-}}\right) = t, \ldots, \left(k^{\mathrm{I}}_{2,0}, N^{n}_{\mathrm{O}^{2-}}/k^{n}_{2,0}, N^{\mathrm{I}}_{\mathrm{O}^{2-}}\right) = f$. Then

$$N_{i,c,m,\ldots,k} = V^{c-1}(UV)^{i} t^{m} \ldots f^{k}. \tag{3.48}$$

Expression (3.48) contains the variables $N^{\mathrm{I}}_{\mathrm{O}^{2-}}, N^{\mathrm{II}}_{\mathrm{O}^{2-}}, \ldots, N^{n}_{\mathrm{O}^{2-}}, N^{\mathrm{I}}_{1,4}$, which can be found by using the following expressions:

$$\sum_{c=0}^{\infty} \sum_{i=c+1}^{\infty} \sum_{m=0}^{i+1-c} \cdots \sum_{k=0}^{i+1-c-m-\cdots-p} N_{i,c,m,\ldots,k} + \sum_{j=1}^{n} N^{j}_{\mathrm{O}^{2-}} = 1; \tag{3.49}$$

$$N_{\mathrm{Me^{I}O}} = \frac{N^{\mathrm{I}}_{\mathrm{O}^{2-}} + \sum_{c=0}^{\infty} \sum_{i=c+1}^{\infty} \sum_{m=0}^{i+1-c} \cdots \sum_{k=0}^{i+1-c-m-\cdots-p} (i+1-c-m-\cdots-k) N_{i,c,m,\ldots,k}}{1 + \sum_{c=0}^{\infty} \sum_{i=c+1}^{\infty} \sum_{m=0}^{i+1-c} \cdots \sum_{k=0}^{i+1-c-m-\cdots-p} (2i-c) N_{i,c,m,\ldots,k}}; \tag{3.50}$$

$$N_{\mathrm{Me^{II}O}} = \frac{N^{\mathrm{II}}_{\mathrm{O}^{2-}} + \sum_{c=0}^{\infty} \sum_{i=c+1}^{\infty} \sum_{m=0}^{i+1-c} \cdots \sum_{k=0}^{i+1-c-m-\cdots-p} m N_{i,c,m,\ldots,k}}{1 + \sum_{c=0}^{\infty} \sum_{i=c+1}^{\infty} \sum_{m=0}^{i+1-c} \cdots \sum_{k=0}^{i+1-c-m-\cdots-p} (2i-c) N_{i,c,m,\ldots,k}}; \tag{3.51}$$

$$N_{\mathrm{Me^{n}O}} = \frac{N^{n}_{\mathrm{O}^{2-}} + \sum_{c=0}^{\infty} \sum_{i=c+1}^{\infty} \sum_{m=0}^{i+1-c} \cdots \sum_{k=0}^{i+1-c-m-\cdots-p} k N_{i,c,m,\ldots,k}}{1 + \sum_{c=0}^{\infty} \sum_{i=c+1}^{\infty} \sum_{m=0}^{i+1-c} \cdots \sum_{k=0}^{i+1-c-m-\cdots-p} (2i-c) N_{i,c,m,\ldots,k}}. \tag{3.52}$$

As in [171], the activities of the principal oxides in the melt were assumed to be equal to the following: $a_{\mathrm{Me^{I}O}} = N^{\mathrm{I}}_{\mathrm{O}^{2-}}$; $a_{\mathrm{Me^{II}O}} = N^{\mathrm{II}}_{\mathrm{O}^{2-}}$; $\ldots a_{\mathrm{Me^{n}O}} = N^{n}_{\mathrm{O}^{2-}}$. The activity of silica was calculated from the equation

$$a_{\mathrm{SiO_2}} = \left(k^{\mathrm{I}}_{2,0}\right)^{2} N^{\mathrm{I}}_{1,4} \left(N^{\mathrm{I}}_{\mathrm{O}^{2-}}\right)^{-2}. \tag{3.53}$$

We will illustrate the method developed in the example cases of the FeO–MnO–MgO–CaO–SiO_2 and FeO–MgO–CaO–SiO_2 systems. For the five-component system, it is difficult to obtain finite expressions for the sums appearing in Eqs. 3.49–3.52, because the final dependences are extremely cumbersome. Therefore, the summation in this case was carried out numerically using Euler's method. Table 3.2 presents the ionic composition and activities of FeO in

Table 3.2 Ionic composition and activity of the components in the $FeO–MnO–MgO–CaO–SiO_2$ system at 1,873 K

Slag composition (wt.%)					$N^{I}_{1,4}$	$N^{I}_{O^{2-}}$	$N^{II}_{O^{2-}}$	$N^{III}_{O^{2-}}$	$N^{IV}_{O^{2-}}$	Exp. a_{FeO} [165]
CaO	SiO_2	FeO	MgO	MnO						
16.78	5.73	67.79	4.71	2.39	2.32×10^{-4}	0.745	0.026	7.44×10^{-2}	4.74×10^{-2}	0.765
0.32	2.52	62.24	1.82	33.01	2.30×10^{-3}	0.633	0.312	2.09×10^{-2}	6.71×10^{-4}	0.581
0.34	13.74	42.48	5.68	37.88	1.01×10^{-2}	0.490	0.248	2.49×10^{-2}	1.65×10^{-4}	0.477
0.28	22.12	28.63	12.76	36.47	4.39×10^{-3}	0.410	0.299	5.49×10^{-2}	1.76×10^{-4}	0.396
Root mean square deviation										0.029

$FeO–MnO–MgO–CaO–SiO_2$ melts at 1873 K. The calculations were performed with the following values of the polymerization constants [171, 175]: $k^{I}_{2,0} = 0.9$, $k^{II}_{2,0} = 0.2$, $k^{III}_{2,0} = 0.03$, $k^{IV}_{2,0} = 0.003$.

System of Eqs. 3.49–3.52 was solved using several iteration methods: Newton's method, the Newton–Kantorovich method, the curve-fitting method, and a multidimensional Halley iteration function. The Newton–Kantorovich function provided the smallest number of iterations.

The determination of the ionic composition and activities of the components for $FeO–MgO–CaO–SiO_2$ melts involves solving the following system of four non-linear equations:

$$\sum_{c=0}^{\infty}\sum_{i=c+1}^{\infty}\sum_{m=0}^{i+1-c}\cdots\sum_{k=0}^{i+1-c-m} N_{i,c,m,k} + N^{I}_{O^{2-}} + N^{II}_{O^{2-}} + N^{III}_{O^{2-}} = 1, \tag{3.54}$$

$$N_{FeO} = \frac{N^{I}_{O^{2-}} + \sum_{c=0}^{\infty}\sum_{i=c+1}^{\infty}\sum_{m=0}^{i+1-c}\cdots\sum_{k=0}^{i+1-c-m}(i+1-c-m-k)\,N_{i,c,m,k}}{1+\sum_{c=0}^{\infty}\sum_{i=c+1}^{\infty}\sum_{m=0}^{i+1-c}\cdots\sum_{k=0}^{i+1-c-m}(2i-c)\,N_{i,c,m,k}}, \tag{3.55}$$

$$N_{MgO} = \frac{N^{II}_{O^{2-}} + \sum_{c=0}^{\infty}\sum_{i=c+1}^{\infty}\sum_{m=0}^{i+1-c}\cdots\sum_{k=0}^{i+1-c-m} mN_{i,c,m,k}}{1+\sum_{c=0}^{\infty}\sum_{i=c+1}^{\infty}\sum_{m=0}^{i+1-c}\cdots\sum_{k=0}^{i+1-c-m}(2i-c)\,N_{i,c,m,k}}, \tag{3.56}$$

$$N_{CaO} = \frac{N^{III}_{O^{2-}} + \sum_{c=0}^{\infty}\sum_{i=c+1}^{\infty}\sum_{m=0}^{i+1-c}\cdots\sum_{k=0}^{i+1-c-m} kN_{i,c,m,k}}{1+\sum_{c=0}^{\infty}\sum_{i=c+1}^{\infty}\sum_{m=0}^{i+1-c}\cdots\sum_{k=0}^{i+1-c-m}(2i-c)\,N_{i,c,m,k}}. \tag{3.57}$$

Table 3.3 presents the ionic composition of $FeO–MgO–CaO–SiO_2$ melts for the following values of the polymerization constants:

$$k^{I}_{2,0} = 0.9, \quad k^{II}_{2,0} = 0.03, \quad k^{III}_{2,0} = 0.003.$$

The values of $k^{i}_{2,0}$ can also be evaluated from thermochemical data for a reaction of the type

$$2Me_iO + SiO_2 = Me_{i2}SiO_4 \tag{3.58}$$

Table 3.3 Ionic composition of slags of the FeO–MgO–CaO–SiO$_2$ system at 1873 K

Slag no.	Slag composition (mole%)				$N^{I}_{O^{2-}}$	$N^{II}_{O^{2-}}$	$N^{III}_{O^{2-}}$	$N^{I}_{1,4}$	$N^{II}_{1,4}$	$N^{III}_{1,4}$
	FeO	MgO	CaO	SiO$_2$						
1	53	10	10	27	0.646	9.3×10^{-3}	1.6×10^{-4}	0.12	0.022	6.6×10^{-4}
2	61	10	10	19	0.760	2.1×10^{-2}	2.8×10^{-3}	0.036	0.025	4.4×10^{-2}
3	30	15	10	45	0.386	8.5×10^{-3}	9.4×10^{-5}	0.096	0.042	5.1×10^{-4}
4	20	10	30	40	0.446	6.9×10^{-3}	4.2×10^{-4}	0.11	0.024	8.8×10^{-3}
5	35	10	30	25	0.673	1.1×10^{-2}	2.1×10^{-4}	0.12	0.029	1.1×10^{-3}
6	80	5	5	10	0.879	1.8×10^{-2}	3.2×10^{-3}	0.021	0.004	2.5×10^{-2}
7	25	10	10	55	0.228	5.5×10^{-3}	7.2×10^{-4}	0.042	0.022	3.8×10^{-2}
8	20	25	15	40	0.350	1.2×10^{-2}	1.1×10^{-3}	0.055	0.058	4.9×10^{-2}
9	15	40	10	35	0.445	1.5×10^{-2}	8.8×10^{-4}	0.073	0.075	2.6×10^{-2}

with the equilibrium constant k^i_{eq}. The assumption made above regarding the absence of a dependence of the polymerization constant on the size of the complexes allows us to write the following relation between $k^i_{eq}$and $k^i_{2,0}$ [171]:

$$k^i_{eq} = \left(k^i_{2,0}\right)^{-2}. \tag{3.59}$$

Since the activities of SiO$_2$, Me$_i$O, and Me$_{i2}$SiO$_4$ have been normalized to the liquid oxides and the liquid silicate, the standard change in the Gibbs free energy of reaction (3.58) at $T = 1{,}873$ K can be evaluated approximately by the relation

$$\begin{aligned}\Delta G^0 = \Delta G_s^{0\mathrm{Me}_{i2}\mathrm{SiO}_4} &- \Delta H_m^{\mathrm{SiO}_2}\left(1 - \frac{T}{T_m^{\mathrm{SiO}_2}}\right) - 2H_m^{\mathrm{Me}_i\mathrm{O}}\left(1 - \frac{T}{T_m^{\mathrm{Me}_i\mathrm{O}}}\right) \\ &+ \Delta H_m^{\mathrm{Me}_{i2}\mathrm{SiO}_4}\left(1 - \frac{T}{T_m^{\mathrm{Me}_{i2}\mathrm{SiO}_4}}\right),\end{aligned} \tag{3.60}$$

where $\Delta G_s^{0\mathrm{Me}_{i2}\mathrm{SiO}_4}$ is the standard change in the Gibbs free energy of reaction (3.58) with the participation of solid reactants, kJ/mole; and ΔH^i_m and T^i_m are, respectively, the enthalpy and melting point of the i-th reactant. We evaluate $k^i_{2,0}$ for the following values of the physicochemical parameters [176]:

$$\Delta H_m^{\mathrm{SiO}_2} = 9 \text{ kJ/mole}, \quad T_m^{\mathrm{SiO}_2} = 1953 \text{ K};$$

$$\Delta H_m^{\mathrm{CaO}} = 79.61 \text{ kJ/mole}, \quad T_m^{\mathrm{CaO}} = 2873 \text{ K};$$

$$\Delta H_m^{\mathrm{MgO}} = 77 \text{ kJ/mole}, \quad T_m^{\mathrm{MgO}} = 3053 \text{ K};$$

$$\Delta H_m^{\mathrm{Ca}_2\mathrm{SiO}_4} = 131.3 \text{ kJ/mole}; \quad \Delta H_m^{\mathrm{Mg}_2\mathrm{SiO}_4} = 64.1 \text{ kJ/mole};$$

$$\Delta G_s^{0\mathrm{Ca}_2\mathrm{SiO}_4} = -125.7 \text{ kJ/mole}; \quad \Delta G_s^{0\mathrm{Mg}_2\mathrm{SiO}_4} = -62.9 \text{ kJ/mole};$$

$$\Delta G_s^{\mathrm{Fe}_2\mathrm{SiO}_4} = -20.9 \text{ kJ/mole}; \quad T_m^{\mathrm{Ca}_2\mathrm{SiO}_4} = 2403 \text{ K}; \quad T_m^{\mathrm{Mg}_2\mathrm{SiO}_4} = 2163 \text{ K}.$$

Table 3.4 Activities of the components of the FeO–MgO–CaO–SiO_2 system at 1873 K

Slag no.	a_{FeO}	a_{FeO} (2.61)	a_{FeO} (2.62)	a^a_{FeO} [172]	a^b_{FeO} [172]	a^{exp}_{FeO} [177]	a_{SiO_2}	$a^a_{SiO_2}$ [172]	$a^b_{SiO_2}$ [172]	$a^{exp}_{SiO_2}$ [178]
1	0.646	0.784	1.09	0.433	0.451	0.69	0.254	0.081	0.077	0.26
2	0.760	0.80	1.28	0.566	0.581	0.83	0.055	0.041	0.039	0.08
3	0.386	0.678	0.244	0.105	0.107	0.44	0.568	0.34	0.33	0.51
4	0.446	0.638	0.246	0.075	0.071	0.42	0.488	0.12	0.11	0.47
5	0.673	0.722	0.664	0.446	0.529	0.70	0.230	0.0057	0.0038	0.10
6	0.879	0.860	0.901	0.746	0.750	0.94	0.024	0.043	0.029	0.05
7	0.228	0.555	0.126	0.047	0.048	0.23	0.713	0.672	0.667	0.70
8	0.350	0.638	0.239	0.084	0.096	0.41	0.396	0.136	0.124	0.36
9	0.445	0.533	0.201	0.106	0.112	0.45	0.325	0.043	0.0377	0.25
Root mean square deviation	0.045	0.17	0.24	0.29	0.27	–	0.056	0.179	0.185	–

[a] Calculation for $k^{I}_{2,0} = 0.9$, $k^{II}_{2,0} = 0.03$, and $k^{III}_{2,0} = 0.003$
[b] Calculation for $k^{I}_{2,0} = 0.5$, $k^{II}_{2,0} = 0.025$, and $k^{III}_{2,0} = 0.0016$ [172]

The calculated values of the constants $k^i_{2,0}$ are equal to $k^{I}_{2,0} = 0.5$, $k^{II}_{2,0} = 0.25$, and $k^{III}_{2,0} = 7.4 \times 10^{-3}$ and are fairly close to the literature data adopted above.

It is seen from Table 3.3 that the content of groups such as Mg_2SiO_4 and Ca_2SiO_4 increases and complexes containing several cations of the type $Fe_{i+1-c-m-k}Mg_m \cdot Ca_kSiO_{3i+1-c}$ appear as the SiO_2 content is increased.

In Table 3.4 the results of the calculations are compared with experimental data. It is noteworthy that data for (MnO + MgO + CaO) + FeO + (SiO_2 + P_2O_5) systems are presented in [177, 178]. This can be a source of discrepancies, especially when there are large amounts of MgO (25 and 40 mol%). The activities of iron monoxide and silicon dioxide were normalized in [177, 178] to pure liquid nonstoichiometric FeO and pure solid SiO_2. Calculations of a_{FeO} and a_{SiO_2} according to the method described in [172] were also presented, and the activity of FeO was calculated according to the theory of regular ionic solutions [165] using the expression

$$\frac{\log a_{FeO}}{\log N_{FeO}} = \frac{1000}{T}\left[2.18 N_{SiO_2} N_{MnO} + 5.9\left(N_{MgO} + N_{CaO}\right) N_{SiO_2}\right] \tag{3.61}$$

and for the (CaO–MgO)–(FeO–Fe_2O_3)–SiO_2 pseudoternary system [176] using the expression

$$\begin{aligned}\frac{\log a_{FeO}}{\log N_{FeO}} = &-\left(0.81 N_{FeO} + 0.2 N_{SiO_2}\right)\left[1 - \left(N_{CaO} + N_{MgO}\right)\right] - \\ &- 0.5 N^2_{SiO_2}\left[1 - 2\left(N_{CaO} + N_{MgO}\right)\right] + 9.7 N_{FeO} N_{SiO_2} \\ &- 19.4 N_{FeO} N^2_{SiO_2}.\end{aligned} \tag{3.62}$$

The calculated values of the activities (Table 3.4) are in fairly good agreement with the experimental data. According to the evaluations, the root mean square deviations of the calculated activities of FeO and SiO_2 from the experimental values, taking into account the composition of the cation environment increases the accuracy of the calculations by a factor of 3–7.

For the purpose of facilitating the calculations, the results are described by correlation equations:

$$a_{FeO} = \exp[1.0834 - 0.845N_{FeO} - 1.014N_{MgO} - 0.346N_{CaO} - 3.852N_{SiO_2}], \quad (3.63)$$

$$R = 0.979, \quad \Delta = 7.0\%$$

$$a_{MgO} = 0.0236N_{FeO}^{0.3}\,N_{MgO}^{0.633}\,N_{CaO}^{0.051}\,N_{SiO_2}^{-0.816}, \quad (3.64)$$

$$R = 0.958, \quad \Delta = 8.2\%$$

$$a_{SiO_2} = 3{,}065N_{FeO}^{-0.788}\,N_{MgO}^{-0.737}\,N_{CaO}^{0.185}\,N_{SiO_2}^{2.0}, \quad (3.65)$$

$$R = 0.979, \quad \Delta = 17.5\%.$$

For CaO it was not possible to devise an equation with acceptable values of the multiple correlation coefficient R and the average relative deviation Δ.

Further improvement of the proposed method is associated with taking into account the presence of fluorides in the melt, the influence of i and c on the polymerization constants, and the nonideality of an ionic melt.

3.3.8.2 Calculation of the Ionic Composition and Activities of the Components in CaO–Al_2O_3 Melts Based on Polymer Theory

From literature data it is known [179] that the structural units of the polymerized anions in the system under consideration are AlO_4^{5-}complexes, which form chains, rings, and other forms of alumo-oxygen anions [175]. Since Al_2O_3 is an amphoteric oxide, the appearance of free Al^{3+} cations [180] or, according to some data, AlO^+ cations [175] in the melt may be expected.

An attempt to describe the ionic composition in oxide systems containing Al_2O_3 was undertaken in [175]. However, the authors confined themselves to consideration of only chainlike anionic structure and did not take into account the appearance of Al^{3+} in the melt or the influence of the AlO^+ cation on the distribution of the complex ions.

The absence of data regarding the actual ionic composition of such melts allows us to assess the correctness of a particular calculation scheme only indirectly, that is, according to the closeness of the calculated and experimental values of the activities of the components. We will apply this approach to the following five models of the constitution of CaO–Al_2O_3 melts or the corresponding calculation schemes [181].

1. The melt consists of Ca^{2+}, O^{2-}, AlO_4^{5-}, $Al_2O_7^{8-}$, and all the other possible (besides the isomeric) alumo–oxygen anions with the general formula $Al_iO_{3i+1-c}^{(3i+2-2c)-}$ (which can be abbreviated as $Al_xO_y^{z-}$).
2. The composition of the melt includes the following: Ca^{2+}, Al^{3+}, O^{2-}, $Al_xO_y^{z-}$.
3. The composition of the melt includes the following: Ca^{2+}, AlO^{+}, O^{2-}, $Al_xO_y^{z-}$.
4. The composition of the melt includes the following: Ca^{2+}, Al^{3+}, AlO^{+}, O^{2-}, $Al_xO_y^{z-}$.
5. The composition of the melt includes Ca^{2+}, Al^{3+}, AlO^{+}, O^{2-}, and $Al_xO_y^{z-}$, and the influence of the composition of the cation environment on the polymerization of the $Al_xO_y^{z-}$ anions is taken into account.

The method developed in [171] was used for the calculations.

We take the following complexation reactions:

$$Al_2O_3 + 5O^{2-} = 2AlO_4^{5-}, \tag{3.66}$$

$$iAlO_4^{5-} = Al_iO_{3i+1-c}^{(3i+2-2c)-} + (i-1+c)O^{2-}, \tag{3.67}$$

$$Al_2O_3 + O^{2-} = Al^{3+} + AlO_4^{5-}, \tag{3.68}$$

$$Al_2O_3 + 2O^{2-} = AlO^{+} + AlO_4^{5-}. \tag{3.69}$$

For the calculation of the ionic composition according to scheme I (Eqs. 3.66 and 3.67), we use the following balance equations, which were derived from the stoichiometry of the reactions that form the various structural units:

$$\sum_{c=0}^{\infty}\sum_{i=c+1}^{\infty} N_{i,c} + N_{O^{2-}} = 1, \tag{3.70}$$

$$2C_{O^{2-}} + \sum_{c=0}^{\infty}\sum_{i=c+1}^{\infty} (3i+2-2c)C_{i,c} = 2C_{Ca^{2+}}, \tag{3.71}$$

where $N_{O^{2-}}$ and $N_{i,c}$ are the ion fractions of the O^{2-} and $Al_iO_{3i+1-c}^{(3i+2-2c)-}$ anions, respectively; $C_{O^{2-}}$, $C_{i,c}$, and $C_{Ca^{2+}}$ are the concentrations of the O^{2-}, $Al_iO_{3i+1-c}^{(3i+2-2c)-}$, and Ca^{2+} ions, respectively (mole/m^3).

Next, using the relations

$$C_{Ca^{2+}} = C_{CaO}, \tag{3.72}$$

$$\sum C^{-} \sum_{c=0}^{\infty}\sum_{i=c+1}^{\infty} iN_{i,c} = 2C_{Al_2O_3}, \tag{3.73}$$

where C_{CaO} and $C_{Al_2O_3}$ are, respectively, the concentrations of CaO and Al_2O_3 (mole/m^3), and $\sum C^{-}$ is the sum of the concentrations of all the anions (mole/m^3). After dividing (3.71) by (3.73), we obtain the system of equations

$$\sum_{c=0}^{\infty}\sum_{i=c+1}^{\infty} N_{i,c} + N_{O^{2-}} = 1,$$

$$\frac{N_{O^{2-}} + \sum_{c=0}^{\infty}\sum_{i=c+1}^{\infty}(1,5i+1-c)\,N_{i,c}}{0,5\sum_{c=0}^{\infty}\sum_{i=c+1}^{\infty} iN_{i,c}} = \frac{N_{CaO}}{N_{Al_2O_3}}. \tag{3.74}$$

Here N_{CaO} and $N_{Al_2O_3}$ are the mole fractions of CaO and Al_2O_3, respectively.

Since the structure of alumo-oxygen ions is similar to the structure of silicate ions [175], the following relations can be adopted for them in analogy to [170]:

$$\sum_{c=0}^{\infty}\sum_{i=c+1}^{\infty} N_{i,c} = \frac{N_{1,4}}{(1-x)(1-y)}, \tag{3.75}$$

$$\sum_{c=0}^{\infty}\sum_{i=c+1}^{\infty} iN_{i,c} = \frac{N_{1,4}-x^3}{(1-x)^2(1-y)^2}, \tag{3.76}$$

$$\sum_{c=0}^{\infty}\sum_{i=c+1}^{\infty} cN_{i,c} = \frac{x^2}{(1-x)(1-y)^2}. \tag{3.77}$$

Here $x = k_p N_{1,4}/N_{O^{2-}}$, and $y = k_p^2 N_{1,4}/(N_{O^{2-}})^2$, where $N_{1,4}$ is the ion fraction of AlO_4^{5-}; k_p is the equilibrium constant of the dimerization reaction

$$2AlO_4^{5-} = Al_2O_7^{8-} + O^{2-}. \tag{3.78}$$

Assuming that the activities of the ions are equal to their ion fractions [173, 182], we find the activities of the components of the melt within scheme I from the expressions

$$a_{CaO} = N_{Ca^{2+}} N_{O^{2-}} = N_{O^{2-}}, \tag{3.79}$$

$$a_{Al_2O_3} = k_p^5 N_{1,4}^2/(N_{O^{2-}})^5. \tag{3.80}$$

Here $N_{Ca^{2+}}$ is the ion fraction of Ca^{2+}, which is equal to 1 in the present case. The equilibrium constant of reaction (3.66) is expressed in terms of k_p raised to a power equal to the number of O^{2-} ions in this reaction, since the change in the Gibbs free energy per mole of "free" oxygen does not depend on the complexity of the anions [177].

The calculation of the ionic composition according to scheme II (reactions (3.67) and (3.68)) involves solving a system of equations that includes Eq. 3.70 and the following equations:

$$C_{Al^{3+}} + \sum_{c=0}^{\infty}\sum_{i=c+1}^{\infty} iC_{i,c} = 2C_{Al_2O_3}, \tag{3.81}$$

$$2C_{O^{2-}} + \sum_{c=0}^{\infty}\sum_{i=c+1}^{\infty}(3i+2-2c)C_{i,c} = 2C_{Ca^{2+}} + 3C_{Al^{3+}} \tag{3.82}$$

where $C_{Al^{3+}}$ is the concentration of Al^{3+} cations (mole/m^3).

After some transformations we obtain

$$\frac{N_{O^{2-}} + \sum_{c=0}^{\infty} \sum_{i=c+1}^{\infty} (1.5i + 1 - c)\, N_{i,c}}{\sum_{c=0}^{\infty} \sum_{i=c+1}^{\infty} iN_{i,c}} = \frac{N_{CaO}}{N_{Al_2O_3}} + 1.5. \tag{3.83}$$

The combined solution of Eqs. 3.70, 3.72, and 3.83 with allowance for (3.75)–(3.77) enables us to calculate the ionic composition of the oxide melt. The activities of the components in this case are equal to

$$a_{CaO} = N_{Ca^{2+}} N_{O^{2-}}, \tag{3.84}$$

$$a_{Al_2O_3} = k_p N_{Al^{3+}} / N_{O^{2-}}, \tag{3.85}$$

where $N_{Al^{3+}}$ is the ion fraction of the Al^{3+} cation.

Applying similar arguments to reaction (3.69), instead of Eq. 3.83 we obtain

$$\frac{N_{O^{2-}} + \sum_{c=0}^{\infty} \sum_{i=c+1}^{\infty} (1.5i + 1 - c)\, N_{i,c}}{\sum_{c=0}^{\infty} \sum_{i=c+1}^{\infty} iN_{i,c}} = \frac{N_{CaO}}{N_{Al_2O_3}} + 0.5. \tag{3.86}$$

Expression (3.86) is used together with (3.70) and (3.72) to calculate the ionic composition of a melt based on scheme III.

We find the activities of the components in a CaO–Al_2O_3 melt for this calculation scheme from expression (3.84) and the following expression

$$a_{Al_2O_3} = k_p^2 N_{1,4} N_{AlO^+} / (N_{O^{2-}})^2, \tag{3.87}$$

where N_{AlO^+} is the ion fraction of the AlO^+ cation.

In scheme IV reactions (3.67), (3.68), and (3.69) should be taken into account in the calculation of the ionic composition. When AlO^+ and Al^{3+} ions form in the melt, the balance equation of type (3.71) takes the form

$$\frac{C_{Al_2O_3}}{\sum_{c=0}^{\infty} \sum_{i=c+1}^{\infty} iN_{i,c}} \left[N_{O^{2-}} + \sum_{c=0}^{\infty} \sum_{i=c+1}^{\infty} (1.5i + 1 - c)\, N_{i,c}\right] = C_{Ca^{2+}} + 1.5 C_{Al^{3+}} + 0.5 C_{AlO^+}, \tag{3.88}$$

where C_{AlO^+} is the concentration of AlO^{3+} ions (mole/m^3).

After transforming Eq. 3.88, we have

$$\frac{1 - \sum_{c=0}^{\infty} \sum_{i=c+1}^{\infty} iN_{i,c}}{\sum_{c=0}^{\infty} \sum_{i=c+1}^{\infty} iN_{i,c}} = \frac{N_{CaO}}{N_{Al_2O_3}} - \frac{N_{O^{2-}}}{N_{O^{2-}} + k_d}. \tag{3.89}$$

Here k_d is the dissociation constant of AlO^+ in the reaction

$$AlO^+ = Al^{3+} + O^{2-}, \tag{3.90}$$

$$k_d = N_{O^{2-}} N_{Al^{3+}} / N_{AlO^+}. \tag{3.91}$$

Equations 3.70, 3.72, 3.89, and 3.91 are used to calculate the ionic composition for scheme IV.

The activities of CaO and Al_2O_3 were found from expressions (3.84) and (3.87).

The literature data on the activities of the components in the CaO–Al_2O_3 system, which were summarized in [177], exhibit considerable divergence, especially in the activities of Al_2O_3. In [179] the data from [183], with which we compared our results, were presented as the most reliable.

For comparison with the experimental data on the activities of the components in the CaO–Al_2O_3 system, where the solid oxides [183] were taken as the standard state, the activities were recalculated using the equation

$$\ln\frac{a_s^i}{a_l^i} = \frac{\Delta H_m^i}{R}\left(\frac{1}{T} - \frac{1}{T_m^i}\right) + \frac{\Delta C_P^i}{R}\ln\frac{T_m^i}{T} - \frac{\Delta C_P^i}{R}T_m^i\left(\frac{1}{T} - \frac{1}{T_m^i}\right), \tag{3.92}$$

Here ΔC_P^i is the change in the specific heat of the *i*th component (J/mole·K); a_l^i and a_s^i are the activities of the *i*th component of the melt normalized to the liquid and solid states of the *i*-th component, respectively; T_m^i is the melting point of the *i*th component; T is the actual temperature of the melt (K); ΔH_m^i is the heat of fusion of the *i*th component (J/mole · K).

The values $\Delta C_P^{CaO} = 27.76$ J/mole · K and $\Delta C_P^{Al_2O_3} = 23.85$ J/mole · K were taken.

Calculation scheme I does not provide activity values that are close to the experimental values.

Schemes II and III do not describe the activities of the two components over the entire range of variation of the concentrations. The best fit with the literature data is achieved for the value $k_p = 0.04$, which was used in the ensuing calculations. At the same time, changes in the polymerization constant in the range from 0.03 to 0.09 do not very strongly alter the values of the activities of the components.

Scheme II gives values of $a_{Al_2O_3}$ that are close to the published values [177] only when $N_{Al_2O_3} \leq 0.45$. The values of a_{CaO} are considerably higher than those in [183], that is, the calculation gives an overestimated concentration of "free" oxygen ions $N_{O^{2-}}$.

Clearly, taking into account reaction (3.69) (scheme III) should lower the calculated values of the activities of CaO, since the number of ions $N_{O^{2-}}$ diverges more strongly in this case than in reaction (3.68). According to the calculations, scheme III accurately describes the variation of the activity of CaO over the entire range of concentrations. At the same time, the values of the activities of Al_2O_3 were overestimated.

The results obtained with allowance for the simultaneous occurrence of reactions (3.68) and (3.69) showed the following. Definite difficulties arose when the calculations were performed because of the lack of values for the dissociation constant of AlO^+ in reaction (3.90). It was initially assumed that $k_d = k_p$, since reaction (3.90) is the sum of reactions (3.68) and (3.69), and the polymerization does not depend on the size of the anion. However, the calculations performed with $k_d = k_p = 0.04$ showed that the values of $a_{Al_2O_3}$ in the range $N_{Al_2O_3} \geq 0.4$ are far from the experimental values. Apparently, the ordinary condition that k_p does not depend on i and c [170] is valid only for reactions involving alumo-oxygen ions in which the aluminum atom has tetrahedral coordination, that is, beginning with AlO_4^{5-}. Therefore, the value $k_d = 0.012$, with which the calculated data accurately describe the activities of the components over the entire range of compositions investigated, was chosen. However, when $N_{CaO} \approx 0.55$, the deviation of $a_{Al_2O_3}$ from the experimental data is still relatively large.

Let us evaluate the influence of the composition of the cation sublattice on the polymerization of complex ions (scheme V). For this purpose we write the reaction

$$iAl_2O_3 + mAlO^+ = Al_iO_{3i+1-c-m}O_m^{(3i+2-2c)-} + mAl^+ \\ + (i - 1 + c + m)O^{2-}. \tag{3.93}$$

In this case the calculation equations will be the following:

$$\sum_{c=0}^{\infty} \sum_{i=c+1}^{\infty} \sum_{m=0}^{i+1-c} N_{i,c,m} + N_{O^{2-}} = 1, \tag{3.94}$$

$$N_{Al_2O_3} = \frac{\sum_{c=0}^{\infty} \sum_{i=c+1}^{\infty} \sum_{m=0}^{i+1-c} N_{i,c,m}}{1 + \sum_{c=0}^{\infty} \sum_{i=c+1}^{\infty} \sum_{m=0}^{i+1-c} (2i - c - m) N_{i,c,m}}, \tag{3.95}$$

$$\frac{N_{CaO} - N_{AlO^+}}{N_{Al_2O_3}} = \frac{1 + \sum_{c=0}^{\infty} \sum_{i=c+1}^{\infty} \sum_{m=0}^{i+1-c} cN_{i,c,m}}{\sum_{c=0}^{\infty} \sum_{i=c+1}^{\infty} \sum_{m=0}^{i+1-c} mN_{i,c,m}}. \tag{3.96}$$

Here

$$N_{i,c,m} = \frac{N_{O^{2-}}}{k_p} \left(\frac{k_p N_{1,4}}{N_{O^{2-}}} \right)^i \left(\frac{k_p}{N_{O^{2-}}} \right)^c \left(\frac{k_p N_{AlO^+}}{k'_p N_{O^{2-}} N_{Al^{3+}}} \right); \tag{3.97}$$

k'_p is the equilibrium constant of the reaction

$$Al_2O_3 + AlO^+ = AlO_4^{5-} + 2Al^{3+}. \tag{3.98}$$

To find the triple sums appearing in Eqs. 3.94–3.96, we introduce the notations

$$U = N_{1,4}; \quad V = \frac{k_p}{N_{O^{2-}}}; \quad t = \frac{k_p N_{AlO^+}}{k'_p N_{O^{2-}} N_{Al^{3+}}}, \tag{3.99}$$

and we use the expressions obtained in [171] for ternary silicate systems.

The values of the activities of CaO and Al_2O_3 were calculated using Eqs. 3.84, 3.85, and 3.94–3.97 for $k'_p = 9.00$. The values of the activities obtained in the calculation and in the experiment differ only slightly over the entire range of variation of the concentrations of the components in the melt. For this reason, when scheme V was used, the ionic composition of the melts of the CaO–Al_2O_3 system was calculated from Eqs. 3.94–3.96.

Table 3.5 Ionic composition of melts of the $CaO–Al_2O_3$ system

N_{CaO}	Ion fractions						
	O^{2-}	AlO_4^{5-}	$Al_2O_7^{8-}$	$Al_3O_9^{9-}$	AlO^+	Al^{3+}	Alumi-nates
0.50	0.036	0.28[a] / 0.016	0.087 / 8.08×10^{-6}	0.0301 / 1.29×10^{-10}	0.38	0.12	0.551
0.55	0.051	0.37 / 0.012	0.107 / 4.11×10^{-6}	0.0244 / 4.03×10^{-11}	0.35	0.10	0.435
0.60	0.086	0.49 / 6.72×10^{-3}	0.117 / $2.67 \times 10 - 7$	0.0118 / 4.74×10^{-12}	0.32	0.08	0.295
0.65	0.180	0.59 / 4.00×10^{-3}	0.077 / 2.67×10^{-7}	0.0023 / 2.96×10^{-13}	0.30	0.05	0.151
0.70	0.350	0.54 / 1.80×10^{-3}	0.033 / 3.60×10^{-8}	0.0002 / 8.00×10^{-15}	0.27	0.03	0.077

[a] The values above the bars are the ion fractions of the anions surrounded only by Ca^{2+} , and the values below the bars are the ion fractions of the anions surrounded only by Al^{3+}

The results of the calculations are presented in Table 3.5.

It is seen that the content of Al^{3+} ions in the melt decreases sharply with increasing mole fraction of CaO, while the content of AlO^+ varies only slightly. Most of the alumo-oxygen anions exist in the form of AlO_4^{5-} ions surrounded by Ca^{2+} cations.

The calculation of the activities of the components in binary silicate melts based on polymer theory was examined in detail in [184].

3.3.8.3 Properties and Structure of $MeO–Al_2O_3–SiO_2$ Melts

To describe the properties and structure of melts of the $MeO–Al_2O_3–SiO_2$ system, $SiO_4^{4-}$and AlO_4^{5-} groups that form chains, rings, and other (besides the isomeric) forms of complex anions were taken as the structural units, according to the literature data.

The ionic compositions and activities of the components of the CaO(22–42%)–Al_2O_3(19–49%)–SiO_2(8–38%) system at 1,873 K were calculated. Satisfactory correspondence between the calculated and experimental data for the activities of all the components of the system under consideration was discovered. The root mean square deviations are: CaO, 0.034; SiO_2, 0.012; Al_2O_3, 0.058.

According to the calculations, the silicon and aluminum in the melts studied appear mainly in the composition of complex alumosilicate anions (60–80%). Only insignificant portions of the Si and Al atoms form simple silicate, aluminate, and alumoslicate anions: SiO_4^{4-}, 10–18%; AlO_4^{5-}, 0.1–5%; $Si_2O_7^{6-}$, 0.1–3%; $Al_2O_7^{8-}$, 0.02–0.8%; $SiAlO_7^{7-}$, 2–10%. A considerable portion of the aluminum (up to 41%) can exist in the form of Al^{3+} cations.

The ionic composition of various welding fluxes was evaluated on the basis of the proposed methods (Table 3.6). To simplify the calculations, their empirical chemical composition was adjusted to a hypothetical four-component system, in which CaO = CaO + MgO + MnO + FeO. As can be seen from Table 3.6, the content of free O^{2-} oxygen ions is low in all the fluxes, indicating their relatively low oxidizing capacity. It is noteworthy that for the fluxes used in electroslag technologies, nearly all the silicon exists in the form of SiO_4^{4-} anions, and nearly all of the aluminum exists in the form of Al^{3+}.

3.3.9 Diffusion of Ions in Molten Slags

3.3.9.1 General Diffusion Laws of Ions

Knowledge of the diffusion characteristics of the components of slag melts can help to resolve questions concerning the mechanism and kinetics of metallurgical processes, since in most cases diffusion in the slag is the rate-limiting step in reactions that occur on the slag/metal and slag/gas interfaces. In liquid slags, as in ionic melts, the transport rate of particles is determined not only by their concentration gradient (chemical potential), but also by the electric potential appearing upon the simultaneous diffusion of different charged particles, for example, a cation C and an anion A. In slags that have ion conduction, the displacement of oxygen ions is accompanied by the motion of calcium, iron, and other cations. An electric field hinders the transport of the most mobile ions and accelerates the displacement of the slow-moving ions. Thus, addition of a tiny amount of F^- ions, which are more mobile than O^{2-} anions, to an oxide melt increases the diffusion coefficients of Ca^{2+} and Fe^{2+} cations by two to three fold [185]. Under the conditions of simultaneous displacement of cations and anions with maintenance of the overall electroneutrality of an ionic melt, the diffusion coefficients D_C and D_A of the individual ions are related to the experimentally determined mutual diffusion coefficient $\widetilde{D}$ by the relation [186]:

$$\widetilde{D} = \frac{D_C z_A u_A c_A + D_A z_C u_C c_C}{z_C u_C c_C + z_A u_A c_C}.$$

If the concentration of the diffusing cation is very small compared with that of the anion, the terms containing c_C can be neglected. In this case the diffusion coefficient of the cation can be equated to the measured value of the mutual diffusion coefficient, i.e., $\widetilde{D} \cong D_C$. In a solution that is highly diluted with respect to the cation C, its activity coefficient scarcely depends on the composition. Then, in agreement with Darken's well-known equation, D_C will be equal to the self-diffusion coefficient of the particular species of particles $^*D_C$, which characterizes their motion in the medium in the absence of a concentration gradient of the other ions. The self-diffusion coefficient is usually measured using radioactive isotopes. In this case it characterizes the simple exchange of places between ions of the radioactive and stable isotopes. Thus, when the concentration of ions in a slag is low, their self-diffusion coefficient can be determined by methods that do not use

Table 3.6 Approximate ionic composition of some welding fluxes

Type of flux	Composition used in calculations (wt.%)				Ionic composition, ion fractions of anions						
	SiO_2	Al_2O_3	CaO	CaF_2	O^{2-}	SiO_4^{4-}	AlO_4^{5-}	Al^{3+}	$SiAlO_7^{7-}$	F^-	Other
ANF-28	22	3	31	44	$4.7 \times \times 10^{-3}$	0.16	1.7×10^{-4}	4.9×10^{-2}	5.5×10^{-5}	0.79	4.5×10^{-2}
AN-348	45	4.5	45.5	5	3.7×10^{-3}	0.35	9.6×10^{-5}	9.1×10^{-2}	8.2×10^{-5}	0.27	0.38
OSts-45	42	5	44	9	4.2×10^{-3}	0.32	1.7×10^{-4}	9.7×10^{-2}	1.3×10^{-4}	0.37	0.30
AN-8	35	13	35	17	2.3×10^{-2}	0.35	3.1×10^{-3}	0.22	3.4×10^{-3}	0.49	0.16
AN-20	25	30	20	25	1.8×10^{-2}	0.16	3.7×10^{-2}	0.36	3.9×10^{-2}	0.65	0.12
AN-22	20	20	35	25	2.3×10^{-2}	0. 24	5.9×10^{-2}	0.19	1.8×10^{-2}	0.61	0.15
AN-26	30	20	20	30	6.4×10^{-3}	0.32	7.6×10^{-4}	0.34	1.3×10^{-3}	0.63	4.16×10^{-2}
ANF-6	2.5	31.5	6.0	60.0	2.5×10^{-2}	4.2×10^{-2}	8.5×10^{-2}	0.28	5.8×10^{-3}	0.86	7.3×10^{-2}
AN-1	38	38	16	8	1.3×10^{-2}	0.46	9.9×10^{-3}	0.26	1.2×10^{-2}	0.30	0.24
FTs-1	47	45	2	6	3.1×10^{-3}	0.30	1.1×10^{-4}	4.0×10^{-2}	7.2×10^{-5}	0.29	0.41
AN-10	39	21	20	20	2.4×10^{-2}	0.28	6.3×10^{-2}	0.18	1.8×10^{-2}	0.53	0.19
AN-15	27	28	25	20	1.8×10^{-2}	0.26	3.2×10^{-2}	0.31	3.5×10^{-2}	0.55	0.19

labeled atoms. In oxide melts, as in metallic melts, the temperature dependence of the diffusion coefficient of the ions generally obeys the Arrhenius equation

$$D = D_0 \exp(-E_D/RT), \tag{3.100}$$

which contains the pre-exponential factor D_0 and the activation energy for diffusion E_D.

3.3.9.2 Methods for Studying the Diffusion of Ions in Slag Melts

The same methods that are used for liquid metals are used to determine the diffusion coefficients in oxides and slags. A review of these methods was given in [186–197]. The *capillary* and *electrochemical* methods have become most widely employed.

In different versions of the *capillary method*, the diffusion coefficient is determined using Fick's equation for diffusion in a body of infinite length. The experiment is set up so that the element whose transport is being investigated would diffuse into a melt found in a capillary. A radioactive isotope of the element is usually employed as the diffusant. The variation of the concentration of the diffusing ion along the length of the capillary is established from the distribution of the radioactivity in the solid.

When there is no radioactive isotope that is suitable for the task, stable isotopes, for example, ^{17}O and ^{18}O, are also used, and their concentration in the slag is determined using a mass spectrometer.

The shortcomings of the capillary method include the impossibility of completely eliminating the convective mixing of the melt in the capillary, the insufficient accuracy of the establishment of the distribution of the concentration of the diffusing element along the length of the capillary, and the possibility of changes in the composition of the diffusion medium as a consequence of its interaction with the material of the capillary.

The *electrochemical methods* for determining diffusion coefficients are based on the laws governing the electrode processes that occur upon alternating-current or direct-current polarization of the electrodes.

When *alternating-current electrode polarization* is employed, the capacitive reactance and total ohmic resistance of the electrodes are found. A component part of the latter is the diffusional resistance, which is related by an equation to the frequency of the alternating current and the diffusion coefficient of a given component. In this method convection in the bulk of the melt has no influence on D, since the diffusional resistance is concentrated in a thin layer near the electrode and the concentration of the potential-determining ions should be small.

When *direct-current electrode polarization* is employed, the diffusion coefficient of the ion discharged on the electrode is evaluated from the value of the limiting current, since polarization has a concentration character in slags. If the thickness of the diffusion layer is unknown, this method can be used to determine only the relative diffusion coefficient [190]. However, a combination of this method with the rotating disk method with controlled convection provides a way to find absolute values of D [191].

In the *galvanostatic method*, which can be employed when the process is limited by the diffusion of ions in the molten slag, the variation of the electrode potential with time is recorded when a current of constant density passes through the electrode/oxide-melt interface. If the current density exceeds the limiting value, we have a special case of the galvanostatic method, namely, chronopotentiometry. When the constant current density is above the limiting value, the concentration of the diffusant on the interface decreases rapidly to 0, causing a jump in the electrode potential. The value of D is determined from the time when a zero concentration is achieved (the transition time), the initial concentration of the analyte, and the current density.

The *rotating disk method* with controlled convection is used to investigate the kinetics of the dissolution of solid pure oxides in molten slags. The diffusion coefficient is determined from the decrease in the mass of the disk, its dissolution time, and its angular rotation rate. However, the accuracy of this method is low. Experimental errors arise because of the inaccurate determination of the mass loss, the roughness of the disk surface, its deformation at high temperatures, the existence of a temperature gradient between the disk surface and the melt due to the presence of a thermal effect upon dissolution of the disk material, etc.

The rotating disk method combined with the use of labeled atoms was employed to study the self-diffusion of iron ions in a molten slag in [192]. An iron disk with a flat face was coated with the radioactive isotope ^{59}Fe, whose concentration in the slag was determined radiometrically.

The diffusion coefficients of hydrogen and nitrogen in liquid slags were studied in [193, 194] by the *method of gas removal from a thin film* of liquid (slag) with one impermeable boundary. Here it was assumed that desorption of the gas is fairly fast and that the rate-limiting step of the process is diffusion of the substance in the liquid.

The following codes are used for the methods in the tables regarding diffusion in liquid slags:

Investigative method	Code
Capillary method followed by chemical analysis	1-1
Capillary method using isotopes	1-2
Chronopotentiometry	3
Rotating disk method with an equally accessible surface	4
Isotope method	5
Alternating-current electrode polarization	6
Galvanostatic method	7
Membrane method with one impermeable boundary (gas removal from a thin film of a slag melt)	8
Porous filter method	9

3.3.9.3 Calculation Methods for Determining Diffusion Coefficients of Ions in Molten Slags

Approximate estimates of diffusion coefficients in liquid slags, as in metals, are obtained using the Stokes–Einstein equation

$$D = kT/\bar{n}\pi r\eta, \tag{3.101}$$

in which η is the dynamic viscosity of the melt, r is the radius of the diffusing particle, $\bar{n}$ is a numerical coefficient, k is Boltzmann's constant, and T is the absolute temperature (K).

However, choosing the radius of the ion is difficult in many cases, because slags are associated liquids and a definite fraction of the ions exist in the form of polyatomic complexes (M_xO_y). The formation of complexes in slags and the percentage of covalent character of the M–O bonds have a significant effect on the activation energy for diffusion E_D, and thus the following relation between D and η is more correct here:

$$D^n \cdot \eta = \text{const.}$$

Here $n = E_\eta/E_D$, where E_η is the activation energy for viscous flow.

If a slag melt has unipolar conduction and the only carriers of electricity are ions of one kind, the value of D_i can be found from the Nernst–Einstein equation

$$\kappa = D_i c_i z_i^2 \cdot F^2/RT, \tag{3.102}$$

where κ is the specific conductance, F is Faraday's number, c_i is the concentration, and z_i is the valence.

The current method for calculating diffusion coefficients, which is based on the rigid-sphere model, has the advantage that the following parameters, which are related to the structure of the melt [195], are used as starting data:

$$D = 0.28 r_m \left(\frac{\pi kT}{M}\right)^{1/2} \cdot \frac{(\bar{\eta}_m/\bar{\eta})^{1/2}}{[(10T_m\rho)/T\rho_m - 1]};$$

$$r_m = [(3/4\pi)V_m]^{1/3};$$

$$\bar{\eta}_m = l^3/r_m^3,$$

where ρ_m, V_m, and $\bar{\eta}_m$ are the atomic density, the atomic volume, and the packing density at the melting point T_m, respectively, and l is the radius of a rigid sphere.

The self-diffusion coefficients calculated from the equations presented above for various ions in oxide melts [196] and fluoride–oxide melts [197] were of the same order of magnitude as the experimental values. The compositions of the melts in this section are given everywhere in weight percentages (wt.%).

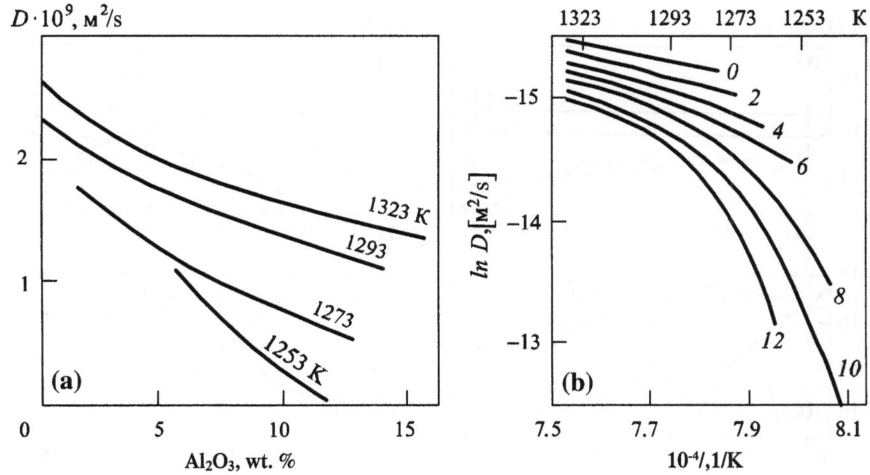

Fig. 3.42 Dependence of the diffusion coefficient of aluminum on the Al_2O_3 content at different temperatures (**a**) and on the temperature (**b**) for different Al_2O_3 contents (numbers on the curves), wt.%. Method 4 [200]

Fig. 3.43 Dependence of E_D of aluminum on the Al_2O_3 content in different temperature ranges (K): *1*—1,253–1,273; *2*—1,293–1,323. Method 4 [200]

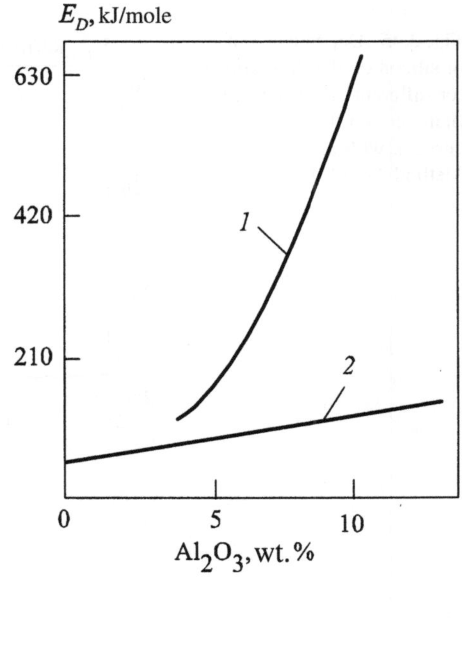

Data on the diffusion in oxide melts are presented in Figs. 3.42, 3.43, 3.44, 3.45, 3.46, 3.47, 3.48, 3.49, 3.50, 3.51, 3.52, 3.53, 3.54, 3.55 and in Tables 3.7, 3.8, 3.9, 3.10, 3.11, 3.12, 3.13, 3.14, 3.15, 3.16, 3.17.

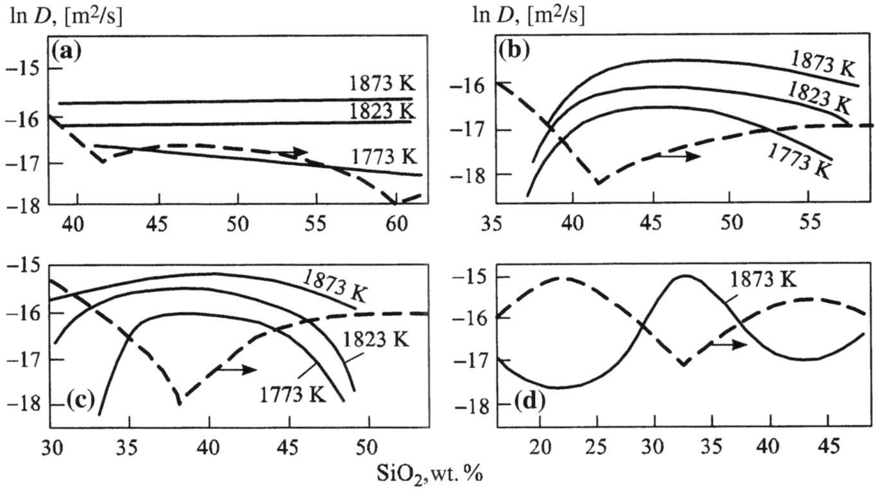

Fig. 3.44 Dependence of the diffusion coefficient of silicon on the SiO_2 content in the presence of 16.5 (**a**), 21 (**b**), 27 (**c**), and 38 wt.% Al_2O_3 (**d**) at different temperatures. *Dashed curve*—liquidus temperature. Method 1-1 [201]

Fig. 3.45 Dependence of E_D of silicon on the SiO_2 content for different Al_2O_3 contents (numbers on the curves), wt.%. Method 1-1 [201]

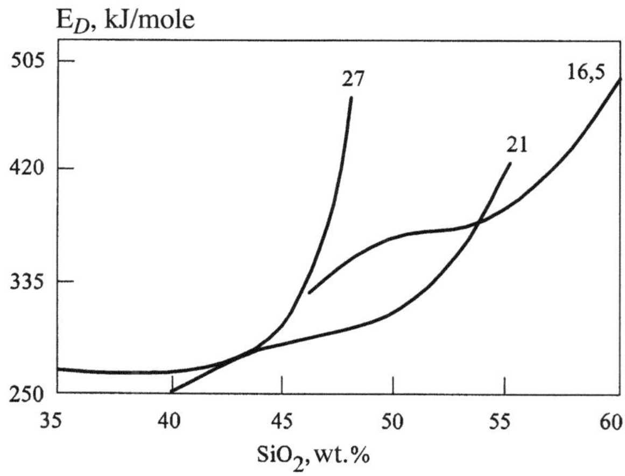

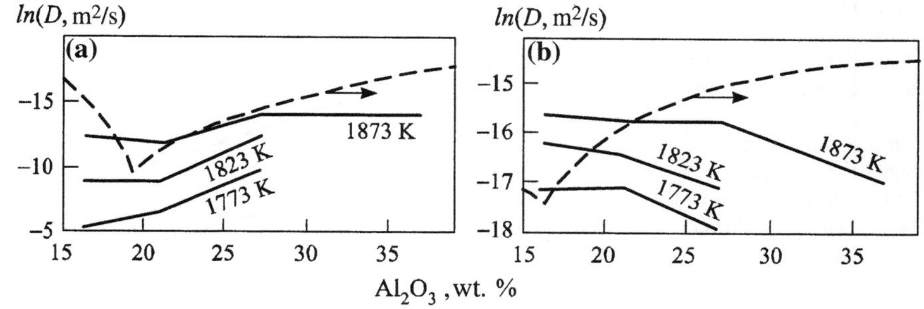

Fig. 3.46 Dependence of the diffusion coefficient of silicon on the Al_2O_3 content for CaO/SiO_2 ratios equal to 0.8 (**a**) and 0.5 (**b**) *Dashed curve*—liquidus temperature. Method 1-1 [201]

Fig. 3.47 Dependence of the diffusion coefficient of ^{59}Fe (1) and ^{45}Ca (2) ions on the temperature of the melt. Composition, wt.%: 20 Al_2O_3, 38 CaO, 42 SiO_2. Method 1-2 [185]

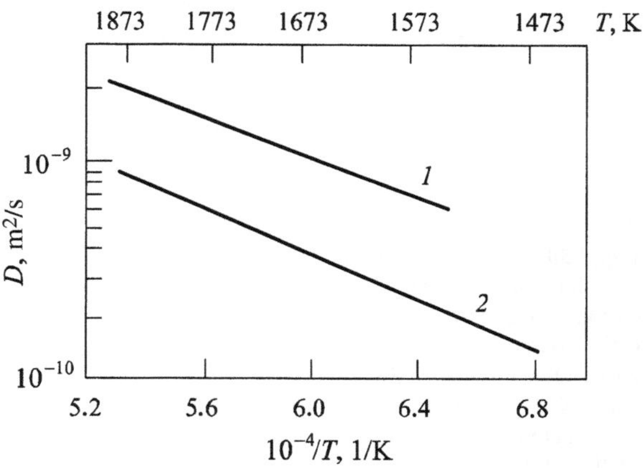

Fig. 3.48 Dependence of the diffusion coefficient of ^{45}Ca ions on the temperature of the melt. Method 1-2 [185]. Composition, wt.%: *1*—50 Al_2O_3, 50 CaO; *2*– 14.5 Al_2O_3, 23.5 CaO, 62 SiO_2

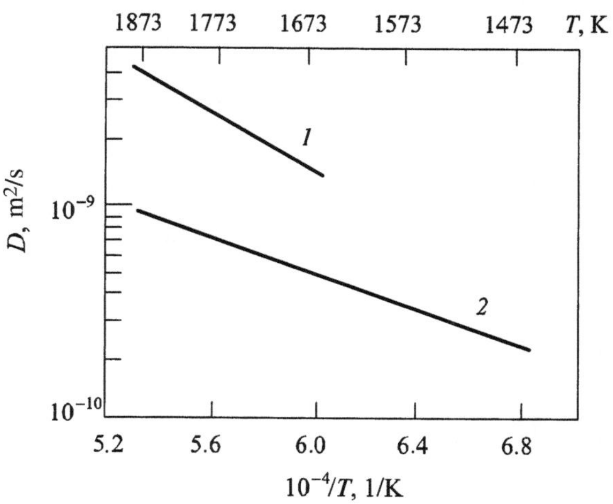

Fig. 3.49 Dependence of the diffusion coefficient of ^{45}Ca (*1*, *2*) and ^{18}F (*3*) ions on the temperature of the melt. Method 1-2 [185]. Composition of the melt, wt.%: 20 Al_2O_3, 38 CaO, 42 SiO_2 without CaF_2 (1) and with an addition of $^{45}Ca^{18}F$ (*2*, *3*)

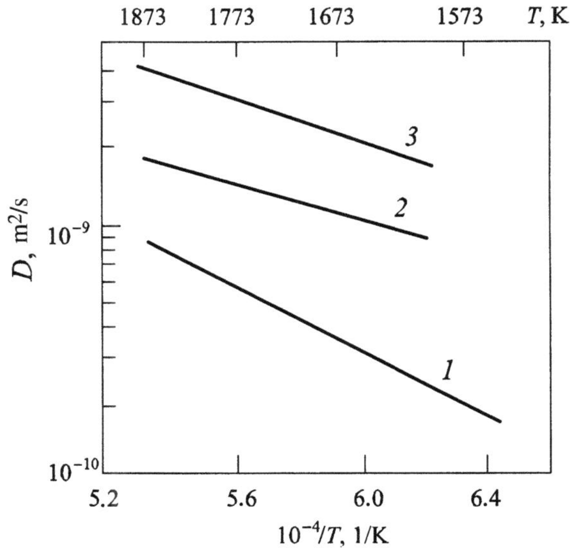

Fig. 3.50 Dependence of the diffusion coefficient of ^{59}Fe on the temperature of the melt. Composition, wt.%: 20 Al_2O_3, 38 CaO, 42 SiO_2 with 1 CaF_2 (*1*), 10 CaF_2 (*2*), and 1 FeO (*3*). Method 1-1 [185]

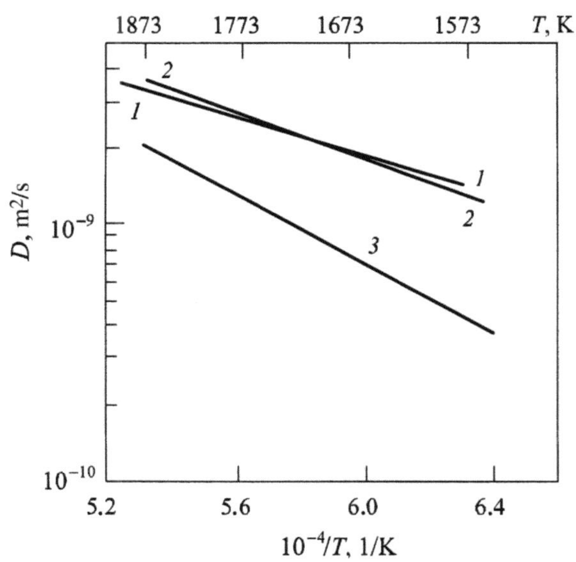

Fig. 3.51 Dependence of the diffusion coefficient of hydrogen on the temperature of the melt. Method 8 [193]. Composition of the melt, wt.%: *1*—16.3 Al_2O_3, 27.3 CaO, 56.4 SiO_2; *2*—41 Al_2O_3, 53 CaO, 6 SiO_2

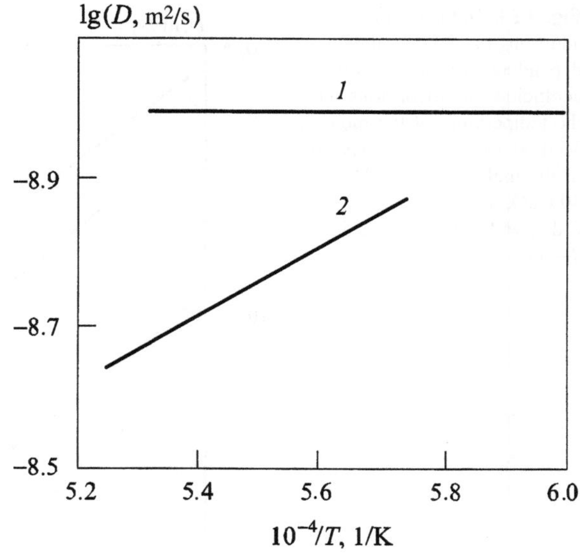

Fig. 3.52 Dependence of the diffusion coefficient of hydrogen on the CaO content at 1,873 K, method 8 [193]: *1*– for a constant SiO_2 content equal to 57 wt.%; *2*—for a constant Al_2O_3 content equal to 28 wt.%

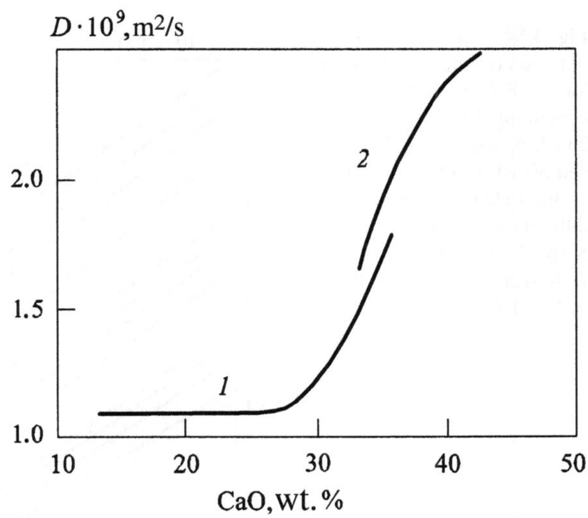

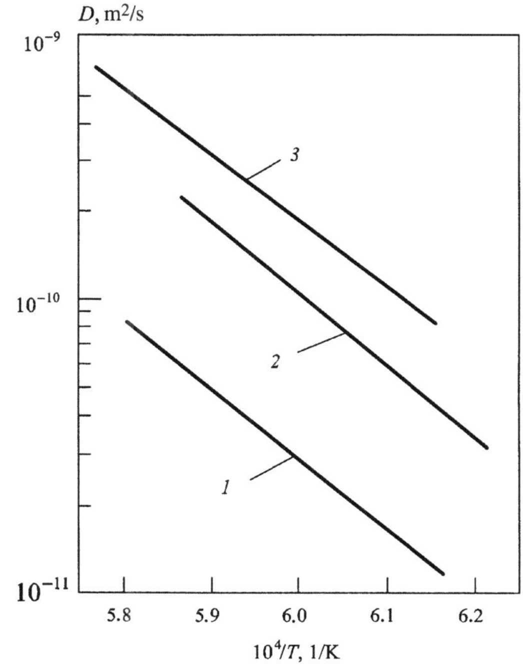

Fig. 3.53 Influence of additions of CaF_2 on the dependence of the diffusion coefficient of silicon ions on the temperature of the melt. Method 7 [202]. Composition of the melt, wt.%: 20 Al_2O_3, 40 CaO, 40 SiO_2. CaF_2, wt.%: *1*—0; *2*—2.5; *3*—4.8

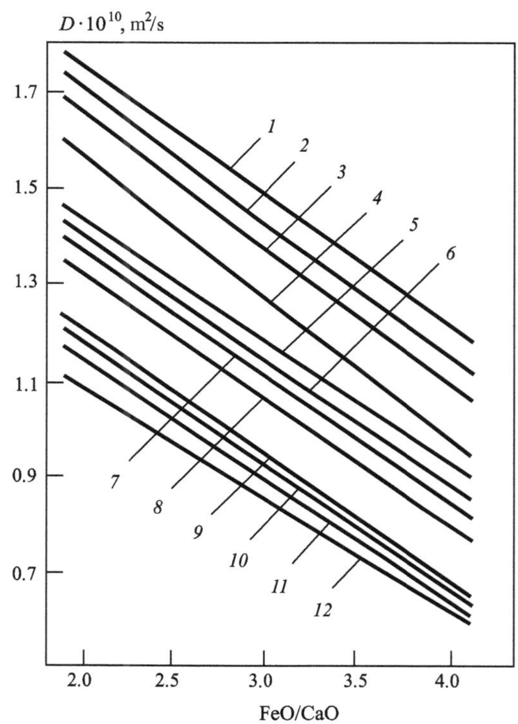

Fig. 3.54 Dependence of the diffusion coefficient of silicon ions in FeO–CaO–SiO_2 melts containing 30 (*1*, *5*, *9*), 25 (*2*, *6*, *10*), 40 (*3*, *7*, *11*), and 50 wt.% SiO_2 (*4*, *8*, *12*) on the FeO/CaO ratio at different temperatures, K (method 4 [203]): *1–4*—1,573; *5–8*—1,523; *9–12*—1,473

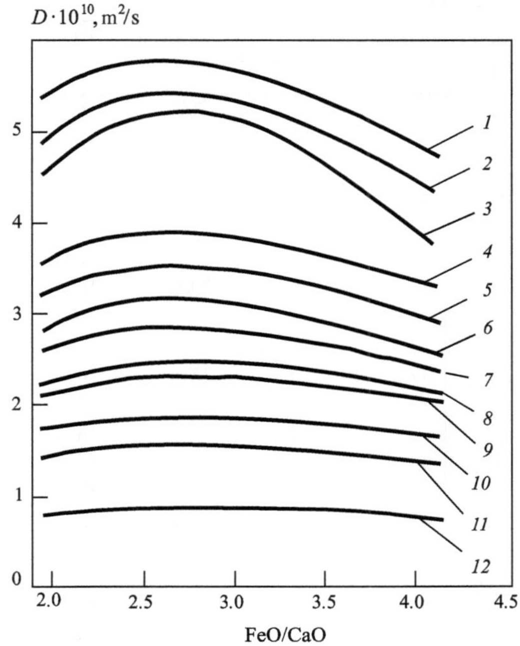

Fig. 3.55 Dependence of the diffusion coefficient of calcium ions in $FeO–CaO–SiO_2$ melts containing 30 (*1, 2, 3*), 35 (*4, 5, 6*), 40 (*7, 9, 10*), and 50 wt.% SiO_2 (*8, 11, 12*) on the FeO/CaO ratio at different temperatures, K (method 4 [203]): *1–4, 7, 8*—1,573; *2, 5, 9, 11*—1,523; *3, 6, 10, 12*—1,473

Table 3.7 Diffusion of ^{26}Al ions in 100 % Al_2O_3. $E = 89$ kJ/mole. Method 5 [204]

T (K)	$D\cdot10^{10}$ (m^2/s)	T (K)	$D\cdot10^{10}$ (m^2/s)
2335	19.5	2475	26.5
2375	21.5	2575	30.5
2425	23.5		

Table 3.8 Diffusion of ^{45}Ca [205] and ^{31}Si [206] ions. Method 1-2

T (K)	$D\cdot10^{10}$ (m^2/s)		T (K)	$D\cdot10^{10}$ (m^2/s)	
	^{45}Ca	^{31}Si		^{45}Ca	^{31}Si
65 SiO_2; $E_D^{Ca} = 146$ kJ/mole			54.7 SiO_2; $E_D^{Ca} = 142$ kJ/mole		
1773	1.7	–	1823	3.4	–
1823	2.1	–	1873	4.7	0.76
1833	2.0	–	1973	6.9	–
1873	2.9	0.32	50.5 SiO_2; $E_D^{Ca} = 142$ kJ/mole		
1884	2.9	–	1823	4.2	–
1923	3.4	–	1833	4.4	–
60.4 SiO_2; $E_D^{Ca} = 125$ kJ/mole			1873	5.4	1.1
1833	2.8	–	1923	6.9	–
1873	3.6	0.51	46.5 SiO_2; $E_D^{Ca} = 121$ kJ/mole		
1923	4.2	–	1773	4.5	–
			1873	6.2	–
			1973	10.2	–

Table 3.9 Diffusion of ions in Al_2O_3–CaO–SiO_2–M_mX_n melts

Melt composition (%)				Diffusing element	T (K)	$D \cdot 10^{10}$ (m^2/s)	E_D (kJ/mole)	Method	Literature source
Al_2O_3	CaO	SiO_2	M_mX_n						
20	39	41	–	^{26}Al	1673	0.26	250 ± 42	1-2	[207]
					1713	0.38			
					1758	0.59			
10	43	47	–	^{26}Al	1713	0.56	354 ± 73	1-2	[207]
					1753	1.1			
					1793	2.2			
20	40	40	–	^{45}Ca	1623	0.33	292 ± 84	1-2	[208]
					1673	0.68			
					1723	1.3			
21	39	40	–	^{45}Ca	1623	0.35	292	1-2	[209]
					1773	2.1			
					1813	3.4			
21	39	40	–	^{31}Si	1638	0.047	292	1-2	[209]
					1703	0.105			
21	42	37	–	^{45}Ca	1573	0.4	167	1-2	[210]
					1673	1.2			
					1723	2.4			
					1773	3.2			
21	42	37	–	^{59}Fe	1573	0.5	167	1-2	[210]
					1673	1.5			
					1773	3.5			
21	42	37	–	^{32}P	1573	1.0	167	1-2	[210]
					1673	1.9			
					1723	4.5			
					1773	4.4			
21	42	37	–	^{35}S	1573	0.3	167	1-2	[210]
					1773	2.5			
20	40	40	–	^{17}O	1645	3.3	355	1-2	[211]
					1686	4.6			
					1748	12			
					1808	34			
20	40	40	–	^{18}O	1645	2.0	355	1-2	[211]
					1686	3.7			
					1748	10.2			
					1808	29			
20	38	42	–	^{45}Ca	1723	4.1	134	1-2	[185]
20	38	42	–	Ca	1723	4.5	104	1-1	[185]
			FeO						
19.8	37.6	41.6	1.0	^{59}Fe	1723	9.6	123	1-2	[185]
19.8	37.6	41.6	1.0	Fe	1723	11.2	88	1-1	[185]
15	30	54	0.8	Fe	1773	2.7	–	6	[212]
20	38	42	–	^{45}Ca	1723	4.1	134	1-2	[213]
					1773	5.5			
20	38	42	–	Ca	1723	4.5	105	1-1	[213]
					1773	6.5			

(continued)

Table 3.9 (continued)

Melt composition (%)				Diffusing element	T (K)	$D{\cdot}10^{10}$ (m^2/s)	E_D (kJ/mole)	Method	Literature source
Al_2O_3	CaO	SiO_2	M_mX_n						
14.5	23.5	62	–	Ca	1723	5.7	75	1-1	[213]
					1773	6.5			
50	50	–	–	Ca	1723	20	151	1-1	[213]
			CaF_2						
19.8	37.6	41.6	1.0	Ca	1723	12	75	1-1	[213]
					1773	14			
19.8	37.6	41.6	1.0	F	1723	25	84	1-1	[213]
					1773	30			
14.5	23.5	62	–	Si	1723	2.2	84	1-1	[213]
					1773	2.7			
20	38	42	–	Si	1723	2.4	250	1-1	[213]
					1773	3.9			
–	38	62	–	Si	1723	1.3	410	1-1	[213]
					1773	1.7			

Table 3.10 Diffusion of nitrogen in Al_2O_3–CaO–SiO_2 melts at 1773 K [194] and in Al_2O_3–CaO–SiO_2–MgO–FeO–M_xO_y melts at 1873 K [214]. Method 8

Slag composition (%)						$D{\cdot}10^{10}$ (m^2/s)	E_D (kJ/mole)
Al_2O_3	CaO	SiO_2	MgO	FeO	M_xO_y		
12.3	47.5	40.2	–	–	–	5.2	135
22.9	48.1	29.0	–	–	–	3.9	117
19.6	30.3	50.1	–	–	–	0.5	–
18.6	44.6	36.5	–	–	–	5.0	–
21.1	53.2	25.7	–	–	–	1.7	130
20.1	32.7	47.2	–	–	–	1.0	–
10	57	24	8	0.2	–	8.8	–
9	56	23	12	0.1	–	7.3	95
8	55	23	14	0.2	–	6.1	112
7	54	22	17	0.1	–	5.2	130
5	50	20	25	0.1	–	4.0	–
					Cr_2O_3		
7	58	23	9	0.1	3	7.6	107
7	56	22	9	0.1	6	5.2	–
7	52	20	9	0.2	12	2.2	140
7	50	19	9	0.2	15	1.1	–
					MnO		
15	20	55	2	3	5	2.0	135
13	17	60	2	3	5	1.2	142
13	15	62	2	3	5	0.8	164
13	12	65	2	3	5	0.5	186
					TiO_2		
50	50	–	–	–	–	5.3	–
45	50	–	–	–	5	6.0	–
40	50	–	–	–	10	7.1	–

Table 3.11 Diffusion (Method 3 in combination with method 4 was used in [191], and method 3 was used in [215].) of nitrogen in Al_2O_3–CaO–SiO_2 and Al_2O_3–CaO–MgO melts containing the oxides M_xO_y of the diffusing elements

M_xO_y content (%)	Diffusant	T (K)	$D \cdot 10^{10}$ (m^2/s)		E_D (kJ/mole)	
			[191]	[215]	[191]	[215]
20 Al_2O_3; 40 CaO; 40 SiO_2 (base melt)						
3.3 CoO	CaO	1623	1.1	–	180	–
		1673	1.8	–		
		1723	2.5	–		
3.17 NiO	Ni	1573	0.64	1.2	230	160
		1623	1.1	2.4		
		1673	1.9	2.8		
7.7 MoO_3	Mo	1623	0.11	–	400	–
		1683	0.29	–		
		1723	0.54	–		
4.15 WO_3	W	1623	0.13	–	400	–
		1673	0.29	–		
		1723	0.74	–		
1.49 FeO	Fe	1623	0.87	2.1	–	–
2.88 Fe_2O_3	Fe	1673	1.9	2.7	250	160
		1723	2.7	4.0		
47 Al_2O_3; 47 CaO; 6 MgO (base melt)						
4.32 SiO_2	Si	1653	0.20	0.14	490	200
		1703	0.57	0.25		
		1743	1.3	0.35		
3.74 Cr_2O_3	Cr	1653	0.78	–	440	–
		1703	2.2	–		
		1743	4.7	–		
2.71 V_2O_5	V	1653	0.39	0.11	420	200
		1703	0.75	0.20		
		1743	2.1	0.35		

Table 3.12 Diffusion of Si ions at 1,823 K (Method 7 [216])

Composition (%)				Ca/SiO_2	$D \cdot 10^{10}$ (m^2/s)
CaO	SiO_2	Al_2O_3	CaF_2		
33.3	47.5	15.0	4.0	0.7	0.068
36.0	45.0	15.0	4.0	0.8	0.093
38.3	42.5	15.0	4.0	0.9	0.15
40.4	40.4	15.0	4.0	1.0	0.22
44.1	36.8	15.0	4.0	1.2	0.45
31.7	45.3	15.0	8.0	0.7	0.21
34.2	42.7	15.0	8.0	0.8	0.37
36.4	40.5	15.0	8.0	0.9	0.52
38.5	38.5	15.0	8.0	1.0	0.66
42.1	35.0	15.0	8.0	1.2	0.91

(continued)

Table 3.12 (continued)

Composition (%)				Ca/SiO_2	$D \cdot 10^{10}$ (m^2/s)
CaO	SiO_2	Al_2O_3	CaF_2		
28.8	42.1	15.0	15.0	0.7	0.29
31.1	38.8	15.0	15.0	0.8	0.62
33.0	36.9	15.0	15.0	0.9	0.90
35.1	35.0	15.0	15.0	1.0	1.2
36.6	33.4	15.0	15.0	1.1	1.4
38.1	31.8	15.0	15.0	1.2	1.6

Table 3.13 Diffusion of ions of alkali and alkaline earth metals (Method 1–2)

Diffusing element and melt composition (wt.%)	$D \cdot 10^{10}$ at 1473 (K m^2/s)	$D_0 \cdot 10^5$ (m^2/s)	E_D (kJ/mole)	Temperature range	Literature source
^{24}Na					
$Na_2O \cdot 2SiO_2$	32	0.018	50	1123–1483	[217]
$Na_2O \cdot Li_2O \cdot 4SiO_2$	29	0.028	56	1123–1483	[217]
$Na_2O \cdot K_2O \cdot 4SiO_2$	26	0.063	67	1123–1483	[217]
$Na_2O \cdot BaO \cdot 4SiO_2$	16	23	145	1373–1573	[217]
$Na_2O \cdot SrO \cdot 4SiO_2$	15	1.1	110	1373–1573	[217]
$Na_2O \cdot CaO \cdot 4SiO_2$	13	0.12	83	1373–1573	[217]
$Na_2O \cdot MgO \cdot 4SiO_2$	14	0.16	86	1373–1573	[217]
^{42}K					
$K_2O \cdot 2SiO_2$	27	0.0087	43	1123–1483	[217]
$K_2O \cdot Li_2O \cdot 4SiO_2$	17	0.087	76	1123–1483	[217]
$K_2O \cdot Na_2O \cdot 4SiO_2$	23	0.05	66	1123–1483	[217]
$K_2O \cdot BaO \cdot 4SiO_2$	8.9	560	192	1373–1573	[217]
$K_2O \cdot SrO \cdot 4SiO_2$	6.8	4.8	136	1373–1573	[217]
$K_2O \cdot CaO \cdot 4SiO_2$	6.6	1.7	124	1373–1573	[217]
$K_2O \cdot MgO \cdot 4SiO_2$	5.2	0.093	92	1373–1573	[217]
^{45}Ca					
$Na_2O \cdot CaO \cdot 4SiO_2$	0.33	132	240	–	[189]
^{24}Na					
$Na_2O \cdot 2SiO_2$	35	0.0023	52	1173–1573	[196]
$Na_2O \cdot 3SiO_2$	20	0.0046	67	1173–1573	[196]
$2Na_2O \cdot 3GeO_2$	39	0.0013	44	1273–1473	[196]
$3Na_2O \cdot 7GeO_2$	23	0.0029	59	1273–1473	[196]

Table 3.14 Diffusion of ions in $CaO–SiO_2–Fe_2O_3$ melts [189]

Melt composition (%)			Diffusing element	$D·10^9$ at 1873 K (m^2/s)	$D_0·10^4$ (m^2/s)	E_D (kJ/mole)	Temperature range	Measurement error	Method
Fe_2O_3	CaO	SiO_2							
40	33	27	^{45}Ca	5	0.1	118	1633 – 1733	±26	1-2
40	33	27	^{59}Fe	18	0.2	109	1633–1733	±26	1-2
38.3	29.8	31.9	^{59}Fe	18	0.15 ± 0.044	105 ± 12	1523–1723	–	1-2
70.5	–	29.5	^{59}Fe	60	0.065 ± 0.019	73 ± 10	1523–1723	–	1-2
25	30	45	P	0.57	29.3	244	1623–1723	±10	1-1
25	30	45	Ti	0.83	16000	336	1623–1723	±10	1-1
25	30	45	Mn	2.5	0.24	143	1623–1723	±10	1-1
25	30	45	Mg	4.9	0.021	97	1623–1723	±10	1-1
25	30	45	Ni	4.1	28	214	1623–1723	±10	1-1

Table 3.15 Diffusion of ^{45}Ca and ^{59}Fe ions in $CaO–SiO_2–FeO$ melts at 1873 K. Method 9 [218]

Melt composition (%)			$D·10^{10}$ (m^2/s)	
CaO	SiO_2	FeO	^{45}Ca	^{59}Fe
32.3	67.7	–	1.78 (+1.13–0.73)	–
28.3	66.2	5.5	1.09 (+0.10–0.16)	13.2 (+9.2–3.7)
27.0	63.5	9.5	1.61 (+0.38–0.29)	8.1 (+3.6–2.6)
21.2	61.1	17.7	–	9.8 (+1.7–2.6)
21.1	60.1	18.8	1.27 (+0.28–0.21)	12.4 (+2.1–2.9)
29.8	32.0	38.2	–	14.5
–	10	90	–	58 ± 13[a]

[a] According to measurements of the mobility of the ^{59} Fe isotope at 1,523 K [219]

Table 3.16 Diffusion of ^{59}Fe, ^{63}Ni, ^{45}Ca, and ^{35}S ions in a melt containing 9.8 Al_2O_3, 21 CaO, 39.9 SiO_2, 10.2 MgO, and 19.1 FeO. Method 1–2 [220]

Temperature (K)	$D·10^{10}$, m^2/s			
	^{59}Fe	^{63}Ni	^{45}Ca	^{35}S
1523	2.4	2.8	2.0	1.6
1573	3.7	4.0	3.2	2.7
1593	–	–	4.2	–
1623	5.2	5.6	4.5	4.1
1643	–	–	–	4.6
1673	7.3	7.8	6.5	6.0
E_D (kJ/mole)	153	146	169	189

Table 3.17 Diffusion of ions of ^{59}Fe ions in a melt containing 16.0 Al_2O_3, 25.8 CaO, 24.0 SiO_2, 25.4 CaF_2, 7.2 FeO, and 1.4 Fe_2O_3. Method 4 in combination with method 5 [192]

T (K)	$D \cdot 10^{10}$ (m^2/s)	$D_0 \cdot 10^5$ (m^2/s)	E_D (kJ/mole)
1593	1.7	0.017	121
1643	2.2	0.017	121
1693	2.9	0.017	121

3.3.9.4 Diffusion Characteristics of Slag Melts

Al_2O_3–CaF_2 melts
Diffusion of Al ions

Al_2O_3 content (wt.%)	T (K)	$D \cdot 10^{10}$ (m^2/s)	Measure-ment error (%)	Method	Literature source
30	1782	85	±20	4	[198]
30	1791	81	±20	4	[198]
20	1773	40	±20	4	[198]
15	1353	7.2	±15	3	[199]
0.5–4	1293	15	±15	3	[199]

CaO–SiO_2 melts
Melts based on the Al_2O_3–CaO–SiO_2–MgO system
CaO–SiO_2–Al_2O_3–CaF_2 melts
Melts of silicates and germinates containing oxides of alkali and alkaline earth metals
Melts containing iron oxides

References

1. Pastukhov, É.A., et al.: Diffraction Studies of the Constitution of High-Temperature Melts (in Russian), 353 pp. Ural Branch of RAS, Ekaterinburg (2003)
2. Bogolyubov, N.N.: Dynamic Theory Problems in Statistical Physics (in Russian), 120 pp. Gostechizdat, Moscow (1946)
3. Fisher, I.Z.: Statistical Theory of Liquids (in Russian), 280 pp. Nauka, Moscow (1961)
4. Croxton, C.A.: Liquid State Physics—A Statistical Mechanical Introduction. Cambridge University Press, London (1974)
5. Vainshtein, B.K.: Diffraction of X-Rays by Chain Molecules. Elsevier, London (1966)
6. Gel'chinskii, B.R., Beskachko, V.P., Ukhov, V.F.: Correlation function method in the theory of liquid metals. In: Reviews of Thermophysical Properties of Substances (in Russian), vol. 2, pp. 1–103. Moscow (1978)
7. Waseda, Y., Suzuki, K.: Characteristics of soft core in pair potential, static structure in liquid metals. Sci. Rep. Res. Inst. Tohoku. Univ. **24**(4), 139–184 (1973)
8. Popel', S.I., Spiridonov, M.A., Zhukova, L.A.: Atomic Ordering in Molten and Amorphous Metals (in Russian), 384 pp. Ural Branch of RAS, Ekaterinburg (1997)

9. Popel', S.I., Manov, V.P., Spiridonov, M.A.: Constitution of liquid iron revealed by diffraction methods. Zh. Fiz. Khim. **51**(6), 1384–1387 (1977)
10. Lashko, A.S.: X-ray diffraction investigation of the structure of several liquid metal systems. Zh. Fiz. Khim. **33**(8), 1730–1738 (1959)
11. Wagner, C.N.J., Holder, N.C.: Atomic distribution and electronic transport properties in liquid alloys studied by X-ray diffraction. Adv. Phys. **16**(2), 241–261 (1967)
12. Belashchenko, D.K., Spektor, E.Z.: Investigation of the structure of liquid metals. Itogi. Nauki. Tekh. Teor. Metallurg. Prots. **3**, 5–55 (1973)
13. Enderby, J.E., North, D.M., Egelstaff, P.A.: The partial structure factors of liquid Cu–Sn. Philos. Mag. **14**(131), 961–970 (1966)
14. Belashchenko, D.K.: Structure of Liquid and Amorphous Metals (in Russian), 192 pp. Gostechizdat, Moscow (1985)
15. Barnes, A.C., Enderby, J.E.: The structural, electrical properties of liquid copper selenide. Philos. Mag. B **58**(5), 497–512 (1988)
16. Derrien, J.Y., Dupuy, J.: Structural analysis of the ionic liquids potassium chloride and cesium chloride by neutron diffraction. J. Phy. **36**(2), 191–198 (1975)
17. Edwards, F.G., Enderby, J.E., Howe, R.A., Page, D.I.: The structure of molten sodium chloride. J. Phys. C Solid. State. Phys. **8**, 3483–3490 (1975)
18. Mitchell, E.W.J., Poncet, P.F.J., Stewart, R.J.: The ion pair distribution functions in molten rubidium chloride. Philos. Mag. **34**(5), 721–732 (1976)
19. Derrien, J.Y., Dupuy, J.: Structure of molten silver chloride. Phys. Chem. Liq. **5**(1), 71–91 (1976)
20. Bratia, A.B., Ratti, V.K.: Partial structure factors and neutron scattering in compound-forming molten alloys. J. Phys. F. Met. Phys. **6**, 927–941 (1976)
21. Waseda, Y., Tamaki, S.: The partial structure factors of liquid Ni–Si alloys. Philos. Mag. **32**(5), 951–960 (1975)
22. Wagner, C.N.J., Halder, N.C., North, D.M.: Partial interference functions of liquid gold–tin alloys. Phys. Lett. A **25**(9), 663–664 (1967)
23. Bühner, H.F., Steeb, S.: Partielle Interferenz und partielle Atomverteilungsfunctionen sowie electrischer Widerstand von Geschmolzenen Magnesium-Zinn-Legirungen. Z. Naturforsch. **25a**(12), 1862–1867 (1970)
24. Bühner, H.F., Steeb, S.: Röntgenographische Structuruntersuchung und Bestimmung des electrischen Widerstandes geschmozener Aluminium-Gold-Legirungen. Z. Metalk. **62**(1), 27–33 (1971)
25. North, D.M., Wagner, C.N.: Atomic distributions in liquid copper–tin alloys. J. Phys. Chem. Liq. **2**(2), 87–113 (1970)
26. Waseda, Y., Ohtani, M., Suzuki, K.: Partial structure factors and atomic distributions in liquid Na–K and Al–Mg alloys. Z. Naturforsch. **28a**(6), 1002–1008 (1973)
27. Waseda, Y., et al.: Partial structure factors and atomic distributions in liquid Hg–In and Hg–Tl alloys. Phys. Status. Solidi. B **52**, 571–579 (1972)
28. Waseda, Y., Tamaki, S.: X-ray diffraction study of molten Te and Tl–Te alloys. Z. Naturforsch. **30a**(12), 1655–1660 (1975)
29. Frenkel, J.: Kinetic Theory of Liquids. Clarendon Press, Oxford (1946)
30. Khrushchev, B.N.: Structure of Liquid Metals (in Russian), 112 pp. FAN, Tashkent (1970)
31. Wilson, J.R.: Structure of liquid metals and alloys. Metall. Rev. **10**, 381–590 (1965)
32. Bloom, H., Bockris, J.O'.M.: Structural aspects of ionic liquids. In: Sundheim, B.R. (ed.) Fused Salts. McGraw-Hill, New York (1964)
33. Dutchak Ya, I: X-ray Diffraction Analysis of Liquid Metals (in Russian), 163 pp. Vyshcha Shkola, L'vov (1977)
34. Gubanov, A.I.: Quantum Electronic Theory of Amorphous Conductors (in Russian), Moscow (1963)
35. Bagchi, S.N.: Kinematic theory of diffraction by matter of any kind and the theory of liquids. Adv. Phys. **19**(77), 119–173 (1970)

36. Bagchi, S.N.: The structure of liquids as revealed by the analysis of their radial distribution functions. Acta. Crystallogr. A **28**, 560–557 (1972)
37. Breitling, G., Handtmann, D., Richter, H.: Verschiedene Verfahren zur Untesuchung der Structur geschmolzener Metalle. Z. Phys. **178**(3), 294–318 (1964)
38. Spektor, E.Z., Khokhlov, S.F.: Interpretation of intensity curves of radiation scattered by a substance in the liquid state. Fiz. Met. Metaloved. **18**(3), 451–454 (1964)
39. Grinberg, B.M., Spektor, E.Z.: X-ray diffraction analysis of the structure of melts. Izv. Akad. Nauk. SSSR Met. (4), 131–135 (1971)
40. Romanova, A.V.: Some questions from the investigation of the structure of liquid metals and alloys. Metallofizika. Kiev. **1**(36), 3–14 (1971)
41. Vatolin, N.A., Pastuhov, E.A.: X-ray structural analysis of liquid alloys of iron and nickel with silicon and carbon. In: The 4th Japan–USSR Joint Symposium on Physical Chemistry of Metallurgical Processes, pp. 1–11. Tokyo (1973)
42. Romanova, A.V., Il'inskii, A.G.: Method for interpreting experimental data from X-ray diffraction investigations of two-component melts. Ukr. Fiz. Zh. **19**(9), 1565–1568 (1974)
43. Pastukhov, É.A., Vatolin, N.A.: Constitution and electrical properties of melts containing transition metal oxides. In: Constitution of Ionic Melts and Solid Electrolytes (in Russian), pp. 36–48. Kiev (1977)
44. Spektor, E.Z., et al.: Constitution of liquid alloys of iron with silicon. Izv. Vyssh. Uchebn. Zaved. Fizika. (10), 68–72 (1971)
45. Ruppersberg, H., Seemann, H.J.: X-ray investigation of liquid Fe. Z. Naturforsch. **21**(6), 820–826 (1966)
46. Evseev, A.M., Voronin G.F.: Thermodynamics and Structure of Liquid Metal Alloys (in Russian), 131 pp. Moscow (1966)
47. Fessler, R.R., Kaplow, R., Averbach, B.L.: Pair correlations in liquid and solid aluminum. Phys. Rev. **150**(1), 34–43 (1966)
48. Lashko, A.C.: Structure of potassium and gold in the liquid state. Ukr. Fiz. Zh. **13**, 728–732 (1968)
49. Lashko, A.C.: Structure of liquid germanium. In: Structure of Metal Alloys (in Russian), pp. 144–148. Kiev (1966)
50. Finkel', V.A.: High-Temperature X-Ray Diffraction Analysis of Metals (in Russian), 204 pp. Moscow (1968)
51. Prins, J.A., Petersen, H.: Theoretical diffraction patterns corresponding to some simple types of molecular arrangement in liquids. Physica **3**, 147–153 (1936)
52. Morozumi, C., Ritter, H.L.: Calculated powder patterns from very small crystals: body-centered cubic cubes. Acta. Crystallogr. **6**(7), 588–590 (1953)
53. Grigson, C.W.B., Barton, E.: The development of the face-centered cubic interference functions as crystals grow. Br. J. Appl. Phys. **18**(2), 175–183 (1967)
54. Sadoc, J.F., Dixmier, J., Guinier, A.: Theoretical calculation of dense random packings of equal and non-equal sized hard spheres applications to amorphous metallic alloys. J. Non-Cryst. Solids **12**(1), 46–60 (1973)
55. Okazaki, H., Iida, K., Tamaki, S.: Pair correlation function of liquid aluminium. J. Phys. Soc. Jpn. **29**(5), 1396 (1970)
56. Vatolin, N.A., Pastukhov, É.A., Sermyagin, V.N.: Influence of the temperature on the structure of liquid aluminum. Dokl. Akad. Nauk. SSSR **222**(3), 641–643 (1975)
57. Levin, E.S., Gel'd, P.V., Yakubchik, V.P.: Investigation of the structure of aluminum near the melting point. Izv. Akad. Nauk. SSSR Met. **2**(5), 80–88 (1975)
58. Vatolin, N.A., Pastukhov, É.A., Kern, É.M.: Influence of the temperature on the structure of molten iron, nickel, palladium, and silicon. Dokl. Akad. Nauk. SSSR **217**(1), 127–130 (1974)
59. Vatolin, N.A., Kern, É.M., Sergin, B.I.: Determination of the coordination numbers and structure of liquid metals from intensity curves. Dokl. Akad. Nauk. SSSR **232**(2), 359–362 (1977)

60. Richter, H., Breitling, G.: Structurkomponenten von geschmolzenem Gold, Silber und Blei. Z. Metallk. **61**(9), 628–636 (1970)
61. Popel', S.I., Ya, Maslennikov.Yu., Spiridonov, M.A.: Investigation of the constitution of Bi–In melts near the liquidus. Izv. Akad. Nauk. SSSR Met. (1), 21–23 (1976)
62. Popel', S.I., Spiridonov, M.A.: Structure of molten In_2Bi. Izv. Akad. Nauk. SSSR Met. (3), 47–50 (1976)
63. Manov, V.P., Popel', S.I.: Diffraction studies of the structural features of liquid metals. Izv. Akad. Nauk. SSSR Met. (2), 68–72 (1978)
64. Lysov, V., Khar'kov, V.I.: Interpretation of X-ray diffraction data regarding the structure of liquid metals within the rigid-sphere model. Ukr. Fiz. Zh. **20**(3), 468–475 (1975)
65. Gel'chinskii, B.R., et al.: Some Experimental and Numerical Methods for Investigating the Structure of the Short-Range Order (in Russian), 222 pp. Frunze, (1987)
66. Vatolin, N.A., Sryvalin, I.T., Evseev, A.M.: Molecular dynamics modeling of liquid metals and alloys. Dokl. Akad. Nauk. SSSR **219**(6), 1394–1396 (1974)
67. Kazimirov, V.P., Smyk, S.Yu.: The reverse Monte Carlo method—a new approach to modeling the structure of unordered systems. 1. Liquid metals. In: Reports to the 9th All-Russia Confererence (in Russian), vol. 1, pp. 25–26. Chelyabinsk (1998)
68. Kazimirov, V.P., Smyk, S.Yu.: The reverse Monte Carlo method—a new approach to modeling the structure of unordered systems. 2. Binary metallic melts. In: Reports to the 9th All-Russia Conference (in Russian), vol. 1, pp. 73–75. Chelyabinsk (1998)
69. Vatolin, N.A., Ukhov, V.F., Esin, O.A.: Application of an oscillatory 3–12 potential to the calculation of the properties of metallic melts. Dokl. Akad. Nauk. SSSR **216**(2), 339–341 (1974)
70. Vatolin, N.A., et al.: Evaluation of the excess volume of a solution using a model of a spherical field with an oscillatory potential. Dokl. Akad. Nauk. SSSR **218**(2), 332–335 (1974)
71. Rushbrooke, G.S.: Equilibrium theories of liquid state. In: Temperley, H.N.V., Rowlinson, J.S., Rushbrooke, G.S. (eds.) Physics of Simple Liquids. North–Holland, Amsterdam (1968)
72. March, N.H.: Liquid Metals. Pergamon, Oxford (1968)
73. Johnson, M.D., Hutchinson, P., March, N.H.: Ion–ion oscillatory potentials in liquid metals. Proc. R. Soc. London Ser. A Math. Phys. Sci. **282**(1389), 283–302 (1964)
74. Waseda, Y., Ohtani, M.: Effective interionic potentials and properties of molten noble and transition metals. Z. Naturforsch. **30a**(4), 485–491 (1975)
75. Waseda, Y., Suzuki, K.: Interionic potentials in liquid metals including liquid noble and transition metals. Phys. Status. Solidi. B **57**(1), 351–367 (1973)
76. Vatolin, N.A., et al.: Calculation of interaction potentials in liquid metals and alloys. Dokl. Akad. Nauk. SSSR **222**(6), 1323–1325 (1975)
77. Vatolin, N.A., et al.: Investigation of the interparticle interaction in liquid metals and alloys. Izv. Akad. Nauk. SSSR Met. (1), 45–49 (1975)
78. Gel'chinskii, B.R., et al.: Calculation of interatomic interaction potentials and their application in the molecular dynamics simulation of metallic melts. In: Physicochemical Investigations of Metallurgical Processes, vol. 4 (in Russian), pp. 4–11. Sverdlovsk (1976)
79. Harrison, W.A.: Pseudopotentials in the Theory of Metals. Benjamin, New York (1966)
80. Fuks, D.L., Bogdanov, V.I., Belan-Gaiko, L.V.: Calculation of the radial distribution function in liquid metals by the pseudopotential method. Fiz. Met. Metalloved. **46**, 924–930 (1978)
81. Kovalenko, N.P., Kuz'mina, L.M.: Effective interionic interaction in liquid copper. Ukr. Fiz. Zh. **22**(8), 1336–1339 (1977)
82. Watts, R.O., McGee I.J: Liquid State Chemical Physics. Wiley, New York (1976)
83. Polukhin, V.A., et al.: Application of molecular dynamics to the evaluation of the properties and structure of Fe–Ni and Al–Zn melts. Izv. Akad. Nauk. SSSR Met. (3), 35–37 (1976)
84. Ukhov, V.Ph., Gelchinskii, B.R., Vatolin, N.A.: Investigation of the properties of liquid metals derived from the data of diffraction experiments. In: Proceedings of the 7th Symposium on Thermophysical Properties, pp. 870–873. New York (1977)

85. Stillinger, F.H.: Equilibrium theory of pure fused salts. In: Blander, M. (ed.) Molten Salt Chemistry, pp. 1–108. Interscience, New York (1964)
86. Woodcock, L.V.: Isothermal molecular dynamics calculations for liquid salts. Chem. Phys. Lett. **10**, 257–261 (1971)
87. Woodcock, L.V., Singer, K.: Thermodynamic and structural properties of liquid ionic salts obtained by Monte Carlo computation. Trans. Faraday Soc. **67**(577), 12–30 (1971)
88. Krebs, H., Lazarev, V.B., Winkler, L.: Structure and properties of metalloids XVIII Separation of atoms in liquid germanium at 980° and 1270°. Z. Anorg. Allg. Chem. **352**(5), 277–284 (1967)
89. Il'inskii, A.G., Romanova, A.V.: Structure and properties of metallic melts. In: Metals, Electrons, and Lattices (in Russian). Kiev (1975)
90. Orton, B.R.: Model of the structure of liquid germanium. Z. Naturforsch. **30A**(11), 1499–1501 (1975)
91. Spektor, E.Z.: Structure of liquid iron and nickel. Dokl. Akad. Nauk. SSSR **190**(6), 1322–1324 (1970)
92. Waseda, Y., Suzuki, K.: Atomic distribution and magnetic moment in liquid iron by neutron diffraction. Phys. Status. Solidi. **39**(2), 669–678 (1970)
93. Schmitz-Pranghe, N., Kohlhaas, R.: Röntgenbeugungsuntersuchungen an Eisen, Cobalt und Nicel im flüssigen Zustand. Z. Naturforsch. **25a**(11), 1752–1754 (1970)
94. Slukhovskii, O.I., Lashpo, A.S., Romanova, A.V.: Structural changes in liquid iron. Ukr. Fiz. Zh. **20**(12), 1961–1965 (1975)
95. Klimenkov, E.A., Gel'd, P.D., Baum, B.A.: Structure of the short-range order in liquid cobalt and nickel. Dokl. Akad. Nauk. SSSR **230**, 71–73 (1976)
96. Vertman, A.A., Samarin A.M.: Properties of Iron Melts (in Russian), 280 pp. Moscow (1969)
97. Nevzorova, É.G., Radovskii, I.Z., Gel'd, P.V.: Magnetic susceptibility of iron at high temperatures. In: New Methods for Investigating Reduction Processes of Ferrous Metals (in Russian), pp. 121–124. Moscow (1974)
98. Turaev, U., Vertman, A.A.: Magnetic susceptibility of liquid metals at high temperatures and in the liquid state. Izv. Akad. Nauk. Tadzh. SSR **1**, 50–55 (1973)
99. Filippov, S.I., Kupriyanov, A.A.: Surface properties and adsorption phenomena of iron–carbon melts. Izv. Vyssh. Uchebn. Zaved. Chern. Metall. **1**, 15–18 (1969)
100. Kupriyanov, A.A., Filippov, S.I.: Density and structural changes in iron and alloys with carbon. Izv. Vyssh. Uchebn. Zaved. Chern. Metall. **9**, 10–15 (1969)
101. Kisun'ko, V.Z., Novokhatskii, I.A., Arkharov, V.I.: Influence of various additions on the temperature of the structural transformation in liquid iron. Izv. Akad. Nauk. SSSR Met. **2**, 176–179 (1975)
102. Korchemkina, N.V., et al.: Viscosity and phase diagram of the iron–niobium system. Rasplavy **5**, 3–9 (2002)
103. Grigorovich, V.K.: Mendeleyev's Periodic Law and the Electronic Structure of Metals (in Russian), 287 pp. Nauka, Moscow (1966)
104. Spektor, E.Z., Khokhlov, S.F.: Structure of copper and nickel melts near the melting point. Izv. Akad. Nauk. SSSR Met. **1**, 70–73 (1967)
105. Waseda, Y., Suzuki, K., Tomaki, S.: Neutron diffraction study of nickel in the liquid state. Phys. Status. Solidi. **39**(1), 181–188 (1970)
106. Pauling, L.: The Nature of the Chemical Bond and the Structure of Molecules and Crystals. Cornell University Press, Ithaca (1939)
107. Krieger, W., Trenkkler, H.: Aussagen über den Zustand von Eisen-Kohlenstoff-Schmelzen bei hohen Temperaturen auf Grund des Viskositätsverhaltens. Arch. Eisenhuttenwes. **42**(10), 685–689 (1971)
108. Gel'd, P.V., Baum, B.A., Petrushevskii, M.S.: Melts in Ferroalloy Production (in Russian), 288 pp. Metallurgiya, Moscow (1973)

109. Shvartsman, L.A., Tomilin, I.A.: Activities of carbon and oxygen in liquid iron. In: Proceedings of the Central Scientific-Research Institute of Ferrous Metallurgy (in Russian), vol. 74, pp. 5–18. Moscow (1971)
110. Schürman, E., Rimkus, H.J.: Verlauf der Activitäten und partiellen energetischen Grüßen des Kohlenstoff in den Schmelzen des Systems Lisen-Kohlenstoff. Geissereiforschung **26**, 175–179 (1974)
111. Spektor, E.Z., et al.: Nature of the interatomic interaction in an iron–carbon melt. Dokl. Akad. Nauk. SSSR **218**, 890–893 (1974)
112. Shvarev, K.M., et al.: Integral radiating power of Fe–C alloys at high temperatures. Izv. Akad. Nauk. SSSR Met. **1**(4), 70–80 (1976)
113. Maier, U., Steeb, S.: Röntgen- und Neutronenbeugungsuntersuchungen an Schmelzen des Systems Fe–C. Phys. Condens. Matter. **17**(1), 11–16 (1973)
114. Waseda, Y., Tokuda, M., Ohtani, M.: X-ray structural investigations of Fe and Fe–C melts. Tetsu-to-Hagane **61**(1), 54–70 (1975)
115. Vatolin, N.A., Veselova, S.I., Kern, É.M.: X-ray diffraction investigation of the characteristics of the short-range order of iron–carbon melts. Fiz. Met. Metalloved. **37**, 181–184 (1974)
116. Grigorovich, V.K.: Electronic Structure and Thermodynamics of Iron Alloys (in Russian), 292 pp. Moscow (1970)
117. Filippov, E.S., Grigorovich, V.K., Samarin, A.M.: Structural transitions in iron–carbon melts. Dokl. Akad. Nauk. SSSR **173**(3), 564–566 (1967)
118. Ruhl, R.C., Cohen, M.: Splat quenching of iron–carbon alloys. Trans. Metall. Soc. AIME **245**(2), 241–251 (1969)
119. Uchaev, A.N., et al.: Constitution and properties of melts based on iron. In: Scientific Reports to the 4th All-Union Conference on the Constitution and Properties of Metallic and Slag Melts. Part 2. Investigations of Metallic Melts (in Russian), pp. 83–85. Sverdlovsk (1980)
120. Avaliani, M.I., et al.: Viscosity of alloys of iron with carbon. Izv. Vyssh. Uchebn. Zaved. Chern. Metall. **2**, 123–126 (1977)
121. Nevzorova, É.G., Radovskii, I.Z., Dovgopol, S.P.: Magnetic susceptibility of alloys of iron with carbon. Fiz. Met. Metalloved. **35**, 737–742 (1973)
122. Kalistratov, O.N., Filippov, S.I.: Investigation of the properties of Fe–C melts using ultrasound. Izv. Vyssh. Uchebn. Zaved. Chern. Metall. **3**, 11–15 (1973)
123. Arsent'ev, P.P., Filippov, S.I., Linitskii, B.S.: Electrical conductivity of iron and iron–carbon melts. Izv. Vyssh. Uchebn. Zaved. Chern. Metall. **3**, 18–22 (1970)
124. Rostovskaya, L.A., et al.: Constitution of iron–carbon melts. Ukr. Fiz. Zh. **31**(9), 1385–1389 (1986)
125. Kurbatov, V.N., et al.: X-ray diffraction investigation of iron–carbon melts. In: Proceedings of the 10th Russian Conference on the Constitution and Properties of Metallic and Slag Melts (in Russian), vol. 2, pp. 166–167. Chelyabinsk (2001)
126. Zelinskaya, et al.: Structure of iron–boron melts close to the eutectic composition. Ukr. Fiz. Zh. **28**(2), 222–226 (1983)
127. Romanov, A.V.: Structure of metallic melts. In: Structure of Real Metals (in Russian), pp. 204–235. Kiev (1988)
128. Zelinskaya, G.M., et al.: Influence of the conditions for obtaining amorphous nonmetallic films of a hypereutectic iron–boron alloy on their structure. Ukr. Fiz. Zh. **28**(9), 1393–1399 (1983)
129. Bazin Yu, A., Baum, B.A.: Short-range order in hypoeutective Fe–B alloys in the liquid state. Neorg. Mater. **33**(1), 46–48 (1977)
130. Gel'd, P.V., Sidorenko, F.A.: Silicides of Fourth-Period Transition Metals (in Russian), 582 pp. Metallurgija, Moscow (1971)
131. Schubert, K.: Kristallstrukturen zweikomponentiger Phasen. Springer, Berlin (1964)
132. Gladyshevskii, E.I.: Crystal Chemistry of Silicides and Germanides (in Russian), 296 pp. Metallurgiya, Moscow (1971)

133. Kita, Y., Zeze, M., Morita, Z.: Structural analysis of molten Fe–Si alloys by X-ray diffraction. Trans. Iron Steel Inst. Jpn. **22**(8), 571–576 (1982)
134. Waseda, Y.: The Structure of Non-Crystalline Materials Liquids and Amorphous Solids, 326 pp. McGraw-Hill, New York (1980)
135. Sokolov, V.I., Popel', S.I., Esin, O.A.: Density and molar volume of slags. Izv. Vyssh. Uchebn. Zaved. Chern. Metall. **2**, 10–15 (1970)
136. Allersma, T., Hakim, R., Kennedy, T.N., Mackenzie, J.D.: Structure and physical properties of solid and liquid vanadium pentoxide. J. Chem. Phys. **46**(1), 154–160 (1967)
137. Henderson, J.: Density of lime–iron oxide–silica melts. Trans. Metall. Soc. AIME **230**(3), 501–504 (1964)
138. Zyazev, V.L., Esin, O.A.: Viscosity and density of the V_2O_5–CaO system and the V_2O_5–MgO system. Izv. Sib. Otd. Akad. Nauk. SSSR (9), 3–9 (1958)
139. Arsent'ev, P.P., Koledov, L.A.: Metallic Melts and Their Properties (in Russian), 376 pp. Metallurgiya, Moscow (1976)
140. Vatolin, N.A., Lisin, V.L., Pastukhov, É.A.: Structure of molten ferrous oxide and vanadium pentoxide. Dokl. Akad. Nauk. SSSR **218**(3), 597–599 (1974)
141. Waseda, Y., Shirai-shi, Y.: Structure of molten FeO at 1,420°C. Trans. Iron Steel Inst. Jpn. **18**(12), 783–784 (1978)
142. Krebs, H.: Fundamentals of Inorganic Crystal Chemistry. McGrawHill, New York (1968)
143. Koch, F., Cohen, J.B.: The defect structure of $Fe_{1-x}O$. Acta. Crystallogr. B **25**(2), 275–287 (1969)
144. Men',A.N., Vorob'ev Yu.D., Chufarov G.I.:Physicochemical Properties of Nonstoichiometric Oxides (in Russian), 223 pp. Khimiya, Leningrad (1973)
145. Pastukhov, É.A., et al.: Structure of molten iron silicates. Dokl. Akad. Nauk. SSSR **226**(5), 1136–1139 (1976)
146. Waseda, Y., Toguri, J.M.: The structure of molten binary silicate systems calcium oxide–silica and magnesium oxide–silica. Metall. Trans. **8B**(4), 563–568 (1977)
147. Waseda, Y., Toguri, J.: The structure of the molten FeO–SiO_2 system. Metall. Trans. **9B**(4), 595–601 (1978)
148. Okada, I.: MD-simulation of molten lithium chloride; self-exchange velocities of lithium isotopes near chloride ions. Z Naturforsch. A Phys. Phys. Chem. Kosmophys. **39**(9), 880–887 (1984)
149. Bertaut, E.F., Blum, P., Sagnieres, A.: Structure du ferrite bicalcique et de la brownmillerite. Acta. Crystallogr. **12**(2), 149–159 (1959)
150. Decker, B.F., Kasper, J.S.: The structure of calcium ferrite. Acta. Crystallogr. **10**(4), 332–337 (1957)
151. Chessin, H., Turkdogan, E.T.: A crystallographic investigation of calcium diferrite. J. Am. Ceram. Soc. **45**(12), 597–599 (1962)
152. Berezhnoi, A.S.: Multicomponent Oxide Systems (in Russian), 544 pp. Kiev (1970)
153. Pastukhov, É.A., Musikhin, V.I., Vatolin, N.A.: Electrical Properties of Nonstoichiometric Oxide Melts (in Russian), 112 pp. Sverdlovsk (1984)
154. Vatolin, N.A., Kern, É.M., Lisin, V.L.: X-ray diffraction investigation of the structure of silicate melts. In: Structure and Physicochemical Properties of Metallic and Oxide Melts (in Russian), p. 38–56. Sverdlovsk (1986)
155. Kern, É.M., et al.: X-ray diffraction investigation of the structure of a magnesium silicate melt. Fiz. Khim. Stekla. **12**(1), 14–19 (1986)
156. Sokol'skii, V.É., et al.: Constitution of melts of the MnO–TiO_2–SiO_2 and MnO–ZrO_2–SiO_2 ternary silicate systems. Rasplavy (6), 34–40 (1987)
157. Dancy, E., Derge, G.: The electrical conductivity of FeO_x–CaO slags. Trans. Metall. Soc. AIME **236**, 1642–1648 (1966)
158. Dickson, W.R., Dismukes, E.B.: The electrolysis of FeO–CaO–SiO_2 melts. Trans. Metall. Soc. AIME **224**, 505–511 (1962)

159. Nikitin, Yu.P., Barmin, L.N., Lopatin, V.M.: Measurement of the electronic conductivity of melts containing iron oxides. In: New Methods for Investigating the Reduction Processes of Ferrous Metals (in Russian), pp. 157–160. Moscow (1974)
160. Toporishchev, G.A., et al.: Electrical conductivity and thermoelectric power of melts of the $CaO–FeO–Fe_2O_3–Al_2O_3–SiO_2$ system. In: Physicochemical Investigations of Metallurgical Processes (in Russian), vol. 6, pp. 38–54. Sverdlovsk (1978)
161. Suchet, J.P.: Chemical Physics of Semiconductors. Van Nostrand, Princeton (1965)
162. Morin, F.J.: Oxides of the 3d transition metals. In: Hannay, N.B. (ed.) Semiconductors, pp. 600–633. Reinhold, New York (1959)
163. Esin, O.A.: Structure of molten silicates. Usp. Khim. **26**(12), 1374–1387 (1957)
164. Mikhailov, G.G., Kozheurov, V.A.: Mechanism of electrical conduction in wüstite. Zh. Fiz. Khim. **39**(3), 775–776 (1965)
165. Kozheurov, V.A.: Thermodynamics of Metallurgical Slags (in Russian), 165 pp. Sverdlovsk (1955)
166. Sryvalin, I.T., Esin, O.A.: Activities of the components of melts of the $CaO–Al_2O_3–SiO_2$ system. Izv. Vyssh. Uchebn. Zaved. Chern. Metall. **8**, 9–16 (1959)
167. Burylev, B.P., Moisov, L.P., Kretov, A.I.: Thermodynamics of multicomponent welding slags. In: Physicochemical Investigations of Metallurgical Processes (in Russian), vol. 10, pp. 47–50. Sverdlovsk (1982)
168. Ponomarenko, A.G.: Questions in the thermodynamics of phases of variable composition with a collective electronic system. Zh. Fiz. Khim. **7**, 1668–1674 (1974)
169. Masson, C.R., Jamieson, W.D., Mason, F.D.: Ionic constitution of metallurgical slag. In: Physical Chemistry of Process Metallurgy, pp. 223–231. London (1974)
170. Esin, O.A.: Application of polymer theory to molten slags. In: Physicochemical Investigations of Metallurgical Processes (in Russian), , vol. 1, pp. 5–17. Sverdlovsk (1973)
171. Esin, O.A.: Distribution of polymers in two- and three-component silicate melts. In: Physicochemical Investigations of Metallurgical Processes (in Russian), vol. 7. pp. 4–13. Sverdlovsk (1979)
172. Novikov, V.K.: Development of a polymer model of silicate melts. Rasplavy **6**, 21–33 (1987)
173. Esin, O.A.: Calculation of the activity of silica using the polymer model. In: Physicochemical Investigations of Metallurgical Processes (in Russian), vol. 3, pp. 19–34. Sverdlovsk (1975)
174. Esin, O.A.: Distribution of anions in molten silicates. Zh. Fiz. Khim. **47**, 2329 (1973)
175. Novikov, V.K., Toporishchev, G.A.: Calculation of the concentration of "free" oxygen ions in multicomponent alumosilicate melts. In: Physicochemical Investigations of Metallurgical Processes (in Russian), , vol. 11,pp. 12–17. Sverdlovsk (1983)
176. Kresnikov, A.N., et al.: Handbook for Calculations of the Equilibria of Metallurgical Reactions (in Russian), 416 pp. Metallurgiya, Moscow (1963)
177. Kulikov, I.S. (ed.). Atlas of Slags (in Russian), 208 pp. Metallurgiya, Moscow (1985)
178. Elliott, J.F., Gleiser, M., Ramakrishna, V.: Thermochemistry for Steelmaking, vol. 2. Addison–Wesley, Reading (1963)
179. Turkdogan, E.T.: Physical Chemistry of High-Temperature Processes (in Russian), 344 pp. Metallurgiya, Moscow (1985)
180. Belyaev, A.I., Zhemchuzhina, E.A., Firsanova, L.A.: Physical Chemistry of Molten Salts (in Russian), 359 pp. Metallurgizdat, Moscow (1957)
181. Zalomov, N.I., Boronenkov, V.N., Shalimov, M.P.: Calculation of the ionic composition and activities of the components in $CaO–A1_2O_3$ melts based on polymer theory. Rasplavy **1**(1), 49–55 (1992)
182. Esin, O.A.: Calculation of the activity of the components of molten silicates. In: Physicochemical Investigations of Metallurgical Processes (in Russian), vol. 11, pp. 5–11. Sverdlovsk (1974)
183. Sharma, R.A., Richardson, F.D.: Activities in lime-alumina melts. J. Iron Steel Inst. **198**, 386–390 (1961)

184. Zalomov, N.I., Zinigrad, M., Boronenkov, V.N., Shanchurov, S.M.: On calculating the activity of components in binary silicate melts. Glass. Phys. Chem. **33**(5), 459–463 (2007)
185. Johnston, R.F., Stark, R.A., Taylor, J.: Diffusion in liquid slags. Ironmak. Steelmak. **1**(4), 220–227 (1974)
186. Esin, O.N., Gel'd, V.P.: Physical Chemistry of Pyrometallurgical Processes, Part 2 (in Russian), 703 pp. Metallurgiya, Moscow (1966)
187. Linchevskii, B.V.: Techniques of Metallurgical Experimentation (in Russian), 334 pp. Metallurgizdat, Moscow (1967)
188. Malkin, V.I., Mogutnov, B.M.: Methods for measuring diffusion coefficients in silicate melts. Zavod. Lab. **28**(3), 291–296 (1962)
189. Wozniaková, B., Kubíček, P., Kursa, M.: Přehled metod studia difúze a difúzních koeficientů v roztavených struskách Sborník vědeckých prací Vysoké školy báňské v Ostravě. Řada. hutnická. **33**(1), 107–142 (1987)
190. Musikhin, V.I., Esin, O.A.: Relative diffusion coefficients in molten slags. Izv. Vyssh. Uchebn. Zaved. Chern. Metall. (12), 3–12 (1959)
191. Boronenkov, V.N., Esin, O.A., Shurygin, P.M.: Kinetics of electrochemical processes on a disk electrode in oxide melts. In: Physical Chemistry of Molten Slags (in Russian), pp. 129–146. Kiev (1970)
192. Han, Ch'ih.-Yung., Grigoryan, V.A., Zhukhovitskii, A.A.: Isotope exchange of iron in a two-phase solid metal–liquid slag system. Izv. Vyssh. Uchebn. Zaved. Chern. Metall. **5**, 5–16 (1961)
193. Novokhatskii, I.A., Esin, O.A., Chuchmarev, S.K.: Diffusion of hydrogen in molten slags. Dokl. Akad. Nauk. SSSR **136**(4), 868–870 (1961)
194. Kamyshov, V.M., Esin, O.A., Chumarev, S.K.: Nitrogen permeability of liquid slags. Izv. Akad. Nauk. SSSR Met. (2), 57–61 (1965)
195. Ascarelly, P., Paskin, A.: Dense-gas formulation of self-diffusion of liquid metals. Phys. Rev. **165**(1), 222–224 (1968)
196. Goto, K., Sasabe, M., Kawakami, M.: Relation between tracer diffusivity and electrical conductivity on multi-component oxide slags at 900° to 1,600°C. Trans. Iron Steel Inst. Jpn. **17**(4), 212–214 (1977)
197. Lepinskikh, B.M., Istomin, S.A.: Electrochemical Alloying and Inoculation of Metals (in Russian), 133 pp. Nauka, Moscow (1984)
198. Mitchell, A., Burel, B.: The solution rate of alumina in CaF_2–Al_2O_3 slags. Metall. Trans. **1**(8), 2253–2256 (1970)
199. Thonstad, J.: Chronopotentiometric measurements on graphite anodes in cryolite–alumina melts. Electrochem. Acta. **14**(2), 127–134 (1969)
200. Desclaux, P., Rolin, M.: Etude du coefficient de diffusion de l'alumine dans les melanges cryolithe-alumine. Rev. Int. Hautes. Temp. Refract. **8**(3–4), 227–236 (1971)
201. Majdic, A., Wagner, H.: Diffusion von Silizium in schmelzflüssigen Kalk-Kieselsäure-Tonerde-Schlacken. Arch. Eisenhuttenw. **41**(6), 529–537 (1970)
202. Mellberg, P.O., Lindström, P.A.: An electrochemical study of silica diffusion in liquid CaO–SiO_2–Al_2O_3–CaF_2 slags. Trans. Iron Steel Inst. Jpn. **21**(8), 591–595 (1981)
203. Satirov, B., et al.: Investigation of the kinetics of the dissolution of solid SiO_2 and CaO in molten slags. Rudodob. Metall. **23**(5), 45–50 (1968)
204. Anisimov Yu, S., Mitin, B.S., Grits, E.F.: Surface tension and density of Al_2O_3–SiO_2 melts. Izv. Akad. Nauk. SSSR Neorg. Mater. **13**(8), 1444–1446 (1977)
205. Keller, H., Schwerdtfeger, K., Hennesen, K.: Tracer diffusivity of Ca^{45} and electrical conductivity in CaO–SiO_2 melts. Metall. Trans. B **10**(1), 67–70 (1979)
206. Keller, H., Schwerdtfeger, K.: Tracer diffusivity of Si^{31} in CaO–SiO_2 melts. Metall. Trans. B **10**(4), 551–554 (1979)
207. Henderson, J., Yang, Ling., Derge, G.: Self-diffusion of aluminum in CaO–SiO_2–Al_2O_3 melts. Trans. Metall. Soc. AIME **221**, 56–60 (1961)
208. Towers, H., Paris, M., Chipman, J.: Diffusion of calcium ion in liquid slag. J. Met. **5**(11), 1455–1458 (1953)

209. Towers, H., Chipman, J.: Diffusion of calcium and silicon in a lime–alumina–silica slag. J. Met. **9**(6), 769–773 (1957)
210. Vorontsov, E.S., Esin, O.A.: Mechanism of diffusion in liquid slags. Izv. Akad. Nauk. SSSR Otd. Tekh. Nauk. (2), 152–155 (1958)
211. Koros, P.J., King, T.B.: The self-diffusion of oxygen in a lime-silica-alumina slag. Trans. Metall. Soc. AIME **224**(2), 299–306 (1962)
212. Nikitin Yu, P., Esin, O.A., Vorontsov, E.S.: Determination of diffusion coefficients in molten oxides. Zh. Fiz. Khim. **33**(6), 1420–1424 (1958)
213. Taylor, J.: Diffusion in liquid slags. In: Chemical Metallurgy of Iron and Steel, pp. 31–34. London (1973)
214. Ershov, G.S.: Diffusion of nitrogen in steelmaking slags. Izv. Vyssh. Uchebn. Zaved. Chern. Metall. **8**, 11–14 (1967)
215. Musikhin, V.I., Esin, O.A.: Diffusion coefficients of ions in molten slags. Dokl. Akad. Nauk. SSSR **136**(2), 388–390 (1961)
216. Gzielo, A.: Effect of CaF_2 concentration on chemical interdiffusivity in liquid $CaO–SiO_2–Al_2O_3–CaF_2$ slags at 1823 K. Arch. Hurnictwa. **32**(4), 613–623 (1987)
217. Malkin, V.I., Mogutnov, B.M.: Self-diffusion of alkali metal ions in silicate melts. Dokl. Akad. Nauk. SSSR **141**(5), 1127–1130 (1961)
218. Keller, H., Schwerdtfeger, K.: Measurement of tracer diffusivities of Ca^{45} and Fe^{59} in silica-saturated $FeO–CaO–SiO_2$ melts with the porous frit technique. Metall. Trans. B **17**(3), 497–501 (1986)
219. Simnad, T.M., Yang, Ling., Derge, G.: Direct measurement of ferrous ion mobility in liquid iron silicate by a radioactive tracer technique. J Met. **1**(5), 690–692 (1965)
220. Vanyukov, A.V., Montil'o, I.: Study of calcium, iron, nickel, and sulfur diffusion in liquid slags. Izv. Vyssh. Uchebn. Zaved. Tsvetn. Metall. **1**(5), 59–68 (1959)

Chapter 4
Modeling and Simulation of High-Temperature Processes

The practical application of modeling methods requires extensive knowledge of the laws governing the processes that occur in a metal–slag–gas system. Therefore, we will turn to well-known thermodynamic and kinetic methods of analysis which provide ways to reveal both the general laws of interaction and the special features of the individual steps of reactions and of the behavior of the components of the system.

4.1 Thermodynamic Methods for Analyzing Multicomponent Systems

As a result of the appearance of powerful computer technology [1], one of the methods currently used in physicochemical investigations of multicomponent systems is complete thermodynamic analysis, on the basis of which thermodynamic modeling has been developed. The term thermodynamic modeling refers to numerical modeling and prediction of the composition and properties of heterogeneous multicomponent systems with allowance for chemical and phase transformations in various, predominantly high-temperature processes [1, 2]. Despite the fact that oxidation–reduction reactions (or redox reactions) are generally far from a state of equilibrium in metallurgical plants, knowledge of the conditions for achieving equilibrium are very important from the practical point of view. On the basis of thermodynamic calculations, the behavior of the components in the complex slag and metal phases in various stages of a steelmaking process can be evaluated from the degree of deviation of the system from equilibrium, the optimal conditions for obtaining a final product of the required quality can be predicted, and the most economic way to carry out the process can be determined.

Thermodynamic modeling allows the solution of the following three major interrelated problems [1]:

M. Zinigrad et al., *Phase Interaction in the Metal–Oxide Melts–Gas System*, Engineering Materials, DOI: 10.1007/978-3-642-22377-8_4,

- determination of the equilibrium composition and properties of a system from known properties of the components comprising it;
- finding the thermodynamic properties of individual components of the system from the known equilibrium state and the values of the thermodynamic functions of the other components of the system;
- evaluation of the equilibrium conditions if the composition of the system and the thermodynamic functions of its components are known.

Practically all thermodynamic calculations are based on various model representations of the constitution of slag and metallic melts.

One of the most widely employed methods for calculating the activities of the components in a metal phase uses the Wagner interaction parameter [3]. The basis of the method is an expansion of an excess thermodynamic function (ΔG_i, ΔH_i, ΔS_i) into a Taylor series near the point corresponding to the pure solvent. If a 1% ideal solution of component i in the solvent is taken as the standard state, the equation for calculating the activity coefficient f_i of the component will have the form

$$\log f_i = \sum_{j=2}^{n} e_i^j[\%j] + \sum_{j=2}^{n} r_i^j[\%j]^2 + \sum_{\substack{j=2 \\ j<k}}^{n} \sum_{k=2}^{m} r_i^{j,k}[\%j][\%\,k], \tag{4.1}$$

where e_i^j, r_i^j and $r_i^{j,k}$ are the Wagner interaction parameters.

The activity of a component in a metal is calculated in this case from the equation

$$a_i = f_i[\%\,i]. \tag{4.2}$$

The presently accumulated material makes it possible to take into account the influence of the temperature and concentration of the other components on the interaction parameters over a broad range of compositions [3–6]. As practical experience has shown, Eq. 4.2 provides fairly high calculation accuracy, which is sufficient for evaluating the solubility of the components in a metal (Fe, Ni, etc.) and their equilibrium concentrations under real conditions.

The following three model approaches have been used most often in recent years to calculate the activities of the components in oxide melts:

- the theory of regular ionic solutions;
- the polymer theory of slags;
- the theory of the electronic structure of condensed phases.

The last of the theories just listed was used by Ponomarenko [7] to describe the characteristics of a slag phase. The slag is treated as a continuous nuclear matrix containing a system of itinerant electrons. Each atom appearing in the composition of the phase introduces electrons in a number equal to the atomic number of the element in Mendeleev's periodic table and its set of electronic levels. According to modern theories [8], the electronic levels of all the atoms forming the phase make up a single quasimechanical system, which is characterized by a specific value of the partial Gibbs free energy of the electrons μ (the Fermi level). It was shown in [7] that the complete expression for the chemical potential of component i of an oxide phase should contain a term that takes into account the characteristics of the electronic system of that phase:

$$\mu_i = \mu_i^0 + RT \ln a_i' + \mu v_i, \tag{4.3}$$

where μ_i^0 is the standard chemical potential of component i, $i_i^0 = k_i^0 C_{O_i}^{1-\alpha} C_{R_i}^{\alpha}$ is the nominal atomic activity, and v_i is the valence of component i.

The following main advantages of this approach can be singled out.

- The solution is found using the methods of mathematical statistics.
- A set of atoms, rather than chemical compounds, is used, allowing us to ignore the form in which the elements exist in the slag and the possible chemical reactions.

However, along with Mikhailets [9] we note that the series of equations which are used for the calculations do not have a theoretical basis and were obtained empirically by adjusting the form of the dependences. Moreover, as was shown in [10], the introduction of the last term into Eq. 4.3 is not thermodynamically rigorous.

The polymer theory of slags is essentially the only theory which enables us to calculate both the activities of the components of oxide and oxyfluoride melts and the activities of their structural components. It underwent extensive development in connection with the experimental confirmation of the presence of complex associates in slags [11–13]. In particular, the application of this theory to the description of a number of multicomponent slag melts was demonstrated in the preceding chapter. It should be noted that when the slag contains more than four components, solution of the problem is possible either by introducing a specific simplification [13] or by using numerical methods on a computer. Therefore, application of the polymer theory of slags is not always worthwhile, since the accuracy of the results is often comparable to the accuracy of other methods and the complexity of the calculations is considerably greater.

One of the widely used methods for describing the thermodynamic properties of slags is based on Kozheurov's theory of regular ionic solutions [14]. Russian and other scientists have subsequently made great contributions to the development of this theory and its practical use for specific oxide systems. In recent years the theory of regular ionic solutions has been developed fairly actively by the efforts of several Japanese scientists.

The following are the fundamental laws that are used in calculations:

$$RT \ln(\gamma_i)_{\text{reg}} = \sum_j^n x_j Q_{ij} + \sum_{j=1}^{n-1} \sum_{k=j+1}^{n} x_j x_k (Q_{ij} + Q_{ik} - Q_{jk}), \tag{4.4}$$

$$RT \ln(\gamma_i)_{\text{real}} = RT \ln(\gamma_i)_{\text{reg}} + I_a, \tag{4.5}$$

where x_i is the ion fraction of the cation, Q_{ij} is the energy of mixing or the energy of interaction of the cations for the (cation *i*)–O–(cation *j*) structure, and I_a is a correction factor for the transition from regular solutions to real solutions.

Introduction of the correction factor provided a way to describe the activities of components in systems whose constitution does not fit the requirements of the theory of regular ionic solutions. In particular, the activities of the components were calculated for slags of the $FeO–MnO–SiO_2–TiO_2$, $Cu_2O–CaO–CoO–FeO–SiO_2$, $FeO–Fe_2O_3–SiO_2–MgO–CaO–P_2O_5$, $FeO–Fe_2O_3–MnO–SiO_2–MgO–CaO$, and other systems, as was the hydrogen capacity of the slags [15, 16].

Taking into account that a relatively simple method for determining the energies of mixing of oxides was presented in [17] and that the data in [18] enable us to evaluate these parameters for systems containing CaF_2, we can conclude that the possibilities of the theory of regular ionic solutions have been expanded significantly and that it can be used to describe an extensive list of real metallurgical melts.

The following criteria are generally used for calculating the equilibrium compositions of the phases, as well as for determining the equilibrium conditions in a metal–slag–gas system:

- the extremum of the entropy or the Gibbs free energy of the system;
- the equality of the chemical potentials of the components in the equilibrium phases (the metal, slag, and gas);
- the dependences of the equilibrium constants of individual reactions on the temperature.

There is presently a fairly large number of publications which are devoted to the calculation of equilibrium in multicomponent systems, but only some of them describe the method as applied to metallurgical melts [3, 19–21]. It can be noted that the existing methods for calculating the equilibrium distribution of the components differ mainly with respect to the choice of the chemical reactions or reactants that are analyzed. At the same time, none of the approaches is universal, and each has its own shortcomings. In particular, it was noted in several reports that the formation not only of CO, but also of CO_2 must be taken into account when equilibria between a metal containing carbon and an oxide melt are calculated [22, 23]. However, the concentration of CO_2 is generally neglected. Therefore, in each specific case the thermodynamic relations must be obtained on the basis of the general laws presented for calculating the equilibrium distribution of the components in a metal–slag–gas system.

4.2 Thermodynamic Characteristics of Interaction Processes in a Metal–Slag–Gas System

The investigation and mathematical modeling of real metallurgical processes requires knowledge of the parameters and characteristics that determine the thermodynamic state of the system.

In this section we will examine the use of calculation and experimental methods for evaluating thermodynamic characteristics in a metal–slag–gas system for a series of components.

4.2.1 Method for Calculating the Equilibrium Distribution of Elements in Multicomponent Systems

The methods for calculating the values of equilibrium concentrations are generally based on using the equilibrium constants of the reactions that are possible in the system under investigation as the starting data [3, 24]. At the same time, the use of equilibrium constants expressed in terms of the concentrations, rather than the activities, of the components can lead to errors in the calculations.

Let us examine the following method [25] as one of the possible methods for calculating the equilibrium concentrations of the elements in a multicomponent carbon-containing metallic melt and a multicomponent slag using the reactions

$$[\mathrm{C}] + 2[\mathrm{O}] = \{\mathrm{CO_2}\} \tag{4.6}$$

$$[\mathrm{C}] + [\mathrm{O}] = \{\mathrm{CO}\} \tag{4.7}$$

$$n[\mathrm{E}_i] + m\,[\mathrm{O}] = (\mathrm{E}_{in}\mathrm{O}_m), \tag{4.8}$$

where E_i denotes Fe, Ni, Al, Si, B, etc.

The choice of deoxidation reactions is attributable to the fact that they have been studied fairly thoroughly and the equilibrium constants of most of the reactions are expressed in terms of the activities of the components [4]:

$$\log K_{P(4.6)} = \log \frac{P_{\mathrm{CO_2}}}{[\mathrm{C}] f_{\mathrm{C}}\, [\mathrm{O}]^2 f_{\mathrm{O}}^2}, \tag{4.9}$$

$$\log K_{P(4.7)} = \log \frac{P_{\mathrm{CO}}}{[\mathrm{C}] f_{\mathrm{C}} [\mathrm{C}] f_{\mathrm{O}}}, \tag{4.10}$$

$$\log K_{P(4.8)} = \log \frac{a_{\mathrm{E}_{in}\mathrm{O}_m}}{[\mathrm{E}_i]^n f_{\mathrm{E}\,i}^n\, [\mathrm{O}]^m f_{\mathrm{O}}^m}. \tag{4.11}$$

Here f_i denotes the Wagner interaction parameter.

We express the temperature dependence of the equilibrium constants of reactions (4.6–4.8) in accordance with the well-known equation

$$\log K_P = A/T + B. \tag{4.12}$$

We perform several transformations with Eqs. 4.9 and 4.10, taking into account that $P_{\mathrm{CO}} = 1 - P_{\mathrm{CO_2}}$. We assume that the equilibrium concentrations of all the elements in the alloy except carbon are insignificant and that their influence on f_{C}

can, therefore, be neglected. Then, we can find the values of $f_C = f_C^C$ and P_{CO} for assigned values of the temperature and the carbon concentration in the metal.

Knowing P_{CO}, we easily obtain

$$\log[O] + \log f_O = \log\frac{1 - P_{CO}}{P_{CO}} - \log K_{P(4.6)} + \log K_{P(4.7)} = D. \quad (4.13)$$

For the reaction of the base of an alloy (for example, iron) with oxygen we write

$$\begin{aligned}\log K_P^{Fe} &= \log\frac{a_{FeO}}{[Fe][O]f_O} = \log a_{FeO} - \log[Fe] - \log[O] - \log f_O \\ &= \log a_{FeO} - \log[Fe] - D.\end{aligned} \quad (4.14)$$

For the calculation of the equilibrium concentrations of the elements we assign the temperature, the composition of the slag with respect to all the components except FeO, and the concentration of carbon in the metal.

Let us examine the sequence of calculations in the example of an iron–carbon–metal/oxide-slag system.

We first find the concentration of iron in the alloy

$$[Fe] = 100 - [C]. \quad (4.15)$$

We determine the values of P_{CO} and D, and we find the activity of ferrous oxide from Eq. 4.14. Taking into account that $\gamma_{FeO} = 2$ [19], we can write

$$N_{FeO} = \frac{a_{FeO}}{2}. \quad (4.16)$$

Since all the remaining concentrations of the components of the original slag are assigned, the mole fraction of FeO can be expressed in the following manner

$$N_{FeO} = \frac{\dfrac{(\text{wt.\%})\text{FeO}}{M_{FeO}}}{\sum\dfrac{(\text{wt.\%})\,\text{E}_{in}\,\text{O}_m}{M_{\text{E}_{in}\text{O}_m}} + \dfrac{(\text{wt.\%})\text{FeO}}{M_{FeO}}}, \quad (4.17)$$

whence the mass percentages are easily found.

We take into account the content of (FeO) and find new values of the concentrations of the remaining components in the slag. Knowing the composition of the slag, we calculate the activities of the components in the slag using the equations of regular solution theory [26]. When the slags contain ions of complex-forming elements, we recalculate the activities using the relations that were presented in [26].

Using equations similar to (4.14), we find the values of all the components in the metal $[E_i]$ except iron and carbon. The Wagner interaction parameters for each of the i components are calculated from the relation

$$f_{E_i} = f_{E_i}^{C} f_{E_i}^{E_i}. \qquad (4.18)$$

From relation (4.13) we find the concentration of oxygen in the metal, assuming that

$$f_O = f_O^O f_O^C \prod f_O^{E_i}. \qquad (4.19)$$

This completes the first cycle of calculations.

The second cycle of calculations begins with determination of the Wagner interaction parameter for carbon with allowance for the influence of oxygen and all the components in the metal.

Then the calculation is continued according to the procedure described. Here the concentration of iron in the metal is determined from the equation

$$[Fe] = 100 - [C] - [O] - \Sigma[E_i]. \qquad (4.20)$$

The calculation is repeated until the values of all the concentrations of the elements in the metal are identical (with an assigned accuracy) in two consecutive iterations.

When this calculation method is used, the accuracy of the results is determined by the completeness of the information regarding the Wagner interaction parameters for different temperatures and melts with different carbon concentrations, including a concentration close to saturation, as well as the heat of mixing of the oxides.

When the data in [17] were used, it became possible to determine the values of Q_{ij} for systems in which these values have not been determined experimentally.

We will next consider the use of the method described for calculating the equilibrium concentrations of the components in examples of specific systems.

4.2.2 Distribution of Boron, Aluminum, and Phosphorus Between Liquid Metal and Slag Phases

For this investigation we selected an oxide phase that is similar to the slags used in the ladle treatment of steel, whose base melt contained the following (wt.%): 51 Al_2O_3, 43 CaO, 6 MgO. Up to 15 wt.% B_2O_3 or up to 10 wt.% P_2O_5 was introduced into it. An Fe–C melt with a carbon content equal to 0.5–3 wt.% was also used as the starting melt.

The equilibrium concentrations of the components in the metal and the slag were calculated according to the method described [25].

Published values of the Wagner interaction parameters [3, 4], heats of mixing [17, 26], equilibrium constants [4, 27], and I_a [15, 16] were used in the calculations.

The results of the calculations reveal the nonmonotonic character of the dependence of the boron and aluminum concentrations in the metal on the carbon content in it. It is attributed to the fact that when the concentration of carbon in the

Table 4.1 Equilibrium concentrations of elements in the metal and slag

T, K	(B_2O_3), wt.%	[C], wt.%	[B], wt.%	[Al], wt.%	$L_B = (B_2O_3)/[B]$	$L_{Al} = (Al_2O_3)/[Al]$
1773	10	1.0	0.015	9.9×10^{-5}	667	4.6×10^3
	15	1.0	0.026	1.2×10^{-4}	577	3.6×10^5
1823	10	0.5	0.014	1.2×10^{-4}	714	3.8×10^5
		1.0	0.031	3.2×10^{-4}	323	1.4×10^5
		1.5	0.046	6.0×10^{-4}	218	7.6×10^4
		2.0	0.058	9.2×10^{-4}	173	4.9×10^4
		2.5	0.060	1.0×10^{-3}	167	4.6×10^4
		3.0	0.057	9.8×10^{-4}	177	4.7×10^4
1873	5	0.5	0.012	2.5×10^{-4}	417	2.1×10^5
		1.0	0.028	7.2×10^{-4}	175	7.5×10^4
		1.5	0.043	1.3×10^{-3}	116	4.1×10^4
		2.0	0.050	2.0×10^{-3}	100	2.7×10^4
		2.5	0.051	2.3×10^{-3}	98	2.3×10^4
		3.0	0.048	2.2×10^{-3}	104	2.4×10^4
	10	0.5	0.028	3.5×10^{-4}	357	1.3×10^5
		1.0	0.065	1.0×10^{-3}	154	4.6×10^4
		1.5	0.096	1.8×10^{-3}	104	2.5×10^4
		2.0	0.114	2.8×10^{-3}	88	1.6×10^4
		2.5	0.118	3.1×10^{-3}	85	1.5×10^4
		3.0	0.111	3.0×10^{-3}	90	1.5×10^4
	15	0.5	0.055	3.7×10^{-4}	273	1.2×10^5
		1.0	0.114	1.1×10^{-3}	132	4.1×10^4
		1.5	0.179	1.9×10^{-3}	84	2.2×10^4
		2.0	0.230	3.0×10^{-3}	65	1.4×10^4
		2.5	0.242	3.7×10^{-3}	62	1.2×10^4
		3.0	0.222	3.8×10^{-3}	64	1.1×10^4
1923	10	0.5	0.057	9.0×10^{-4}	175	5.1×10^4
		1.0	0.132	2.5×10^{-3}	76	1.8×10^4
		1.5	0.198	4.7×10^{-3}	51	9.7×10^3
		2.0	0.235	7.2×10^{-3}	43	6.3×10^3
		2.5	0.247	8.0×10^{-3}	40	5.7×10^3
		3.0	0.242	7.7×10^{-3}	41	5.9×10^3
1973	5	1.0	0.106	5.1×10^{-3}	47	9.5×10^4
	10		0.261	6.6×10^{-3}	38	7.0×10^3
	15		0.480	7.4×10^{-3}	81	5.8×10^3

metal is greater than 2 wt.%, it is a stronger deoxidizing agent than boron and aluminum in corresponding concentrations. Since the concentration of oxygen passes through a minimum at $[C] \geq 2$ wt.%, the concentration of boron and aluminum accordingly passes through a maximum. According to the data in [4], the concentration of carbon that ensures the minimum oxygen content in iron at 1873 K is equal to 2.25 wt.%. This value is fairly close to the value that we obtained. The experimental data are presented in Table 4.1.

Let us analyze the dependence of the distribution coefficients of boron and aluminum between the metal and the slag, which are equal, respectively, to

$$L_B = \frac{(B_2O_3)}{[B]} \text{ and } L_{Al} = \frac{(Al_2O_3)}{[Al]}, \qquad (4.21)$$

on various parameters.

They depend on the temperature and concentration of carbon in the metal. In addition, as in [27], we assume that they obey the following linear dependences on the concentration of the oxide in the slag:

$$L_B = a + b\,(B_2O_3), \; L_{Al} = a_1 + b_1(Al_2O_3). \qquad (4.22)$$

As a result of the treatment of the data obtained, we determined the equations for calculating L_B and L_{Al} and the limits of their use

$$L_B = \left[\exp\left(\frac{47830}{T} - 20.23\right) - \exp\left(\frac{42310}{T} - 20.96\right)(B_2O_3)\right]\frac{1}{[C]}, \qquad (4.23)$$

$$L_{Al} = \left[\exp\left(\frac{72800}{T} - 30.8\right)(Al_2O_3) - \exp\left(\frac{73200}{T} - 27.6\right)\right]\frac{1}{[C]^{1.5}}. \qquad (4.24)$$

These equations enable us to calculate the values of L_B and L_{Al} in the temperature range from 1723 to 1973 K, the range of concentrations of B_2O_3 in the slag up to 25 wt.%, and the range of concentrations of Al_2O_3 in the slag up to 55 wt.%.

It should be noted that similar results with respect to the values of the equilibrium concentrations of aluminum and their dependence on the carbon content in the metal were obtained for slag and metallic systems that do not contain boron in [28].

At low values of (B_2O_3) in the slag, Eq. 4.23 is transformed slightly. When $(B_2O_3) \leq 3$ wt.%, L_B scarcely depends on the concentration of boron oxide in the slag and can be calculated from the following expression:

$$L_B = \left[\exp\left(\frac{47830}{T} - 20.23\right)\right]\frac{1}{[C]}. \qquad (4.25)$$

In order to test the calculated data, the equilibrium distribution coefficients of boron were determined experimentally. The compositions of the metal and the slag in the experiments were similar to those taken in the calculations. The metals were prepared by melting together powders of iron carbonyl, spectroscopically pure graphite, and amorphous boron, and the slag was prepared from the chemically pure oxides CaO, Al_2O_3, MgO, and B_2O_3. The concentration of carbon in the metal was 1.0, 1.5, or 2.0 wt.%, and the concentration of boron ranged from 0 to 0.3 wt.%.

To ensure the approach to the equilibrium position from both sides, three series of experiments were performed. In the first series the boron content in the metal significantly exceeded the equilibrium concentration, and there was no B_2O_3 in the original slag. Conversely, in the second series the metal did not contain boron, and the concentration of B_2O_3 in the slag was 5, 10, or 15 wt.%. In the third series boron was present in both the metal and the slag.

Weighed portions of the preliminarily ground metal and slag were taken in equal amounts by weight and mixed together. Thus, the contact surface between the phases was maximized in order to shorten the time needed by the system to achieve the equilibrium state and to eliminate the influence of the mass of the metal and slag phases on the distribution coefficient of boron. The mixture was placed in an isothermal zone of a carbon resistance furnace and held at a constant temperature for 60 min. To rule out the possibility of an interaction between the components of the metal and the gas phase during the heating and melting, the heating zone was isolated from the carbon heating element by a corundum cover and purged with argon. The purging was stopped after the phases melted, because it is known [28] that an air atmosphere does not have an appreciable effect on the composition of the reacting melts.

The experimental data allow us to conclude that equilibrium with respect to boron was not always successfully established under the conditions of our experiments. However, the character of the variation of its concentrations in the metal and the slag is such that the appearance of experimental points on the calculated curve can be expected.

Therefore, it may be concluded that the procedure described can be used to calculate the equilibrium concentrations in an iron–carbon-melt/oxide-slag system.

Expressions for calculating the ratios between the equilibrium concentrations for the following reactions were also derived on the basis of the data obtained:

$$2/3(\mathrm{BO}_{1.5}) + [\mathrm{C}] = 2/3[\mathrm{B}] + \{\mathrm{CO}\}, \quad (4.26)$$

$$2/3(\mathrm{BO}_{1.5}) + [\mathrm{Fe}] = 2/3[\mathrm{B}] + (\mathrm{FeO}), \quad (4.27)$$

$$2/3(\mathrm{AlO}_{1.5}) + [\mathrm{C}] = 2/3[\mathrm{Al}] + \{\mathrm{CO}\}, \quad (4.28)$$

$$2/3(\mathrm{AlO}_{1.5}) + [\mathrm{Fe}] = 2/3[\mathrm{Al}] + (\mathrm{FeO}), \quad (4.29)$$

$$\log K_{P(4.26)} = -\frac{13413}{T} + 5.676, \quad (4.30)$$

$$\log K_{P(4.27)} = -\frac{4012}{T} - 2.013, \quad (4.31)$$

$$\log K_{P(4.28)} = -\frac{21005}{T} + 8.077, \quad (4.32)$$

$$\log K_{P(4.29)} = -\frac{12340}{T} + 0.793. \quad (4.33)$$

Expressions (4.30–4.33) were used in our research both for performing kinetic analyses of multicomponent systems containing boron and aluminum and for developing mathematical models of technological processes.

Calculations of the distribution of phosphorus between a metal and a slag were performed in a similar manner. In this case the following reaction was considered for the oxidation of phosphorus:

Table 4.2 Equilibrium concentrations of the components in the metal and the slag

T, K	(P_2O_5), wt.%	[C], wt.%	[P], wt.%	$L_P = (P_2O_5)/[P]$
1873	5	1.0	0.27	18.5
	10	0.5	0.15	66.7
		1.0	0.49	20.4
		1.5	0.75	13.3
1923	10	1.0	0.92	10.9

$$2/5[P] + (FeO) = 1/5(P_2O_5) + [Fe], \tag{4.34}$$

whose equilibrium constant is described by the expression [4, 19]

$$\log K_{P(4.34)} = \log \frac{a_{P_2O_5}[Fe]}{[P]^{2/5} f_{P^{2/5}} a_{FeO}} = -\frac{22910}{T} + 15.414. \tag{4.35}$$

The results are presented in Table 4.2.

The data obtained allow us to draw the following conclusions.

The distribution ratio depends weakly on the value of (P_2O_5) in the slag, in agreement with the data in [29]. At the same time, L_P depends to a considerable extent on the concentration of carbon in the metal. A decrease in the concentration of carbon results in an increase in L_P to values close to the values obtained in [29] for a carbon-free metal and in virtual constancy of the value of the distribution ratio of phosphorus.

In order to test the calculated data, experiments were performed to determine the distribution ratio of phosphorus between the metal and the slag. In particular, it was found for $T = 1973$ K and a carbon content in the metal equal to 0.3 wt.% that $L_P = 28.6$ when equilibrium is approached from the slag side. When equilibrium is approached from the metal side, $L_P = 4.3$. The calculated value of the distribution ratio of phosphorus for these conditions is equal to 25.5.

Treatment of the results enabled us to obtain the following expressions for the distribution ratio of phosphorus between the metal and the slag:

for (P_2O_5) > 0.5 wt.%,

$$L_P = \left[\exp\left(\frac{42250}{T} - 19.79\right) - \exp\left(\frac{49930}{T} - 27.57\right)(P_2O_5)\right] \frac{1}{[C]^{1.5}}; \tag{4.36}$$

for (P_2O_5) ≤ 0.5 wt.%,

$$L_P = \left[\exp\left(\frac{40485}{T} - 18.17\right)(P_2O_5)\right] \frac{1}{[C]^{1.5}}. \tag{4.37}$$

Expressions for calculating the ratios between the equilibrium concentrations for the following reactions involving phosphorus were also obtained:

$$2/5(PO_{2.5}) + [C] = 2/5[P] + \{CO\}, \tag{4.38}$$

$$\log K_{P(4.38)} = -\frac{8120}{T} + 3.81, \tag{4.39}$$

$$2/5(PO_{2.5}) + [Fe] = 2/5[P] + (FeO), \tag{4.40}$$

$$\log K_{P(4.40)} = -\frac{15740}{T} - 11.29. \tag{4.41}$$

4.2.3 Equilibrium Distribution of Tungsten Between Liquid Metal and Slag Phases

When a tungsten-containing metal is in contact with an oxide melt, the main process determining the distribution of tungsten between the interacting phases can be represented by the reaction

$$1/3[W] + (FeO) = 1/3(WO_3) + [Fe], \tag{4.42}$$

The direction of the reaction for assigned phase compositions can easily be determined from the value of its equilibrium constant ($K_{P(4.42)}$). In addition, knowledge of its value is needed for calculations of the chemical composition of the metal and the slag during their interaction with the participation of tungsten. However, there are no experimental values of this equilibrium constant for phase compositions that are close to the process of the oxidative remelting of tungsten-containing scrap.

It is practically impossible to use calculation methods to find the equilibrium concentrations of the components, because the available literature data on the thermodynamic parameters of reaction (4.42) are very limited and contradictory. According to the data in [30, 31], the difference between the values of $K_{P(4.42)}$ found by calculation and by experiment reaches 150%. Therefore, we will use only experimental methods to determine the equilibrium concentrations for this system.

It should be noted that the presently available data on the distribution of tungsten between a metal and a slag [30–35] refer to oxide systems containing at least 30 wt.% ferrous oxide. At the same time, there are no data on the distribution of tungsten between a metal and a slag for FeO concentrations below 20 wt.%, which are characteristic of steelmaking slags.

The equilibrium distribution of tungsten between a carbon-containing metal and a slag was studied at 1773–1873 K in [36] according to the method described.

The starting materials used were a slag based on the low-melting eutectic containing (wt.%) 51 Al_2O_3, 43 CaO, and 6 MgO with additions of 0.5–20.0 wt.% FeO and 5–20 wt.% WO_3 and a metal based on iron containing 0.5–3 wt.% carbon and up to 6 wt.% tungsten.

In order to determine the holding time of the metal with the slag needed for the system to achieve an equilibrium state, experiments that provide a way to evaluate

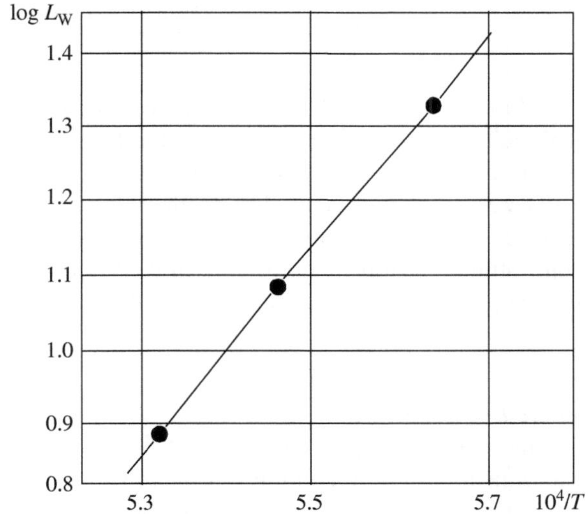

Fig. 4.1 Influence of the temperature on the equilibrium concentration ratio for reaction (4.42)

the dependence of the reactant concentrations on the interaction time of the melts were performed.

It was shown experimentally that variation of the composition of the metal and the slag with time scarcely occurs after 10–15 min of holding at 1823 K for initial concentrations of the ferrous oxide in the slag and of carbon and tungsten in the metal equal to 10, 1.1, and 6 wt.%, respectively. This holding time is apparently sufficient for the system to reach a state close to the equilibrium state. Experiments with other phase compositions were performed with a holding time equal to 30 min.

It follows from the results obtained that an increase in the concentration of FeO in the final slag leads to significant lowering of the values of $L_W = [W]/(WO_3)$. When (FeO) > 7.0 wt.%, variation of the temperature and an increase in the ferrous oxide content in the slag have practically no effect on L_W. Fairly similar results were obtained in [31, 33, 34]. At the same time, values of L_W for concentrations of FeO in the slag below 3 wt.% were obtained for the first time. Under these conditions L_W is strongly dependent on the temperature, and the values of the distribution coefficient are considerably higher than the generally accepted values.

It should be noted that the variation of the concentrations of the components in the experiments is due to the interaction with carbon and FeO. The experimental data allow us to theorize that reaction (4.42) is in a state close to equilibrium at the end of an experiment. In fact, it follows from the experimental data that the value of $K_{P(4.42)}$ remains nearly constant at an assigned temperature in the region of the phase compositions obtained after an experiment. The dependence of the average values of $K_{P(4.42)}$ on the phase interaction temperature was evaluated from the experimental data. It is linearized in log $K_{P(4.42)}$ versus $1/T$ coordinates (Fig. 4.1) and is described by the expression

$$\log K_{P(4.42)} = \log \frac{(WO_3)^{1/3}[Fe]}{[W]^{1/3}(FeO)} = \frac{14860}{T} - 7.05. \quad (4.43)$$

It follows from Eq. 4.43 that an increase in temperature causes displacement of the equilibrium of reaction (4.42) to the left, in agreement with the conclusions drawn in [31, 33, 34] regarding the exothermal character of the process studied.

There are practically no data on the influence of carbon on the distribution of tungsten in the literature. It is only known [31] that an increase in the carbon concentration results in an increase in the value of $K_{P(4.42)}$. The experimental data in [31] give the following dependence of L_W on the carbon content in the metal

$$L_W = A[C]^b, \quad (4.44)$$

where $b = 0.7–0.85$.

The experimental dependence of L_W on [C] obtained from the data from our experiments is described by Eq. 4.44, in which $b = 0.5–0.6$.

In our opinion, the following dependence of L_W on the initial concentration of carbon in the metal, which can be used for approximate calculations, is interesting:

$$L_W = \frac{A_1}{[C]_0}, \quad (4.45)$$

where $[C]_0$ is the initial concentration of carbon in the metal (wt.%). The value of A_1 can be evaluated from the following data.

- If the initial value of (FeO) in the slag is greater than 7.0 wt.%, $A_1 = 8.88$ in the temperature range from 1673 to 1973 K.
- If the initial value of (FeO) in the slag is less than 4.0 wt.%, A_1 can be calculated from the expression

$$\ln A_1 = -\frac{10310}{T} + 8.88. \quad (4.46)$$

4.2.4 *Calculation and Experimental Determination of the Equilibrium Distribution Coefficients of Sulfur*

The equilibrium distribution of sulfur between immiscible oxide and metal phases is often described by the ratio between the equilibrium concentrations

$$L_S = (S)/[S]. \quad (4.47)$$

When the composition of the interacting phases is constant, the coefficient L_S is proportional to the equilibrium constant of the desulfurization processes.

The desulfurization reaction in an iron–carbon-melt/oxide-slag system is written [37] in the form

$$[\mathrm{FeS}] + (\mathrm{CaO}) + [\mathrm{C}] = (\mathrm{CaS}) + [\mathrm{Fe}] + \{\mathrm{CO}\}, \qquad (4.48)$$

or in the ionic form

$$[\mathrm{S}] + (\mathrm{O})^{2-} + [\mathrm{C}] = (\mathrm{S})^{2-} + \{\mathrm{CO}\}. \qquad (4.49)$$

The expression for the equilibrium constant of process (4.49) can be written in the form

$$K_{P(4.49)} = \frac{a_{(\mathrm{S})^{2-}} P_{\mathrm{CO}}}{a_{[\mathrm{S}]} a_{[\mathrm{C}]} a_{(\mathrm{O})^{2-}}}. \qquad (4.50)$$

We set the equilibrium carbon monoxide pressure (P_{CO}) equal to unity, which is valid if CO bubbles form on the metal/oxide-melt boundary when the external pressure is equal to 1 atm. In addition, we assume that the activity coefficients of sulfur and oxygen in the slag are equal to 1, and we express the activities of sulfur and carbon in the metal using the Wagner interaction parameters. Then Eq. 4.50 is simplified.

We express the equilibrium distribution coefficient of sulfur L_S using formula (4.50).

We assume that $N_{(S^{2-})} = N_{(S)}$, which is fully permissible when the sulfur content in the slag is low (up to 3 wt.%). Then the expression for L_S can be written in the form

$$L_S = (\mathrm{S})/[\mathrm{S}] = AK_{\mathrm{eq}}[\mathrm{C}]\, f_C f_S N_{(O^{2-})}, \qquad (4.51)$$

where A is a constant that takes into account the switch from $N_{(S^{2-})}$ to (S), K_P is the equilibrium constant of reaction (4.48), [C] is the concentration of carbon in the metal (wt.%), f_C is the activity coefficient of carbon in the metal, f_S is the activity coefficient of carbon in the metal, and $N_{(O^{2-})}$ is the ion fraction of "free" oxygen in the slag.

The value of K_P was determined from the equation [37]

$$\log K_{P(4.48)} = -\frac{3800}{T} + 3.71. \qquad (4.52)$$

The activity coefficients of carbon and sulfur were calculated according to a method that uses Wagner interaction parameters [3, 4].

The difference between the present case and all the preceding variants of the calculation of L_S based on expression (4.51) is the fact that the ion fraction of "free" oxygen in the oxide phase $N_{(O^{2-})}$was determined using the polymer theory of liquid slags. Taking into account that the MgO content in the slag investigated is insignificant and that the properties of calcium oxide and magnesium oxide are similar, we used the method that we developed for the Al_2O_3–CaO system to

Table 4.3 Dependence of the calculated equilibrium distribution coefficients of sulfur on the temperature and the concentration of carbon in the metal

T, K	[C], wt.%											
	0.2	0.3	0.4	0.5	0.6	0.9	1.0	1.1	1.3	1.6	1.9	2.0
	Equilibrium distribution coefficient of sulfur, L_S											
1773	18	22	33	46	52	85	99	110	132	175	225	250
1823	20	27	38	52	65	110	112	124	160	215	280	301
1873	22	33	45	60	75	115	128	157	188	265	350	375

determine the ion fraction of "free" oxygen appearing in Eq. 4.51. In our opinion, this approach leads to refinement of the calculated values of L_S and better agreement with the experimental data.

The results of calculations of L_S for different temperatures and concentrations of carbon in the metal using Eq. 4.51 are presented in Table 4.3. As follows from the table, the extent of desulfurization increases with increasing temperature and carbon content in the metal phase.

The main deficiency of most of the experimental studies on the determination of L_S was the approach to equilibrium from only one side. We attempted to take into account this circumstance [25, 28, 100].

The values of L_S for the metal containing less than 1.5 wt.% carbon, which is close to the conditions of the ladle treatment of steel by synthetic slags, are of greatest interest. Therefore, a metal containing 0.5–1.9 wt.% carbon was used in the experiments. The procedure for preparing the melts and carrying out the experiments was described previously. In order to ensure the approach to the equilibrium state from both sides, three series of experiments were performed at each carbon concentration: (1) the concentration of sulfur in the original metal was higher than the equilibrium concentration obtained in the calculation, (2) the content of sulfur corresponded to the equilibrium concentration, and (3) there was no sulfur in the original metal.

In each series of experiments, the sulfur concentration in the oxide melt was varied: 0.5, 1.0, 2.0, and 3.0 wt.%.

The metal was analyzed for its carbon and sulfur content, and the slag was analyzed for its sulfur content. The value of [C] remained nearly constant during the experiments. It follows from the experimental data that equilibrium with respect to sulfur between the metal and the oxide melt is achieved after approximately 30 min when the interfacial surface is sufficiently developed.

Figure 4.2 presents the experimental results for the distribution of sulfur between metallic and oxide melts. The points indicate the initial and final sulfur concentrations, and the arrows show the direction of their variation.

A comparison of the calculated and experimental data (Fig. 4.2) allows us to conclude that the assumptions made in the calculation do not distort the values of L_S obtained.

The influence of the carbon concentration on the value of L_S is illustrated in Fig. 4.3. As follows from the figure, L_S increases with increasing [C]. The experimental values of L_S were found to be close to the calculated values (Fig. 4.3).

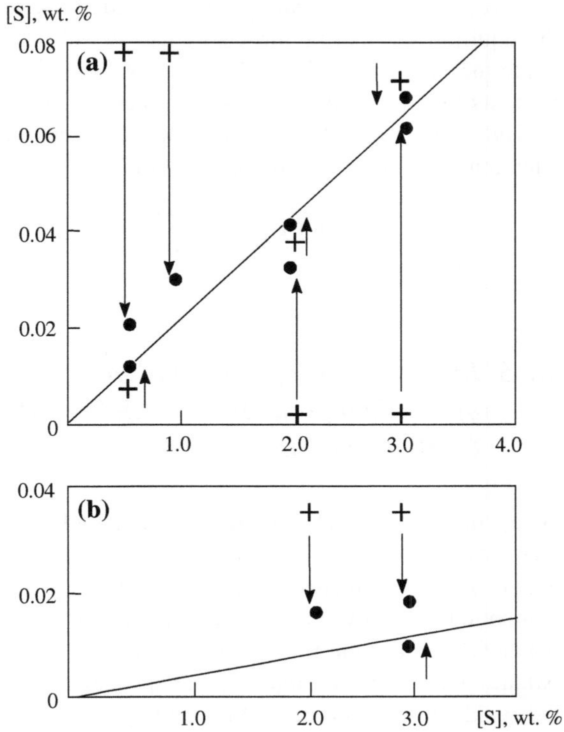

Fig. 4.2 Equilibrium distribution of sulfur between oxide and iron-containing melts: a—[C] = 0.5 wt.%; b—[C] = 1.9 wt.%. Lines—calculation; points—experiment; +—initial concentrations; ●—final concentrations

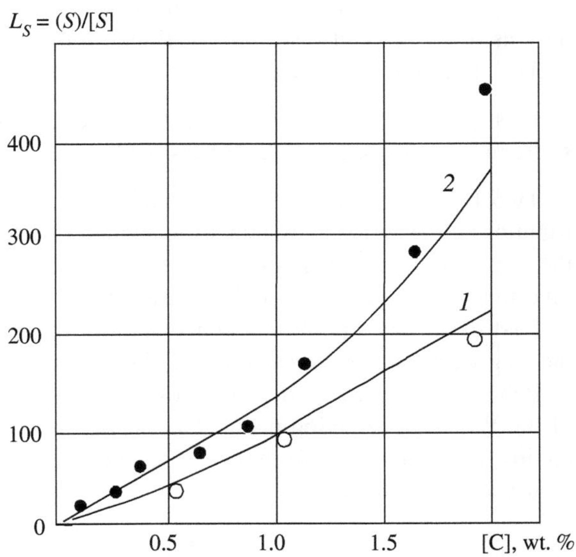

Fig. 4.3 Influence of the concentration of carbon in the metal on the equilibrium distribution of sulfur between metallic and oxide melts. Lines—calculation: *1*—T = 1773 K; *2*—T = 1873 K. Points—experiment: ○—data from [72] at T = 1773 K; ●—our data at T = 1873 K

Therefore, Eq. 4.51 can be recommended for determining numerical values of the equilibrium distribution coefficient of sulfur between Fe–C–S melts and an oxide melt containing 51 wt.% Al_2O_3, 43 wt. % CaO, and 6 wt. % MgO, as well as melts that are similar to it with respect to the content of "free" oxygen.

The treatment of the experimental data enabled us to obtain the following equation for calculating the distribution coefficient of sulfur:

$$\log L_S = -\frac{3700}{T} + 4.10 + 1.25\log[\mathrm{C}]. \qquad (4.53)$$

4.2.5 *Investigation of the Equilibrium Concentrations of the Components in a Nickel-Alloy/Oxyfluoride-Slag System*

When nickel and nickel alloys are remelted or welded, special oxide–fluoride mixtures (fluxes) are used. In the molten form, they protect the metals from the action of the gas phase and prevent oxidation of the alloying components. The compounds CaO, MgO, Al_2O_3, and CaF_2 are generally employed as flux components. At the same time, elements such as Cr, Ti, and Co, which can undergo oxidation and pass into the slag, are specially added to nickel alloys. Inasmuch as the welding materials used for welding nickel and its alloys contain transition metal carbides, which are easily oxidized to the corresponding oxides, the slag in contact with the liquid metal can contain oxides of chromium, titanium, etc.

For this reason, we conducted studies to determine the equilibrium concentrations of Ni, Ti, and Cr in the metal and in the oxyfluoride melt.

The starting phases selected were:

- the slag (wt.%): 50 Al_2O_3, 30 CaO, 15 CaF_2, 5 MgO;
- the metal (wt.%): Ni as the base, 0.1 C.

Two types of experiments were performed. In one type the equilibrium distribution of the elements was investigated after up to 15 wt.% TiO_2 was added to the slag with the composition indicated, and in the other type the distribution was investigated after up to 20 wt.% Cr was added to the metal.

First of all, the equilibrium concentrations of the component in the metal and the slag were calculated according to the procedure described.

Published temperature dependences of the equilibrium constants [4, 27], values of the Wagner interaction parameters [4, 27], heats of mixing [18, 26], and correction factors for going from regular solutions to real solutions [15, 16] were used.

It was also assumed that

$$a_{\mathrm{NiO}} = N_{\mathrm{NiO}}. \qquad (4.54)$$

The results of the calculations are presented in Table 4.4.

Table 4.4 Equilibrium concentrations of the components in a nickel-alloy/oxyfluoride–slag system

[C], wt.%	*T*, K	$(NiO)_{eq}$, wt.%	$(TiO_2)_{eq}$, wt.%	$[Ti]_{eq}$, wt.%	$L_{Ti} = (TiO_2)/[Ti]$
0.1	1773	13.0	2.0	4.1×10^{-4}	4878
		11.7	15.0	4.0×10^{-3}	3750
	1873	8.7	2.0	1.9×10^{-3}	1053
		7.6	10.0	0.011	909
		7.0	15.0	0.021	714
	1973	5.6	2.0	7.6×10^{-3}	263
		5.5	10.0	0.037	270
		4.8	15.0	0.073	205
	T, K	$(NiO)_{eq}$, wt.%	$(Cr_2O_3)_{eq}$, wt.%	$[Cr]_{eq}$, wt.%	$L_{Cr} = (Cr_2O_3)/[Cr]$
	1773	13.3	5.0	0.65	7.7
		12.6	10.0	1.00	10.0
		11.5	20.0	1.50	13.3
		10.8	30.0	1.85	16.2
	1873	7.1	1.4	1.0	1.4
		6.9	5.8	2.0	2.9
		6.7	12.8	3.0	4.3
		639	25.0	4.0	6.3

In addition to the calculations, experiments were performed to study the equilibrium distribution of the components between the metal and the slag with the compositions indicated above. The procedure for preparing and performing the experiments was previously described. We only note that in this case the metal was prepared by melting together powders of NP-2 nickel with the alloying components. Metallic powders of titanium and chromium served as the alloying components. The isothermal holding time of the crucibles with the melts was about 60 min.

The results of the chemical analysis of the metal and the slag after the experiments showed that equilibrium could not be achieved when it was approached either from the metal side or from the slag side. In our opinion, there are several possible reasons for this.

First, the equilibrium concentrations of titanium in the metal are not very high even when the content of its oxide in the slag is high. Therefore, it is fairly difficult to approach the equilibrium state from the metal side, especially within a relatively short time.

Second, the slag of the composition investigated has a very small oxidation potential; therefore, the rates of the interactions on the nickel-alloy/oxyfluoride–slag boundary are low. In addition, the reactions practically cease when a significant part of the CaF_2 is removed from the slag by vaporization.

For this reason, the correctness of the calculated data obtained can be confirmed only indirectly by comparing them with published data in the literature.

In particular, it is known [38, 39] that the final slag obtained after welding nickel and its alloys under molten and ceramic fluxes contains 6–8 wt.% NiO, in agreement with our data (Table 4.4).

The calculations of the activity of TiO_2 in multicomponent slag systems [40, 41] exhibit considerable negative deviations from ideality. In particular, when the titanium dioxide content is in the range from 34.0 to 44.0 wt.%, its activity at 1873 K varies form 0.041 to 0.106, depending on the composition of the slag. Our calculations exhibit a similar dependence. As the TiO_2 concentration in the slag varies from 2.0 to 15.0 wt.%, its activity at 1873 K increases from 0.002 to 0.02.

In addition, the results of similar calculations that were presented in the preceding sections exhibit fairly good agreement with experiment.

On the basis of these data, estimates of the ratios between the equilibrium concentrations were obtained for the following reactions:

$$1/2[\mathrm{Ti}] + (\mathrm{NiO}) = 1/2(\mathrm{TiO_2}) + [\mathrm{Ni}], \tag{4.55}$$

$$2/3[\mathrm{Cr}] + (\mathrm{NiO}) = 1/3(\mathrm{Cr_2O_3}) + [\mathrm{Ni}]. \tag{4.56}$$

Their temperature dependences are given by the equations

$$\log K_{P(4.55)} = \frac{4440}{T} + 0.22, \tag{4.57}$$

$$\log K_{P(4.56)} = \frac{1000}{T} + 0.68. \tag{4.58}$$

Dependences (4.57) and (4.58) were used in the mathematical modeling of metallurgical processes in similar systems.

4.2.6 Equilibrium Distribution of Silicon, Manganese, and Titanium Between the Metal and Oxyfluoride Melts

The literature contains practically no information about the equilibrium distribution of elements between the metal and multicomponent oxyfluoride slags of the SiO_2–CaO–CaF_2–Al_2O_3–MgO–FeO–TiO_2 system. Moreover, we note that most of the existing data for simpler systems refer to the 1750–1950 K temperature range, while higher temperatures, up to 2300–2500 K, are realized in some technological processes. The absence of reliable data on the activities of the components under such conditions makes thermodynamic evaluation of the equilibrium difficult. For this reason, the equilibrium of the reactions

$$n/m[\mathrm{E}_i] + (\mathrm{FeO}) = n/m\left(\mathrm{E}_i\mathrm{O}_{m/n}\right) + [\mathrm{Fe}] \tag{4.59}$$

with the participation of silicon, manganese, and titanium at 1800–2250 K was studied experimentally.

When the variations in the composition of the phases are relatively small and the activity coefficients can be assumed to be constant, the apparent equilibrium constant for a reaction of type (4.59) is given by the formula

Table 4.5 Initial composition of the slags

Slag No.	Concentration of components, wt.%								
	SiO_2	MnO	TiO_2	Cr_2O_3	Al_2O_3	CaO	CaF_2	MgO	FeO
1	35.6	6.1	–	0.80	22.3	5.1	17.3	11.2	1.21
2	34.2	6.0	10.1	–	20.5	5.1	16.2	11.6	4.53
3	22.3	2.0	10.1	0.15	22.2	5.2	25.1	10.1	2.04
4	23.0	0.5	–	0.16	31.5	6.2	24.1	10.2	1.50
5	24.2	2.6	1.4	0.96	26.2	6.0	22.3	9.8	0.55
6	24.7	1.1	1.7	0.50	27.3	7.3	24.3	10.8	0.71
7	22.6	1.0	–	0.10	22.3	6.7	28.5	11.0	6.22
8	23.0	1.1	–	0.11	24.9	5.1	27.3	11.2	5.43
9	23.7	2.6	3.3	0.18	28.0	14.0	17.6	10.5	0.52
10	23.7	2.6	–	0.19	27.5	14.1	20.0	11.0	0.61
11	23.0	0.9	–	–	27.0	10.8	26.1	11.6	0.83
12	20.2	4.2	2.7	0.25	25.9	9.2	28.8	8.7	0.52

$$K_{P(4.59)} = \frac{\left(\mathrm{E}_i\mathrm{O}_{m/n}\right)^{n/m}}{[\mathrm{E}_i]^{\,n/m}(\mathrm{FeO})}. \tag{4.60}$$

To eliminate contamination of the slag by components of the refractories, a melt in the suspended state was employed [42, 43]. The small relative mass of the slag in the case of a melt in the suspended state leads to considerable variation of its composition in response to a comparatively small change in the concentration of the elements in the metal. In addition, during prolonged holding, vaporization of the components results in appreciable changes in the compositions of the metallic and slag melts. The investigations performed enabled the development of an experimental procedure that ensures the achievement of equilibrium for different metal and slag compositions. A sample of the metal with a mass of 3–5 g was melted in a high-frequency electromagnetic field (440 kHz) in an argon atmosphere at a pressure of 0.1–0.3 MPa and heated to the assigned temperature, and a powdered flux was supplied to the metal. The samples were quenched in a heavy copper mold. The mass of the slag interacting with the metal was 0.05–0.2 g. The temperature of the metal was measured using an optical pyrometer with accuracy equal to ±10 K.

The slags were analyzed by x-ray fluorescence analysis using a VRA-20R spectrometer and by emission spectral analysis using an ISP-30 spectrograph. The silicon content in the slag was additionally determined by laser microspectral analysis on an LMA-10 analyzer. The metal was analyzed on the ISP-30 spectrograph. The relative error for x-ray fluorescence analysis was 1.5–2.7%, the relative error for emission spectral analysis was 3.5–6.6%, and the relative error for laser microscpectral analysis was 8–10%.

The initial compositions of the slags and metals were selected so that equilibrium would be approached "from above" and "from below," i.e., from the directions of oxidation and reduction of the elements. The initial compositions of the slags are presented in Table 4.5.

Table 4.6 Data for calculating equilibrium constants

Reaction equation	A	B	Original slag No.
$1/2[Si] + (FeO) = 1/2(SiO_2) + [Fe]$	6420	−2.65	1, 2
	4020	−1.54	3–12
$[Mn] + (FeO) = (MnO) + [Fe]$	9080	−4.33	1, 2
	6550	−3.32	3–12
$2/3[Cr] + (FeO) = 2/3(CrO_{1.5}) + [Fe]$	8270	−4.5	3–12
$1/2[Ti] + (FeO) = 1/2(TiO_2) + [Fe]$	11940	−5.23	3–12
$2/3[Al + (FeO)] = 2/3(AlO) + [Fe]$	15250	−5.63	3–12

An analysis of the experimental data shows that equilibrium could not be achieved in all of the experiments. In particular, data on the distribution of manganese which do not fit equilibrium were obtained in a few experiments. This is attributed to the rapid vaporization of manganese. For this reason, those results were disregarded in the further analysis. At the same time, experiments showed that the data obtained for the less volatile components (Si, Ti, Cr, and Al) with different variants for supplying the slag are in good agreement with each other. The coefficients in Eq. 4.12, which relate the apparent equilibrium constant to the temperature, were determined by statistical treatment of the results of the experiments (Table 4.6).

The values of $K_{P(4.59)}$ differ markedly from the known literature values for basic and acidic oxide melts because of the significant variation of the activity coefficients of the components upon transition from one slag system to another.

4.3 Methods for the Kinetic Analysis of Multimolecular and Simultaneously Occurring Reactions

The interaction of metal, slag, and gas in metallurgical plants encompasses a large set of interrelated reactions. Several methods for mathematically describing the kinetics of both parallel heterogeneous processes with common reactants and individually taken reactions are known in the literature.

4.3.1 Analysis of the Kinetics of an Interaction Under Steady-State Conditions

The use of the law of mass action for describing the kinetics of processes in metallurgical plants is fairly widespread [44–50].

The general form of the equation for the rate of a process can be described in the following manner:

$$V = k_0 e^{-\frac{E_j}{RT}} \prod_{j=1}^{n} C_j^{P_j}, \tag{4.61}$$

where V is the rate of the process, E_j is the activation energy of the reaction involving component j, R is the universal gas constant, T is the temperature, C_j is the concentration of the j-th component, P_j is the order of the reaction with respect to the j-th component, n is the number of reactants, and k_0 is the pre-exponential factor. The parameters E_j, P_j, and k_0 can be evaluated from the experimental data. It should be noted that the values of these parameters generally remain constant only in relatively narrow ranges of variation of the temperature of the system and concentrations of the reactants. In addition, the values of these parameters can vary upon transition from one system to another. For this reason this approach is fairly effective in the analysis of the kinetics of an individually taken reaction, but its use for mathematically describing the interaction in complex metallurgical systems ultimately leads to ordinary statistical laws, which are suitable for a specific system, specific conditions, and a specific plant.

In our opinion, the approach to the description of the interaction processes in a metal–slag–gas system based on fundamental physicochemical laws is more promising.

Kinetic equations for the diffusion-controlled and mixed regimes of a large number of reversible heterogeneous reactions with an arbitrary number of common reactants and different forms of the dependence of the rate of the chemical act on the concentration were obtained in [44].

The authors considered the occurrence of a large number of phase boundary reactions, the intensity of each of which dependence on the rate constant K^n and the mass-transfer constant K^d of all the reactants and products of the specific reaction. Then, assuming that there is no mutual influence of the reactions on the values of K^n and K^d, they considered a set which includes X coupled reactions:

$$\begin{gathered} m_{11}A_{11} + m_{21}A_{21} + \cdots + m_{k1}A_{k1} = n_{11}B_{11} + n_{21}B_{21} + \cdots + n_{k1}B_{k1}, \\ \cdots\cdots\cdots\cdots\cdots\cdots\cdots\cdots\cdots\cdots\cdots\cdots\cdots\cdots\cdots\cdots\cdots\cdots\cdots\cdots, \\ m_{1x}A_{1x} + m_{2x}A_{2x} + \cdots + m_{kx}A_{kx} = n_{1x}B_{1x} + n_{2x}B_{2x} + \cdots + n_{kx}B_{kx}, \end{gathered} \tag{4.62}$$

where m_{ij} and n_{ij} are the stoichiometric coefficients.

For a purely diffusion-controlled regime, where a state close to the equilibrium state is realized near the reaction surface for each reaction, they obtained a system of nonlinear algebraic equations of the form

$$\frac{\prod_{i=1}^{l}\left(1+\frac{\sum_{j=1}^{x} n_{ij}J_j-\sum_{j=1}^{x} m_{ij}J_j}{J_{B_{ij}}^{\lim}}\right)^{n_{ij}}}{\prod_{i=1}^{K}\left(1-\frac{\sum_{j=1}^{x} m_{ij}J_j-\sum_{j=1}^{x} n_{ij}J_j}{J_{A_{ij}}^{\lim}}\right)^{m_{ij}}}=K_{Pj}\cdot\frac{\prod\gamma_{A_{ij}}^{m_{ij}}\cdot C_{A_{ij}}^{m_{ij}}}{\prod\gamma_{B_{ij}}^{n_{ij}}\cdot C_{B_{ij}}^{n_{ij}}}. \quad (4.63)$$

Here J_i is the rate of reaction j, mole/cm^2·s; $J_{R_{ij}}^{\lim}$ is the limiting diffusion flux of the reactant, mole/cm^2·s, $K_{P,j}$ is the equilibrium constant of the reaction of j; $\gamma_{R_{ij}}$ is the activity coefficient of the reactant; $C_{R_{ij}}$is the concentration of the reactant.

The value of $J_{R_{ij}}^{\lim}$ is determined from the relation

$$J_{R_{ij}}^{\lim}=\beta\cdot D_{R_{ij}}^{\alpha}\cdot C_{R_{ij}}, \quad (4.64)$$

where $D_{R_{ij}}$is the diffusion coefficient of the reactant (cm^2/s) and β is the convection constant, which is identical for all reactants and depends on the flow parameters of the liquid and its viscosity. Usually [45, 46], $\alpha \cong \frac{2}{3}$ is taken for a solid/liquid (solid/gas) interface, and $\alpha \cong \frac{1}{2}$ is taken for a liquid/liquid interface.

If the activity coefficients vary only slightly upon transition from the bulk concentration to the surface concentration, then, taking into account Eq. 4.64, for one reaction we can obtain

$$\frac{\prod\left(1+\frac{n_iJ}{\beta D_{B_i}^{\alpha}C_{B_i}}\right)^{n_i}}{\prod\left(1-\frac{m_iJ}{\beta D_{A_i}^{\alpha}C_{A_i}}\right)^{m_i}}=K_C\frac{\prod C_{A_i}^{m_i}}{\prod C_{B_i}^{n_i}}. \quad (4.65)$$

Here K_C is the equilibrium constant expressed in terms of concentrations.

Using expressions (4.64) and (4.65), we can find the concentration dependence of the rate of the process without resorting to empirical constants. In this case the reaction rate is expressed in the form of the ratio J/β. To find the absolute values of J, we must have one experimental value of the convection constant.

To illustrate this, let us consider the kinetic dependences for the reaction of carbon and oxygen dissolved in iron, assuming a diffusion-controlled regime for process (4.7).

Plots of the dependence of the rate of reaction (4.7) on the concentration of carbon in the metal for different values of the oxygen content are presented in Fig. 4.4. The plots obtained show that up to a carbon content of 0.3–0.4 wt.% the values of the reaction rate increase linearly with the carbon concentration, but they remain practically constant when [C] is increased further. This result is consistent with industrial observations [47]. A comparison of the calculated and experimental data [47, 48] (Fig. 4.4) reveals the closeness between them. This finding points to the possibility of diffusional hindrances when reaction (4.7) takes place.

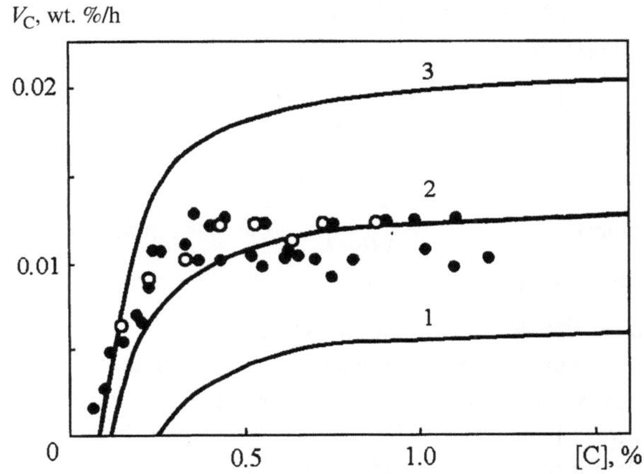

Fig. 4.4 Dependence of the decarbonization rate on the carbon concentration in an open-hearth bath: *1–3*—calculation based on Eq. 4.65 for [O] equal to 0.01, 0.02, and 0.03 wt.%; ○, ●—data from [47] and [48], respectively

The approach described here has been employed in the kinetic analysis of oxidation reactions of impurities in iron by a molten slag. The interaction of the impurities in molten iron with a liquid slag containing iron oxides occurs in steelmaking plants during the production of ferroalloys and several nonferrous metals.

According to the electrochemical theory of the interaction of a metal with a slag, the oxidation of metal impurities by ferrous oxide

$$n/m\,[\mathrm{Me}] + (\mathrm{FeO}) = [\mathrm{Fe}] + 1/m\,(\mathrm{Me}_n\mathrm{O}_m) \tag{4.66}$$

consists of two steps, viz., a cathodic step and an anodic step. The cathodic step is the discharge of Fe^{2+} ions, and the anodic step is the ionization of the impurity atoms. Thus, the reaction under consideration is a complex heterogeneous process that consists of several stages:

- diffusion of Fe^{2+} (O^{2-}) ions and Me atoms from the bulk of the phases to the metal/slag interface;
- adsorption followed by the discharge of Fe^{2+} ions to Fe atoms and the ionization of Me atoms to Me^{x+} in the present case;
- diffusion of the reaction products Fe and Me^{x+} (or $Me_xO_y^{z-}$) from the reaction surface into the bulk of the metal and the slag.

Accordingly, four kinds of diffusional hindrances (according to the number of reactants in the reaction) and two kinds of kinetic hindrances (according to the number of particles that change their charge) can be singled out. For some reactions there can be more factors that hinder the overall reaction, for example, the diffusion of O^{2-} ions. In addition, the act of ion discharge can consist of several stages and can be accompanied by the diffusion of intermediate particles into the bulk of the slag. Diffusion control is realized fairly often in oxidation reactions, where the rate is limited only by diffusion of the reactants, and discharge of the ions occurs relatively fast.

The following equation was obtained from Eqs. 4.64 and 4.65 after some relatively simple transformations:

$$[\mathrm{Me}] = aV + \frac{[(\mathrm{Me}_n\mathrm{O}_m) + cV]^{1/n}}{K_P^{m/n}[(\mathrm{FeO}) - bV]^{m/n}}, \tag{4.67}$$

where $a = \frac{nM_{\mathrm{Me}}}{md_{\mathrm{Me}}}\left(\frac{D_{\mathrm{FeO}}}{D_{\mathrm{Me}}}\right)^{\alpha}$, $b = \frac{M_{\mathrm{FeO}}}{d_{\mathrm{slag}}}$, and $c = \frac{M_{\mathrm{Me}_n\mathrm{O}_m}}{md_{\mathrm{slag}}}\left(\frac{D_{\mathrm{FeO}}}{D_{\mathrm{Me}_n\mathrm{O}_m}}\right)^{\alpha}$, $\mathrm{cm}^3/\mathrm{mole}$; $V = \frac{J\cdot 10^2}{\beta D_{\mathrm{FeO}}^{\alpha}}$, is the reduced rate, mole wt.%/cm^3; M_i is the molecular weight, g/mole; and d_i is the density of the phase, g/cm^3.

Equation 4.67 was used in [49] to analysis the kinetics of the oxidation processes of silicon, manganese, phosphorus, chromium, and vanadium dissolved in liquid iron at 1873 K. A slag of the open-hearth type containing (wt.%) 45 CaO, 25 SiO_2, 10 FeO, 9 MgO, 5 Al_2O_3, 5 MnO, and 1 P_2O_5, to which 5 wt.% Cr_2O_3 and 5 wt.% V_2O_3 were added, served as the base component.

The results of the calculations show that at low concentrations of the impurities their oxidation rates are directly proportional and that at 1–5 wt.% [Me] they are weakly dependent on the concentration of the impurity, especially for silicon and phosphorus. This is due to the fact that at low concentrations the oxidation rate is limited by the diffusion of the impurities in the metal, while at high concentrations it is limited by the diffusion of FeO in the slag. The switching of diffusion control from one starting substance to another occurs under the condition of equality between the limiting fluxes of the respective substances: $J_{\mathrm{FeO}}^{\mathrm{lim}} = m/s\, J_{\mathrm{Me}}^{\mathrm{lim}}$. The reactions considered are reversible; therefore, such switching from [Me] to (FeO) is completed for them in a certain concentration range. The formal reaction order with respect to the content of [Me] is fractional and variable in this case. At impurity concentrations below the equilibrium concentrations, we have $V < 0$, which corresponds to the occurrence of reaction (4.66) in the reverse direction.

The kinetic laws considered above apply to the oxidation of each of the impurities in iron separately. If the metal contains several impurities, their oxidation occurs simultaneously. Ferrous oxide diffusing to the metal is consumed at once in several reactions, ensuring their occurrence with the rates $J_1, J_2, \ldots J_i$. In this case, the following equations were obtained for the diffusion-controlled regime:

$$\left(1 - \frac{J_i}{mJ_{\mathrm{Me}_n\mathrm{O}_m}^{\mathrm{lim}}}\right)^{1/m} = \frac{K_{P,i}}{\prod_i}\left(1 - \frac{\sum J_i}{J_{\mathrm{FeO}}^{\mathrm{lim}}}\right)\left(1 - \frac{nJ_i}{mJ_{\mathrm{Me}}^{\mathrm{lim}}}\right)^{n/m}, \tag{4.68}$$

$$J_i = \frac{\frac{K_{P,i}}{\prod_i} - 1}{\frac{1}{m^2 J_{\mathrm{Me}_n\mathrm{O}_m}^{n}} + \frac{K_{P,i}}{\prod_i}\left(\frac{1}{J_{\mathrm{FeO}}^{\mathrm{lim}}}\frac{\sum J_i}{J_i} - \frac{n^2}{m^2 J_{\mathrm{Me}}^{n}}\right)}. \tag{4.69}$$

The simultaneous oxidation of Mn, Cr, V, P, and Si at $T = 1873$ K in a diffusion-controlled regime with a concentration of each of the elements in the metal

Table 4.7 Reduced oxidation rates of impurities in a diffusion-controlled regime

Oxidation conditions	Cr	Mn	V	P	Si	$\sum V_i$
Separate oxidation	0.048	0.175	0.26	0.33	0.37	–
Simultaneous oxidation						
Without Me_nO_m	4.2×10^{-5}	0.056	4.2×10^{-5}	4.2×10^{-5}	0.31	0.37
In the presence of Me_nO_m	−0.049	−0.01	−0.04	0.069	0.38	0.35

equal to 1 wt.% was considered. Two series of calculations were performed. In one series the slag contained oxides of the metals in the ranges indicated above. In the other series, among the participants in the oxidation reactions, only FeO and SiO_2 were present in the slag. The results of the calculations are presented in Table 4.7.

Initially, when there are no oxides of the corresponding elements in the slag, silicon is oxidized almost exclusively. The oxidation of manganese is also noticeable. With time, as the concentration of silicon in the metal decreases, the rapid oxidation of phosphorus and then of vanadium and the other elements may be expected. As equilibrium is approached, the rates become close, and the successive oxidation of the impurities changes to parallel.

When the oxides Me_nO_m are present in the slag, the impurities with a low oxygen affinity are initially reduced, rather than oxidized. For example, in the case of chromium, the reduction rate is close to the limiting rate of diffusion of Cr_2O_3. Conversely, the oxidation of silicon becomes more intense owing to the oxides of Cr, V, and Mn. Over the course of the process, the concentration of Cr in the metal and, to lesser extent, the concentrations of V and Mn at first increase and then begin to decrease. This supports the possibility of selectively oxidizing impurities in the metal by adjusting the composition of the slag and the optimal time for carrying out the process.

For the mixed regime, denoting the rate constants of the reactions in the forward direction (the direct reactions) by $K_1^d, K_2^d, \ldots, K_X^d$ and the rate constants in the reverse direction by $K_1^r, K_2^r, \ldots, K_X^r$, we obtained a system of equations for determining J_j. For reactions with a reaction order that is equal to the molecularity, the system of equations was obtained in an explicit form:

$$J_j = K_j^d \left[\prod C_{A_{ij}}^{m_{ij}} \left(1 - \frac{\sum m_{ij} J_j - \sum n_{ij} J_{ij}}{J_{A_{ij}}^{\lim}} \right)^{m_{ij}} - \frac{K_j^r}{K_j^d} \prod C_{B_{ij}}^{n_{ij}} \left(1 + \frac{\sum n_{ij} J_j - \sum m_{ij} J_{ij}}{J_{A_{ij}}^{\lim}} \right)^{n_{ij}} \right]. \tag{4.70}$$

We note that Eq. 4.70 is applicable only when there is no mutual influence of the reactions on their rate constants.

When the values of the rate constants are large, Eq. 4.70 transform into expression (4.63) for the diffusion-controlled regime.

In [50] the dependences found were applied to an analysis of the kinetics of the reduction of silicon from liquid silicates and slags by carbon in an Fe–C_{sat} melt, which takes place in a domain process during the production of ferroalloys and the like. The results obtained in the calculations are in satisfactory agreement with independent experimental data.

The main shortcoming of the method described is its complexity—it gives a solution in the form of a system of high-degree algebraic equations, which is not convenient for calculations.

In [51] a relatively simple method for analyzing the kinetics of reactions that take place between a metal and a slag was developed. The method is based on writing all the oxidation reactions in terms of the common reactant (FeO) and takes into account the mutual influence of the reactions and the diffusion of all the reactants in the metal and the slag.

The theoretical basis of the method consists of two assumptions:

- In a diffusion-controlled regime the concentration ratio on the interface for each reaction is close to the equilibrium value.
- The rate of transfer of the reactants to the interface or from it is proportional to the difference between the concentrations in the bulk and on the interface between the metallic and oxide melts.

The occurrence of each reaction of the type

$$1/m[\mathrm{E}_i] + (\mathrm{FeO}) = [\mathrm{Fe}] + 1/m\ (\mathrm{E}_i\mathrm{O}_m) \tag{4.71}$$

is accompanied by the consumption of a definite quantity of (FeO); therefore, their rates J_i are equal to part of the total diffusion flux of (FeO) that is consumed to oxidize impurity i:

$$J_{\mathrm{FeO}} = \sum_{i=1}^{n} J_i = \sum_{i=1}^{n} mJ_{\mathrm{E}_i}. \tag{4.72}$$

The expression for the rate of diffusion of (FeO) to the metal/slag interface [19]

$$J_{\mathrm{FeO}} = J_{\mathrm{FeO}}^{\mathrm{lim}}\left(1 - \frac{x}{(\mathrm{FeO})}\right) \tag{4.73}$$

and the ratio of the equilibrium concentrations of the reactants for reaction (4.71)

$$K_{\mathrm{E}_i} = \frac{(\mathrm{E}_i\mathrm{O}_m)_{\mathrm{eq}}^{1/m}}{(\mathrm{FeO})\cdot[\mathrm{E}_i]_{\mathrm{eq}}^{1/m}} \tag{4.74}$$

were used to obtain [51] expressions for calculating the rates of the individual reactions:

$$\left(1 + \frac{J}{mJ_{\mathrm{E}_i\mathrm{O}_m}^{\mathrm{lim}}}\right)^{1/m} = K_{P,\mathrm{E}_i}\frac{[\mathrm{E}_i]^{1/m}x}{(\mathrm{E}_i\mathrm{O}_m)^{1/m}}\left(1 - \frac{J}{mJ_{\mathrm{E}_i}^{\mathrm{lim}}}\right)^{1/m}. \tag{4.75}$$

Here x is the content of (FeO) on the reaction surface, wt.%; $[E_i]$ and (E_iO_m) are the current concentrations of the components in the metal and the slag, wt.%; $J = J_{FeO}$.

Thus, the method enables us to calculate the oxidation rates of all impurities as a function of only one argument, namely, the concentration of (FeO) on the interface.

Further development of the method [52] provided a way to take into account the kinetic hindrances for the individual reactions.

The problem with this method is the fact that the derivation of the equations for taking into account the kinetic hindrances requires a priori knowledge of the components for which they are significant. In addition, in some technological processes the rates of natural or forced convection are high enough to cause the interaction regime to approach the kinetic regime. In this case the use of this method is also not entirely proper.

In as much as the interaction between a metal and a slag has an electrochemical character, methods that take into account the conditions for electroneutrality of the phases are fairly promising.

One of the latest attempts to create a model description of the interaction in a metal–slag system based on consideration of the electrode process

$$O_i^{n+} + ne = R_i \tag{4.76}$$

was undertaken in [23].

By performing some simple transformations Zhancheng et al. [23] obtained the following equation for the rate of reaction (4.15):

$$J_i = n_i F(k_{j,i} C_{O_i}^b - k_{b,i} C_{R_i}^b) \cdot \exp(\beta_i^2 \tau) \cdot \mathrm{erfc}(\beta_i \sqrt{\tau}). \tag{4.77}$$

Here $C_{O_i}^b$ and $C_{R_i}^b$ are the concentrations of the oxidized and reduced forms of element i in the bulk of the respective phase; β_i is a parameter, which is found from the relation

$$\beta_i = \frac{k_{j,i}}{\sqrt{D_{O_i}}} + \frac{k_{b,i}}{\sqrt{D_{R_i}}}; \tag{4.78}$$

where D_{O_i} and D_{R_i} are the diffusion coefficients of the components in the slag and the metal, respectively; τ is the time; F is Faraday's constant; $k_{j,i}$ and $k_{b,i}$ are parameters for the forward and reverse reactions, respectively, which are defined by the following formulas:

$$k_{j,i} = k_{j,i}^0 \exp\left(\frac{\alpha_i n_i FE}{RT}\right), \tag{4.79}$$

$$k_{b,i} = k_{b,i}^0 \exp\left(-\frac{(1-\alpha_i) n_i FE}{RT}\right). \tag{4.80}$$

Here $k_{j,i}^0$ and $k_{b,i}^0$ are the rate constants of the forward and reverse reactions, respectively, α_i is the transfer coefficient, and E is the compromise or steady-state potential.

The following relation holds for a system of N electrode reactions:

$$\sum_{i=1}^{N} J_i = 0. \tag{4.81}$$

Finding the value of E presents the greatest difficulties in the case of a system in which several electrode process take place, because it depends on the current and equilibrium concentrations or, more precisely, on the activities of the components. For this reason, the method under discussion does not have an explicit solution. Therefore, it was suggested in [23] that the value of the compromise or steady-state potential should be found from the values of the Gibbs free energy of the system and that the rate constants of the reactions should be found from the value of the standard potential for reactions (4.76) using published experimental data and the equation

$$E_i^0 = \frac{RT}{n_i F} \ln \frac{k_{j,i}^0}{k_{b,i}^0}. \tag{4.82}$$

Over the course of many years, studies aimed at creating methods for analyzing reactions in multicomponent systems on the basis of the laws of electrochemical kinetics have been successfully carried out by members of the Ural school of physical chemists, which was created under the guidance of Esin.

In [53, 54] a method for analytically describing the kinetics of simultaneously occurring reactions of type (4.76) was developed using the equation for the discharge rate of an ion on a metal/electrolyte boundary in the mixed regime [55]

$$i_i = i_i^0 \left[\left(1 - \frac{i_i}{i_{O_i}^{\lim}}\right) e^{\frac{\alpha_i n_i F}{RT}\eta_i} - \left(1 + \frac{i_i}{i_{R_i}^{\lim}}\right) e^{-\frac{(1-\alpha_i) n_i F}{RT}\eta_i} \right]. \tag{4.83}$$

Using the notation

$$x_i = \exp\left(-\frac{n_i F}{RT}\eta_i\right), \tag{4.84}$$

we obtain [56, 57]

$$i_i = \frac{x_i - 1}{\frac{x_i^{1-\alpha_i}}{i_i^0} + \frac{x_i}{i_{O_i}^{\lim}} + \frac{1}{i_{R_i}^{\lim}}}. \tag{4.85}$$

Here i_i^0 is the ion-exchange current on the metal/slag interface; η_i is the polarization; $i_{O_i}^{\lim}$ and $i_{R_i}^{\lim}$ are the limiting diffusion currents of the oxidized and reduced forms of the reactant in the respective phases.

To solve Eq. 4.85 we must specify a reference point, relative to which the polarization values will be determined.

In [53] the equilibrium potential of carbon was chosen as the reference point, and in [54] the equilibrium potential of iron was chosen. The latter is convenient for melts that are not saturated with carbon. At high concentrations of iron in the metal, its activity is practically constant, and the equilibrium potential depends mainly on the activity of Fe^{2+} in the slag at a constant temperature.

Since the derivation of the calculation equations in [53, 54] are similar, we present the concise mathematical manipulations made in [54].

For reaction (4.76) the polarizations of the cathodic (η_c) and anodic (η_a) steps are related by the following relation:

$$\eta_c - \eta_a = \frac{RT}{2F} \ln K_{P(4.71)} \frac{a_{E_i}^{1/m} a_{FeO}}{a_{Fe} a_{E_iO_m}^{1/m}} = \frac{RT}{2F} \ln \frac{K_{P(4.71)}}{\prod a_i}. \tag{4.86}$$

After several transformations, the following equation for calculating the rate of passage of any element through the metal/slag interface was obtained in [54]:

$$i_i = \frac{y^{n_i} \left(\dfrac{\prod a_i}{K_{P(4.71)}} \right)^{n_i/2} - 1}{\dfrac{y^{n_i(1-\alpha_i)} \left(\dfrac{\prod a_i}{K_{P(4.71)}} \right)^{n_i(1-\alpha_i)/2}}{i_i^0} + \dfrac{y^{n_i} \left(\dfrac{\prod a_i}{K_{P(4.71)}} \right)^{n_i/2}}{i_{O_i}^{\lim}} + \dfrac{1}{i_{R_i}^{\lim}}}. \tag{4.87}$$

The equation takes into account the mutual influence of the reactions, the diffusion of the particles in the respective phases, and the slowness of the discharge of the ions on the interface and provides a way to find the rates of the interaction in all the regimes of the process: kinetic, diffusion-controlled, and mixed.

Substituting (4.87) into expression (4.81), we obtain an equation containing one unknown quantity, namely, y. After solving this equation for y, the rates and the directions of passage of each element through the metal/slag boundary can be found with the aid of expression (4.87). This method also enables us to take into account the multiple-step character of the discharge of some multiply charged ions, such as Si^{4+}, Cr^{3+}, and P^{5+}.

This approach is very promising as a basis for mathematical models of interaction processes that occur in metallurgical plants. In addition, its further development is possible, for example, by taking into account the non-steady-state conditions for the occurrence of the oxidation–reduction processes, which are caused by various factors: short interaction times, non-steady-state diffusion, application of an external current of various nature, etc. Similar conditions can arise in some high-temperature technological processes (droplet refining of a metal in a slag layer, alternating-current (ac) electroslag technologies, etc.). In addition, this method can also be used to theoretically analyze non-steady-state electrochemical phenomena when the Faradaic current is caused by the simultaneous occurrence of more than two electrode processes.

4.3.2 Development of a Method for Analyzing the Kinetics of Simultaneously Occurring Electrode Reactions Under Non-Steady-State Conditions

The methods described in the preceding section are based on the use of the steady-state diffusion equation

$$V = K^d(C - C') = \frac{D}{\delta}(C - C'), \tag{4.88}$$

where C and C' are the concentrations of the reactant in the bulk and on the interface, respectively (mole/m^3), K^d is the mass-transfer constant (m/s), D is the diffusion coefficient (m^2/s), and δ is the thickness of the diffusion layer (m).

However, in some cases the time of formation of constant diffusion layers is greater than the time of variation of the external action, that is, the mass-transfer rates cannot be calculated from steady-state diffusion equation (4.88). Such conditions can arise in high-temperature technological processes when the time of interaction of the metal droplet and the ionic melt is short or when the variation of the external current is sufficiently fast, as in the case of a pulsed or ac current. The analysis is also complicated by the fact that in the initial state the metal and the electrolyte are often not at equilibrium and redox reactions also occur in the absence of an external current, significantly altering the concentrations of the components near the interface.

In this section we present a method for quantitatively describing the kinetics of an arbitrary number of electrode processes when they occur simultaneously on the metal/electrolyte boundary with allowance for the slowness of the discharge and diffusion of all particles under non-steady-state conditions for any form of the time dependence of the external current [56].

For electrode process (4.76) the rate (A/m^2) is described in the general case by the well-known equation [57, 58]

$$i_i = i_i^0 \left[\frac{C'_{O_i}}{C_{O_i}} e^{\frac{\alpha nF}{RT}\eta_i} - \frac{C'_{R_i}}{C_{R_i}} e^{-\frac{(1-\alpha)nF}{RT}\eta_i} \right], \tag{4.89}$$

where $i_i^0 = k_i^0 C_{O_i}^{1-\alpha} C_{R_i}^{\alpha}$ is the exchange current, C'_i and C_i are the concentrations of the oxidized (O_i^{n+}) and reduced (R_i) forms of the particles on the metal/electrolyte interface (beyond the absorption layer) and in the bulk of the phases, respectively, η_i is the polarization (V), and F is Faraday's constant.

The diffusion of each particle is described by the equation

$$\frac{\partial C}{\partial \tau} = D\frac{\partial^2 C}{\partial X^2}. \tag{4.90}$$

If i electrode processes occur simultaneously, they are thermodynamically and kinetically interrelated. This relationship is stipulated by the single value of the

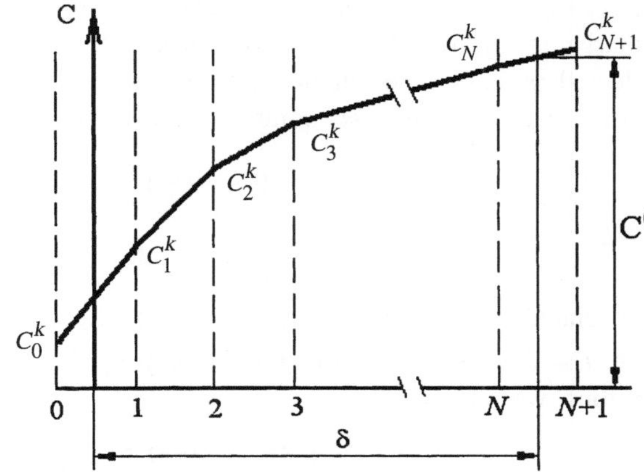

Fig. 4.5 Calculation scheme. Here δ is the thickness of the diffusion layer, C' is the surface concentration of the substance, and C^0 is the bulk concentration of the substance

electrochemical potential of the metal, that is, by interrelationship between the boundary conditions for C'_{O_i} and C'_{R_i} and the condition of electroneutrality

$$\sum i - i(\tau) = 0, \tag{4.91}$$

where $i(\tau)$ is the density of the external current (A/m^2) (a cathodic current is assumed to be positive).

Thus, in the present case the following system of equations of type (4.90) must be solved:

$$\begin{cases} \dfrac{\partial C_{O_i}}{\partial \tau} = D_{O_i} \dfrac{\partial^2 C_{O_i}}{\partial X^2}, & i = 1, 2, \ldots, i \\ \dfrac{\partial C_{R_i}}{\partial \tau} = D_{R_i} \dfrac{\partial^2 C_{R_i}}{\partial X^2}, & i = 1, 2, \ldots, i. \end{cases} \tag{4.92}$$

The numerical finite-difference method was used. For the explicit difference scheme, Eq. 4.90 takes the form [59]

$$C_n^{k+1} = C_n^k + \frac{D\Delta\tau}{\Delta X^2}\left(C_{n+1}^k + C_{n-1}^k - 2C_n^k\right), \tag{4.93}$$

where $k = 0, 1, 2, \ldots$ and $n = 1, 2, \ldots, N$ are the cell indices with respect to the time and the coordinate, and $\Delta\tau$ and ΔX are the dimensions of the cells.

The calculation scheme is presented in Fig. 4.5.

It is seen from Eq. 4.93 that to find the concentrations at any moment in time, beginning from the first ($k = 0$), C_0^k must be determined in the first and subsequent cells ($n = 1, 2, \ldots, N$). Using several relatively simple transformations, we reduce the solution of the system of equations to the solution of one equation that enables us to take into account an arbitrary number of electrode reactions.

Let us examine the essence of the proposed method in the example of an iron-based-melt/oxide-electrolyte system. All the possible chemical reactions involving the elements E_i that are dissolved in the iron can be described by a set of reactions of type (4.71) with a common reactant, for example FeO [19]. We express the polarization η_i for all the electrode processes in terms of the polarization of one selected process. The anodic (η_a) and cathodic ($\eta_c = \eta_{Fe}$) polarizations for the partial electrode processes comprising reaction (4.71)

$$[E_i] = (E_i)^{2m+} + 2me, \tag{4.94}$$

$$(Fe)^{2+} + 2e = [Fe] \tag{4.95}$$

are related by Eq. 4.86 [19]. Substituting (4.86) into (4.89), taking into account that $n = 2m$, using the notation $X = \exp\left(\frac{F\eta_{Fe}}{RT}\right)$, and going over to mass concentrations, we obtain

$$\begin{aligned} i = k_i^0 \Big[& (E_iO_m)' X^{-2\alpha m} K_{P(4.71)}^{-\alpha m} (FeO)^{-\alpha m} \\ & - [E_i]' X^{2(1-\alpha)m} K_{P(4.71)}^{(1-\alpha)m} (FeO)^{(1-\alpha)m} \Big], \end{aligned} \tag{4.96}$$

where k_i^0 is the constant in the equation $i_i^0 = k_i^0 (E_iO_m)^{1-\alpha} [E_i]^{\alpha}$ and $K_{P(4.71)}$ is the equilibrium ratio between the mass concentrations for reaction (4.71). For iron $K_P/\Pi = 1$ [19]; therefore, neglecting its diffusional hindrances in the metal because of the high concentrations, we obtain

$$i_{Fe} = k_{Fe}^0 [Fe]^{\alpha} \left[(FeO)' (FeO)^{-\alpha} X^{-2\alpha} - (FeO)^{(1-\alpha)} X^{2(1-\alpha)} \right]. \tag{4.97}$$

We introduce the notations

$$(FeO) = F_n^k, \quad (E_iO_m) = Q_{i,n}^k, \quad [E_i] = R_{i,n}^k, \quad [Fe] = Fe,$$

$$\frac{D_{Fe^{2+}} \Delta\tau}{\Delta X_{Fe^{2+}}^2} = A_F, \quad \frac{D_{E_i^{2m+}} \Delta\tau}{\Delta X_{E_i^{2m+}}^2} = A_{O_i}, \quad \frac{D_{E_i} \Delta\tau}{\Delta X_{E_i}^2} = A_{R_i}, \tag{4.98}$$

$$\frac{M_{FeO} \Delta x_{Fe^{2+}} 100}{\rho_{slag} 2F2D_{Fe^{2+}}} = B_F, \quad \frac{M_{E_iO_m} \Delta x_{E_i^{2m+}} 100}{\rho_{slag} n_i F2D_{E_i^{2m+}}} = B_{O_i}, \quad \frac{M_{E_i} \Delta x_{E_i} 100}{\rho_{slag} n_i F2D_{E_i}} = B_{R_i},$$

where M_i is the molar weight of component i, ρ_{slag} and ρ_{Me} are the density of the ionic melt and the metal, respectively (kg/m^3), and F = $(FeO)^0$ is the concentration of ferrous oxide in the bulk of the melt.

After several transformations we find

$$i_i^k = \frac{Q_{i,1}^k - R_{i,1}^k X^{2m} K_{P(4.71)}^m F^m}{\frac{X^{2\alpha m} K_{P(4.71)}^{\alpha m} F^{\alpha m}}{k_i^0} + B_{R_i} X^{2m} K_{P(4.71)}^m F^m + B_{O_i}}, \tag{4.99}$$

$$i_F^k = \frac{F_1^k - Fx^2}{\frac{F^{\alpha} X^{2\alpha}}{k_F^0 Fe^{\alpha}} + B_F}. \tag{4.100}$$

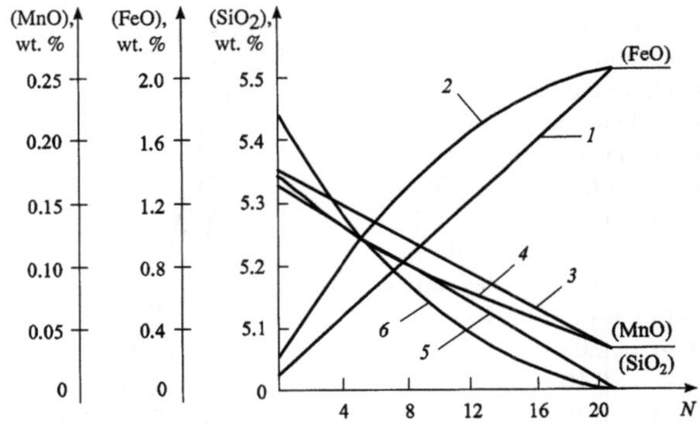

Fig. 4.6 Distribution of the concentrations of $\underline{(FeO)}$, $\underline{(MnO)}$, and $\underline{(SiO_2)}$ in a slag within the diffusion layer: *1*, *3*, *5*—$\tau = 1$ s; *2*, *4*, *6*—$\tau = 0.3$ s

Substituting i_i^k and i_F^k into electroneutrality Eq. 4.91, which takes the form

$$\sum i_i^k + i_F^k - i(\tau) = 0 \tag{4.101}$$

in the present case, we obtain an equation containing only the one unknown X. Solving this equation enables us to find i_i^k and i_F^k followed by the concentrations of all the elements in the hypothetical layer. Then we can calculate the compositions of the phases for $n = 1, 2, \ldots, N$ from Eq. 4.93.

From the stability conditions of the explicit difference scheme [59]

$$\frac{D\Delta\tau}{\Delta x^2} \leq A \quad \text{and} \quad A_{cr} = 1/4 \tag{4.102}$$

it follows that different values of Δx and N should be chosen for different particles. For the sake of convenience in performing calculations, the same minimum value of Δx from condition (4.102) can be taken (which gives different N_i) or the same maximum value of N can be taken (which corresponds to different Δx_i).

To evaluate the possibilities of this method, as well as the reliability of the results obtained, test calculations were performed, and the data obtained were compared with the existing literature data obtained by independent methods.

First, calculations for conditions that are typical of electroslag remelting were performed. The composition of the metal was (wt.%): Fe, 67; C, 1.5; Si, 0.53; Mn, 1.56; V, 1.12; Cr, 18; Ni, 10; Al, 0.01; P, 0.026; S, 0.012; O, 0.005. The composition of the slag was (wt.%): FeO, 2.0–10.0; SiO_2, 5.0; MnO, 0.03; $A1_2 0_3$, 5.0; P_2O_5, 0.03; S, 0.06; CaF_2, the rest. The temperature was 1873 K.

The results of the calculations were compared with the data presented in [60]. The comparison reveals that the values of the rates and the distribution of the concentrations of the components within the diffusion layer are practically equal for all the elements. The only exceptions are carbon and silicon, apparently due to the more exact way in which the kinetic hindrances are taken into account in the proposed method.

It was shown that the times needed for the rates to achieve a steady state, as well as for the formation of a constant concentration gradient, are ~ 1 s and are also in agreement with the available data [19, 57] (Fig. 4.6).

Fig. 4.7 Variation of the concentration of manganese oxide in the slag when a pulsed current passes through the metal–oxide melt system (δ is the thickness of the diffusion layer): 1—$\tau = 0$ (before the current pulse); 2—$\tau = \Delta\tau$ (after the current pulse) 3—$\tau = 0.04$ s; 4—$\tau = 0.08$ s; 5—$\tau = 0.10$ s; 6—$\tau = 0.30$ s; 7—$\tau = 1.10$ s

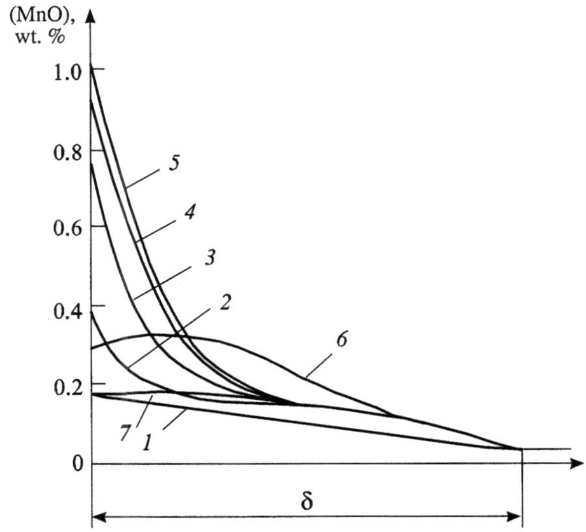

The possibility of taking into account the influence of the reaction rate and the distribution of the elements in the diffusion layer for a pulsed current was analyzed on the same system for a concentration of FeO equal to 2.0 wt.%. The parameters of the current were as follows:

- signal shape, rectangular;
- pulse amplitude, 3 A/cm^2;
- $\tau_{pulse} = 0.1$ s;
- on–off ratio of the pulses, 10.

The results of the calculations for (MnO) are shown in Fig. 4.7.

It is seen that the current influences both the absolute values of the concentration of (MnO) within the diffusion layer and the course of the curves. The dependences obtained graphically demonstrate that under such conditions Eq. 4.88 should not be used to calculate the reaction rates.

Figure 4.8 shows the variation of the surface concentrations of (MnO) and (FeO) with time. The character of the dependences obtained is similar to that of the dependences presented in [61] and corresponds to the form of the function of a transient process under a given perturbation [62]. The reaction rates also vary nonmonotonically, and for manganese a change in the sign of the reaction rate is observed, that is, it begins to be reduced, after a current pulse ends. The process becomes oxidative again only at the end of the pause, but the value of the reaction rate (-3.9×10^{-8} mol/cm^2/s) differs significantly from the value obtained when $I_{in} = 0$ (-1.5×10^{-7} mol/cm^2/s).

The possibility of using the method under discussion to evaluate the kinetic features of electrode processes under steady-state conditions was analyzed in the example of an Fe–O/Al_2O_3–CaO–MgO–FeO metal–slag system.

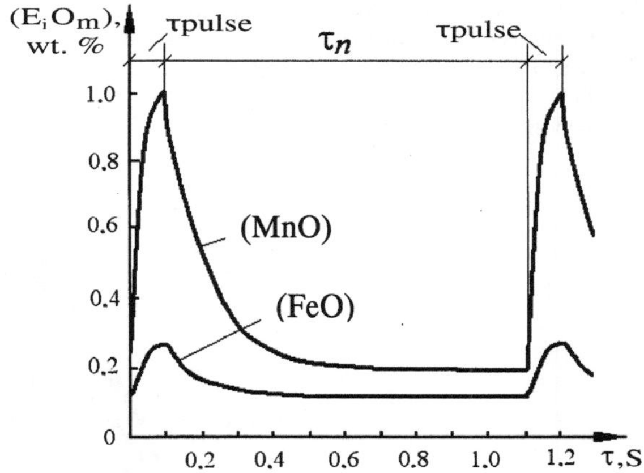

Fig. 4.8 Variation of the concentrations of (MnO) and (FeO) on the metal/oxide-melt interface with time upon passage of a pulsed current

The following empirical relations were presented in [19] for the limiting diffusion currents of ferrous oxide and oxygen, respectively:

$$i^{\lim}_{(\mathrm{FeO})} = 0.08(\mathrm{FeO}), \tag{4.103}$$

$$i^{\lim}_{[\mathrm{O}]} = 16\ [\mathrm{O}]. \tag{4.104}$$

By substituting various values of$i$ into the equation [19]

$$\eta = \frac{RT}{nF} \ln\left(1 + \frac{i}{i^{\lim}}\right), \tag{4.105}$$

it can be used to reconstruct part of the polarization curve and to compare this curve with the calculated polarization curve obtained by the method described here.

Calculations were preformed for the following conditions: $T = 1823$ K, (FeO) = 4 wt.%, [O] = 0.024 wt.%.

The results of the calculations show that equilibrium is established in the system under consideration after a time corresponding to the holding time of the melts in contact before the electrochemical experiment and that this equilibrium is characterized by the absence of concentration gradients for any of the reactants within the diffusion layer and by values of the reaction rates that are smaller than 1×10^{-4} A/cm^2.

Figure 4.9 presents the results obtained using the method described here, as well as results calculated from Eq. 4.105. The satisfactory agreement between them may be noted.

Thus, the method developed enables us to computationally obtain steady-state polarization curves in a metal–slag system.

The test calculations performed showed that the method described provides a way to analyze the kinetics of reactions in a metallic-melt/oxide-electrolyte system with sufficient accuracy under both steady-state and non-steady-state conditions when an external current of diverse nature is applied.

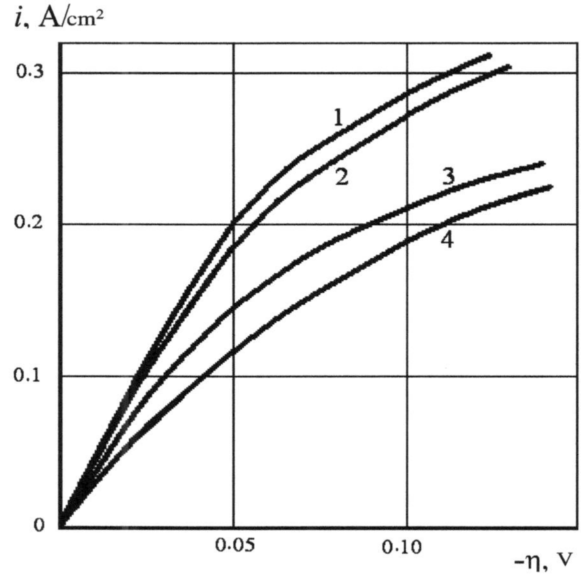

Fig. 4.9 Polarization curves for oxygen and iron oxide at $T = 1{,}823$ K: 1, 2—[O]; 3, 4—(FeO); 1, 4—our data; 2, 3—according to Eq. 4.35

4.3.3 Procedure for Evaluating the Kinetic Laws Governing Electrochemical Processes on a Metal/Polymeric-Oxide-Melt Boundary

Slag melts that have oxides of silicon, aluminum, and boron in their composition are known to contain polymerized anions [63]. It has been shown [64] that such anions influence the kinetics of processes to a certain extent even without allowance for the electrochemical laws. A quantitative description of electrochemical phenomena has been thoroughly developed in theoretical metallurgy, but the laws governing the electrochemical kinetics for polymerized ionic melts have not been discussed in the literature.

We will examine this question in the example of the electrochemical oxidation of silicon [65].

The process of the anodic oxidation of silicon on an interface can be represented by the set of interrelated reactions

$$[\mathrm{Si}] + 4(\mathrm{O})^{2-} = (\mathrm{SiO_4})^{4-} + 4e \tag{4.106}$$

$$2[\mathrm{Si}] + 7(\mathrm{O})^{2-} = (\mathrm{Si_2O_7})^{6-} + 8e \tag{4.107}$$

$$3[\mathrm{Si}] + 10(\mathrm{O})^{2-} = (\mathrm{Si_3O_{10}})^{8-} + 12e \tag{4.108}$$

$$3[\mathrm{Si}] + 9(\mathrm{O})^{2-} = (\mathrm{Si_3O_9})^{6-} + 12e \tag{4.109}$$

etc., or in the general form

$$i[\mathrm{Si}] + (3i + 1 - c)(\mathrm{O})^{2-} = (\mathrm{Si_iO}_{3i+1-c})^{2(i+1-c)-} + 4ie. \tag{4.110}$$

We describe the method for calculating the electrode polarization when reactions (4.106–4.110) occur simultaneously, that is, with allowance for all the possible (apart from the isomeric) species of silicon–oxygen anions. We confine ourselves to the diffusion-controlled regime of electrode reactions (4.106–4.110). In the steady-state regime of reaction (4.110), the diffusion fluxes of the reactants are related by the expression

$$I_{i,c} = \frac{I'_{\mathrm{Si}}}{i} = \frac{I'_{\mathrm{O^{2-}}}}{3i + 1 - c}. \tag{4.111}$$

Here I'_{Si} and $I'_{\mathrm{O^{2-}}}$ are, respectively, the fractions of the diffusion fluxes of Si and $\mathrm{O^{2-}}$ that provide for the occurrence of reaction (4.110) with the rate of formation $I_{i,c}$ of $\mathrm{Si}_i\mathrm{O}_{3i+1-c}^{2(i+1-c)-}$ anions. Summation of the partial fluxes for all the reactions (4.106–4.110) enables us to calculate the total diffusion fluxes of silicon and "free" oxygen ions:

$$I_{\mathrm{Si}} = \sum_{c=0}^{\infty} \sum_{i=c+1}^{\infty} I'_{\mathrm{Si}} = K^d_{[\mathrm{Si}]}\left(C_{[\mathrm{Si}]} - C'_{[\mathrm{Si}]}\right), \tag{4.112}$$

$$I_{\mathrm{O^{2-}}} = \sum_{c=0}^{\infty} \sum_{i=c+1}^{\infty} I'_{\mathrm{O^{2-}}} = K^d_{(\mathrm{O^{2-}})}\left(C_{(\mathrm{O^{2-}})} - C'_{(\mathrm{O^{2-}})}\right). \tag{4.113}$$

In Eqs. 4.112 and 4.113 K_i^d is the mass-transfer constant, m/s; C_i and C'_i are the concentrations of the components in the bulk of the phase and on the metal/slag interface, respectively, mole/m^3.

During anodic dissolution of the metal under the diffusion-controlled regime, the near-boundary electrolyte layer is in a state that is infinitely close to equilibrium with the metal. Therefore, the polarization η appearing when reactions (4.106–4.110) occur can be written in the general form

$$\eta = \frac{RT}{4iF} \ln \frac{a'_{i,c} a^i_{\mathrm{Si}} a^{3i+1-c}_{\mathrm{O^{2-}}}}{a_{i,c} (a^i_{\mathrm{Si}})' (a^{3i+1-c}_{\mathrm{O^{2-}}})'}, \tag{4.114}$$

where a and a' are the activities of the components in the bulk of the phase and on the interface, respectively.

Treating the slag melt as an ideal ionic solution, in which the activities of the ions are equal to their ion fractions N_i, we can write equation (4.114) in a form that is more convenient for further mathematical manipulations:

$$\eta = \frac{RT}{4iF} \ln \frac{\sum C^{-} C'_{i,c} a^i_{\mathrm{Si}} N^{3i+1-c}_{\mathrm{O^{2-}}}}{\sum (C^{-})' C_{i,c} (a^i_{\mathrm{Si}})' (N^{3i+1-c}_{\mathrm{O^{2-}}})'}, \tag{4.115}$$

where ΣC^- and $\Sigma (C^-)'$ are the sums of the concentrations of all the anions in the bulk of the slag and on the surface, respectively (mole/m^3).

After several transformations we obtain

$$I_{\mathrm{Si}} = K_{i,c}^{d}\left(\sum (C^{-})' \frac{N_{\eta} - x_{\eta}^{3}}{(1 - x_{\eta})^{2}(1 - y_{\eta})^{2}} - \sum C^{-} \frac{N - x^{3}}{(1 - x)^{2}(1 - y)^{2}}\right). \quad (4.116)$$

Here

$$N_{\eta} = \frac{N_{1,4} a_{\mathrm{Si}}' (N_{\mathrm{O}^{2-}}')^{4}}{a_{\mathrm{Si}} (N_{\mathrm{O}^{2-}})^{4}} e^{\frac{4iF}{RT}}, \quad (4.117)$$

$$x_{\eta} = K_{2,0} N_{1,4} \frac{a_{\mathrm{Si}}' (N_{\mathrm{O}^{2-}}')^{2}}{a_{\mathrm{Si}} (N_{\mathrm{O}^{2-}})^{4}} e^{\frac{4iF}{RT}}, \quad (4.118)$$

$$y_{\eta} = K_{2,0}^{2} N_{1,4} \frac{a_{\mathrm{Si}}' (N_{\mathrm{O}^{2-}}')^{2}}{a_{Si} (N_{\mathrm{O}^{2-}})^{4}} e^{\frac{4iF}{RT}}, \quad (4.119)$$

$$x = K_{2,0} \frac{N_{1,4}}{N_{\mathrm{O}^{2-}}}, \quad y = K_{2,0}^{2} \frac{N_{1,4}}{(N_{\mathrm{O}^{2-}})^{2}}. \quad (4.120)$$

As we see, Eq. 4.116 contains unknown quantities that determine the concentrations of the components in the surface layer. To find them we use equations (4.112) and (4.113) along with the following equation:

$$I_{\mathrm{Si}} = K_{i,c}^{d}\left(\sum_{c=0}^{\infty} \sum_{i=c+1}^{\infty} iC_{i,c}' - \sum_{c=0}^{\infty} \sum_{i=c+1}^{\infty} iC_{i,c}\right). \quad (4.121)$$

After some relatively simple transformations, we find

$$a_{\mathrm{Si}}' = \gamma_{\mathrm{Si}}' \frac{C_{\mathrm{Si}}}{\sum C'}\left(1 - \frac{I_{\mathrm{Si}}}{C_{\mathrm{Si}} K_{\mathrm{Si}}^{d}}\right), \quad (4.122)$$

$$a_{\mathrm{Si}} = \gamma_{\mathrm{Si}} \frac{C_{\mathrm{Si}}}{\sum C}, \quad (4.123)$$

$$\frac{I_{\mathrm{Si}}}{K_{i,c}^{d}} = \frac{C_{1,4} - (\alpha')^{3} \sum (C^{-})'}{(1 - \alpha')^{2}(1 - \beta')^{2}}, \quad (4.124)$$

$$K_{\mathrm{O}^{2-}}^{d}(C_{\mathrm{O}^{2-}} - C_{\mathrm{O}^{2-}}') = 3I_{\mathrm{Si}} + K_{i,c}^{d}\left(\begin{array}{c} \dfrac{C_{1,4}'(1 - \beta')^{2} - (\alpha')^{2} \sum (C^{-})'}{(1 - \alpha')^{2}(1 - \beta')^{2}} \\ - \dfrac{C_{1,4}(1 - y)^{2} - x^{2} \sum C^{-}}{(1 - y)^{2}(1 - x)^{2}} \end{array}\right). \quad (4.125)$$

Here

$$\alpha' = K_{2,0} \frac{C_{1,4}'}{C_{\mathrm{O}^{2-}}'}; \quad \beta' = K_{2,0}^{2} \frac{C_{1,4}' \sum (C^{-})'}{(C_{\mathrm{O}^{2-}}')^{2}}. \quad (4.126)$$

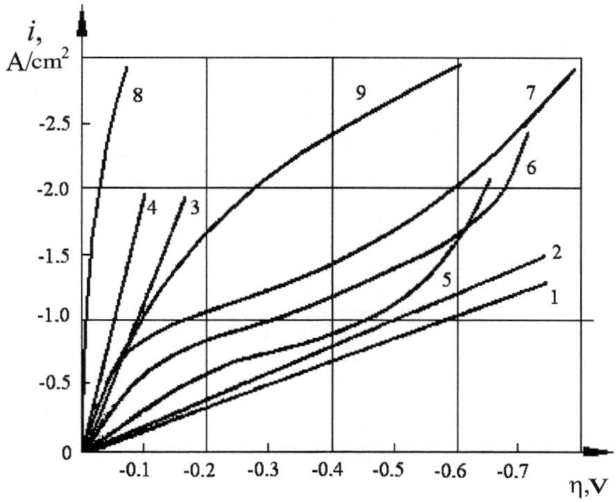

Fig. 4.10 Cathodic polarization curves in a CaO–SiO_2/Fe–Si system: 1–4—experiment [66] for the following system (wt.%): slag—35 CaO, 10 MgO, and 55 SiO_2; metal—77.5 Fe and 22.5 Si; 1—$T = 1753$ K; 2—$T = 1773$ K; 3—$T = 1873$ K; 4—$T = 1893$ K; 5–9—calculation for $T = 1873$ K; 5–7—according to the method described; 8—according to Eq. 4.63; 9—according to Eq. 4.64; 5—slag containing (wt.%) 55 CaO and 45 SiO_2; 6—slag containing (wt.%) 50 CaO and 50 SiO_2; 7–9—slag containing (wt.%) 45 CaO and 55 SiO_2

System of Eqs. 4.116–4.120 and 4.122–4.126 provide a way to calculate the values of the polarization and the concentration of all the ions on the reaction surface for assigned oxidation or reduction rates of silicon.

Figures 4.10, 4.11, 4.12 and 4.13 present the results of calculations based on the proposed method at 1873 K for the Fe–Si/CaO–SiO_2 system. The calculated cathodic branches of the polarization curves (Fig. 4.10, curve 7) closely agree with the experimental curves [66] (Fig. 4.10, curve 3). For comparison, the figure also shows polarization curves that were calculated without taking into account the polymerization using the ordinary equation for concentration polarization (Fig. 4.10, curve 8)

$$\eta = \frac{RT}{4F}\ln\left(1 - \frac{i}{i^{\mathrm{lim}}}\right), \tag{4.127}$$

and with allowance for the change in the activity coefficient of SiO_2 using the following equation (Fig. 4.10, curve 9)

$$\eta = \frac{RT}{4F}\ln\left(\frac{\gamma'_{\mathrm{SiO_2}}}{\gamma_{\mathrm{SiO_2}}}\left(1 - \frac{i}{i^{\mathrm{lim}}}\right)\right). \tag{4.128}$$

The activity coefficients of silicon in the slag were calculated from Kozheurov's equation for the CaO–SiO_2 system.

As the calculations show, taking into account the change in $\gamma_{\mathrm{SiO_2}}$ and the polarization leads to displacement of the polarization curve toward higher values of η.

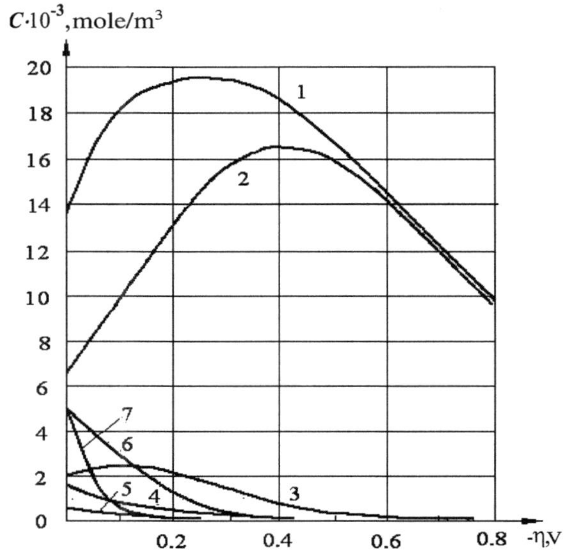

Fig. 4.11 Dependence of the concentrations of various anions at the metal/slag interface on the polarization for 55% CaO : 1 – $-\sum iC_{i,0}$; 2 – $-SiO_4^{4-}$; 3 – $-Si_2O_7^{6-}$; 4 – $-Si_2O_6^{4-}$; 5 – $-Si_3O_9^{6-}$; 6 – $-\sum iC_{i,1}$; 7 – $-\sum iC_{i,c>1}$

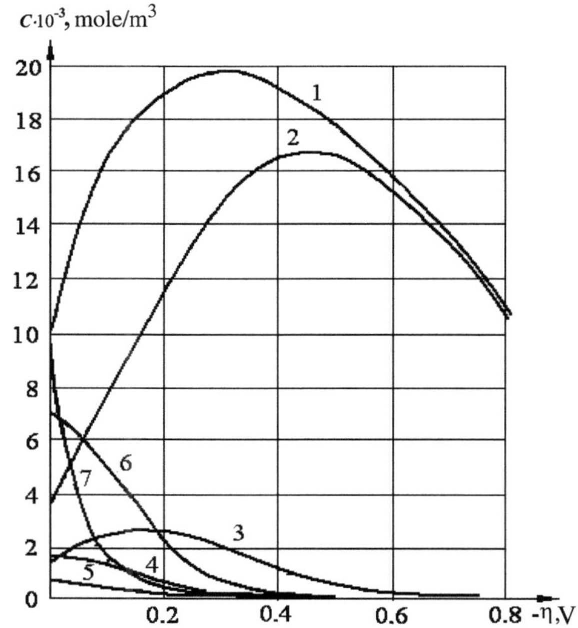

Fig. 4.12 Dependence of the concentrations of various anions at the metal/slag interface on the polarization for 50% CaO : 1 – $-\sum iC_{i,0}$; 2 – $-SiO_4^{4-}$; 3 – $-Si_2O_7^{6-}$; 4 – $-Si_2O_6^{4-}$; 5 – $-Si_3O_9^{6-}$; 6 – $-\sum iC_{i,1}$; 7 – $-\sum iC_{i,c>1}$

The large values of η for polymerized melts are attributed to counter fluxes of various silicon–oxygen anions. For this reason, a stronger negative shift of the electrochemical potential is needed to create the total diffusion flux. A similar conclusion was also reached in [64]. The calculation showed that at specific values of the polarization (for example, $\eta = -0.618$ V) the cathodic reduction of silicon occurs

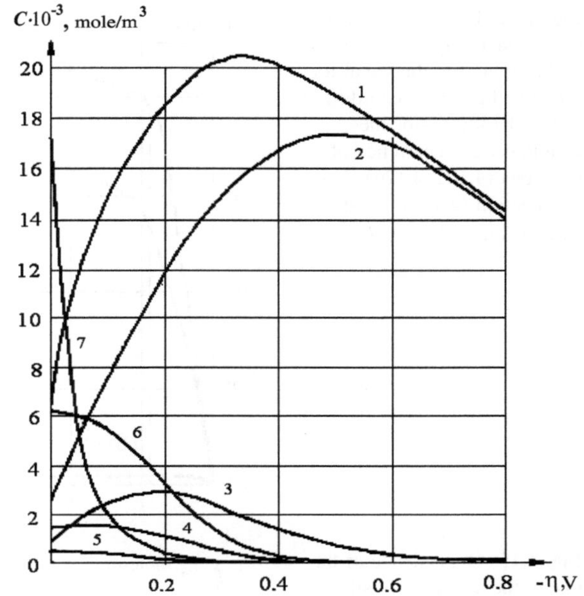

Fig. 4.13 Dependence of the concentrations of various anions at the metal/slag interface on the polarization for 45% CaO : 1 – – $\sum iC_{i,0}$; 2 – – SiO_4^{4-}; 3 – –$Si_2O_7^{6-}$; 4 – $Si_2O_6^{4-}$; 5 – –$Si_3\,O_9^{6-}$; 6 – – $\sum iC_{i,1}$; 7 – – $\sum iC_{i,c>1}$

with a higher rate from melts with a smaller concentration of SiO_2. This is associated with the fact that at the polarization value indicated the chainlike anions in a melt consisting of 50 mol% CaO and 50 mol% SiO_2 transport silicon from the surface into the bulk (Fig. 4.12), thereby slowing the rate of the reaction. Conversely, in a melt consisting of 55 mol% CaO and 45 mol% SiO_2 the silicon atoms are not transported by such anions (Fig. 4.11). However, other complex species transport particles to the reaction surface. It should be noted that the ionic composition of the bulk of the melt corresponds to $\eta = 0$.

Taking into account that for the conditions of the calculation and the experiments in [66] the limiting diffusion currents of SiO_2 in the slag are approximately an order of magnitude smaller than the silicon ion-exchange currents [67], we can expect that the reaction regime of the cathodic deposition of silicon is close to the diffusion-controlled regime. Therefore, the use of the method described above is justified.

Let us consider the application of this method to the oxidation of silicon by an iron-containing slag according to the reaction

$$[Si] + 2(FeO) = (SiO_2) + [Fe]. \qquad (4.129)$$

This process can be represented by a set of interrelated reactions of type (4.106–4.109) that take into account the actual structural units in a polymerized silicate melt. We write them in the general form

$$i[Si] + (3i + 1 - c)(O)^{2-} + 2i\left(Fe^{2+}\right) = \left(Si_iO_{3i+1-c}\right)^{2(i+1-c)^-} + 2i[Fe]. \qquad (4.130)$$

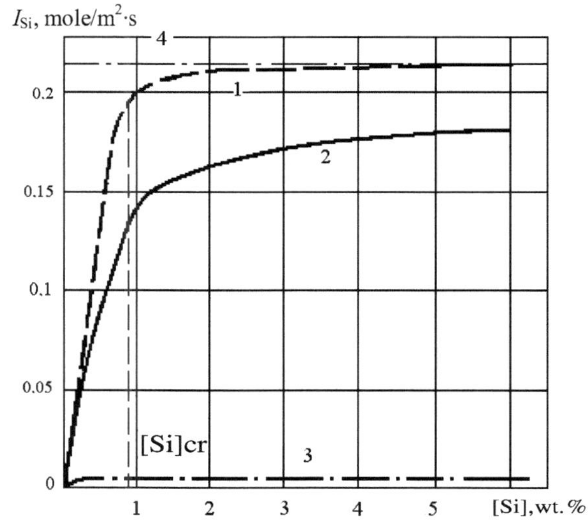

Fig. 4.14 Dependence of the rate of the oxidation of silicon on its content in the metal for 30 wt.% FeO: 1—according to Eq. 4.138; 2—for the simultaneous occurrence of reactions (4.122–4.126); 3—for reaction (4.122); 4—the straight line $I^{\text{lim}} = 0.5I^{\text{lim}}_{\text{Fe}^{2+}}$

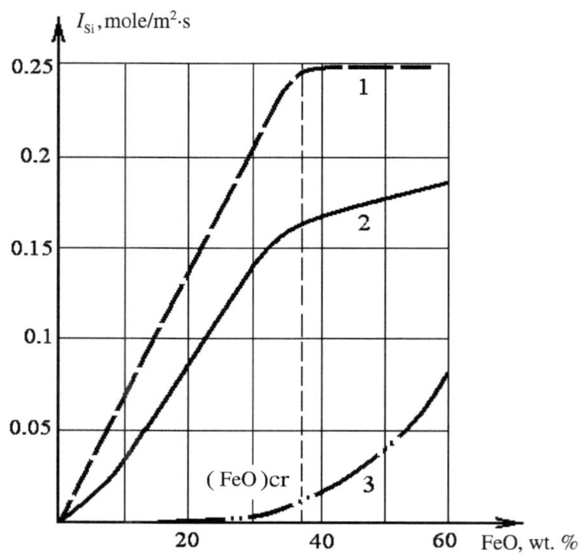

Fig. 4.15 Dependence of the rate of the oxidation of silicon on the concentration of FeO in the oxide melt for 1 wt.% [Si]: 1—according to Eq. 4.138; 2—for the simultaneous occurrence of reactions (4.122–4.126); 3—for reaction (4.122)

Accordingly, the equation for the diffusion flux of iron must be presented:

$$I_{\text{Fe}^{2+}} = \sum_{c=0}^{\infty} \sum_{i=c+1}^{\infty} I'_{\text{Fe}^{2+}} = K^{d}_{(\text{Fe}^{2+})}\left(C_{(\text{Fe}^{2+})} - C'_{(\text{Fe}^{2+})}\right). \qquad (4.131)$$

After several transformations that take into account the presence of iron oxides in the slag, system of Eqs. 4.116–4.120 and 4.122–4.126 provided a way to obtain results for $FeO–CaO–SiO_2$ and Fe–Si melts at 1873 K. The data from these calculations are presented in Figs. 4.14, 4.15 and 4.16.

The oxidation rates of silicon in Figs. 4.14 and 4.15 are presented for a typical value of the mass-transfer constant equal to 1 $s^{-0.5}$.

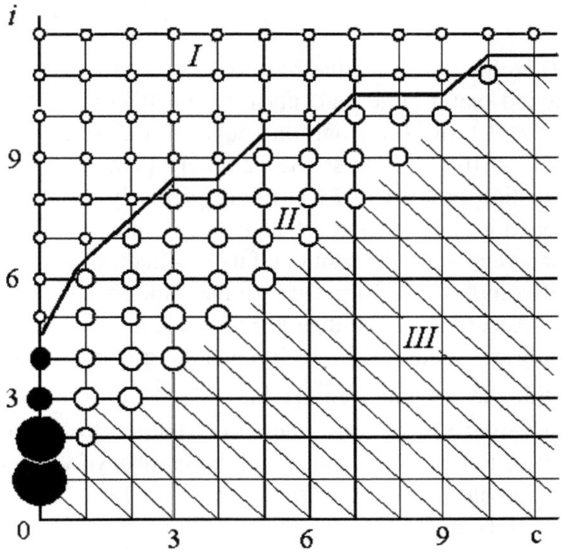

Fig. 4.16 Distribution of silicon–oxygen anions according to their direction of motion and participation in the interaction with the metal for 1 wt.% [Si]

The plots of the dependence of the oxidation rates of silicon on its content in the metal (Fig. 4.14) have the characteristic form for such reactions, i.e., a nearly linear relation between I_{Si} and [Si] at low concentrations, where the process is limited by the diffusion of silicon in the metal, and a weak relation at values of [Si] greater than the critical value, where the reaction is limited by the diffusion of FeO in the slag. The switch between limiting steps is revealed especially clearly when the oxidation rate is calculated without taking into account the polymeric constitution of the slag. Taking into account the polymeric constitution of the melt introduces several distinctive features. First and foremost, our attention is attracted by the extremely low oxidation rates of silicon under the condition that the appearance of only the simplest $(SiO_4)^{4-}$ ions is taken into account (Fig. 4.14, curve 3). Here, because of the low concentrations of free oxygen ions, the reaction is limited by their diffusion to the metal/slag interface. Taking into account other anions results in a sharp increase in I_{Si} (Fig. 4.14, curve 2).

Increasing the concentration of FeO to 40 wt.% causes an increase in the oxidation rate of silicon (Fig. 4.15). When (FeO) > 40 wt.%, the process is limited by the diffusion of silicon in the metal. The significant increase in the oxidation rate upon an increase in the FeO content in the slag is associated with the increase in the concentration of free oxygen ions.

The passage of silicon from the metal into the slag must be accompanied by the binding of the O^{2-} anions that enter the tetrahedral environment of an Si^{4+} cation; therefore, a source of supply of O^{2-} ions that provide for the increase in I_{Si} must exist. Such a source may include $(SiO_4)^{4-}$ and $(Si_2O_7)^{6-}$ ions, as well as other ions that are formed when these are converted into larger anions. The following reaction is an example of a resultant reaction that is obtained without taking into account the O^{2-} ions:

$$[Si] + 7(SiO_4)^{4-} + 2(Fe^{2+}) = 4(Si_2O_7)^{6-} + 2[Fe]. \qquad (4.132)$$

Therefore, some of the reactions proceed in the reverse direction. In fact, it was found that a large percentage of the silicon–oxygen anions contained in the slag ($\approx$80%) diffuse from the bulk of the slag to its boundary with the metal (Fig. 4.16). We stress the definitely paradoxical nature of this finding, i.e., the reaction product of SiO_2 (more precisely, of $(Si_xO_y)^{z-}$ ions) moves toward the reaction surface, rather than away from it. However, this contraction is only apparent, since the resultant flux of silicon atoms within anions of different degrees of complexity is directed into the bulk of the slag.

According to the data in Fig. 4.16, in the case of 1 wt.% Si and 30 wt.% FeO, only chainlike ions ($c = 0$) containing from one to five silicon atoms ($i_0 = 1$–5) with the general formula $Si_{i_0}O_{3i_0+1}^{2(i_0+1)-}$ diffuse toward the metal. If we confine ourselves to reactions whose contribution to the overall oxidation rate of silicon exceeds 1%, their number and the number of varieties of silicon–oxygen anions is far greater: 37 (Fig. 4.16).

Thus, under these conditions the reactions that are actually responsible for the passage of silicon from the metal into the slag are

$$\begin{aligned}&(i - i_0 + i_0c)[Si] + (3i + 1 - c)(Si_{i_0}O_{3i_0+1}^{2(i_0+1)-}) + 2(i - i_0 + i_0c)(Fe^{2+})\\ &= (3i_0 + 1)(Si_iO_{3i_0+1-c}^{2(i+1-c)-}) + 2(i - i_0 + i_0c)[Fe]\end{aligned}, \qquad (4.133)$$

where $i_0 = 1$–5 and $c \geq 1$.

We note that the diffusion of such cumbersome ions hardly proceeds by an ordinary activation mechanism. If it is taken into account that the stable structural unit of all the complex anions is the SiO_4^{4-} tetrahedron, the process of the counter diffusion of ions of different complexity may be interpreted as the exchange of SiO_4^{4-} between them. Such a description of the mechanism of the diffusion of silicon in the slag allows us to draw a conclusion that the diffusion coefficients of all the silicon–oxygen anions in the slag are equal.

It should be noted that the overall oxidation rate of silicon obtained with allowance for the polymeric constitution of the melt is fairly close to the rate obtained from molecular descriptions. In some cases this allows the use of simpler methods for determining the reaction rates [19, 49].

At the same time, it must be noted that further development of the method is possible both by taking into account kinetic factors in the calculation equations and by using more exact thermodynamic and kinetic constants.

4.3.4 Kinetic Parameters of the Passage of Elements Through the Interface in a Metal/Oxide-Melt System

The interaction of metallic and slag melts consists of a set of heterogeneous reactions, whose distinguishing features include their complexity and multiple-step character. If the electrochemical character of the interaction is taken into account [68], the reactions on a metal/slag interface should be treated as a set of electrode processes, which include a minimum of one or more mass-transport steps (the diffusion steps) and an electrochemical discharge–ionization step (the kinetic step). The slowest step in a heterogeneous reaction determines the rate of the process as a whole, i.e., is the rate-limiting step when all the steps occur in succession. In order to determine the slowest step of an overall electrode reaction, the influence of various factors on the kinetics of the process being studied must be investigated under steady-state conditions, under which its rate is determined by the laws governing the rate-limiting step. If the rate of a reaction is determined by diffusional hindrances, the kinetic features of the fast stage of an electrochemical process can be revealed after they are eliminated.

The kinetic characteristics of the passage of elements through a metal/slag interface have been obtained using electrochemical methods [19, 69–71]. The reliability of the data obtained is also determined by the method used to treat the experimental dependences.

4.3.4.1 Character of the Polarization and Kinetic Parameters of Transport Processes of Boron Under Steady-State Conditions

Alloying steel with boron significantly raises the thermal stability, heat resistance, stability under the action of aggressive media, and wear resistance. A fairly large amount of experimental data regarding the kinetics of the passage of several elements through a metal/slag boundary has been accumulated. Let us examine the laws governing the passage of boron through this interface. On the basis of the electrochemical character of the redox reactions in a metal/oxide-melt system [68], the passage of boron can be represented by the following overall electrode reaction

$$[\mathrm{B}] = \left(\mathrm{B}^{3+}\right) + 3e. \qquad (4.134)$$

The kinetic laws governing the interaction of Fe–B–C melts with oxide slags were studied using a galvanostatic method with compensation for the ohmic resistance of the electrolyte.

Melts that are close in composition to actually employed metallurgical slags [72] and welding fluxes [73] were chosen for the investigation. A slag containing (wt.%) 51 Al_2O_3, 43 CaO, and 6 MgO and an Fe–B–C metal containing (wt.%) 0.04–0.12 B and 1–3 C were taken as the base.

A series of values of i and the values of η corresponding to them were obtained by varying the polarizing current and were used to construct the polarization curves. The influence of the temperature and of the concentrations of boron and carbon in the metal on the rate of process (4.134) was studied in [74, 75]. A few of the polarization curves obtained are presented in Figs. 4.17 and 4.18. They have characteristic points of inflection, which are caused by the existence of limiting currents. Such a form of the curves may be a consequence of the diffusional hindrances appearing in the system or the slowness of one of the chemical steps

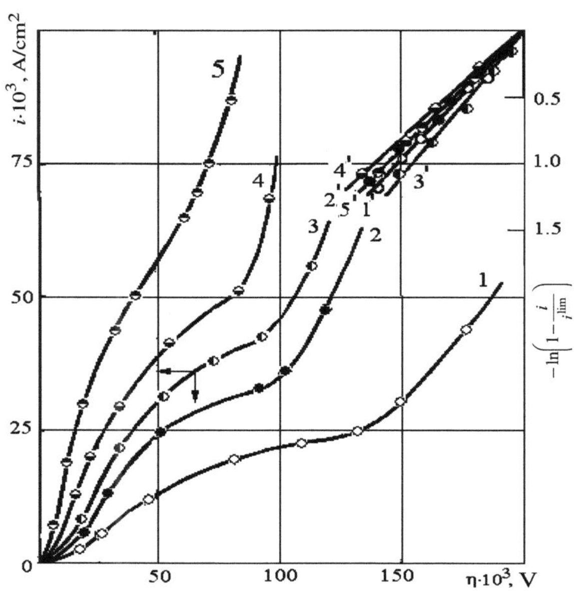

Fig. 4.17 Anodic polarization of an Fe–B–C melt in an oxide melt when [B] = 0.065 wt.% and [C] = 2 wt.%: 1—T = 1723 K; 2—T = 1773 K; 3—T = 1803 K; 4—T = 1833 K; 1—T = 1873 K; 1'–5'—same in semilogarithmic coordinates

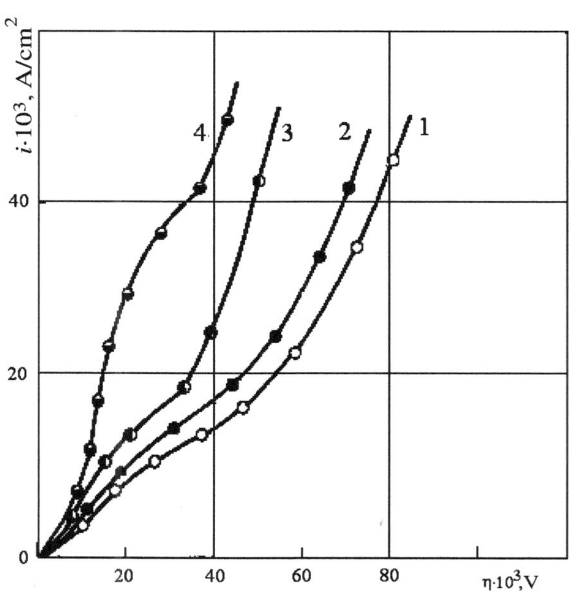

Fig. 4.18 Anodic polarization of an Fe–B–C melt in an oxide melt at 1723 K when [C] = 3 wt.% in all the melts: 1—[B] = 0.045 wt.%; 2—[B] = 0.065 wt.%; 3—[B] = 0.080 wt.%; 4—[B] = 0.120 wt.%

comprising the overall electrode reaction. We will first analyze the data obtained using conventional treatment methods.

During the anodic polarization of an Fe–B–C melt, along with reaction (4.134), the occurrence of reaction (4.95) and the reaction

$$[C] + (O)^{2-} = \{CO\} + 2e \tag{4.135}$$

Table 4.8 Kinetic parameters of the anodic polarization of Fe–B–C melts

Melt No.	[C], wt.%	[B], wt.%	*T*, K	$i^{lim}\bullet 10^3$, A/cm^2	*N*	*E*, kJ/mole
1	3.0	0.12	1673	45	2.8	240
			1723	85	2.9	
			1773	120	3.1	
2	3.0	0.08	1673	25	2.8	280
			1723	45	2.8	
			1773	85	3.2	
3	3.0	0.065	1673	20	3.2	270
			1723	35	2.8	
			1773	65	2.8	
4	3.0	0.045	1673	15	2.9	280
			1723	30	2.8	
			1773	55	3.0	
5	2.0	0.12	1723	46	2.9	270
			1773	74	3.0	
			1803	90	3.0	
			1833	132	3.2	
6	2.0	0.08	1723	30	2.8	250
			1773	50	3.0	
			1803	65	2.9	
			1833	85	2.9	
			1873	120	3.1	
7	2.0	0.064	1723	25	3.0	250
			1773	40	2.9	
			1803	50	3.2	
			1833	65	2.8	
			1873	98	3.0	
8	1.0	0.12	1773	40	3.1	250
			1803	60	3.1	
			1833	85	2.9	
			1873	105	3.1	
9	1.0	0.08	1773	30	2.8	240
			1803	45	3.0	
			1833	60	3.0	
			1873	70	3.2	

is possible. We will assign the values of the limiting currents obtained (Table 4.8) to specific processes.

It was noted in [76, 77] that half-reaction (4.135) takes place in a slow-discharge regime and has polarization values that significantly exceed the observed values. The contribution of reaction (4.95) to the total current of the anode process was ignored, because constancy of the iron content in the slag during an experiment was established by chemical analysis. The experimental data were used to determine the number of electrons participating in the electrode process, which turned out to be close to 3 (Table 4.8). This number is consistent with reaction (4.134). It follows from the data in Figs. 4.18 and 4.19 that the value of the

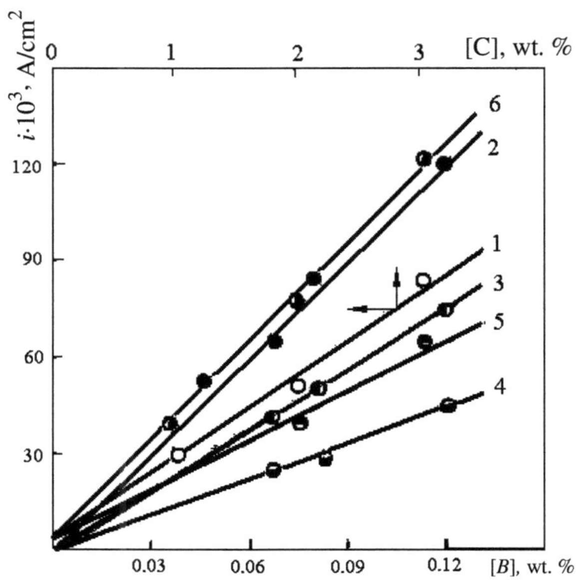

Fig. 4.19 Influence of the carbon (1, 5, 6) and boron (2–4) concentrations in the metal on the limiting current density: 1—$T = 1773$ K and [B] = 0.080 wt.%; 2—$T = 1773$ K and [C] = 3.0 wt.%; 3—$T = 1723$ K and [C] = 2.0 wt.%; 4—$T = 1773$ K and [C] = 2.0 wt.%; 5—$T = 1773$ K and [B] = 0.065 wt.%; 6—$T = 1773$ K and [B] = 0.120 wt.%

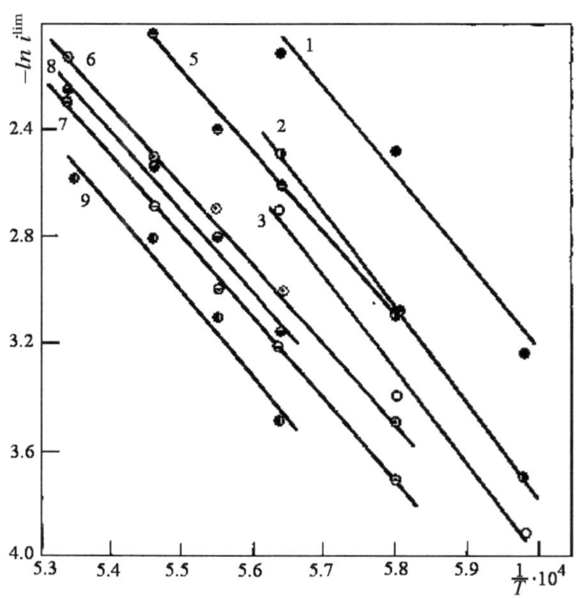

Fig. 4.20 Temperature dependence of the limiting current density. The numbers on the curves correspond to the numbers of the experiments in Table 4.1

limiting current is proportional to the concentration of boron in the metal. This would seem to allow us to conclude that the polarization curves obtained should be assigned to process (4.134). They are described well by the concentration polarization equations when there are diffusional hindrances in the metal phase. Based on the data in Fig. 4.20 and Table 4.8, we write

$$i_{\mathrm{B}}^{\mathrm{lim}} = 1.6 \times 10^7 [\mathrm{C}][\mathrm{B}] \exp\left(-\frac{31290}{T}\right). \tag{4.136}$$

In accordance with the data presented in Fig. 4.19, for a metal containing less than 0.2 wt.% carbon, instead of Eq. 4.136 we have

$$i_{\mathrm{B}}^{\mathrm{lim}} = 3.2 \times 10^6 [\mathrm{B}] \exp\left(-\frac{31290}{T}\right). \tag{4.137}$$

However, several points compel us to examine the data obtained in greater detail, including with the use of the method described in Sect. 4.3.2. We first note the following.

- Evaluation of the diffusion coefficient of boron in the metal under the assumption that process (4.134) takes place according to the equation

$$i_{\mathrm{B}}^{\mathrm{lim}} = nF\beta D_{[B]}^{0.5} \frac{[\mathrm{B}]\rho_{\mathrm{Me}}}{100M_{\mathrm{B}}}, \tag{4.138}$$

  where β is the convection constant ($s^{-0.5}$), [B] is the concentration of boron in the metal (wt.%), ρ_{Me} is the density of the metal (g/cm^3), and M_{B} is the Bohr atomic weight (g/mole), gives values ranging from 1×10^{-6} to 1×10^{-7} cm^2/s. They are approximately 1–2 orders of magnitude smaller than the known values of the diffusion coefficients in iron for most elements [78].
- The values of the limiting currents are approximately an order of magnitude smaller than the values for elements with similar properties, viz., Al and Si [65, 79], at similar concentrations and temperatures.
- The data in Fig. 4.19 show that extrapolation of linear plots 1, 5, and 6 to a concentration of carbon in the metal equal to 0 gives approximately identical values of the limiting current at boron concentrations differing by almost two fold.

We will illustrate the new approach to the analysis of the data obtained in the following example: Fe–B(0.065 wt.%)–C(2 wt.%) as the metal, $T = 1833$ K.

We will consider all the chemical reactions that are possible in this system:

$$[\mathrm{C}] + (\mathrm{FeO}) = [\mathrm{Fe}] + \{\mathrm{CO}\}, \tag{4.139}$$

$$2/3[\mathrm{Al}] + (\mathrm{FeO}) = [\mathrm{Fe}] + 2/3(\mathrm{AlO}_{1.5}), \tag{4.140}$$

$$2/3[\mathrm{B}] + (\mathrm{FeO}) = [\mathrm{Fe}] + 2/3(\mathrm{BO}_{1.5}), \tag{4.141}$$

$$(\mathrm{FeO}) = [\mathrm{Fe}] + [\mathrm{O}], \tag{4.142}$$

which can consist of partial electrode processes (4.95), (4.134), (4.135), and

$$[\mathrm{Al}] = \left(\mathrm{Al}^{3+}\right) + 3e, \tag{4.143}$$

$$[\mathrm{O}] + 2\mathrm{e} = (\mathrm{O}^{2-}). \tag{4.144}$$

In accordance with (4.99) are (4.100), the expressions for the rates of passage of the elements equal

$$i^k_{[\mathrm{C}]} = \frac{P_{\mathrm{CO}} - [\mathrm{C}]^k_1 x^2 K_{P_\mathrm{C}} F}{\frac{P_{\mathrm{CO}}^{1-\alpha-\beta} x^{2\alpha} K^{\alpha}_{P_\mathrm{C}} F^{\alpha}}{K^0_\mathrm{C}} + B_{[\mathrm{C}]} x^2 K_{P_\mathrm{C}} F}, \tag{4.145}$$

$$i^k_{[\mathrm{Al}]} = \frac{(\mathrm{AlO}_{1.5})^k_1 - [\mathrm{Al}]^k_1 x^3 K^{1.5}_{P_\mathrm{Al}} F^{1.5}}{\frac{(\mathrm{AlO}_{1.5})^{1-\alpha-\beta} x^{3\alpha} K^{1.5\alpha}_{P_\mathrm{Al}} F^{1.5\alpha}}{K^0_\mathrm{Al}} + B_{[\mathrm{Al}]} x^3 K^{1.5}_{P_\mathrm{Al}} F^{1.5} + B_{(\mathrm{AlO}_{1.5})}}, \tag{4.146}$$

$$i^k_{[\mathrm{B}]} = \frac{(\mathrm{BO}_{1.5})^k_1 - [\mathrm{B}]^k_1 x^3 K^{1.5}_{P_\mathrm{B}} F^{1.5}}{\frac{(\mathrm{BO}_{1.5})^{1-\alpha-\beta} x^{3\alpha} K^{1.5\alpha}_{P_\mathrm{B}} F^{1.5\alpha}}{K^0_\mathrm{B}} + B_{[\mathrm{B}]} x^3 K^{1.5}_{P_\mathrm{B}} F^{1.5} + B_{(\mathrm{BO}_{1.5})}}, \tag{4.147}$$

$$i^k_{[\mathrm{O}]} = \frac{[\mathrm{O}]^k_1 - x^2 K_{P_\mathrm{O}} F^k_1}{\frac{x^{2\alpha} K^{\alpha}_{P_\mathrm{O}} F^{\alpha}}{K^0_\mathrm{Al}} + B_{(\mathrm{FeO})} x^2 K_{P_\mathrm{O}} + B_{[\mathrm{O}]}}, \tag{4.148}$$

$$i^k_{[\mathrm{Fe}]} = \frac{F^k_1 - x^2 F}{\frac{x^{2\alpha} F^{1-\beta}}{K^0_\mathrm{Fe} [\mathrm{Fe}]^{\alpha}} + B_{(\mathrm{FeO})}}. \tag{4.149}$$

Let us analyze the processes that occur between the metal and the slag when the system is held before performing the electrochemical measurements. Calculations showed that during a time close to 1 h, the system reaches a state that is characterized by low values of the rates of passage of the elements through the interface and a nearly constant value of the potential, but is far from equilibrium with respect to the component studied (boron). This is verified by the high polarization values and the considerably higher concentrations of B_2O_3 on the interface than in the bulk of the slag. In addition, it should be noted that an appreciable quantity of aluminum appears in the metal due to the occurrence of the reaction

$$[\mathrm{B}] + (\mathrm{AlO}_{1.5}) = [\mathrm{Al}] + (\mathrm{BO}_{1.5}). \tag{4.150}$$

Accordingly, it may be assumed that electrode process (4.143) is also possible upon anodic polarization of the metal.

Galvanostatic steady-state measurements were next simulated. The results are presented in Table 4.9 and in Fig. 4.21. The following data were used for the calculations: $m_{\mathrm{Me}} = 7$ g, $m_{\mathrm{slag}} = 150$ g, $\beta = 1\ \mathrm{s}^{-0.5}$, $S_{\mathrm{Me-slag}} = 1\ \mathrm{cm}^2$. The calculations show that the processes of the anodic passage of aluminum and carbon, rather than of boron, were actually studied experimentally. The values of the limiting currents obtained most likely belong to process (4.143). This is also verified by the evaluation of these parameters from the data in [79]: $i^{\mathrm{lim}}_{\mathrm{Al}} = 0.01$ to $0.30\ \mathrm{A/cm}^2$.

Table 4.9 Results of calculations that simulate galvanostatic steady-state measurements at 1833 K

I, A/cm^2	η, V	i_{Fe}, A/cm^2	i_C, A/cm^2	i_B, A/cm^2	i_{Al}, A/cm^2	i_O, A/cm^2
−0.005	0.0063	-3.7×10^{-5}	0.0	−0.0003	−0.0046	-1.0×10^{-5}
−0.010	0.0133	-8.1×10^{-5}	0.0	−0.0006	−0.0092	-1.9×10^{-5}
−0.015	0.0207	-1.3×10^{-4}	0.0	−0.0010	−0.0136	-2.9×10^{-5}
−0.020	0.0268	-1.8×10^{-4}	−0.0010	−0.0013	−0.0174	-3.9×10^{-5}
−0.030	0.0377	-2.7×10^{-4}	−0.0042	−0.0020	−0.0234	-5.9×10^{-5}
−0.050	0.0610	-5.1×10^{-4}	−0.0115	−0.0036	−0.0343	-1.1×10^{-4}
−0.100	0.1184	-1.5×10^{-3}	−0.0341	−0.0097	−0.0543	-3.5×10^{-4}
−0.150	0.1693	-3.2×10^{-3}	−0.0624	−0.0186	−0.0650	-7.5×10^{-4}

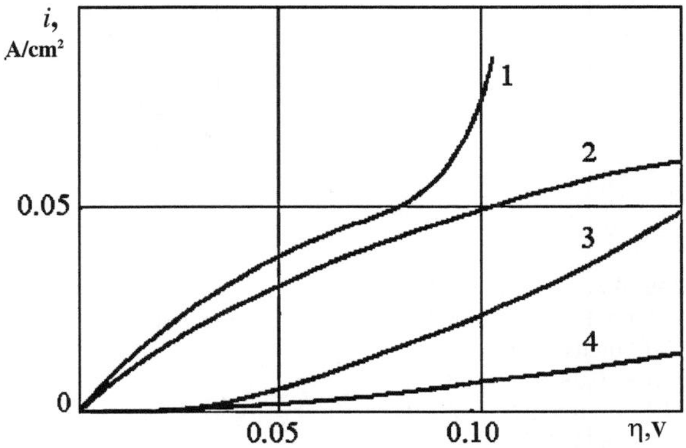

Fig. 4.21 Anodic polarization of an Fe–B–C–Al melt in a slag at $T = 1833$ K: *1*—experiment; *2*–*4*—calculation: *2*—[Al]; *3*—[C]; *4*—[B]

It should be noted that the question of the reasons for the abrupt rise on the polarization curves in the experiment remains open. A similar phenomenon occurs on calculated polarization curves at considerably higher polarizations (Fig. 4.21).

Data corresponding to galvanostatic steady-state measurements of the system studied with boron concentrations close to the equilibrium value were also obtained mathematically. The calculations were performed for the following conditions: T = 1773 K; a metal containing (wt.%) 2 carbon, 0.08 boron, and the rest iron; a slag containing (wt.%) 40.8 Al_2O_3, 34.4 CaO, 20 B_2O_3, and 4.8 MgO. The results of the calculations are presented in Fig. 4.22.

When the conventional treatment method is used and the errors in the experimental determination of the polarization values are taken into account, a conclusion that there are diffusional hindrances with a limiting current close to 0.120 A/cm^2 can be drawn from the polarization curve obtained (Fig. 4.22, curve 1). However, an analysis of the partial polarization curves (Fig. 4.22, curves 2–4) shows that such a conclusion would be erroneous.

Thus, the data obtained allow us to state that setting up steady-state electrochemical experiments to study multicomponent metallurgical systems can be effective in conjunction with modern methods of kinetic analysis.

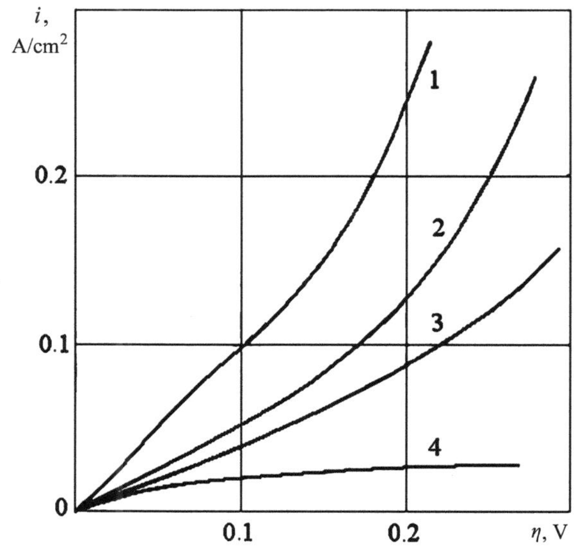

Fig. 4.22 Calculated anodic polarization curves of an Fe–B–C–Al melt in a slag at $T = 1773$ K: *1*—overall polarization curve; *2*—[O]; *3*—[B]; *4*—[Al]

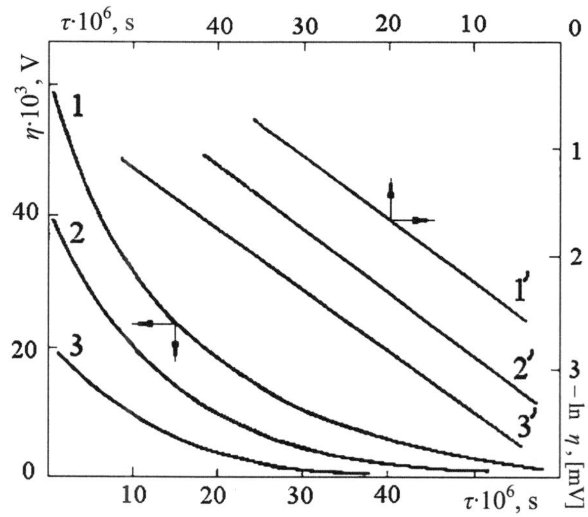

Fig. 4.23 Variation of the polarization with time for Fe–B–C melts when [B] = 0.08 wt.% and [C] = 2.0 wt.%: *1, 1'*—$T = 1673$ K; *2, 2'*—$T = 1723$ K; *3, 3'*—$T = 1773$ K

4.3.4.2 Rate of a Process in the Absence of Diffusional Hindrances

The study of the rates of the steps of reaction (4.134) that occur without diffusional hindrances is of considerable interest. The features of the interaction were studied experimentally using a coulostatic method [80].

The kinetic parameters of the reactions on the boundary between Fe–B–C melts and a slag containing (wt.%) 51 Al_2O_3, 43 CaO, and 6 MgO in the 1673–1873 K temperature range are presented in Table 4.10 and in Fig. 4.23.

Table 4.10 Kinetic parameters of the interaction of boron-containing iron–carbon and oxide melts

Melt No.	[C], wt.%	[B], wt.%	T, K	C_0, μF/cm^2	R_0, $\Omega\cdot$cm^2	C_0R_0, μF$\cdot\Omega$
1	3.0	0.12	1673	8.6	0.944	8.12
			1723	23.0	0.374	8.60
			1773	72.0	0.142	10.22
			1803	112.6	0.084	9.46
			1833	163.8	0.054	8.85
			1873	365.4	0.026	9.50
2	3.0	0.08	1673	7.2	1.244	8.96
			1728	18.2	0.502	9.14
			1773	45.4	0.204	9.26
			1803	68.6	0.124	8.51
			1833	134.4	0.066	8.87
			1873	300.0	0.030	9.00
3	3.0	0.05	1673	5.8	1.766	10.24
			1723	15.4	0.628	9.67
			1773	43.4	0.244	10.59
4	3.0	0.04	1673	3.2	0.256	8.08
			1723	6.0	1.106	6.64
			1773	19.0	0.424	8.06
5	3.0	0.03	1673	1.2	4.900	5.38
			1723	5.2	1.354	7.04
			1773	8.8	0.672	5.91
6	2.0	0.12	1723	17.0	0.552	9.38
			1773	38.2	0.226	8.63
			1803	59.2	0.142	8.41
			1833	105.0	0.082	8.61
			1873	276.8	0.038	10.52
7	2.0	0.08	1723	8.0	1.010	8.08
			1773	23.4	0.312	7.30
			1803	59.0	0.154	9.09
			1833	92.8	0.094	8.72
			1873	209.3	0.046	9.63
8	2.0	0.05	1723	2.8	0.962	8.08
			1773	21.8	0.366	7.98
			1803	27.6	0.242	6.68
			1833	59.8	0.136	8.13
			1873	122.8	0.069	8.35
9	1.0	0.12	1803	20.2	0.410	8.28
			1833	39.2	0.242	9.48
			1873	65.0	0.144	9.36
10	1.0	0.18	1803	12.8	0.800	10.24
			1833	15.4	0.632	9.73
			1873	33.8	0.302	10.21

An analysis of the data obtained shows that the decrease in the resistance R_0 is accompanied by an increase in the capacitance C_0, so that the product of these quantities remains nearly constant. According to [71], the existence of such a relationship between R_0 and C_0 is stipulated by the slow relaxation of the electrical double layer. The impedance of the electrical double layer can be represented in this case in the form of the following components: the discharge resistance (R_{disch}), the relaxation resistance (R_1), and the Helmholtz capacitance (C_H). The following relations can be obtained using well-known relations:

$$C_0 \cdot R_0^2 = C_H \cdot R_{disch}^2, \tag{4.151}$$

$$i_0 = \frac{1}{R_{disch}} \cdot \frac{RT}{nFS}. \tag{4.152}$$

The values $C_H = 2.25$ μF/cm^2 and $R_1 = 4.16$ Ω cm^2, which are close to the values obtained in [71], were found from the data in Table 4.10 using the least-squares method. The kinetic parameters calculated from equations (4.151) and (4.152) with allowance for relaxation of the electrical double layer are presented in Table 4.11. The activation energy of the processes was calculated from the temperature dependence of the exchange current (Fig. 4.24).

Let us analyze the data obtained with consideration of the composition of the metal and the slag that are formed after holding for 1 h before performing the electrochemical experiments.

Taking into account the position of boron in Mendeleev's periodic table, we compared the values of the exchange current of the fast step for B, Al, and Si at similar concentrations of the components in the metal and the slag on the basis of the data in [65, 77]. For Al, the values of the exchange current lie in the range 0.001–0.004 A/cm^2 in the 1673–1873 K temperature range when the carbon content in the metal is 2 wt.%, and for Si, $i_0 = 0.004$–0.008 A/cm^2. Comparison with the data in Table 4.11 reveals that the exchange current values that we obtained for boron are approximately an order of magnitude greater.

Calculations performed according to the procedure described in Sect. 4.3.2, which simulate galvanostatic experiments with interaction times from 1×10^{-8} to 1×10^{-6} s, show that approximately 10% of the total current is used for reaction (4.134). Then the data obtained (Table 4.11) can be interpreted in the following manner. We assume that in the equivalent electrical circuit a series of resistances that correspond to each electrode process are connected in parallel instead of the discharge resistance (R_{disch}) and that C_H remains nearly constant because it does not depend on the composition of the electrolyte [71]. Then the value of R_{disch} for reaction (4.134) will be an order of magnitude greater than the values presented in Table 4.11. In this case the calculation based on Eq. 4.152 gives values of the exchange currents that are an order of magnitude smaller than the values in Table 4.11. We assign just these values to process (4.134). The relationship between the exchange current and the parameters of the process, viz., the

Table 4.11 Kinetic parameters of the interaction of Fe–B–C and an oxide melt determined with allowance for the slow relaxation of the electrical double layer

Melt No.	[C], wt.%	[B], wt.%	T, K	R_H, Ω·cm^2	i_0, A/cm^2	E, kJ/mole
1	3.0	0.12	1673	1.87	0.026	225
			1723	1.21	0.041	
			1773	0.81	0.063	
			1803	0.60	0.086	
			1833	0.47	0.113	
			1873	0.34	0.159	
2	3.0	0.08	1673	2.24	0.021	240
			1728	1.44	0.034	
			1773	0.93	0.056	
			1803	0.69	0.075	
			1833	0.52	0.102	
			1873	0.35	0.154	
3	3.0	0.05	1673	2.87	0.017	245
			1723	1.66	0.030	
			1773	1.08	0.047	
4	3.0	0.04	1673	3.04	0.015	245
			1723	1.83	0.027	
			1773	1.24	0.041	
5	3.0	0.03	1673	3.79	0.013	230
			1723	2.07	0.027	
			1773	1.35	0.038	
6	2.0	0.12	1723	1.53	0.032	240
			1773	0.94	0.054	
			1803	0.74	0.070	
			1833	0.57	0.093	
			1873	0.43	0.126	
7	2.0	0.08	1723	1.93	0.026	260
			1773	1.02	0.050	
			1803	0.80	0.065	
			1833	0.61	0.086	
			1873	0.45	0.120	
8	2.0	0.05	1723	1.87	0.026	245
			1773	1.15	0.044	
			1803	0.85	0.060	
			1833	0.71	0.074	
			1873	0.51	0.106	
9	1.0	0.12	1803	1.24	0.042	215
			1833	1.02	0.052	
			1873	0.78	0.069	
10	1.0	0.18	1803	1.94	0.026	225
			1833	1.67	0.032	
			1873	1.18	0.045	

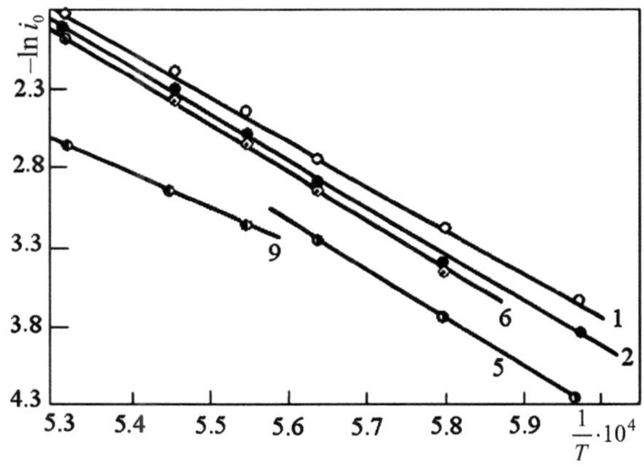

Fig. 4.24 Dependence of the exchange current on the temperature. The numbers on the curves correspond to the numbers of the experiments in Tables 4.9 and 4.10

temperature and the concentrations of boron in the metallic and slag phases, was represented in the form

$$i_0 = 1.15 \times 10^6 \cdot [\mathrm{C}][\mathrm{B}]^{0.5}(\mathrm{BO}_{1.5})^{0.5}\exp(-28300/T). \qquad (4.153)$$

4.3.4.3 Investigation of the Fast Step in the Passage of Sulfur through an Interface

The measurements were performed by a coulostatic method [81] in the 1673–1873 K temperature range. The compositions of the iron–carbon–sulfur melts are presented in Table 4.12, and the oxide melt contained (wt.%) 51 Al_2O_3, 43 CaO, and 6 MgO.

The treatment of the experimental polarization versus time dependences enabled us to determine the parameters of the interface: the capacitance C_0, the resistance R_0, and the apparent exchange current of the reaction i_0' whose values for various temperatures and metal compositions are presented in Table 4.12.

The large capacitance values, which reach 1000 μF/cm^2 or more in several cases, cannot be identified with the capacitance of the electrical double layer. The product C_0R_0 varies weakly as the temperature and the composition of the melt are varied.

Therefore, slow reorganization of the particles in the electrolyte cloud of a variable-polarity electrical double layer also occurs in an Fe–C–S/oxide-melt system [82]. Another, more complex equivalent electrical circuit of the interface is valid for this case. The experimental results for all the metal compositions and temperatures investigated are plotted on a straight line in $C_0^{0.5}$ versus $C_0^{-0.5}R_0^{-1}$ coordinates (Fig. 4.25), confirming the validity of such an equivalent circuit.

The capacitance of the Helmholtz layer (C_H) found in accordance with the equation

$$C_0^{0.5} = C_\mathrm{H}^{0.5} + R_1 C_\mathrm{H} R_0^{-1} C_0^{-0.5} \qquad (4.154)$$

Table 4.12 Data from coulostatic measurements

Melt No.	[C], wt.%	[S], wt.%	T, K	C_0, μF/cm^2	$R_0 \cdot 10^2$, Ω•cm^2	C_0R_0, μF·Ω	i_0', A/cm^2	R_{disch}, Ω•cm^2	i_0, A/cm^2	E, kJ/mole
1	2.0	0.75	1673	1010	0.50	5.0	14.5	0.092	0.78	150
			1723	1800	0.35	6.3	22.5	0.081	0.92	
			1773	2360	0.19	4.5	40.0	0.053	1.44	
2	2.0	0.48	1673	840	0.063	5.3	11.5	0.105	0.69	180
			1773	1970	0.24	4.9	32.0	0.061	1.25	
			1823	2280	0.14	3.2	34.5	0.046	1.7	
			1873	3800	0.08	3.0	100.0	0.028	2.88	
3	2.0	0.30	1723	1600	0.38	6.0	20.0	0.088	0.84	150
			1773	2240	0.24	5.4	32.0	0.065	1.17	
			1823	2820	0.18	5.1	44.0	0.055	1.42	
4	2.0	0.13	1673	760	0.80	6.1	9.0	0.127	0.57	180
			1723	1190	0.60	7.1	12.4	0.120	0.62	
			1773	1650	0.27	4.5	27.8	0.076	1.0	
			1823	2050	0.21	4.3	37.0	0.059	1.33	
			1873	2150	0.14	3.0	58.0	0.038	2.12	
5	2.0	0.02	1673	250	2.70	6.8	2.7	0.248	0.29	160
			1773	690	0.94	6.4	8.1	0.142	0.54	
			1823	1140	0.58	6.6	13.5	0.113	0.70	
			1823	1520	0.37	5.6	22.0	0.083	0.97	
6	3.0	0.02	1673	360	1.92	6.9	3.8	0.211	0.34	190
			1723	580	1.14	6.6	6.5	0.159	0.47	
			1773	1060	0.61	6.4	12.3	0.115	0.66	
			1823	1560	0.35	5.5	22.0	0.080	0.98	
			1873	1980	0.22	4.4	37.0	0.057	1.42	
7	0.5	0.02	1773	330	2.51	8.4	3.0	0.256	0.29	190
			1823	500	1.78	9.0	4.4	0.230	0.34	
			1873	700	0.93	6.6	8.7	0.142	0.57	

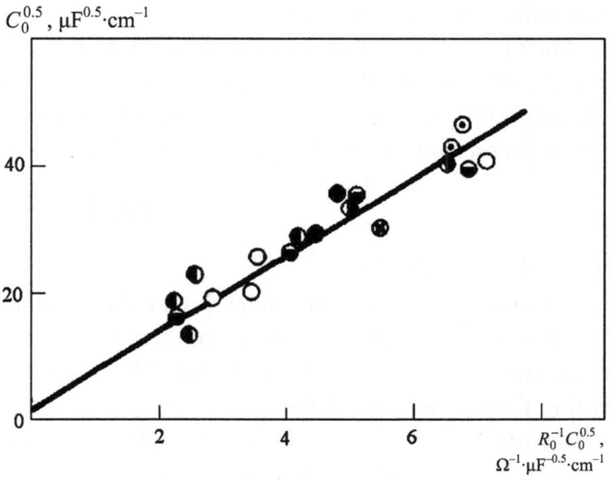

Fig. 4.25 Relationship between the interface parameters obtained by the coulostatic method in an (Fe–C–S)/oxide-melt system at 1673–1873 K: ⊙—1; ⊗—2; ◑—3; ●—4; ◒—5; ○—6; ◐—7 (the numbers correspond to the numbers of the metal compositions in Table 4.12)

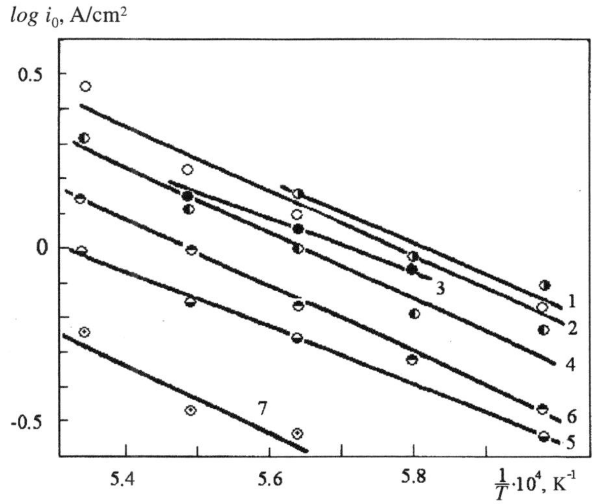

Fig. 4.26 Influence of the temperature on the exchange current in an Fe–C–S/oxide-melt system (the numbers correspond to the numbers of the metal compositions in Table 4.12)

was equal to 3.0 ± 0.1 μF/cm². After determining the value of the discharge resistance (R_{disch}) from relation (4.151), we calculated the exchange current (i_0^S) of the fast step in the reaction involving the passage of sulfur through the interface. The activation energy of the process was determined from the temperature dependence of the exchange current in $\log i_0$ versus $1/T$ coordinates (Fig. 4.26). The kinetic parameters obtained are listed in Table 4.12.

In accordance with the theories of electrochemical kinetics, we write an equation for the exchange current of the fast step of the reaction involving the passage of sulfur through the interface [70, 83, 84]

$$i_0^S = 2FK_1[S]^\alpha(S)^{1-\alpha}, \quad (4.155)$$

where K_1 is a constant, which takes into account the exponential dependence of i_0 on the temperature, and [S] and (S) are the concentrations of sulfur in the metallic and oxide phases, respectively.

The insignificant change in the concentration of sulfur in the oxide melt, which is associated with the relatively large mass of the latter (which exceeds the mass of the metal by approximately two orders of magnitude), allows us to assume that the value of (S) is constant in a first approximation. Then relation (4.155) takes the form

$$i_0^S = 2FK_2[S]^\alpha, \quad (4.156)$$

where $K_2 = K_1(S)^{1-\alpha}$.

An analysis of the experimental data showed that the exponent α in Eq. 4.156 decreases with increasing sulfur content in the metal phase at a constant concentration of carbon and temperature. The values of α were determined by analyzing the dependence of the exchange current on the concentration of sulfur in the iron–carbon melt (Fig. 4.27, curves 1–3). The variation of α with increasing S content and 2.0 wt.% carbon in the metal is shown in Fig. 4.27 (linear plot 4).

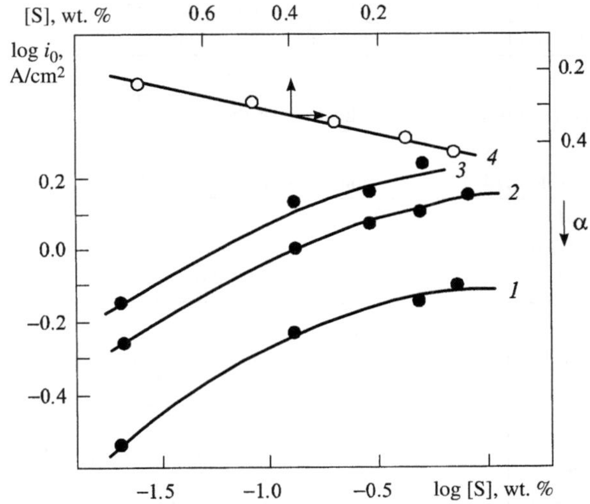

Fig. 4.27 Dependence of the exchange current on the concentration of sulfur in an Fe–C–S melt when [C] = 2 wt.%: 1–3—at 1673, 1773, and 1823 K, respectively; 4—influence of the sulfur concentration on the value of the exponent α

Sulfur has high capillary activity; therefore, the values of its concentration in the bulk and in the surface layer of the metallic melt can differ significantly [85]. The extent of adsorption (Γ_S) is specified by the difference between the surface (N_S^{ω}) and bulk (N_S) concentrations

$$\Gamma_S \sim (N_S^{\omega} - N_S), \tag{4.157}$$

where ω denotes the molar surface.

Since the dimensions of the interfacial surface are finite, when its degree of filling is significant, the bulk concentration will increase more rapidly than the surface concentration. Therefore, as the sulfur content in the bulk of the metal is increased, lowering of the rate of increase of the exchange current, which is expressed quantitatively by a decrease in α in Eq. 4.156, can be expected (Fig. 4.27).

An analysis of the dependence of i_0 on N_S^{ω} is of unquestionable interest, because this relationship reflects the real influence of the concentration for surface-active sulfur.

We will determine the concentration of sulfur in the surface layer of a metal in contact with an oxide melt. To evaluate N_S^{ω} we use the equation [86]

$$N_i^{\omega} = \frac{N_i (C_i'/C_i)^{\alpha_i}}{\sum\limits_{i=0}^{n} N_i (C_i'/C_i)^{\alpha_i}}. \tag{4.158}$$

For the interface between the melts investigated, we have

$$N_S^{\omega} = \frac{N_S\left(\frac{C_S'}{C_S}\right)^{\alpha_S}}{N_S\left(\frac{C_S'}{C_S}\right)^{\alpha_S} + N_{Fe}\left(\frac{C_{Fe}'}{C_{Fe}}\right)^{\alpha_{Fe}} + N_C\left(\frac{C_C'}{C_C}\right)^{\alpha_C} + N_O\left(\frac{C_O'}{C_O}\right)^{\alpha_O}}. \tag{4.159}$$

Table 4.13 Composition of the contacting phases

$C_{Fe,}$ mole/cm^3	C'_{Fe}, mole/cm^3	$C_{C,}$ mole/cm^3	$C'_{C,}$ mole/cm^3	$C_{O,}$ mole/cm^3	$C'_{O,}$ mole/cm^3
0.12	1.0×10^{-4}	1.0×10^{-2}	5.0×10^{-5}	8.0×10^{-6}	7.0×10^{-2}

Note. The prime sign indicates that the quantity refers to the oxide phase

Table 4.14 Concentrations of sulfur in the surface layer and in the bulk of the metal

[S]	0.02	0.13	0.30	0.48	0.75
$N_S \cdot 10^3$	0.3	2.3	4.9	6.8	10.1
$N_S^{\omega} \cdot 10^2$	0.7	5.0	10.0	14.0	21.0

The values of α_i were assumed to be equal to 0.25 [86]. The concentrations of Fe, C, and O in the metal and the oxide melt were calculated from the relation

$$C_i = (\text{wt. }\%\ i)d/100M_i, \tag{4.160}$$

where (wt.% i) is the mass percentage of the component in the particular phase based on the results of chemical analysis, d is the density of the metal or the oxide melt, and M_i is the atomic weight of the element.

The densities of the metal (6.65 g/cm^3) and the oxide melt (2.85 g/cm^3) were taken from [87] and [88]. The ratio between the concentrations of sulfur was chosen in accordance with its equilibrium distribution coefficient (L_S), since the concentrations of S in the liquid phases near the interface are close to the equilibrium values. When [C] = 2.0 wt.%, the average value of L_S for the compositions of the iron–carbon and oxide melts investigated was assumed to be equal to 250 [89].

In accordance with the results of chemical analysis, the following compositions of the contacting phases were taken in the calculation (Table 4.13).

The values of the surface concentration of sulfur for various values of its content in the metal are presented in Table 4.14.

When the temperature dependence of the exchange current is taken into account, the influence of the surface concentration of sulfur in the metal is conveyed by the equation

$$i_0^S = K(N_S^{\omega})^{\alpha} \exp(-20460/T). \tag{4.161}$$

The experimental data are described well by Eq. 4.161. The dependence of i_0 on N_S^{ω} is linear in $\log i_{exch}$ versus i_i^k and i_F^k coordinates (Fig. 4.28). The value of α determined at different temperatures was equal to 0.3.

For the compositions of the metallic and oxide phases investigated, Eq. 4.161 ultimately takes on the form

$$i_{exch}^S = 2.5 \cdot 10^{-5} (N_S^{\omega})^{0.3} \exp(-20460/T). \tag{4.162}$$

We note that the average value $E = 170 \pm 20$ kJ/mole was taken in Eq. 4.162.

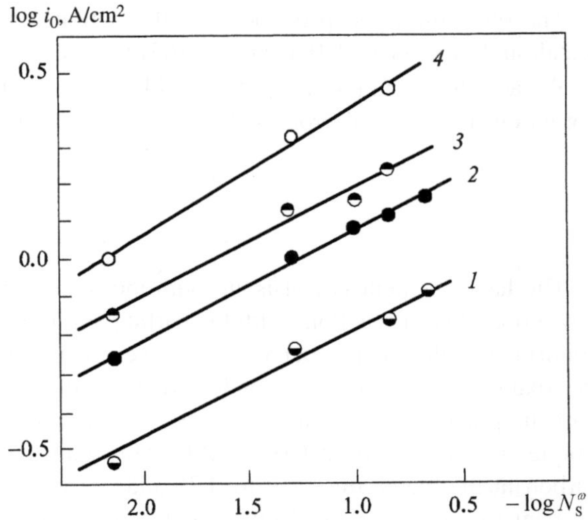

Fig. 4.28 Influence of the surface concentration of sulfur on the value of the exchange current in an Fe–C–S/oxide-melt system: *1*–*4*—at 1673, 1723, 1773, and 1873 K, respectively

The experimentally determined values of the exchange current for the fast step (i_0^S) are, on the average, 5–10 times greater than the values of the limiting diffusion currents of sulfur in the metal (i_S^{lim}) [90].

The initial segments of the polarization curves obtained under steady-state conditions [90] characterize the slow transport of sulfur in the metal phase. The rate of the faster process that takes place in the absence of diffusional hindrances was measured by a coulostatic method. When the polarization has a concentration character, drawing a conclusion regarding the reaction mechanism is known to be impossible [57]. A systematic analysis of the steady-state and relaxation measurements verifies the multiple-step character of the reaction in which sulfur passes from the metal into the oxide phase

$$[\mathrm{S}] + 2e = (\mathrm{S}^{2-}). \tag{4.163}$$

One of the steps in this process is the adsorption of sulfur on the interface from the side of the metal, as is indicated by the experimental dependence of the exchange current on the value of N_S^{ω} (Fig. 4.28).

We represent reaction (4.163) as a reaction consisting of the following consecutive steps:

- diffusion of sulfur from the bulk of the metal to the interface with the oxide melt;
- adsorption of sulfur on the interface from the side of the metal phase

$$[\mathrm{S}] = \mathrm{S}_{\mathrm{ads}}; \tag{4.164}$$

- ionization of the adsorbed sulfur atoms

$$\mathrm{S}_{\mathrm{ads}} + 2e = \mathrm{S}^{2-}. \tag{4.165}$$

The rate-limiting step of the overall reaction is the diffusion of sulfur in the metal, and processes (4.164) and (4.165) take place reversibly.

We assign the values of i^{S}_{exch} to step (4.165), since the adsorption step (4.164) is not an electrochemical process. Process (4.165) can take place in steps:

$$S_{ads} + e = S^{-}, \tag{4.166}$$

$$S^{-} + e = S^{2-}. \tag{4.167}$$

The last statement is consistent with the opinion in [57, 91, 92] regarding the low probability of reactions with the participation of more than one electron and is confirmed by the weak dependence of the exchange current on the sulfur content in the oxide phase. Specially conducted experiments verify that the value of i_0 remains practically constant when the concentration of sulfur in the oxide electrolyte is increased from 0.05 to 0.19 wt.% and the concentration of sulfur and carbon in the metal are equal to 0.75 and 2.0 wt.%, respectively. Therefore, the exchange currents obtained by the coulostatic method characterize step (4.166). The addition of the second electron (4.167) apparently takes place so rapidly that it is not detected under the conditions of the experiments.

4.3.4.4 Influence of the Pressure and the Surface Concentrations of Elements on the Anodic Polarization of Fe–C Melts

When the kinetics of reaction (4.135) were studied in an Fe–C/oxide-melt system [76, 77, 93–95], chemical polarization was discovered over the comparatively broad temperature range from 1333 to 1793 K. It was shown on the basis of an analysis of the basic experimental data in [96] that the discharge of O^{2-} ions on liquid metal anodes which are capable of dissolving nascent oxygen is described by reaction (4.144).

According to the data in [164], reaction (4.135) takes place in two steps:

$$[C] + (O^{2-}) = C(O)^{-}_{ads} + 1e, \tag{4.168}$$

$$C(O)^{-}_{ads} = \{CO\} + 1e. \tag{4.169}$$

The rate-limiting step of reaction (4.135) is electrochemical desorption (4.169), and step (4.168) takes place relatively fast. The expression for the exchange current in reaction (4.169) has the form

$$i_0 = kP^{\alpha}_{CO}C^{(1-\alpha)}_{C(O)^{-}_{ads}}. \tag{4.170}$$

Replacing the concentration of the adsorbed particles from the equilibrium condition of reaction (4.168), we obtain

$$i_0 = k'P^{\alpha}_{CO}[C]^{(1-\alpha)}C^{(1-\alpha)}_{C(O)^{-}_{ads}}. \tag{4.171}$$

Table 4.15 Kinetic characteristics of the anodic oxidation of carbon dissolved in iron

	T, K	P, atm	$i_0 \cdot 10^3$, A/cm^2	$k_{exch} \cdot 10^3$, A/cm^2	α	E, kJ/mole
Fe–C$_{sat}$ melt	1673	0.013	7.3	24.0	0.36	250
		0.25	12.2			
		0.5	18.7			
	1723	0.5	28.0	38.5	0.5	
		1.0	38.5			
		1.5	46.0			

Since the concentration of oxygen ions in the slag and the concentration of carbon in the metal do not depend on the pressure, then, according to Eq. 4.171, the exchange current should increase with pressure proportionally to P^{α}_{CO} ($\alpha = 0.4$–0.6). The value of α should increase with increasing temperature.

The experiments were performed at pressures equal to 0.013–1.5 atm and temperatures equal to 1673–1723 K with an Fe–C$_{sat}$ melt and a slag with the following composition (wt.%): 40 CaO, 40 Al_2O_3, 6 MgO. The composition of the slag was chosen such that it would have sufficient fluidity at the temperatures of the experiments and would not contain silicon dioxide, which can be reduced by pig iron. The appearance of Si in the metal can distort the results as a consequence of its anodic dissolution.

The results of the experiments are presented in Table 4.15.

The following equation relating the variation of the exchange current to the pressure and temperature was obtained:

$$\log i_0 = 5.57 - \left(\frac{4450}{T} - 2.96\right) \log P - \frac{13000}{T}. \tag{4.172}$$

Thus, the experimental data confirm the conclusions drawn during the analysis of expression (4.171) both with respect to the form of the dependence and with respect to the values of the exponent.

As has already been noted, additional information regarding the mechanism of the process can be obtained from relaxation measurements. The data from such measurements [77] confirm the multiple-step character of reaction (4.134), since the values of the exchange currents obtained exceed the values found by steady-state methods by more than an order of magnitude.

Let us analyze the influence of several factors on the kinetic parameters of step (4.168) [97] in the example of the data in [77]. The results in [77] were obtained in the temperature range from 1603 to 1808 K for an Fe–C metallic melt containing carbon in concentrations from 0.22 wt.% up to the saturation concentration and the following series of oxide melts (wt.%): 40 CaO, 40 SiO_2, 20 Al_2O_3 (I); 51 Al_2O_3, 43 CaO, 6 MgO (II); 48 CaO, 42 Al_2O_3, 5 SiO_2, 5 MgO (III).

First of all, we note the influence of the temperature and the carbon content in the metal (Fig. 4.29) [77]. It is seen that the bulk concentration of carbon has practically no influence on the values of the exchange current for step (4.168).

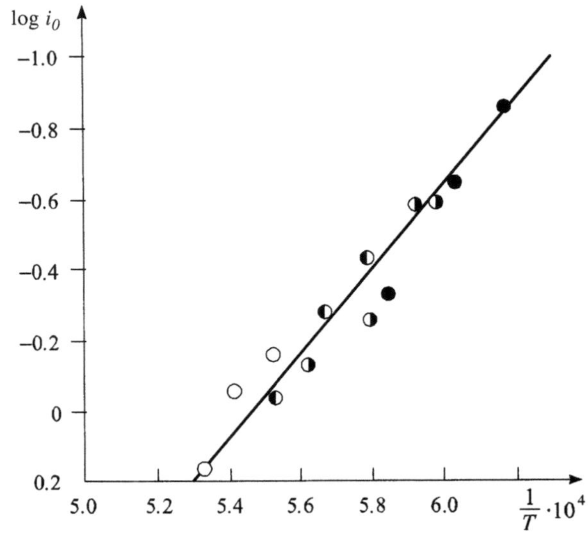

Fig. 4.29 Dependence of the exchange current for reaction (4.168) on the temperature: ● —[C] = $[C]_{sat}$; ◖— [C] = 1.6 wt.%; ◑ — [C] = 0.46 wt.%; ○ — [C] = 0.22 wt.%

The activation energy was found to be close to 250 kJ/mole. Taking into account that step (4.168) takes place on the slag/metal interface, we will analyze the relationship between the values of the exchange current of the fast step and the surface concentrations of the components.

To calculate he surface concentrations of the components of the metallic melt, we will use the Popel'–Pavlov thermodynamic-model Eq. 4.158 [19].

In [77] only the values of the concentrations of carbon in the metal were presented. We will evaluate the content of the remaining components mathematically.

For alloys that are saturated with carbon, the values of its concentration were calculated from the equation [98]

$$[C]_{sat} = 1.34 + 2.54 \cdot 10^{-3} t. \qquad (4.173)$$

Here t is the temperature of the melt, °C.

Subsequently, the calculation procedure was as follows.

Assuming that a state close to the equilibrium state is achieved during the holding time of the metal and the slag before performing the electrochemical measurements, we calculated the concentrations of the components in the contacting phases according to the procedure described in Sect. 4.2.1. The data obtained in this manner served as the starting material for finding the surface concentrations of the components according to expression (4.158).

It should be noted that the following fairly specific data are needed for calculations based on Eq. 4.158: the concentrations of carbon and oxygen ions in the slag. The former quantity was calculated using the following expression, which was presented in [99]:

$$
\begin{aligned}
\ln(\%C) = &-\frac{100879}{T} + 55.45 + \ln a_{CaO} \\
&+ \left(-13.8875 + 1.5882 \cdot 10^{-7}T - 4.546 \cdot 10^{-6}T^2\right)K_P^0 \\
&+ \left(3.9279 \cdot 10^{-1} - 4.4054 \cdot 10^{-4}T + 1.2353 \cdot 10^{-7}T^2\right) \cdot \left(K_P^0\right)^2.
\end{aligned}
\tag{4.174}
$$

Here T is the temperature, K; a_{CaO} is the activity of CaO in the slag investigated; K_P^0 is a constant defined by the expression

$$
K_{eq}^0 = \frac{26.106 \cdot a_{CaO}^0}{0.755 \cdot B - 0.04}, \tag{4.175}
$$

where a_{CaO}^0 is the activity of CaO in the $CaO–SiO_2–Al_2O_3$ system when the concentration of SiO_2 in the slag is >20 wt.% or a_{CaO}^0 is the activity of CaO in the $CaO–SiO_2$ system when the concentration of SiO_2 in the slag is $\leq$20 wt.% and

$$
B = \frac{(\%CaO)}{(\%SiO_2)} + (\%Al_2O_3). \tag{4.176}
$$

The values of $N_{O^{2-}}$ were calculated from the following equations, which were presented in [99]:

$$
\begin{aligned}
\ln \overline{K^{polym}} = &\frac{N_{SiO_2}}{1 - \sum_{j=0}^{k} N_j} \sum_{j=0}^{k} \frac{N_j}{\sum_{j=0}^{k} N_j} \ln K_{j-SiO_2}^{polym} \\
&+ \frac{N_{Al_2O_3}}{1 - \sum_{j=0}^{k} N_j} \sum_{j=0}^{k} \frac{N_j}{\sum_{j=0}^{k} N_j} \ln K_{j-Al_2O_3}^{polym},
\end{aligned}
\tag{4.177}
$$

$$
\overline{K^{polym}} = \frac{\beta\left[1 - 3\left(1 - \sum_{j=0}^{k} N_j\right) + 2\beta\left(1 - \sum_{j=0}^{k} N_j\right)\right]}{\left(1 - \sum_{j=0}^{k} N_j\right) \cdot (1 - \beta)^2}, \tag{4.178}
$$

$$
N_{O^{2-}} = \frac{1 - 3\left(1 - \sum_{j=0}^{k} N_j\right) + 2\beta\left(1 - \sum_{j=0}^{k} N_j\right)}{1 - 2\left(1 - \sum_{j=0}^{k} N_j\right) + \beta\left(1 - \sum_{j=0}^{k} N_j\right)}. \tag{4.179}
$$

Fig. 4.30 Dependence of the exchange current for reaction (4.168) on the surface concentration of oxygen: ● —$T = 1713$ K; ○ — $T = 1658$ K; ◑ — $T = 1623$ K. The numbers at the points correspond to the number of the slag

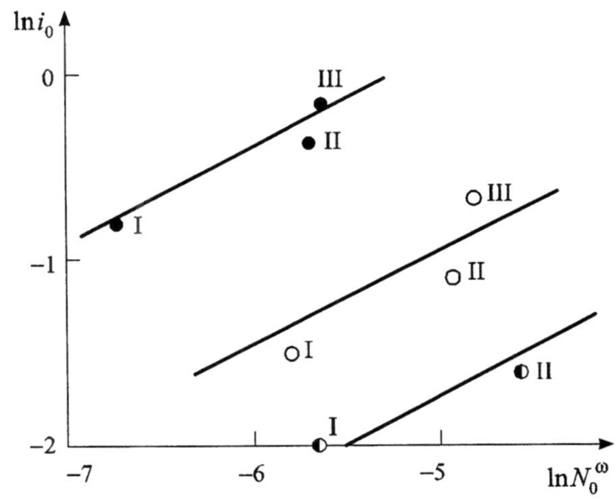

Here N_j is the mole fraction of each of the k basic oxides; $K_{j-SiO_2}^{polym}$ and $K_{j-Al_2O_3}^{polym}$ are the polymerization constants for the MeO–SiO$_2$ and MeO–Al$_2$O$_3$ binary systems, respectively.

The values of the polymerization constants at 1873 K were taken as the following:

$$K_{FeO-SiO_2}^{polym} = 0.9; \quad K_{MgO-SiO_2}^{polym} = 0.03; \quad K_{CaO-SiO_2}^{polym} = 0.003;$$
$$K_{FeO-Al_2O_3}^{polym} = 0.9; \quad K_{MgO-Al_2O_3}^{polym} = 0.2; \quad K_{CaO-Al_2O_3}^{polym} = 0.044.$$

Despite the approximate character of the calculations, the following equation for the exchange current of the fast step (4.168) was obtained on the basis of their results:

$$i_0 = 4.17 \cdot 10^9 N_C^{\omega} (N_O^{\omega})^{0.5} \exp(-30700/T), \tag{4.180}$$

where N_C^{ω} and N_O^{ω} are the surface concentrations of carbon and oxygen, respectively.

Thus, the dependence of the exchange current of the fast step on the surface concentration of oxygen has been revealed (Fig. 4.30). A similar dependence of the rate of the fast step of the passage of sulfur through a sulfide-melt/liquid-slag interface was found in the preceding section. This may be evidence that ion-exchange processes on interfaces between metallic and oxide melts have a common nature. The specific dependence of the exchange current on $N_{O^{2-}}$ could not be revealed. This is most likely due to the fact that the concentration of "free" oxygen ions near the metal/slag boundary, rather than in the bulk, must be taken into account. As was noted in Sect. 4.3.3, these values can differ strongly from one another.

The dependence of the exchange current on the surface concentration of carbon obtained (Fig. 4.31) shows that for identical initial compositions of the metal

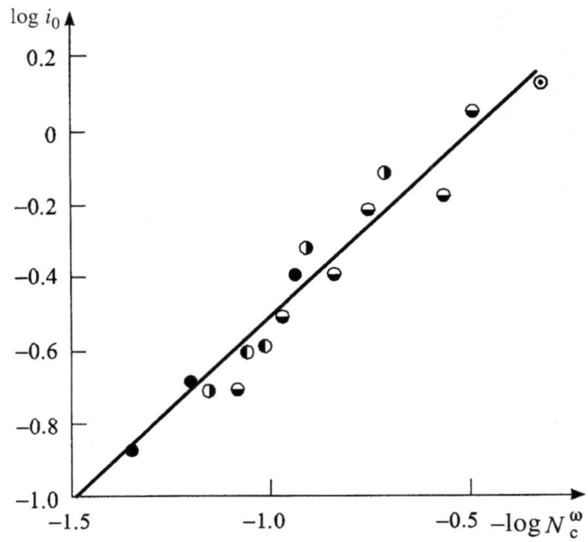

Fig. 4.31 Dependence of the exchange current for reaction (4.190) on the surface concentration of carbon

phase ([C] = $[C]_{sat}$), the temperature and composition of the slag significantly alter the surface concentration of carbon.

4.4 Conclusion

Extensive experimental material regarding the kinetics of the interaction between oxide and metallic melts, as well as the reduction of metals by solid and gaseous reactants, has been accumulated. Not only general laws, but also individual steps of the process, as well as the influence of factors such as the composition of the phases, the temperature, the pressure, the stirring intensity, etc., have been investigated.

The equilibrium distribution of most of the components appearing in the composition of steels between the metal and the slag has been studied in fairly great detail. Equations relating the distribution coefficients of elements to the temperature, the composition of the slag phase, and the pressure have been obtained.

Without dwelling on the special features of the behavior of the individual components of the metal and the slag, we note the main laws that were discovered.

The interaction in an oxide-melt/solid–carbon system takes place mainly according to two schemes: a two-step mechanism based on the reactions

$$MeO_n + nCO = Me + nCO_2, \quad (4.181)$$

$$C + CO_2 = 2CO \quad (4.182)$$

and an electrochemical mechanism, in which the reduction process consists of two electrode steps, namely, the anodic oxidation of carbon (4.135) and the cathodic deposition of the metal

Table 4.16 Most probable kinetic laws for the reduction of metals from slags by solid carbon at 1573–1773 K

Strength of the Me–O bond	Concentration of MeO, wt.%	Rate-limiting step	Formal reaction order
Strong:	<1–3	Diffusion in the slag to graphite	1
MnO, V_2O_3, Cr_2O_3	>5–10	Anodic oxidation of carbon	≤0.5
Intermediate: FeO	<1–2	Diffusion in the slag to graphite	1–2
		Anodic oxidation of carbon	≤0.5
	5–20	Diffusion in the slag to the gas/slag interface	2
		Diffusion in the slag to graphite	1–3
		Anodic oxidation of carbon	≤0.5
	>30	Diffusion in the slag to the gas/slag interface	2
		Diffusion in gaseous CO_2 to graphite	1/m = 0.2–1
		Chemical act on the gas/slag interface	>3?
Weak: NiO, CuO, PbO, etc.	<5–10	Diffusion in the slag to the gas/slag interface	2
		Diffusion in the slag to graphite	1–3
	>20	Chemical act on the gas/slag interface	≤2
		Diffusion in gaseous CO_2 to graphite	
		Diffusion in the slag to the gas/slag interface	

$$Me^{n+} + ne = Me. \tag{4.183}$$

The percentages of the direct and indirect reduction of oxides from liquid slags by solid carbon, as well as the rate-limiting steps of the process, depend on several factors: the concentration of the oxide in the slag, the temperature, the amount of convection, and the pressure. The dependence of the rate on the concentration of the oxide in the slag is different for different reduction mechanisms, and within a single mechanism, it depends on the regime and the rate-limiting step of the reaction. In addition, the regime of the evolution of gases, the mass-transfer conditions, and the ratio between the areas of the slag/graphite interface and the slag/gas interface influence this dependence.

The mechanism and rate of the reduction of metals from slags depend on the strength of their oxides. All metals can be divided by convention into three groups according to the strength of their bonds with oxygen: strong, intermediate, and weak. This division can be made quantitatively both according to the value of the dissociation pressure of the oxide (P_{O_2}) and according to the value of the equilibrium constant of the corresponding indirect reduction reaction $\left(K_{P(4.181)}\right)$. It has been shown that if $P_{O_2} < 10^{-9}$atm or $K_{P(4.181)} << 1$, the oxides are strong, and they can be reduced only electrochemically; when $P_{O_2} > 10^{-9}$ atm or $K_{P(4.181)} > > 1$, the oxides readily dissociate, and reduction takes place mainly through the gas phase; finally, when $P_{O_2} \approx 10^{-9}$ atm or $K_{P(4.181)} \approx 1$, reduction proceeds with similar rates according to both mechanisms. As the temperature is increased and the concentration of oxides in the slag decreases, the percentage of electrochemical reduction increases, and the regime approaches diffusion control. The literature data are summarized in Table 4.16.

An analysis of the interaction of oxide melts with carbon-containing liquid metals enabled us to reveal some special features of these processes.

Because of the better wetting of pig iron by a slag, the percentage of indirect reduction can be smaller for Me–C melts than for solid carbon, and, conversely, the reaction rates in a diffusion-controlled regime should be higher. In addition, the diffusion processes of the elements in the metal should be taken into account in this case.

The data in Table 4.16 allow us to draw several conclusions of a methodical nature.

1. A conclusion regarding the rate-limiting step and the regime of the reduction process drawn on the basis of indirect data, namely, the value of the activation energy, the reaction order with respect to the concentration of the reduced oxide, and the influence of stirring on the rate of the process, can be equivocal.
2. The approach based on comparison of the experimental reaction rates with reaction rates calculated from independent data can be more convincing for determining the reduction scheme.
3. To substantiate a conclusion regarding the regime and the rate-limiting steps of the process, the kinetics of the reduction process should be studied over a broad range of concentrations, temperatures, and pressures.

References

1. Moiseev, G.K., Vatolin, N.A.: Thermodynamic modeling: concentration, application, examples. Rasplavy **5**, 15–40 (1990)
2. Marshuk, L.A., et al.: Use of thermodynamic modeling for devising theoretical principles for obtaining new ferroalloys. In: Physicochemical Principles of Metallurgical Processes. Part III (in Russian), pp. 179–180. Chermetinformazia, Moscow (1991)
3. Grigoryan, V.A., Belyanchikov, L.P., Stomakhin, A.Ya.: Theoretical Foundations of Electro-Steelmaking Processes (in Russian), p. 272. Metallurgiya, Moscow (1987)
4. Kulikov, I.S.: Deoxidation of Metals (in Russian), p. 504. Metallurgiya, Moscow (1975)
5. Sabirzyanov, T.G.: Activity and solubility of carbon in iron melts. Izv Vyssh Uchebn Zaved Chern Metall **1**, 1–4 (1989)
6. Mikhailov, G.G.: Thermodynamic principles of the optimization of the deoxidation of steel and the modification of nonmetallic inclusions. Doctoral dissertation, Moscow (1985)
7. Ponomarenko, A.G.: Questions in the thermodynamics of phases of variable composition with an itinerant electronic system. Zh Fiz Khim **7**, 1668–1674 (1974)

8. Kireev, P.S.: Physics of Semiconductors (in Russian), p. 584. Vysshaya Shkola, Moscow (1975)
9. Mikhailets, V.N., et al.: Physicochemical models for predicting the distribution of components between a metal and a slag under oxidative conditions. In: Physicochemical and Technological Databases for Optimizing Metallurgical Technologies (in Russian), pp. 154–161. Dnepropetrovsk (1988)
10. Laptev, D.M.: Analysis of inaccuracies in the literature on the thermodynamics of solutions. Izv Vyssh Uchebn Zaved Chern Metall **12**, 5–11 (1987)
11. Yanagase, T., Suginahara, Y.: Studies on constitution of vitreous silicates by infrared absorption spectra. Trans Jpn Inst Metals **11**, 4000–4003 (1970)
12. Ezikov, V.I., Pasishnik, S.V., Chuchmarev, S.K.: Application of trimethylsilylation and rapid quenching to the investigation of the constitution of anions in silicate melts. In: Abstracts of Reports to the 6th All-Union Conference on the Constitution and Properties of Metallic and Oxide Melts. Part 3 (in Russian), pp. 229–230, Sverdlovsk, 1986
13. Stolyarova, V., et al.: High temperature mass spectrometric study of multicomponent silicate systems. In: Proceedings 4th International Conference on Molten Slags and Fluxes, pp. 185–188. Sendai, Japan (1992)
14. Kozheurov, V.A.: Thermodynamics of Metallurgical Slags (in Russian), p. 163. Metallurgizdat, Sverdlovsk (1995)
15. Ban-Ya, S.: Mathematical expression of slag-metal reactions in steelmaking process by quadratic formalism based on the regular solution model. In: Proceedings 4th International Conference on Molten Slags and Fluxes, pp. 8–13. Sendai, Japan (1992)
16. Jahanshaki, S., Wright, S.: Aspects of the regular solution model and its application to metallurgical slags. In: Proceedings 4th International Conference on Molten Slags and Fluxes, pp. 61–66. Sendai, Japan (1992)
17. Snitko, Yu.P., Surova, Yu.N., Lyakishev, N.P.: Method for determining the heats of mixing of oxides. In: Abstracts of Reports to the 5th All-Union Conference on the Constitution and Properties of Metallic and Slag Melts. Part 3 (in Russian), pp. 39–41. Sverdlovsk, 1983
18. Gyae, H., et al.: A statistical thermodynamics model of slags: applications to systems containing S, P, P_2O_5, and Cr oxides. In: Proceedings 4th International Conference on Molten Slags and Fluxes, pp. 103–108. Sendai, Japan (1992)
19. Popel', S.I., Sotnikov, A.I., Boronenkov, V.N.: Theory of Metallurgical Processes (in Russian), p. 403. Metallurgiya, Moscow (1986)
20. Sinyarev, G.B., et al.: Application of Computers to Thermodynamic Calculations of Metallurgical Processes (in Russian), p. 263. Nauka, Moscow (1982)
21. Mochalov, S.P., et al.: Method for the mathematical modeling and calculation of the equilibrium conditions in complex steelmaking systems. In: Modeling of Physicochemical Systems and Technological Processes in Metallurgy (in Russian), pp. 132–133. Novokuznetsk Institute of Metal Press, Novokuznetsk (1991)
22. Gasik, M.M.: Carbon–oxygen equilibrium in liquid iron during vacuum–carbon deoxidation. Izv Vyssh Uchebn Zaved Chern Metall **10**, 9–13 (1991)
23. Zhancheng, G., et al.: A model on slag-metal reaction kinetics. In: Proceedings 4th International Conference on Molten Slags and Fluxes, pp. 308–313. Sendai, Japan (1992)
24. Kazachkov, E.A.: Calculations Based on the Theory of Metallurgical Processes (in Russian), p. 288. Metallurgiya, Moscow (1988)
25. Shalimov, M.P., et al.: Calculation of the equilibrium distribution of elements between metallic and slag melts. In: Structure and Properties of Slag Melts (in Russian), pp. 62–66. Kurgan Mach Build Institute Press, Kurgan (1984)
26. Ban-Ya, S., Dong-Shim, Y.: Application of the regular solution model to steelmaking slags. In: Proceeding of the 10th Soviet-Japanese Symposium on the Physicochemical Principles of Metallurgical Processes (in Russian), pp. 21–41. Metallurgiya, Moscow (1983)
27. Turkdogan, E.T.: Physical Chemistry of High-Temperature Processes (in Russian), p. 344. Metallurgiya, Moscow (1985)

28. Zinigrad, M.I.: Kinetics and mechanism of the interaction of metals and oxide melts. Doctoral dissertation, p. 419, Sverdlovsk (1981)
29. Medzhibozhskii, M.Ya., Konoplya, V.G., Plokhikh, G.A.: Some laws governing the distribution of phosphorus between a metal with a high phosphorus content and a slag. In: Electrochemistry and Melts (in Russian), pp. 201–207. Nauka, Moscow (1974)
30. Chuiko, N.M., Zaozernyi, M.T.: Distribution of tungsten between a metal, a slag, and a gas phase during the smelting of high-speed steels. Metally **6**, 20–26 (1974)
31. Zamoruev, V.M.: Tungsten in Steel (in Russian), p. 199. GNTI, Moscow (1962)
32. Adel'shin, Yu.G., et al.: Thermodynamic analysis of the oxidation of tungsten in a steelmaking arc furnace. Izv Vyssh Uchebn Zaved Chern Metall **11**, 51–54 (1984)
33. Perevalov, N.M., Mogutnov, B.M., Shvartsman, L.A.: Influence of the basicity of a slag on the oxidation of elements of the chromium subgroup dissolved in liquid iron. Dokl Akad Nauk SSSR **124**(1), 150–152 (1959)
34. Perevalov, N.M., Mogutnov, B.M., Shvartsman, L.A.: Influence of calcium oxide on the distribution of tungsten between liquid iron and a slag. Izv Akad Nauk SSSR Metall Topl **1**, 22–28 (1959)
35. Merkulov, V.F., et al.: Technology for recovering tungsten from boring bit scrap. Stal' **10**, 30–32 (1984)
36. Okol'zdaev, A.G., et al.: Influence of the oxidative capacity of the slag on the distribution of tungsten between the metallic and oxide phases. In: Abstracts of Reports to the 6th All-Union Conference on the Constitution and Properties of Metallic and Slag Melts, Part 3 (in Russian), pp. 238–239. Sverdlovsk, 1986
37. Yavoiskii, V.I.: Theory of Steel Production Processes (in Russian), p. 792. Metallurgizdat, Moscow (1967)
38. Bagryanskii, K.V.: Electric-Arc Welding and Surfacing under Ceramic Fluxes (in Russian), p. 184. Tekhnika, Kiev (1976)
39. Ignatov, M.N.: Improving the quality of welded joints of nickel and nickel alloys by using transition metal carbides in the electrode coatings. Candidate dissertation, p. 145, Perm′ Polytechnic Institute, Perm′ (1987)
40. Burylev, B.P., Kretov, A.I., Moisov, L.P.: Thermodynamic activity of components of welding fluxes. Avtom Svarka **2**, 67–69 (1978)
41. Burylev, B.P., Kretov, A.I.: Influence of the thermodynamic activity of the slag components on the properties of the metal. In: Physicochemical Investigations of Metallurgical Processes (in Russian), vol. 8, pp. 88–90. Ural Polytechnic Institute Press, Sverdlovsk (1980)
42. Hideaki, S., Ryo, J.: Manganese equilibrium between molten iron and MgO-saturated CaO–Fe_tO–SiO_2–MnO–P_2O_5 slags. Trans. Iron Steel Inst. Jpn. **24**(4), 257–265 (1984)
43. Globovskii, V.G., Burtsev, V.T.: Melting Metals and Alloys in the Suspended State (in Russian), p. 176. Metallurgiya, Moscow (1974)
44. Boronenkov, V.N., Esin, O.A.: Kinetics of the simultaneous occurrence of several heterogeneous reactions with common reactants. Zh Fiz Khim **8**, 2022–2027 (1970)
45. Frank-Kamenetskii, D.A.: Diffusion and Heat Transfer in Chemical Kinetics, 2nd edn. Plenum Press, New York (1969)
46. Levich, V.G.: Physicochemical Hydrodynamics. Prentice-Hall, Englewood Cliffs (1962)
47. Medzhibozhskii, M.Ya., Zinov'ev, V.T., Geineman, A.V.: Influence of several factors on the rate of carbon boil in an open-hearth bath. Izv Vyssh Uchebn Zaved Chern Metall **6**, 47–53 (1960)
48. Kazakov, N.I., Filippov, S.I.: Kinetics of the oxidation of carbon by liquid steel with electromagnetic stirring. Izv Vyssh Uchebn Zaved Chern Metall **11**, 15–21 (1961)
49. Boronenkov, V.N.: Kinetic analysis of the oxidation of impurities in iron by a molten slag. In: Physicochemical Investigations of Metallurgical Processes: Proceedings of Institutions of Higher Education of the Russian Federation, No. 1 (in Russian), pp. 18–34. Sverdlovsk (1973)
50. Boronenkov, V.N., Esin, O.A.: Kinetic equation of a multicomponent reaction in a diffusion-controlled regime. Izv Vyssh Uchebn Zaved Chern Metall **9**, 17–20 (1970)

51. Boronenkov, V.N., Shanchurov, S.M., Zinigrad, M.I.: Kinetics of the interaction of a multicomponent metal with a slag in a diffusion-controlled regime. Izv Akad Nauk SSSR Met **6**, 21–27 (1979)
52. Boronenkov, V.N., Salamatov, A.M.: Mathematical estimation of the kinetics of the interaction of a multicomponent metal with the slag during submerged-arc welding. Avtomat Svarka **8**, 19–24 (1985)
53. Churkin, A.S., Toporishchev, G.A., Esin, O.A.: Kinetic features of the desulfurization of high-silicon pig iron by a flux of the $CaO–SiO_2–Al_2O_3$ system. Izv Akad Nauk SSSR Met **1**, 37–44 (1971)
54. Boronenkov, V.N., Pozdnyakov, A.D.: Mathematical model of the kinetics of the simultaneous oxidation of impurities in iron by a molten slag. In: Physicochemical Principles of Metallurgical Processes. No. 7 (in Russian), pp. 75–83. Ural Polytechnic Institute Press, Sverdlovsk (1979)
55. Damaskin, B.B., Petrii, O.A.: Introduction to Electrochemical Kinetics (in Russian), p 416. Vyshaya Shkola, Moscow (1975)
56. Boronenkov, V.N., Shalimov, M.P., Shanchurov, S.M.: Method for analyzing the kinetics of simultaneously occurring electrode reactions under non-steady-state conditions. Rasplavy **5**, 12–17 (1994)
57. Vetter, K.: Electrochemical Kinetics. Academic, New York (1967)
58. Delahay, P.: Double Layer and Electrode Kinetics. Interscience, New York (1965)
59. Turchak, L.I.: Fundamentals of Numerical Methods (in Russian), p. 320. Nauka, Moscow (1987)
60. Boronenkov, V.N., Shanchurov, S.M.: Kinetics of reactions between a metal and a slag under a non-steady-state diffusion regime. Metally **3**, 5–11 (1992)
61. Kostin, N.A., Kublanovskii, V.S., Zabludovskii, V.A.: Pulsed Electrolysis (in Russian), p. 167. Naukova Dumka, Kiev (1989)
62. Tsymbal, V.P.: Mathematical Modeling of Metallurgical Processes, p. 240. Metallurgiya, Moscow (1986)
63. Esin, O.A.: Application of polymer theory to molten slags. In: Physicochemical Investigations of Metallurgical Processes (in Russian), vol. 1, pp. 5–17. Ural Polytechnic Institute Press, Sverdlovsk (1973)
64. Zalomov, N.I., Boronenkov, V.N., Shanchurov, S.M.: Method for kinetic analysis of the oxidation of silicon by polymeric oxide melts. Rasplavy **2**(1), 12–17 (1988)
65. Zalomov, N.I., Shalimov, M.P., Boronenkov, V.N.: Kinetics of electrochemical processes on a metal/polymeric-oxide-melt interface. In: Constitution and Properties of Metallic and Slag Melts, Part 2 (in Russian), Vol. 3, pp. 199–203. Chelyabinsk Polytechnic Institute Press, Chelyabinsk (1990)
66. Esin, O.A., Gavrilov, L.K.: Electrode polarization at high temperatures. Zh Fiz Khim **29**(3), 566–575 (1955)
67. Novikov, V.K.: Influence of the composition of the molten metal and the molten slag on the kinetics of the transport of silicon and sulfur through the interface between them. Candidate dissertation, p. 175, Sverdlovsk (1972)
68. Esin, O.A., Gel'd, P.V.: Physical Chemistry of Pyrometallurgical Processes (in Russian), p. 704. Metallurgizdat, Moscow (1966)
69. Delahay, P.: New Instrumental Methods in Electrochemistry. Interscience, New York (1954)
70. Damaskin, B.B.: Principles of Modern Methods for Studying Electrochemical Reactions (in Russian), p. 104. Moscow State University Press, Moscow (1965)
71. Sotnikov, A.I.: Structure of the metal/oxide-melt interface and special features of electrochemical methods in metallurgical systems. In: Physicochemical Investigations of Metallurgical Processes, vol. 2, pp. 40–49. Ural Polytechnic Institute Press, Sverdlovsk (1974)
72. Voinov, S.G., Shalimov, A.G., Kosoi, L.F., Kalinnikov, E.S.: Refining Steel by Synthetic Slags (in Russian), p. 186. Metallurgiya, Moscow (1970)

73. Fefelov, A.S.: Metallurgical and technological features of alloying a metal with boron. Candidate dissertation, p. 190, Ural Polytechnic Institute Press, Sverdlovsk (1982)
74. Zinigrad, M.I., et al.: Kinetics of the interaction of a boron-containing metallic melt with an oxide electrolyte. Élektrokhimiya **13**(1), 74–78 (1986)
75. Flyagin, A.A., Zinigrad, M.I.: Kinetics of the passage of aluminum and boron through a liquid-steel/oxide-melt interface. Izv Akad Nauk SSSR Met **1**, 50–55 (1986)
76. Sotnikov, A.I., Esin, O.A., Nikitin, Yu.P.: Chemical polarization at high temperatures. Dokl Akad Nauk SSSR **152**(5), 1173–1176 (1963)
77. Melamud, S.G.: Investigation of the kinetics of high-temperature electrode processes involving carbon, silicon, and chromium in a metal–oxide-melt system. Candidate dissertation, p. 185, Sverdlovsk (1970)
78. Lepinskikh, B.M., Kaibichev, A.V., Savel'ev, Yu.A.: Diffusion of Elements in Liquid Iron-Group Metals (in Russian), p. 192. Nauka, Moscow (1974)
79. Flyagin, A.A.: Features of the passage of aluminum through a metal/slag interface during the treatment of steel with a synthetic slag. Candidate dissertation, p. 167, Sverdlovsk (1980)
80. Barmin, L.N., et al.: Application of a coulostatic method to the study of the rate of ion exchange between liquid manganese and an oxide melt. In: Physical Chemistry and Electrochemistry of Molten Salts and Slags. Part II (in Russian), pp. 99–102. Institute Electrochemistry Press, Kiev (1969)
81. Panov, S.P., Zinigrad, M.I., Barmin, L.M.: Investigation of the fast step of the desulfurization of an iron–carbon melt by a slag. In: Scientific Reports to the 6th All-Union Conference on the Constitution and Properties of Metallic and Slag Melts: Investigations of Slag Melts (in Russian), vol. 3, pp. 60–63. Sverdlovsk (1980)
82. Sotnikov, A.I.: Structure of the metal/oxide-melt interface and features of electrochemical relaxation methods of investigation in metallurgical systems. In: Physicochemical Investigations of Metallurgical Processes: Proceedings of Institutions of Higher Education of the Russian Federation (in Russian), vol. 1, pp. 40–54 (1974)
83. Frumkin, A.N.: Kinetics of Electrode Processes (in Russian), p. 318. Izd MGU, Moscow (1952)
84. Damaskin, B.B., Petrii, O.A.: Foundations of Theoretical Electrochemistry (in Russian), p. 239. Vysshaya Shkola, Moscow (1978)
85. Popel', S.I.: Investigation of phenomena on phase boundaries in the steelmaking process. Doctoral dissertation, pp 347, Sverdlovsk (1959)
86. Pavlov, V.V., Popel', S.I., Esin, O.A.: Dependence of the interfacial tension on the composition and temperature. In: Surface Phenomena in Melts and Solid Phases Derived from them (in Russian), pp. 136–141. Kabardino Balkar Publishing, Nal'chik (1965)
87. Yakobashvili, S.B.: Surface Properties of Welding Fluxes and Slags (in Russian), p. 207. Technika, Kiev (1970)
88. Andronov, V.N., Chekin, B.V., Nesterenko, S.V.: Liquid Metals and Slags. Handbook Edition (in Russian), p. 128. Metallurgiya, Moscow (1977)
89. Zinigrad, M.I., et al.: Investigations of the distribution of sulfur between an Fe–C–S melt and a $CaO–Al_2O_3–MgO$ slag. In: Physicochemical Investigations of Metallurgical Processes, vol. 9, pp. 60–64. Ural Polytechnic Institute Press, Sverdlovsk (1981)
90. Zinigrad, M.I., et al.: Kinetic features of the desulfurization of an iron–carbon melt by a slag. Izv Vyssh Uchebn Zaved Chern Metall **2**, 4–6 (1981)
91. Losev, V.V., Gorodetskii, R.V.: Stepwise occurrence of the discharge and ionization of metals. Élektrokhimiya **3**(9), 1061–1070 (1967)
92. Shantarin, V.D., Esin, O.A., Boronenkov, V.N.: Chemical component of the anodic polarization of molten iron saturated with carbon. Élektrokhimiya **3**(6), 775–778 (1967)
93. Buler, P.I., Esin, O.A., Nikitin, Yu.P.: Influence of the composition on the anodic polarization of iron alloys with carbon and nickel. Élektrokhimiya **3**(3), 288–293 (1967)
94. Fugman, G.I.: Kinetics of metallurgical reactions involving titanium and carbon. Candidate dissertation, p. 185, Sverdlovsk (1972)

95. Sotnikov, A.I.: Kinetics of Electrode Processes on a Metal/Oxide-Melt Interface (in Russian), p. 80. Ural Polytechnic Institute Press, Sverdlovsk (1981)
96. Shalimov, M.P., Zinigrad, M.I.: Kinetic features of ion exchange on an iron–carbon-melt/slag interface. In: Physicochemical Principles of Metallurgical Processes. Part II (in Russian), pp. 7–9. Chermetinformazia, Moscow (1991)
97. Medzhibozhskii, M.Ya., et al.: Dynamic computer simulation of the decarbonization of steel. Izv Vyssh Uchebn Zaved Chern Metall **7**, 158–161 (1977)
98. Sevic, D., Cureija, D.: Temperaturna ovisnost rastvorljivosti uglika u troskama. Zelez Zb **16**(4), 123–126 (1982)
99. Novikov, V.K.: Development of the polymer model of silicate melts. Rasplavy **1**(6), 21–33 (1987)
100. Zinigrad, M.I., et al.: Investigation of the distribution of sulfur between Fe–C–S and CaO–Al_2O_3–MgO melts. In: Physicochemical Investigations of Metallurgical Processes, No. 9 (in Russian), pp. 60–64. Ural Polytechnic Institute Press, Sverdlovsk (1981)

Chapter 5
Modeling and Simulation of Technological Processes for Producing and Refining Steel

The production of metals with required chemical composition in steelmaking plants is a very difficult problem, because it is associated with the occurrence of complex physicochemical processes. The greatest contribution to shaping the composition of the products of a melting process is made by the chemical reactions that occur on the various phase boundaries in the metal–slag–gas system. A comprehensive study of their laws is needed to perfect existing technological processes, to develop new technological processes, and to introduce them into practice. Prediction of the chemical composition of the metal and the slag based on mathematical modeling of the phase interaction not only opens up additional routes for improving the economic parameters of technological processes, but also allows their optimization in the design stage. Kinetic analysis methods that are known and that we developed, thermodynamic and kinetic parameters that have been obtained, and data concerning the mechanism of individual steps of heterogeneous reactions were used to create mathematical models that enabled us to predict the composition and mass of the phases for different variants of metal production and refinement processes.

Most of the existing methods for predicting the composition of liquid phases are based on analyzing statistical data and obtaining semiempirical dependences. This sharply reduces their areas of application.

This chapter is devoted to developing mathematical models of the physicochemical processes in a metal–slag–gas system based on fundamental laws of thermodynamics, chemical kinetics, and heat and mass transfer as applied to various technologies.

5.1 Mathematical Modeling of Steel Smelting Processes

The smelting of steel in steelmaking electric arc furnaces (EAF) is one of the most widespread technological processes for obtaining steel and alloys of various grades. High-quality, high-alloy steels, such as corrosion-resistant, transformer,

M. Zinigrad et al., *Phase Interaction in the Metal–Oxide Melts–Gas System*, Engineering Materials, DOI: 10.1007/978-3-642-22377-8_5,

Table 5.1 Furnace loss of alloying elements [2]

Group of steels or alloys	Melting method	Furnace loss (rel.%)						
		Mn	Si	Cr	W	V	Mo	Ni
Chromium and chrome–nickel structural steels	With oxygen	80	100	20	–	60	0	0
	Without oxygen	20	40	10	–	20	0	0
Chrome–nickel–tungsten structural steels	With oxygen	80	100	40	5	–	0	0
	Without oxygen	20	40	15	10	–	0	0
Tool steels	With oxygen	60	80	15	7	11	0	0
	Without oxygen	20	40	10	10	7	0	0
High-speed steels	With oxygen	40	80	8–10	4–5	10.5	0	0
	Without oxygen	20	40	6–7	6–7	9	0	0
Nichrome alloys	With oxygen	80	100	15	–	–	0	4
	Without oxygen	0	40	10	–	–	0	4

bearing, tool, and other steels, are currently smelted in alternating current EAFs. EAFs of especially high capacity (more than 100 tons) are also used for smelting ordinary carbon steels [1]. The charge materials used in EAFs include scrap from local metallurgical production processes, old scrap, ferroalloys of various types, metallized raw materials, and solid pig iron.

With consideration of the extensive list of smelted steels and alloys, the efficiency of an EAF depends to a considerable extent on perfecting the melting production process. The economic advantage of solutions based on utilizing modern advances of science and technology take on great importance in this case.

The following principal stages of the smelting of steel in an EAF were chosen for mathematical modeling: melting of the charge and subsequent periods up to tapping of the metal into a ladle.

The main goal of the simulation of the charge melting process was to predict the composition of the primary metal and slag. For this purpose, the data presented in [2], which were based on statistical treatment of the results of a large number of melting operations, were used. In particular, Table 5.1 [2] presents the furnace loss of alloying elements during the smelting of steel with and without the use of oxygen. The passage of elements into the slag and the gas was taken into account in a similar manner on the basis of the data from [2]. This provided a way to calculate the mass and composition of the metal and the slag in the melting period.

A preliminary calculation is performed for 100 kg of charge.

We first calculate the mass and composition of the metal in the melting period with allowance for the furnace loss of the elements

$$M_{\mathrm{Me}} = \sum_{i=0}^{n} M_i^0 (1 - a_i), \tag{5.1}$$

where M_{Me} is the mass of the metal in the melting period without taking into account the oxygen (kg), M_i^0 is the mass of the ith component in the charge (kg), and a_i is a coefficient that takes into account the furnace loss of element i.

The following relation was used to take into account the oxygen content. If the mass of carbon in the metal in the melting period (per 100 kg of charge) is greater than 0.1 kg, we assume that the mass of oxygen in the metal is 0.01 kg. If the mass of carbon is less than 0.1 kg, we set the mass of oxygen equal to 0.02 kg. Then the mass of the metal in the melting period will be equal to

$$M^0_{\mathrm{Me}} = M_{\mathrm{Me}} + M_{\mathrm{O}}, \tag{5.2}$$

where M^0_{Me} is the mass of the metal in the melting period (kg) and M_{O} is the mass of oxygen in the melting period (kg).

These data were used to calculate the composition of the metal in the melting period

$$[\mathrm{E}_i] = \frac{M^0_i(1-a_i)}{M^0_{\mathrm{Me}}} \cdot 100, \tag{5.3}$$

where $[\mathrm{E}_i]$ is the concentration of component i in the metal in the melting period (wt.%).

Then we determined the mass of the slag in the melting period

$$M^0_{\mathrm{slag}} = \sum_{i=0}^{n} M^0_i a_i b_i + \sum_{j=0}^{m} \frac{100 - \sum_1^N y_n}{100} M^0_j + \frac{100 - \%\mathrm{Fe_2O_3} - \sum_1^M y_m}{100} M_{\mathrm{ore}} + \sum_1^Z M_z a_z + \sum_1^X M_x a_x. \tag{5.4}$$

Here M^0_{slag} is the mass of the slag in the melting period, kg; b_i is the fraction of the oxide that formed as a result of oxidation of the metal in the slag; M_j is the mass of the slag formers in the charge, kg; y_n is the concentration of the volatile components in the j-th slag former in the charge, wt.%; M_{ore} is the mass of the ore in the charge, kg; $\%\mathrm{Fe_2O_3}$ and y_m are the concentrations of $\mathrm{Fe_2O_3}$ and the volatile components in the ore, wt.%; M_z is the consumption of the inner surfaces of the furnace in one smelting cycle, kg; a_z is the fraction of M_z that passes into the slag in the melting period; M_x is the mass (consumption) of carbon-containing materials, kg; a_x is the fraction of ash in them, wt.%.

After determining the mass of each slag component originating from all the sources enumerated, we find its concentration in the slag. Thus, the composition of the slag in the melting period was calculated.

These data are the starting data for further calculations. The actual masses of the metal and slag in the melting period were obtained with consideration of the mass of the charge loaded.

To evaluate the variation of the concentrations of the components in the metal and the slag over the course of the melting, the method for kinetic analysis of simultaneously occurring processes with a common reactant described in Sect. 4.3.2 was employed.

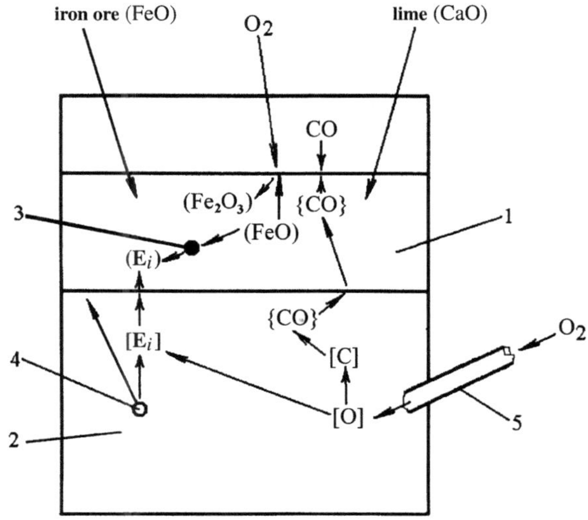

Fig. 5.1 Diagram of the phase interactions that occur during the smelting of steel in an EAF: *1*—oxide melt; *2*—metal pool; *3*—metal droplet in the slag; *4*—slag droplet in the metal; *5*—tuyere for injecting oxygen

In the general case, when steel is smelted in an EAF, interactions can occur on the following interfaces (Fig. 5.1) with the surface area S_j:

1. The metal/slag interface. The reaction area S_1 is for the boundary of the metal pool with the slag; S_2 is for the metal-droplet/slag boundary; S_3 is for the slag-droplet/metal boundary.
2. The metal/gas interface (when oxygen is blown through the metal) with the surface area S_4. Oxidation of components of the metal, primarily carbon, occurs on this interface. The vigorous evolution of CO bubbles results in stirring of the phases with averaging of their temperature and chemical composition.
3. The slag/gas interface with the surface area S_5. The conversion of lower oxides (FeO, MnO, etc.) into higher oxides, the oxidation of sulfur by oxygen in the gas phase, etc. occur here.

Let us examine the processes occurring on these interfaces in greater detail.

The data concerning the conditions of convection of the phases and the fractional composition of the droplets when the metal interacts with the slag are insufficient for quantitatively evaluating the relative involvement of the processes on each of the surfaces in the alteration of the composition of the phases. For this reason we assumed that the interaction between the metal and the slag takes place in one hypothetical step on some effective interface with the area S_{eff}. In accordance with the procedure proposed in [3], we determined the product $\beta \cdot S_{\text{eff}}$ (for further details, see Sect. 5.2). It was found that this product varies from $6S_{\text{geom}}$ in the oxygen blowing period to $1.5S_{\text{geom}}$ in other periods of the smelting process. Here S_{geom} is the geometric area of the metal/slag interface.

Finding the rates of the interaction reactions in the period when oxygen is blown through the metal presents certain difficulties. The concentration of oxygen dissolved in the metal is known to increase considerably in this period of the process, and according to the estimates in [4, 5] it reaches 0.15 wt.%. At the same

Table 5.2 Iron oxide content in the slag during the period of blowing oxygen through the metal

Smelting cycle no., unit 21, Izhstal Corporation	T (K)	FeO content in the slag (wt.%)	
		Determined by chemical analysis	Calculated from Eq. 5.6
8259	2063	23.48	25.88
8261	2068	24.53	25.47
8264	2028	31.58	28.89
8376	2013	22.65	30.34
8378	2033	23.23	28.44
8739	2023	28.47	29.40
8742	2033	29.13	28.44

time, the rate of the decarbonization reaction rises significantly. This results in increases in the temperature and the stirring intensity [5]. It can be assumed that phase interaction conditions under which some reactions reach a state close to equilibrium are created in this period of the smelting process. It was assumed that one of these reactions may be the reaction

$$(\mathrm{FeO}) = [\mathrm{Fe}] + [\mathrm{O}]. \tag{5.5}$$

The validity of this assumption is confirmed by the data that we obtained as a result of chemical analysis of samples of the slag from several industrial smelting cycles (Izhstal Corporation). The content of iron oxide content in the slag when oxygen was blown through the metal was close to its equilibrium concentration (see Table 5.2), which was determined from the relation [6]

$$\log K_{P(5.5)} = \log \frac{(\mathrm{FeO})}{[\mathrm{O}]} = \frac{5750}{T} - 0.55. \tag{5.6}$$

To calculate the rates of the chemical processes that occur when oxygen is blown through the metal, the oxygen concentration was assumed to have a constant value equal to 0.15 wt.%, and the concentration of iron oxide in the slag was calculated from Eq. 5.6.

The concentrations of the components in the metal and the slag can vary not only due to chemical reactions, but also due to the dissolution of various additions (pellets, ore, lime, etc.), as well as, as was previously noted, dissolution of the inner surfaces of the furnace. We assumed that the dissolution of oxides from the walls, roof, hearth, and banks of the furnace occurs uniformly within the oxidation and reduction periods, respectively. The rate of solution of lime and bauxites in the slag determined from the data in [7, 8] was equal to $\sim$10 kg/s. The rate of solution of all the ferroalloys introduced in the metal was assumed to be equal to 7 kg/s. When ferrosilicium was introduced as a deoxidizer in the reduction period, it was assumed that 50% of the Si passes into the metal and that 50% is oxidized to SiO_2 and assimilated by the slag. When ferroalloys are introduced as alloying components, the assimilation of the components is nearly always close to 100%. One exception is titanium, whose assimilation from a ferroalloy by the metal is at the 65% level. When iron-ore pellets were dissolved in the liquid phases, it was assumed [9] that the oxides comprising them pass equally into the metallic and oxide melts with the overall rate of this process being equal to ~25 kg/s. The rate of solution of iron ore and the slag skimming rate estimated from literature data were assumed to be equal to 15 and 5 kg/s, respectively.

The initial temperature was assumed to be 10 K above the liquidus temperature for the metal, whose composition was obtained after melting of the charge. The liquidus and solidus temperatures were calculated using the following relations, which were obtained in [10]:

$$\begin{aligned} T_{\text{liq}} = 1539 &- 78[\text{C}] - 12[\text{Si}] - 5[\text{Mn}] - 1.5[\text{Cr}] - 4[\text{Ni}] - 5[\text{Cu}] \\ &- 2([\text{Mo}] + [\text{V}] + [\text{Al}] + [\text{Co}]) - 1.5[\text{W}] - 10[\text{Ti}] - 6[\text{Zr}] \\ &- 30[\text{P}] - 25[\text{S}] - 80[\text{O}] - 90[\text{N}] - 1300[\text{H}], \end{aligned} \tag{5.7}$$

$$T_{\text{sol}} = T_{\text{liq}} - \Delta T_{\text{cryst}}, \tag{5.8}$$

$$\Delta T_{\text{cryst}} = \Delta T_{[\text{C}]} - \Delta T_{[\text{E}]}, \tag{5.9}$$

$$\begin{aligned} \Delta T_{[\text{E}]} = 4.5[\text{Si}] &+ 2.5[\text{Mn}] + [\text{Cr}] + 4.5[\text{Ni}] + [\text{Co}] + 5[\text{Ti}] \\ &+ 3([\text{Cu}] + [\text{Mo}] + [\text{V}] + [\text{W}] + [\text{Al}]), \end{aligned} \tag{5.10}$$

$$\Delta T_{[\text{C}]} = 13 + 140[\text{C}] \text{ for } 0.6 > [\text{C}] \geq 0.18, \tag{5.11}$$

$$\Delta T_{[\text{C}]} = 21 \text{ for } [\text{C}] < 0.18. \tag{5.12}$$

Here T_{liq} is the liquidus temperature of the metal, °C; T_{sol} is the solidus temperature of the metal, °C; ΔT_{cryst} is the crystallization interval, °C; $\Delta T_{[\text{C}]}$ and $\Delta T_{[\text{E}]}$ are the "contributions" of carbon and the other alloying elements, respectively, to the crystallization interval; $[\text{E}_i]$ is the concentration of the ith component in the metal, wt.%.

These equations give good results for carbon, low-alloy, and medium-alloy steels. For high-carbon and high-alloy steels, Eq. 5.8–5.12 can have only an approximate character.

The variation of the temperature during the oxidation period of a smelting cycle is associated mainly with the evolution of heat upon the oxidation of carbon and iron, as well as with the absorption of heat upon dissolution of the various additions. In particular, the following equation for calculating the variation of the temperature of the metal in the oxidation period during the smelting of steel in an EAF was obtained in [11]:

$$\begin{aligned} \frac{dT_{\text{Me}}}{d\tau} = \frac{1}{\bar{c}_{\text{Me}} G_{\text{Me}} + \bar{c}_{\text{slag}} G_{\text{slag}}} \Bigg[P_g - 0.9P_{\text{ad}} + (Q_{\text{CO}} + 0.5Q_{\text{CO}_2}) \frac{G_{\text{Me}}}{100} \frac{d[\text{C}]}{d\tau} \\ + 0.75Q_{\text{FeO}} \frac{G_{\text{slag}}}{100} \frac{d(\text{FeO})}{d\tau} - 0,5Q_{\text{ad}} G_{\text{ad}} \left(1 - e^{-\alpha(\tau - \tau_g)}\right) (\tau - \tau_g) \Bigg]. \end{aligned} \tag{5.13}$$

Here $\bar{c}_{\text{Me}}$ and $\bar{c}_{\text{slag}}$ are the mean specific heats of the metal and the slag, respectively, J/(kg°C); G_{Me} and G_{slag} are the masses of the metal and the slag, respectively, kg; Q_{CO}, Q_{CO_2}, and Q_{FeO} are the thermal effects of the oxidation of carbon to CO, the oxidation of carbon to CO_2, and the oxidation of iron to FeO, respectively, J/kg; Q_{ad} is the thermal effect of the dissolution of the additive, J/kg; G_{ad} is the mass of the additive, kg; α is the characteristic rate of solution of the additive; τ is the current melting time, s; τ_g is the time of introduction of the additive, s.

At the same time, the experimental investigations that we performed at the Izhstal Corporation, as well as the data in [11], show that the rate of the increase in temperature during the oxidation period can be assumed to be 1 K/min.

The content of any element in the metal $[\text{E}_i]$ at the time τ can be found from the expression

$$[\text{E}_i]^\tau = [\text{E}_i]^0 + \frac{100 M_{\text{E}_i}}{M^0_{\text{Me}}} \sum_{i=1}^{n} \int_0^\tau \frac{i_i^k}{nF} S_{\text{eff}} d\tau + \sum_1^Z [\text{E}_i]_Z \int_0^\tau V_Z a_Z d\tau. \tag{5.14}$$

The concentration of the oxide of the same element (E_iO_m) equals

$$(\text{E}_i\text{O}_m)^\tau = \frac{1}{M^\tau_{\text{slag}}} \left[(\text{E}_i\text{O}_m)^0 M^0_{\text{slag}} + \sum_1^{Z_1} (\text{E}_i\text{O}_m)_{z_1} \int_0^\tau V_{z_1} d\tau \right.$$
$$\left. - \frac{100 M_{\text{E}_i\text{O}_m}}{m} \sum_{i=1}^{n} \int_0^\tau \frac{i_i^k}{nF} S_{\text{eff}} d\tau \right], \tag{5.15}$$

where $[\text{E}_i]^0$ and (E_iO_m) are the concentrations of the element in the metal and the slag in the melting period (wt.%), M^0_{Me} and M^0_{slag} are the masses of the metal and the slag in the melting period (kg), $[\text{E}_i]_z$ and $(\text{E}_i\text{O}_m)_{z_1}$ are the concentrations of the element in technological additive z and of its oxide in technological additive z_1 (wt.%), V_z and V_{z_1} are the rates of solution of technological additive z in the metal and of additive z_1 in the slag, respectively (kg/s), i_i is the rate of passage of the element through the metal/slag interface (A/cm^2), M^τ_{slag} is the mass of the slag at the time τ (kg), a_z is the conversion coefficient of the element, and M_{E_i} and $M_{\text{E}_i\text{O}_m}$ are the atomic weight of the element E_i and the molecular weight of its oxide E_iO_m (kg/mole).

To obtain the character of the variation of the composition of the metal and the slag over time, the entire process was divided into short time intervals ($\Delta\tau$), during which the reaction rates can be considered constant. Then the concentrations of the elements and their oxides at the time $k\Delta\tau$ equal

$$[\text{E}_i]^k = \left\{ [\text{E}_i]^{k-1} M^{k-1}_{\text{Me}} + \sum_1^Z [\text{E}_i]_z V_z a_z \Delta\tau + 100 M_{\text{E}_i} \Delta\tau \sum_{i=1}^{n} \left(\frac{i_i^k}{nF} \right) S_{\text{eff}} \right\} \frac{1}{M^0_{\text{Me}}}, \tag{5.16}$$

$$(\mathrm{E}_i\mathrm{O}_m)^k = \left\{ (\mathrm{E}_i\mathrm{O}_m)^{k-1} M_{\mathrm{slag}}^{k-1} + \sum_{1}^{Z_1} (\mathrm{E}_i\mathrm{O}_m)_{z_1} V_{z_1} \Delta\tau - \frac{100 M_{\mathrm{E}_i\mathrm{O}_m}}{m} \Delta\tau \sum_{i=1}^{n} \left(\frac{i_i^k}{nF} \right) S_{\mathrm{eff}} \right\} \frac{1}{M_{\mathrm{slag}}^0}, \tag{5.17}$$

$$M_i^k = M_i^{k-1} \pm \Delta M_i^{k-1} + \sum_{1}^{Z} V_z \Delta\tau, \tag{5.18}$$

where ΔM_i^{k-1} is the change in the masses of the metal and the slag due to the occurrence of the chemical reactions, kg.

In addition, it was taken into account that the carbon in liquid steel is oxidized not only by oxygen from the slag, but also by oxygen from the gas [5]. Therefore, the contribution of the latter process to the change in the concentration of carbon in the metal was taken into account by the empirical dependence [6]

$$\Delta[\mathrm{C}] = k[\mathrm{C}], \tag{5.19}$$

where k is an empirical coefficient that is equal to 0.5–0.65.

System of Eqs. 5.1–5.4 and 5.7–5.19 and the expression for the reaction rates comprise a mathematical model of the smelting of steel in an EAF, which can be used to predict the composition of the metal and the slag at any time during a smelting cycle.

The calculation is performed in successive steps. First, the rates are determined for the initial compositions. Next, the compositions and masses of the phases at the beginning of the next time interval are found. After this the calculation cycle is repeated.

Only a relatively small number of physicochemical and technological parameters of the process must be known to predict the composition of the metal and the slag.

5.2 Development of a Mathematical Model of the Ladle Refinement of a Metal by a Synthetic Slag

The essence of the technological scheme for the ladle treatment of steel by synthetic slags is as follows. A synthetic slag having a temperature of 1953–2023 K is poured onto the bottom of a steel-pouring ladle, into which metal having a temperature of 1853–1893 K is then tapped from the furnace. The required ferroalloys, which are most often FeSi or FeMn, are also added at the bottom of the ladle before tapping of the metal. The metal is deoxidized by aluminum. The tapping time varies from 4 to 10 min. The mass of the slag is chosen in the range from 3 to 5% of the mass of the metal. The concentrations of the elements in the metal vary both due to the occurrence of chemical reactions in the metal–slag–gas system and due to dissolution of the solid ferroalloys in the metal.

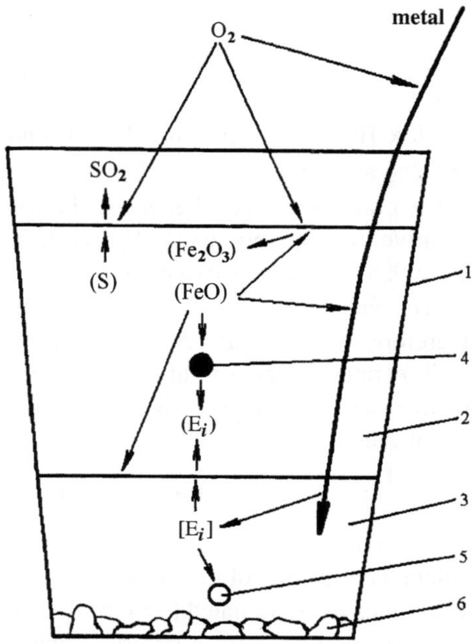

Fig. 5.2 Diagram of the phase interactions that occur during the ladle treatment of steel by a synthetic slag. *1*—steel-pouring ladle; *2*—synthetic slag; *3*—liquid metal; *4*—metal droplet; *5*—slag droplet; *6*—solid ferroalloys

In the general case, when a liquid metal is poured into a ladle with a synthetic slag, an interaction can take place on the following interfaces (Fig. 5.2).

1. The metal-stream/gas interface. Secondary oxidation of the metal by oxygen from the air and the dissolution of nitrogen mainly occur here.
2. The metal/slag interface. If the stirring and emulsification of the phases in the various stages of the steel treatment process are taken into account, several interfaces can be distinguished: metal-stream/slag, metal-in-ladle/slag, metal-droplet/slag, and slag-droplet/metal. Refinement of the metal from sulfur, oxygen, and nonmetallic inclusions takes place on these boundaries.
3. The slag/gas interface. Processes involving oxygen from the air, such as the oxidation of sulfur and the formation of higher oxides of iron, manganese, etc., occur here.
4. The metal/solid ferroalloy interface. Dissolution of the components of the ferroalloys occurs here.

When this is taken into account, the concentration of a specific element $[E_i]$ in the metal at the time τ can be found from the equation

$$[E_i]^{\tau} = \frac{[E_i]^0 \int_0^{\tau} V_{Me} d\tau + 100 M_{E_i} \sum_{i=1}^{n} \int_0^{\tau} \left(\frac{i_i^k}{nF}\right)_j S_j d\tau + \sum_1^Z \int_0^{\tau} [E_i]_{ad}^z V_z d\tau}{\int_0^{\tau} V_{Me} d\tau + \sum_1^Z \int_0^{\tau} V_z d\tau}. \quad (5.20)$$

The concentration of the same element in the slag (E_iO_m) equals

$$(\mathrm{E}_i\mathrm{O}_m)^{\tau} = (\mathrm{E}_i\mathrm{O}_m)^0 - \frac{100}{M_{\mathrm{slag}}}\frac{M_{\mathrm{E}_i\mathrm{O}_m}}{m}\sum_{i=1}^{n}\int_0^{\tau}\left(\frac{i_i^k}{nF}\right)_j S_j d\tau. \tag{5.21}$$

Here $[\mathrm{E}_i]^{\tau}$ and $(\mathrm{E}_i\mathrm{O}_m)^{\tau}$ are the concentrations of the element E_i in the metal and the slag at the time τ, respectively, wt.%; $[\mathrm{E}_i]^0_{\mathrm{Me}}$, $(\mathrm{E}_i\mathrm{O}_m)^0$, and $[\mathrm{E}_i]^z_{\mathrm{ad}}$ are the initial concentrations of the element E_i in the metal, the slag, and the technological additive, respectively, wt.%; V_{Me} is the pouring rate of the metal passing into the ladle, kg/s; S_j is the surface area.

The values of V_{Me}, V_{z}, i_i^k, and S_j can vary over the course of the process; therefore, Eqs. 5.20 and 5.21 contain integral dependences.

The rates of the simultaneously occurring reactions can be found using the method described in Sect. 4.3.2, in which the phase interaction is described by a set of reactions of the form

$$1/m[\mathrm{E}_i] + (\mathrm{FeO}) = 1/m(\mathrm{E}_i\mathrm{O}_m) + [\mathrm{Fe}]. \tag{5.22}$$

The rates of solution of the technological additives, which are needed to calculate the variation of the masses and compositions of the metallic and slag phases with time, were estimated from the data in [12, 13]. Assuming that the lumps of the alloying additions have a spherical shape and diameters up to 10 cm, we used the data in [12, 13] to calculate the decrease in the mass of the ferroalloys per unit time, which was found to be equal to 10–12 kg/s. In addition, according to the data in [14], we assumed that the rate of the incorporation of aluminum into the metal is equal to its rate of solution.

Apart from the interaction processes described by equations of type (5.22), we took into account the secondary oxidation of the metal by atmospheric oxygen and the removal of sulfur from the slag in the reaction

$$(\mathrm{S}) + \{\mathrm{O}_2\} = \{\mathrm{SO}_2\}. \tag{5.23}$$

According to the data in [15], during the time of contact of the metal with the atmosphere when the metal is poured from the furnace into the ladle, the oxygen concentration increases on the average by 0.01 wt.%. This value was taken into account in setting the initial concentrations of the elements in the metal. The rate of process (5.23) was calculated from the expression [16]

$$V_1^{\mathrm{S}} = 6.53 \times 10^{-2} P_{\mathrm{O}_2}^{0.55} (\%\mathrm{S})^{1.2} \exp(-15200/T). \tag{5.24}$$

Here the rate of the removal of sulfur from the slag into the gas phase is expressed in %S/(cm^2 s).

We express the change in the sulfur concentration in the slag due to the interaction with atmospheric oxygen as follows:

$$\Delta(\mathrm{S}) = V_1^{\mathrm{S}} S_{\mathrm{slag/gas}} \Delta\tau. \tag{5.25}$$

As has already been noted, the reactions accompanying the interaction between the metal and the slag can occur simultaneously on several interfaces. Unfortunately, the reliable quantitative data regarding the convection conditions and the fractional composition of the droplets are insufficient for evaluating the proportion of the interaction on each of the interfaces. The following laws, in particular, are known. The size of the interfacial surface (m^2/ton metal) for a constant amount of slag participating in the treatment is proportional to the flow rate of the metal and the drop height of the stream, but is inversely proportional to the viscosity of the slag [17]. Using specially set up experiments, Voinov et al. [18] arrived at the conclusion that 40–50% of all the slag found in the ladle is emulsified into the metal during the treatment. The slag droplets have dimensions ranging from hundredths of a millimeter to 2–3 mm, and their surface area reaches 60 m^2/ton steel [18]. The time for separation of the slag and the metal in the ladle after completion of the pouring of the steel is about 2 min [18]. However, it was noted in [17] that the slag particles entering the metal during the treatment manage to float up to the surface practically completely before the beginning of teeming.

We note the special features of taking into account the interaction in the stage of emulsification, for example, of the slag in the metal.

The following approach is expedient here.

For each droplet fraction, the changes in composition with respect to all the components during the average time of contact with the metal (the droplet lifetime τ_{dr}) should be calculated according to the procedure described in Sect. 4.3.2. The value of $\Delta(E_iO_m)_{dr}$ matches the corresponding change for the respective component in the metal $\Delta[E_i]_{dr}$. When all the droplet fractions are taken into account, the total change in the concentration of the respective component in the metal can be found.

If $\tau_{dr} < \Delta\tau$, the composition of the metal may be assumed to be constant in calculations of the slag droplet interaction rate. If $\tau_{dr} \gg \Delta\tau$, the number of time intervals into which the entire treatment process is divided must be increased accordingly.

Thus, information regarding the number and size of the metal and slag droplets, their lifetime, and their rates of motion is needed. These parameters must be known as functions of the treatment time.

Taking into account that such information is generally not known *a priori*, we will simplify the physical picture somewhat.

We represent the interaction between the metal and the slag as a hypothetically one-step interaction, i.e., an interaction of assigned masses of the metal and the slag that occurs on one interface with a certain effective value of the rate constant, which is equal to the product of the effective values of the mass-transfer constant and the reaction surface

$$K_{E_i}^{eff} = \left(K_{E_i}^{d}\right)_{eff} S_{eff} = \beta D_{E_i}^{0.5} S_{eff} = \sum_{j=1}^{n} \left(K_{E_i}^{d}\right)_j S_j. \qquad (5.26)$$

Finding the values of β and S_{eff} presents the greatest difficulty. According to the data in [3], $S_{eff} = 100 S_{geom}$ and $\beta = 3.3\ s^{-1/2}$ for a time of pouring the metal into

Table 5.3 Composition of the metal after ladle treatment

Melting cycle no.	Tapping time (s)	Steel grade	Metal composition (wt.%): above the bar, experimental; below the bar, calculated								
			C	Mn	Si	P	Cr	Ni	Cu	S	O
625	137	30KhGSA	0.32	1.10	1.07	0.019	1.10	0.18	0.18	0.009	0.0079
			0.33	1.04	1.00	0.018	1.10	0.18	0.18	0.011	0.0050
647	180		0.31	0.83	0.96	0.024	1.10	0.18	0.18	0.008	0.0057
			0.30	0.78	1.00	0.026	1.09	0.18	0.18	0.012	0.0040
651	145		0.32	1.01	1.05	0.025	1.00	0.17	0.20	0.007	0.0073
			0.32	0.96	1.00	0.022	0.99	0.17	0.20	0.008	0.0060
741	125		0.33	0.87	1.02	0.027	1.03	0.23	0.19	0.008	0.0040
			0.33	0.95	1.00	0.025	1.03	0.23	0.19	0.010	0.0045
649	240	10	0.11	0.43	0.25	0.017	0.15	0.21	0.19	0.020	0.0065
			0.11	0.41	0.24	0.015	0.14	0.21	0.19	0.018	0.0060
688	170	40Kh	0.41	0.67	0.26	0.018	1.02	0.21	0.19	0.009	0.0061
			0.41	0.65	0.25	0.014	1.02	0.21	0.19	0.010	0.0056
606	180		0.43	0.58	0.26	0.013	0.88	0.17	0.20	0.013	0.0115
			0.43	0.63	0.25	0.011	0.87	0.17	0.20	0.012	0.0070
617	155		0.45	0.64	0.24	0.017	0.98	0.16	0.15	0.009	0.0019
			0.45	0.62	0.26	0.020	0.98	0.16	0.15	0.012	0.0030

the ladle equal to 300 s under similar conditions. At the same time, it is obvious that the shorter is the time during which tapping of the metal into the ladle occurs, the greater should be the value of the effective mass-transfer constant $\left(K^{d}_{\mathrm{E}_i}\right)_{\mathrm{eff}}$. According to the data in [19], the value of the coefficient β is directly proportional to the amount of convection. After some relatively simple transformations we found

$$\beta = 3.3(300/\tau)^{1/2}. \tag{5.27}$$

The changes in the convection constant and the true surface area from the initial values during the period of residence of the metal in the ladle to the final values, which were equal to $\beta = 1\ \mathrm{s}^{-1/2}$ and $S = S_{\mathrm{geom}}$, were also taken into account.

Equations 5.20, 5.21, 5.24, 5.25, and 5.27 and the expression for the rates of the corresponding reactions comprise a mathematical model of the process of the ladle treatment of steel by a synthetic slag and enables us to predict the composition of the metal during its interaction with the slag and the gas. To perform the calculations, the entire process was divided into time intervals of short duration. The calculation was performed in successive steps. First, the rates for the initial compositions were determined, and the compositions and masses of the phases at the beginning of the next time interval were found on their basis. Then the calculation cycle was repeated until completion of the time of the process.

To verify the faithfulness of the mathematical model developed, calculations of the final composition of the metal after ladle treatment by a synthetic slag were performed for a series of melting cycles of steels of grades 30 KhGSA, 40 Kh, and 10, and the results were compared with experimental data (Table 5.3).

Table 5.4 Composition of the metal (wt.%) before ladle treatment by a synthetic slag

Steel	C	Mn	Si	Cr	P	S	O
10 steel	0.07–0.14	0.05–0.08	0.0	<0.15	<0.025	0.040–0.045	0.03–0.04
40Kh	0.36–0.41	0.50–0.80	0.12–0.15	0.8–1.1	<0.035	0.035–0.040	0.03–0.04
35KhGSA	0.32–0.39	0.80–1.10	0.12–0.15	1.1–1.4	<0.025	0.035–0.040	0.03–0.04

The average initial compositions of the steel grades just cited are presented in Table 5.4. It may be concluded from a comparison of the initial and final compositions of the metal that desulfurization and deoxidization occur when these steels are subjected to ladle treatment by a synthetic slag. The concentrations of the remaining elements, with the exception of those which enter the metal from the ferroalloys, remain practically constant. As a whole, the data obtained indicate that this method is sufficiently reliable for predicting the composition of steel after ladle treatment by a synthetic slag.

The composition and growth rate of the nonmetallic inclusions in the steel can be estimated on the basis of the mathematical model presented.

The growth of a nucleated inclusion with a radius exceeding the critical value occurs as a result of the diffusion of oxygen and other components of the steel to its surface followed by their interaction. The composition of the inclusion is shaped in accordance with the ratio between the oxidation rates of the components:

$$(\mathrm{E_1O}_{m1}) : (\mathrm{E_2O}_{m2}) : \ldots = J_{\mathrm{E}_1} : J_{\mathrm{E}_2} : \ldots . \quad (5.28)$$

It should be kept in mind that Eq. 5.28 is not suitable for evaluating the composition of a charged inclusion, which is shaped by the condition that its work of formation is at a minimum. However, the initial composition is rapidly "lost" during the growth of an inclusion, because the initial quantities of oxides will amount to $\sim 1/1000$ of the mass of a nonmetallic inclusion (NI) even when the radius of the inclusion increases by a factor of 10. Therefore, to simplify the calculations, we assume that the initial composition of a nonmetallic inclusion corresponds to relation (5.28). To find it, we assume that the initial mass of the particle m_0 does not vary and that the products of the oxidation reactions of the components enter it in accordance with expression (5.28). The composition of the inclusion is averaged. The constancy of m_0 is ensured by the continuous removal of the excess part of the inclusion with the averaged composition at the rate

$$\dot{M} = S\Big(\sum J_{\mathrm{E}_i} A_{\mathrm{E}_i} + J_{\mathrm{O}} A_{\mathrm{O}} - J_{\mathrm{C}} A_{\mathrm{O}}\Big), \quad (5.29)$$

where S is surface area of the nonmetallic inclusion, and A_{E_i} and A_{O} are the atomic weights of the component E_i and of oxygen. It was taken into account in expression (5.29) that the oxidation of carbon with the rate J_{C} causes a decrease in the mass of the nonmetallic inclusion due to the loss of oxygen in the form of CO.

Table 5.5 Values of $K^d_{E_i}$ for nonmetallic inclusions of various size

R (cm)	$K^d_{E_i}$	
	Molecular diffusion	Convective diffusion
10^{-6}	100	0.001
10^{-5}	10	0.003
10^{-4}	1	0.01
10^{-3}	0.1	0.03
10^{-2}	0.01	0.1

The composition of a growing inclusion at the time $k\Delta\tau$ can be described by the equation

$$(E_iO_m)^{k+1} = \frac{(E_iO_m)^k + J^k_{E_i} M_{E_iO_m} S^k 100\Delta\tau}{m^k + \dot{M}^k\Delta\tau}. \tag{5.30}$$

The change in the composition of the metal can be found from the expression

$$[E_i]^{k+1} = [E_i]^k - N\frac{J^k_{E_i} A_{E_i} S^k 100\Delta\tau}{\rho_{Me}}, \tag{5.31}$$

where N is the number of inclusions in a unit volume of the metal.

The discharge rates of the ions were calculated using equations from Sect. 4.3.2.

The rate of mass transfer between the metal and an inclusion is determined by molecular and convective diffusion. It was shown in [20] that under given conditions the mass-transfer constant for molecular diffusion equals

$$K^d_{E_i} = \frac{D_{E_i}}{R}. \tag{5.32}$$

Here R is the radius of the nonmetallic inclusion.

For the mass-transfer coefficient for convective diffusion to a floating particle, we have [21]

$$K^d_{E_i} = \left(\frac{4D_{E_i}(\rho_{Me} - \rho_{slag})}{9\pi\eta_{Me}} gR\right)^{1/2}. \tag{5.33}$$

The values of $K^d_{E_i}$ calculated from Eqs. 5.32 and 5.33 are presented in Table 5.5.

An analysis of the data in Table 5.5 shows that the mass-transfer rate is determined mainly by molecular diffusion when $R < 10^{-3}$ cm and by convective diffusion when $R > 10^{-2}$ cm. Although the data in Table 5.5 are estimates, it can still be seen that the overall values of $K^d_{E_i}$ are >0.1 cm/s, i.e., are 10–100 times greater than the values usually observed in a metal–slag system. According to our data [20], at such values of $K^d_{E_i}$ most reactions involving the oxidation of iron impurities by a slag take place under a nearly kinetic regime.

The method described was used to estimate the growth dynamics and composition of nonmetallic inclusions in ShKh15 steel at 1823 K. The composition of the

Table 5.6 Calculated composition of nonmetallic inclusions

[Al] (wt.%)	R (cm)	τ (s)	Composition of inclusions (wt.%)				
			FeO	MnO	SiO_2	Al_2O_3	Cr_2O_3
0.16	10^{-6}	0	0.94	1.40	41.0	56.8	0.06
	10^{-5}	0.5	0.94	1.40	41.0	56.8	0.06
	10^{-4}	6.5	0.90	1.35	40.8	56.9	0.05
	10^{-3}	64.5	0.88	1.28	43.9	53.9	0.04
	10^{-2}	774.5	0.80	1.20	44.0	54.0	0.04
0.016	10^{-6}	0	1.00	1.70	90.0	7.0	0.30

metal was assumed to be constant with time. The parameters needed for the calculations were taken from literature sources.

Table 5.6 presents the calculated data. It is seen that the inclusions consist mainly of SiO_2 and Al_2O_3 and that the content of FeO, MnO, and Cr_2O_3 is insignificant. The time needed for an inclusion to grow to $R = 10^{-2}$ cm is comparatively large ($\sim$13 min). Therefore, it has not been ruled out that the frequently observed absence of large inclusions [22] is attributable not only to their high floating rates, but also to the short holding time of the metal, during which the inclusions do not manage to grow. The weak dependence of the composition of a nonmetallic inclusion on its size is associated with the small change in the relative contributions of the kinetic and diffusional hindrances as inclusions grow.

We note that the accuracy of the currently available data regarding the equilibrium constants, diffusion coefficients, and especially the exchange currents is poor. The advantage of the method described is the possibility of revealing the parameters of the mathematical model which influence the result to the greatest extent, i.e., which should be refined first.

5.3 Predicting the Composition of the Metal in the Head Part of an Ingot During the Teeming of Steel Under Siliceous Heat-Generating Mixtures

Heat-generating and thermally insulating mixtures are employed to reduce the cropping of the head part of an ingot. When coverings consisting of silicocalcium and scale are used, a high-silicon metal layer forms on the surface of the steel (Fig. 5.3). When it is mixed with the base metal, heat associated with the dissolution of silicon is released. Owing to the generation of heat, the lowered melting point of the silicon alloy, and the thermal insulation provided by the slag and the insulating covering, the metal in the upper part does not solidify for a long time. This decreases the volume of the shrinkage cavity or eliminates its formation entirely and significantly increases the yield of suitable metal [23].

Knowledge of the laws governing the formation of a ferrosilicium layer, its behavior over time, and the rate of heat generation enables us to optimize the

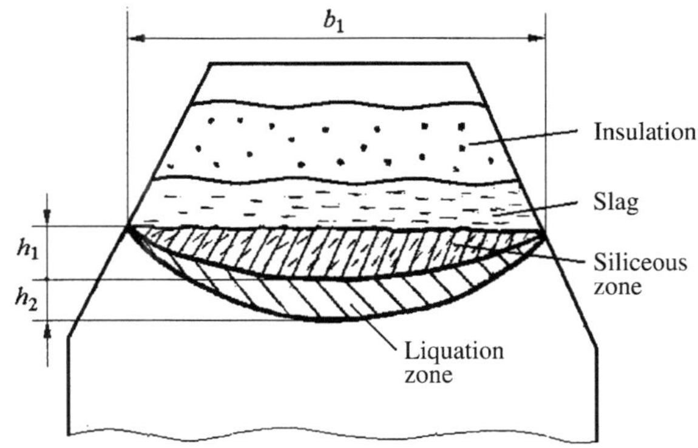

Fig. 5.3 Diagram of the head part of an ingot

process and to increase the yield of suitable metal by reducing cropping, as well as to determine the optimal parameters of the head.

We will analyze the physical and chemical processes for the purpose of developing a physical and mathematical model of the technological process under discussion.

Our calculation based on the material balance of silicon shows that a ferroalloy with a silicon content equal to about 35.5 wt.% appears on the surface of the metal.

A comparison with the data in [23] regarding the concentration of silicon in the siliceous and liquation zones of an ingot (Fig. 5.3) shows that after the "burning" period of the mixture, silicon is retained in the upper head part of the ingot and scarcely passes into the bulk of the liquid metal. The emerging ferrosilicium layer, whose composition corresponds to the homogeneous region of the liquid solution on the iron–silicon phase diagram, does not have an interface with the liquid steel. Therefore, at first glance, it seems unlikely that this layer is not washed away by the flows of metal, whose velocities reach 10–50 cm/s. However, our analysis of the material balance and the fact that there are high-silicon zones in the ingot convince us that the opposite is true.

The possible causes are as follows.

1. The layer of the Fe–Si melt has a low density. When $b_1 = 53$ cm, its thickness h is equal to 0.58 cm.
   The slag has a viscosity that is hundreds of times greater; therefore, the thin ferroalloy layer "sticking" to it may fail to be entrained by the flows of steel.
2. After the mixture is "burned," droplets of the ferroalloy can develop in the highly viscous slag and gradually settle, passing into the metal and maintaining a high concentration of silicon in it. In the case under consideration, this cause is unlikely, since the slag formed, which consists of 36 wt.% CaO and 64 wt.% SiO_2, is close to the eutectic with a melting point of 1703 K. The viscosity of such a slag is relatively low, and the settling velocity of the metal particles in it reaches 5–10 cm/s [13].
3. When the temperature is lowered, the steel begins to crystallize under the layer of the ferroalloy, which has a low melting point. The dissolution of silicon and

the generation of heat accompanying it prevent crystallization, during which heat is also generated. However, the presence of a solid phase (a solid–liquid state) that raises the viscosity and prevents mixing of the ferroalloy with the steel is possible. This cause is supported by the small thickness of the silicon-containing layers on the periphery, where the dissipation of heat is faster and the amount of the solid phase is increased.

The situation just described allows us to adopt the following physical picture (model) of the process.

- The "burning" of the heat-generating mixture results in the formation of a layer of a liquid ferroalloy, which contains all of the silicon that has not reacted, and a slag layer.
- Silicon passes from the ferroalloy into the liquid steel by molecular and convective diffusion with the latter having a local character in the form of small-scale circulation events. The resultant mass transfer from such diffusion, which may be called turbulent, can be described by the equations of mass transfer in a motionless liquid, but with appropriately increased diffusion coefficients.
- The amount of heat generated during such diffusion is determined by the concentration field, i.e., each concentration distribution pattern has its own corresponding heat content. In a first approximation the quantity of heat is proportional to the quantity of silicon that has passed from the ferroalloy into the steel.

The simplifications and assumptions were used during development of the mathematical model.

1. The "burning" of the (SiCa + FeO) mixture occurs rapidly compared with the solidification of the ingot. Therefore, a ferrosilicium layer with a silicon concentration $[\mathrm{Si}]_{\mathrm{init}}$ and a thickness h_0 forms at once on the surface of the metal. According to our data, $[\mathrm{Si}]_{\mathrm{init}} = 35.5$ wt.%, and $h = 0.58$ cm. In the general case part of the silicon can enter the slag in the form of beads and settle into the metal over time with a certain rate $\dot{m}_{\mathrm{Si}}$.
2. Silicon from the ferrosilicium layer diffuses downward, into the body of the head. In the general case, the contributions of both molecular and convective diffusion to the transport of silicon into the steel should be considered.
3. Inasmuch as the silicon concentration varies by two orders over a small distance (Table 5.7), the dependence of the diffusion coefficient of silicon on its concentration must be taken into account.

To obtain a solution, we use the finite-difference method in the one-dimensional approximation.

The equation

$$\frac{\partial C}{\partial \tau} = D\frac{\partial^2 C}{\partial x^2} \tag{5.34}$$

is transformed into the system of equations

Table 5.7 Composition of the metal along the axis of the ingot during the teeming of grade 45 steel under siliceous heat-generating mixtures

Distance from the upper boundary of the head (mm)	[Si] (wt.%)	
	Experiment	Calculation
0	26.70	26.77
20	4.60	4.54
60	0.84	0.81
80	0.38	0.39
100	0.31	0.31
120	0.30	0.30
145	0.30	0.30
Ladle sample	0.30	

$$C_n^{k+1} = C_n^k + \frac{D\Delta\tau}{\Delta x^2}\left(C_{n+1}^k + C_{n-1}^k - 2C_n^k\right), \tag{5.35}$$

where k is the number of the cell with respect to time and n is the number of the cell along the x coordinate.

According to the condition for stability of the solution of the explicit difference scheme, we have

$$\frac{D\Delta\tau}{\Delta x^2} \le A. \tag{5.36}$$

The calculation scheme is presented in Fig. 5.4.

The boundary conditions in this case will be as follows:

$$\text{at}\ \tau = 0 \quad \begin{array}{c} C_1^1 = C_2^1 = \ldots = C_{n_0}^1 = [\mathrm{Si}]_{\mathrm{alloy}} \\ \ldots\ldots\ldots\ldots\ldots\ldots\ldots\ldots\ldots \\ C_{n_0+1}^1 = C_{n_0+2}^1 = \ldots = C_n^1 = [\mathrm{Si}]_{\mathrm{steel}} \end{array}; \tag{5.37}$$

$$\text{at}\ x = 0 \quad C_0^k = C_1^k \quad \text{when} \quad \dot{m}_{\mathrm{Si}} = 0. \tag{5.38}$$

Then, for the first layer at any time we have

$$C_1^{k+1} = C_1^k + \frac{D\Delta\tau}{\Delta x^2}\left(C_2^k + C_0^k - 2C_1^k\right). \tag{5.39}$$

To take into account the arrival of silicon from the slag layer with the rate $\dot{m}_{\mathrm{Si}}$, we make the following assumptions:

1. The silicon alloy formed enters the metal uniformly with time over the course of the entire ingot crystallization period.
2. The silicon supplied is distributed uniformly over the zeroth layer (Fig. 5.4).

In this case, instead of Eq. 5.39, we will have the following:

$$C_n^{k+1} = C_n^k + \frac{D\Delta\tau}{\Delta x^2}\left(C_{n+1}^k + C_{n-1}^k - 2C_n^k\right) + \frac{\dot{m}_{\mathrm{Si}}\Delta\tau}{\rho\Delta x}. \tag{5.40}$$

Here $\dot{m}_{\mathrm{Si}}$ is the arrival rate of silicon, g Si/(cm^2 s); ρ is the density of the metal, g/cm^3.

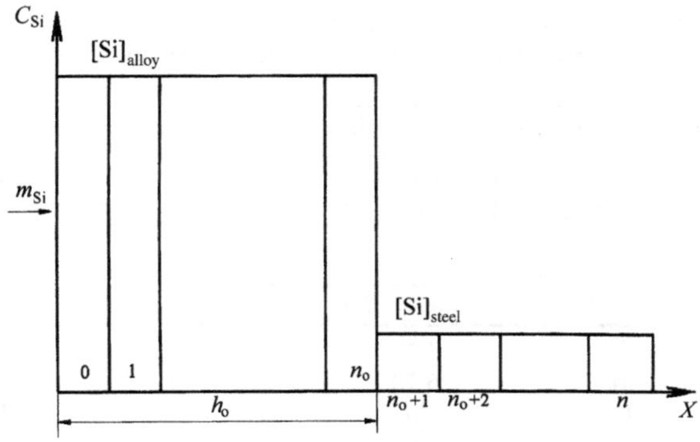

Fig. 5.4 Calculation scheme: $\dot{m}_{Si}$—entry rate of the silicon alloy formed from the slag layer; h_0—thickness of the layer of the silicon alloy; 0, 1, ..., n_0—cells along the *X* coordinate at the time $\tau = 0$ in the layer of the silicon alloy; $n_0 + 1$, $n_0 + 2$, ..., n cells along the *X* coordinate at the time $\tau = 0$ in the steel; $[Si]_{alloy}$—initial concentration of silicon in the alloy formed; $[Si]_{steel}$—concentration of silicon in the steel (ladle sample)

Thus, Eqs. 5.35–5.40 comprise a mathematical model of the process of the distribution of silicon in the head part of an ingot [24].

For adaptation of the mathematical model, calculations were performed for the distribution of silicon along the axis of an ingot during the teeming of grade 45 steel in an ingot weighing 3.5 tons under the 50SK500 heat-generating mixture, which contains 40 wt.% silicocalcium and 50 wt.% scale.

The following were chosen as the principal variable parameters: the amount of the silicon alloy remaining in the slag and the diffusion coefficients of silicon.

Initially, the calculations were performed with one assigned constant value of the diffusion coefficient and different percentages of the silicon alloy remaining in the slag. The latter parameter was varied from 0 to 15% of the total silicon alloy formed. Variation of the diffusion coefficient from 3×10^{-4} to 5×10^{-4} cm^2/s did not enable us to obtain values of the silicon concentration along the axis of the ingot that are close to the experimental values.

In the second stage of the calculations, we distinguished between the "silicon-rich" zone, in which the diffusion coefficient of silicon was assumed to be close to that for ferrosilicium, and the "silicon-poor" zone, in which the diffusion coefficient was approximately an order of magnitude greater than in the steel. The data from the calculations are presented in Table 5.8.

The data in Table 5.8 show that the calculation and experiment are in good agreement under the conditions indicated. This allowed us to recommend this scheme for describing the silicon distribution process.

In the next stage of the calculations, five segments, in which the diffusion coefficients differed from one another but were constant within each segment, were distinguished in accordance with the calculation scheme. The data obtained, which are presented in Table 5.7, demonstrate good correspondence between the calculated and experimental values. Thus, it may be stated that this approach enabled us

Table 5.8 Calculated (values above the bar) and experimental (values below the bar) data on the concentration of Si during the teeming of grade 45 steel under the 50SK500 heat-generating mixture

Distance from the upper boundary of the head (mm)	Concentration of Si along the axis of the ingot (wt.%)			
	$D_{FeSi} = 2.2 \times 10^{-5}$ cm²/s, $D_{steel} = 2.5 \times 10^{-4}$ cm²/s		$D_{FeSi} = 3 \times 10^{-5}$ cm²/s	
			D_{steel}	
			3.3×10^{-4} cm²/s	2.0×10^{-4} cm²/s
	$N_{Si} = 1.0$	$N_{Si} = 0.9$	$N_{Si} = 0.9$	
0	21.70/26.70	27.80/26.70	24.80/26.70	24.40/26.70
20	5.40/4.60	5.20/4.60	6.30/4.60	4.30/4.60
60	0.72/0.84	0.69/0.84	1.10/0.84	0.51/0.84
80	0.34/0.38	0.34/0.38	0.43/0.38	0.31/0.38
100	0.30/0.31	0.30/0.31	0.30/0.31	0.30/0.31
120	0.30/0.30	0.30/0.30	0.30/0.30	0.30/0.30
145	0.30/0.30	0.30/0.30	0.30/0.30	0.30/0.30

Note N_{Si} is the fraction of the silicon alloy that formed on the metal

to describe the distribution of silicon along the axis of an ingot during the teeming of grade 45 steel under the 50SK500 heat-generating mixture.

The computer-aided design (CAD) subsystem Optimal Selection of Head Parameters for Ingots of Round Cross Section was developed on the basis of the mathematical model and equations obtained as a result of the statistical treatment of the experimental data in [23].

The following values served as input parameters: the composition of the steel, the ratio of the height of the ingot (*H*) to its diameter (*D*), the density of the liquid steel (ρ), the mass of the ingot (m_{ingot}), the fraction of silicon forming the ferro-alloy layer, the conicity of the head (k_{head}), and the safety factor of the head (σ). Data regarding the conicity of the ingot (*k*) and the crystallization time (τ_{cryst}) are also used as input information. For estimates these parameters can be calculated from the following relations:

$$\tau_{cryst} = (R_{av}/\kappa). \tag{5.41}$$

Here R_{av} is the average radius of the ingot, cm; κ is an empirical coefficient, which depends on the degree of alloying of the steel (0.22–0.36), cm/s.

The conicity of the ingots (*k*) is assigned in accordance with the mass of the ingot:

$$m_{ingot} \leq 10 \text{ ton} \quad k = 0.03,$$

$$10 \text{ ton} < m_{\text{ingot}} \leq 50 \text{ ton} \quad k = 0.06,$$

$$50 \text{ ton} < m_{\text{ingot}} \leq 100 \text{ ton} \quad k = 0.12.$$

The geometric parameters of the ingot are calculated from equations which were obtained on the basis of the data in [23]:

$$R_{\text{av}} = \sqrt[3]{\frac{3V_{\text{body}}}{2\pi\alpha(3 + n^2)}}. \tag{5.42}$$

Here V_{body} is the volume of the ingot body, cm^3; $\alpha = H/D$; $n = \alpha k$. The volume of the ingot body is calculated from the formula

$$V_{\text{body}} = \frac{100 - (24.9 - 1.8\alpha)}{100\rho}. \tag{5.43}$$

Then we calculate the required mass of the 50SK500 mixture for the specific ingot geometry:

$$m_{\text{mixture}} = 2.2 \times 10^{-3}\pi R_{\text{av}}\left(1 + n^2\right). \tag{5.44}$$

Using Eqs. 5.35 and 5.40 we calculate the distribution of silicon along the axis of the ingot until the condition $\tau = \tau_{\text{cryst}}$ holds (τ is the current time), and we evaluate the depth of the silicon and liquation zones (h_{Si}). This value was used to calculate the parameters of the head. The height of the head is estimated from the equation

$$H_{\text{head}} = h_{\text{Si}}\sigma. \tag{5.45}$$

Setting the maximum diameter of the ingot equal to the maximum diameter of the head, we calculate the minimum diameter of the head

$$R_{\text{head}}^{\min} = R_{\text{head}}^{\max} - H_{\text{head}}k_{\text{head}}. \tag{5.46}$$

Then we estimate the relative volume of the head

$$V_{\text{head}} = \frac{H_{\text{head}}}{3}\left(S_{\text{upper}} + S_{\text{lower}} + \sqrt{S_{\text{upper}}S_{\text{lower}}}\right), \tag{5.47}$$

$$V_{\text{head}}^{\text{rel}} = \frac{V_{\text{head}}}{V_{\text{head}} + V_{\text{body.}}}, \tag{5.48}$$

where V_{head} is the volume of the head (cm^3), and S_{upper} and S_{lower} are the cross-sectional areas of the upper and lower bases of the head, respectively (cm^2).

This CAD subsystem was delivered to the Izhstal Corporation for use in production.

5.4 Development of Mathematical Models of Refining Remelts

The use of refining remelts provides a way to significantly and purposefully alter and the composition and structure of a metal obtained in a steelmaking plant. However, for these technological processes there is no mathematical description of the kinetics of the reactions in the form of sufficiently detailed mathematical models that take into account the technological factors. The complexity of these technologies precludes finding an optimal regime solely on the basis of experimental data. For this reason, the development and adaptation of mathematical models are formidable theoretical and practical undertakings.

5.4.1 Mathematical Model of the Alloying of a Metal with Nitrogen During Plasma Arc Remelting

The behavior of nitrogen during plasma arc remelting (PAR) has been considered in detail is numerous monographs and articles [25–39]. The increased incorporation of nitrogen into the metal to levels that exceed its equilibrium solubility by several fold at the respective partial pressure of nitrogen in the absence of an arc discharge has been established.

On the basis of an analysis of the literature data [25–39], we adopt the following scheme for the behavior of nitrogen during PAR (Fig. 5.5).

Nitrogen dissociates in a plasma jet and is ionized to equilibrium according to the reactions

$$1/2\{N_2\} = \{N\} \tag{5.49}$$

$$\{N\} = \{N^+\} + e^-. \tag{5.50}$$

The gas contains Ar, N_2, N, N^+, and e^-.

The deionization of nitrogen occurs in the diffusion layer near the surface of the metal, and all the atoms combine to form excited N_2^* molecules. Evaluation of the extent of development of such processes in the diffusion layer using the Damköhler number [26] showed that the formation of nitrogen molecules takes place fairly rapidly.

The N_2^* molecules formed diffuse to the metal and dissolve in it according to the reaction

$$½\{N_2^*\} = [N]. \tag{5.51}$$

The equilibrium constant of this reaction K_N^* [25]

$$\log K_N^* = \log\frac{f_N[N]}{P_{N_2^*}^{1/2}} = \frac{2264}{T} - 1.16 \tag{5.52}$$

Fig. 5.5 Diagram of plasma arc refining: *1*—intermediate product (electrode); *2*—liquid metal film; *3*—plasmatron; *4*—pool surface covered by the plasma torch (arc); *5*—free surface of the pool; *6*—metal pool; *7*—mold; *8*—gas bubbles; *9*—ingot

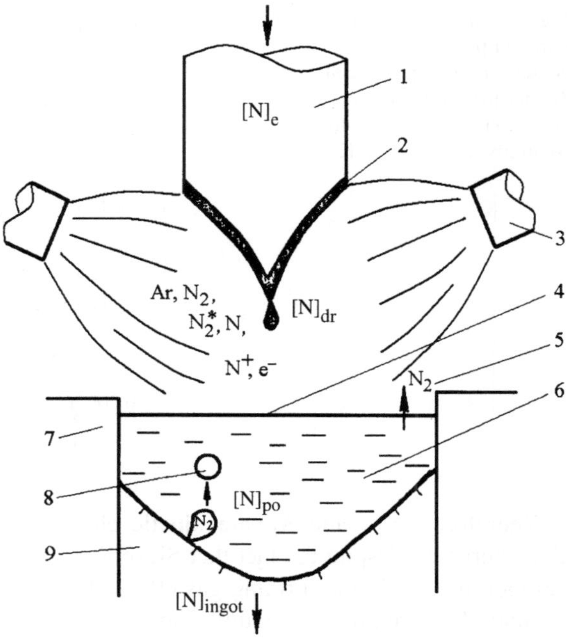

is significantly higher than the value of K_N [26]

$$\log K_N = \log \frac{f_N[N]}{P_{N_2}^{1/2}} = -\frac{293}{T} - 1.16 \qquad (5.53)$$

for the reaction

$$\frac{1}{2}\{N_2\} = [N]. \qquad (5.54)$$

Consequently, the concentration of nitrogen in the surface layer of the metal exceeds the equilibrium concentration according to reaction (5.54).

The diffusion and circulation of the metal transport this "super-equilibrium" nitrogen into the bulk of the metal pool, and it remains in the ingot after crystallization.

On the part of the pool surface that is free of contact with the plasma, nitrogen can be desorbed into the gas according to a reaction that is the reverse of reaction (5.54).

When the critical degree of supersaturation of the metal with nitrogen is achieved, bulk "boiling" of the pool begins. The rate of the removal of nitrogen from the metal is then so high that further supersaturation of the metal above the critical value does not occur (Fig. 5.6). The value of $[N]_{cr}$ depends on the heating method, the total pressure over the pool, and the properties of the diluting gas [26]. As P_{total} increases, $[N]_{cr}$ increases, but the slope of the straight line decreases, i.e., the value of K_N^* decreases. The replacement of argon with helium causes a strong drop in K_N^*, which is inexplicable on the basis of thermodynamics.

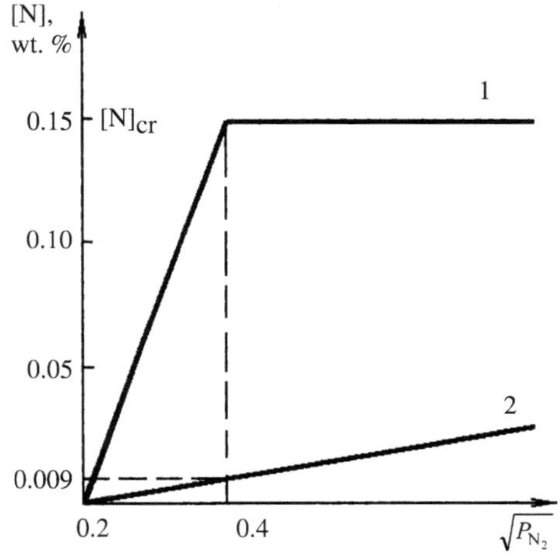

Fig. 5.6 Influence of the nitrogen pressure on its content in the liquid metal: *1*—plasma arc remelting conditions; *2*—equilibrium solubility at $T = 2073$ K

Therefore, $[N]_{cr}$ and K_N^* are kinetic characteristics, rather than equilibrium characteristics, despite the fact that Sievert's law holds at small concentrations of nitrogen in the plasma-forming gas ($P_{N_2} < 0.1$ atm).

Since K_N^* is not the true equilibrium constant, the question of the correctness of including the activity coefficient of nitrogen f_N in expression (5.52) remains open. Apparently, the most reliable method for determining K_N^* is to perform a direct experiment for each grade of metal. The opinion [25–27, 32–36] that the rate-limiting step in the overall process is the slow diffusion of nitrogen in the metal is widely accepted. However, we showed [40] that diffusion in both the metal and the gas in two successive stages, viz., an electrode stage and a pool stage, must be taken into account. A conclusion regarding the principle limiting step under specific conditions can be drawn based on the solution of a complete system of equations that takes into account the factors enumerated above.

First of all, we will examine the processes which take place in the gas phase when the plasma forms. The composition of the plasma is determined by the equilibrium conditions of reactions (5.49) and (5.50) for an assigned initial gas composition, pressure, and temperature. The ionization of argon was neglected in the calculation of the plasma composition. When the total pressure in the diffusion layer is constant, the concentration of excited molecules ($P_{N_2}^*$) can be found from the relation

$$P_{N_2^*} = \frac{P_N + P_{N^+}}{2\left(1 - \frac{P_N + 3P_{N^+}}{2P}\right)}, \tag{5.55}$$

where P is the total pressure.

The correction within the parentheses in the numerator takes into account the decrease in volume due to the deionization of nitrogen atoms and the formation of nitrogen molecules from them. The calculation showed (Table 5.9) that at $T \geq 6000$ K the partial pressure of the active particles which determine the

Table 5.9 Composition of the plasma, P_i (atm)

T (K)	$\frac{\%Ar}{\%N_2}$	P_{tot}	P_N	$P_{N^+} = P_{e^-}$	$P_{N_2^*}$	P_{Ar}	P_{N_2}
8000	$\frac{39}{61}$	1.2	0.819	0.0025	0.627	0.3070	0.0695
	$\frac{70}{30}$		0.4893	0.0019	0.308	0.6821	0.0248
	$\frac{85}{15}$		0.296	0.0015	0.170	0.8920	0.0091
8000	$\frac{39}{61}$	0.6	0.428	0.0018	0.246	0.1500	0.0190
		0.8	0.562	0.0021	0.437	0.2010	0.0328
		1.5	1.002	0.0028	0.758	0.3880	0.1040
6000		1.2	0.197	3.2×10^{-5}	0.107	0.4300	0.5730
10000			0.562	0.0230	0.724	0.2860	0.00395
12000			1.002	0.0920	0.730	0.2700	0.00041
14000			1.002	0.2190	0.730	0.2390	4.84×10^{-5}

passage of nitrogen into the metal is close to the partial pressure of nitrogen in the plasma-forming gas (P_{N_2}) and can be approximated by the equation

$$P^*_{N_2} = 0.7 P_{N_2}. \tag{5.56}$$

Similar results were obtained in [41], in which it was found, in particular, that mainly nitrogen atoms and ions are found in the plasma at $T = 10500$ K.

The nitrogen balance (kg/s) in the metal droplets on the electrode can be described in the following manner:

$$\dot{m}\frac{[N]_0}{100} + V_e S_e M_N = \dot{m}\frac{[N]_{dr}}{100}. \tag{5.57}$$

Here $[N]_{dr}$ and $[N]_0$ are the concentrations of nitrogen in the droplets and in the intermediate product, wt.%; $\dot{m}$ is the melting rate, kg/s; V_e is the rate of reaction (5.51) in the electrode stage, mole N/(m^2·s); S_e is the surface area of the electrode on which reaction (5.51) takes place, m^2; M_N is the molecular weight of nitrogen.

Assuming that the nitrogen adsorption process is diffusion-controlled and that its rate is described by the equation

$$V_e = \frac{K_N^d \rho_{Me}}{100 M_N}\left([N]'_{dr} - [N]_{dr}\right), \tag{5.58}$$

we can obtain

$$[N] = \frac{[N]_0 + C[N]'_{dr}}{1 + C}. \tag{5.59}$$

Here $[N]'_{dr}$ is the concentration of nitrogen in the surface layer of the metal, wt.%; $C = \frac{S_e K_N^d \rho_{Me}}{\dot{m}}$; K_N^d is the mass-transfer constant of nitrogen in the metal, m/s; ρ_{Me} is the density of the metal, kg/m^3.

Thus, if the concentration of nitrogen in the surface layer of the metal can be determined, its rate of adsorption and its concentration in the metal droplets entering the pool from the intermediate product can be found.

In a steady-state regime the diffusion rates of nitrogen in the gas and the metal, as applied to reaction (5.51), are identical:

$$V_e = \frac{K^d_{N_2}}{2RT}\left(P_{N_2^*} - P'_{N_2^*}\right) = \frac{K^d_N \rho_{Me}}{100 M_N}\left([N]'_{dr} - [N]_{dr}\right). \tag{5.60}$$

If we take into account that

$$P'_{N_2^*} = \left(\frac{[N]'_{dr} f'_N}{K^*_N}\right), \tag{5.61}$$

then after some transformations we obtain

$$[N]'_{dr} = \frac{1}{2ab}\left(\sqrt{1 + 4ab([N]_{dr} + aP_{N_2^*})} - 1\right), \tag{5.62}$$

where

$$a = \frac{K^d_{N_2} 100 M_N}{2RTK^d_N \rho_{Me}}; \quad b = \left(\frac{f'_N}{K^*_N}\right), \tag{5.63}$$

$K^d_{N_2}$ is the mass-transfer constant of nitrogen in the gas (m/s) and f_N is the activity coefficient of nitrogen in the metal.

The equation of the nitrogen balance in the pool has the form

$$\dot{m}\frac{[N]_{dr}}{100} + V_{po} S_{po} M_N \alpha = \dot{m}\frac{[N]_{ingot}}{100} + V_{des} S_{po} M_N (1-\alpha) + \frac{m_{po} d[N]_{po}}{100 d\tau}. \tag{5.64}$$

The terms on the left-hand side of Eq. 5.64 represent the rates of the entry of nitrogen with the droplets and from the plasma into the pool (α is the fraction of the pool surface area S_{po} that is in contact with the plasma), and the terms on the right-hand side represent, respectively, the rates of the departure of nitrogen to the crystallizing metal, its desorption into the gas from the part of the surface of the pool that is not in contact with the plasma, and its accumulation in the pool. In a steady-state regime the last term can be neglected. In the latter case the system of calculation equations takes on the following form:

$$\left([N]'_{po}\right)^2 + \frac{\kappa_N + C - \alpha C}{ab(\kappa_N + C)}[N]'_{po} - \frac{P_{N_2^*}}{b} - \left(\frac{C(1-\alpha)[N]'_{des} + [N]_{dr}}{ab(\kappa_N + C)}\right) = 0, \tag{5.65}$$

$$\left([N]'_{des}\right)^2 + \frac{\kappa_N + \alpha C}{a_{des} b_{des}(\kappa_N + C)}[N]'_{des} - \frac{P_{N_2}}{b} - \left(\frac{C\alpha[N]'_{dr} + [N]_{dr}}{a_{des} b_{des}(\kappa_N + C)}\right) = 0, \tag{5.66}$$

$$[\mathrm{N}]_{\mathrm{po}} = \frac{C[\mathrm{N}]'_{\mathrm{po}} + C(1-\alpha)[\mathrm{N}]'_{\mathrm{des}} + [\mathrm{N}]_{\mathrm{dr}}}{\kappa_{\mathrm{N}} + C}. \quad (5.67)$$

Here $[\mathrm{N}]'_{\mathrm{po}}$ and $[\mathrm{N}]'_{\mathrm{des}}$ are the concentrations of nitrogen in the surface layer of the pool metal in the adsorption and desorption zones, respectively, wt.%; a_{des} and b_{des} are the parameters calculated using Eq. 5.63 for the desorption zone.

The concentrations of nitrogen in the pool and in the ingot are related by the equation

$$[\mathrm{N}]_{\mathrm{ingot}} = \kappa_{\mathrm{N}}[\mathrm{N}]_{\mathrm{po}}, \quad (5.68)$$

where κ_{N} is the liquation coefficient of nitrogen.

The liquation coefficient can be found from the equation [26]

$$\kappa_{\mathrm{N}} = \frac{\kappa_{\mathrm{N}}^{0}}{\kappa_{\mathrm{N}}^{0} + (1-\kappa_{\mathrm{N}}^{0})\exp(-U/K^{d})}, \quad (5.69)$$

where $\kappa_{\mathrm{N}}^{0} = 0.29$ is the equilibrium liquation coefficient of nitrogen [26], U is the linear crystallization rate (m/s), and K^{d} is the mass-transfer constant of nitrogen on the boundary between the solid and liquid metal.

To take into account the appearance of bulk boiling of the metal pool, relations obtained in [39] were used:

$$[\mathrm{N}]_{\mathrm{cr}}^{\mathrm{steel}} = \frac{[\mathrm{N}]_{\mathrm{cr}}^{\mathrm{Fe}}}{f_{\mathrm{N}}}, \quad (5.70)$$

where $[\mathrm{N}]_{\mathrm{cr}}^{\mathrm{steel}}$ and $[\mathrm{N}]_{\mathrm{cr}}^{\mathrm{Fe}}$ are the threshold concentrations of nitrogen in the alloy and in pure iron, respectively, wt.%. The latter concentration can be determined from the formula

$$[\mathrm{N}]_{\mathrm{cr}}^{\mathrm{Fe}} = K_{\mathrm{N}}\sqrt{1 + \frac{\Delta P}{P}}. \quad (5.71)$$

Here P is the pressure in the furnace chamber, atm; $\Delta P = 3$ atm is the experimental value.

In addition, on the basis of the literature data, an attempt was made to take into account the dependence of the surface temperature of the metal pool on the composition of the gas and the working current of the plasmatrons, as well as the nonuniformity of the distribution of the temperature over the surface of the metal pool.

On the basis of an analysis of the available data [30, 31], we selected a basic variant, relative to which the temperature of the metal bath was calculated for different gas compositions and plasmatron currents: melting current, 400 A; gas composition, 100% argon; metal pool completely covered by the plasma torch.

The surface of the metal pool was divided into four regions with areas (S_i) equal to 2, 20, 35, and 43% of the total surface area with average temperatures (T_i^{Ar}) equal to 2350, 2250, 2050, and 1975 K under the conditions enumerated above.

Table 5.10 Calculated nitrogen content in PAR ingots of grade Kh21G11AN6 steel (wt.%)

Sample no.	Calculation conditions				$[N]'_{dr}$	$[N]_{ingot}$
	$K^d_{N_2}$	K^d_N	K^*_N	$(1-\alpha)$		
1	0.5	0.02	1.13	0.0	0.451	0.382
2	1.0	0.02	1.13	0.0	0.479	0.401
3	5.0	0.02	1.13	0.0	0.708	0.554
4	10.0	0.02	1.13	0.0	0.990	0.743
5	50.0	0.02	1.13	0.0	3.080	2.140
6	100.0	0.02	1.13	0.0	5.010	3.430
7	10.0	0.01	1.13	0.0	1.120	0.742
8	10.0	0.05	1.13	0.0	0.916	0.743
9	10.0	0.02	0.20	0.0	0.882	0.670
10	10.0	0.02	0.50	0.0	0.974	0.731
11	10.0	0.02	1.13	0.05	0.882	0.649
12	10.0	0.02	1.13	0.10	0.800	0.573
13	10.0	0.02	1.13	0.30	0.612	0.416
14	10.0	0.02	1.13	0.50	0.519	0.356
Experiment						0.42

In order to take into account the dependence of the temperature of the pool surface on the composition of the gas, the data in [31] were approximated by the equations

$$T_i^{\mathrm{Ar+N_2}} = T_i^{\mathrm{Ar}} + 6(\%\mathrm{N_2}) \quad \text{for} \quad \mathrm{N_2} \leq 25\%, \tag{5.72}$$

$$T_i^{\mathrm{Ar+N_2}} = T_i^{\mathrm{Ar}} + 150 + 3.3(\%\mathrm{N_2} - 25) \quad \text{for} \quad \mathrm{N_2} > 25\%. \tag{5.73}$$

The influence of the current was estimated in a first approximation using the following relationship: when the current is increased by 100 A, the temperature increases by 50 K [31]. In our opinion, this dependence should be more complex, since significant roles are played by the dimensions of the pool, the placement of the plasmatrons, and their number. Apparently, the variation of the current density in the pool must be taken into account.

The equations presented comprise a mathematical model of the behavior of nitrogen during PAR.

For adaptation of the model, calculations were preformed for the PAR of grade Kh21G11AN6 steel in a mold with a 435 × 435 mm cross section at a rate of 3.5–4.5 mm/min. The diameter of the intermediate product was equal to 300 mm.

Table 5.10 and Figs. 5.7, 5.8 and 5.9 present the results of calculations based on Eqs. 5.65–5.69 for various values of the parameters appearing in them.

The results of the calculations allowed us to draw several conclusions regarding the kinetics of the adsorption of nitrogen from the plasma in the electrode stage.

The diffusional hindrances in the metal are small, since the value of K^d_N weakly influences the nitrogen content in the droplets (Fig. 5.7). Conversely, the diffusional hindrances in the gas are significant (Fig. 5.8). The composition of the metal also has a significant influence on the nitrogen content in the metal.

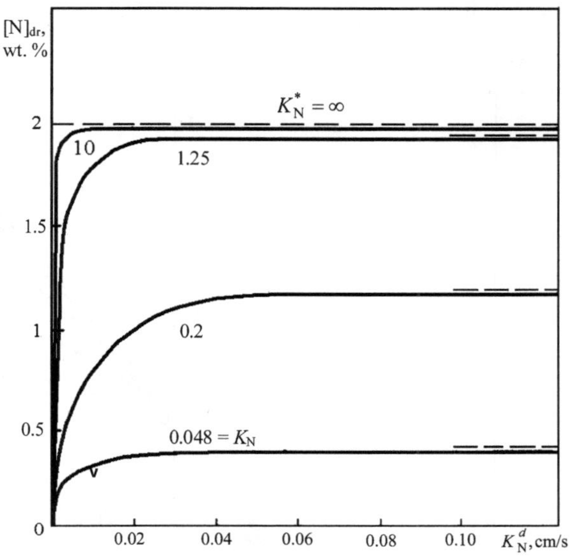

Fig. 5.7 Influence of the mass-transfer constant in the metal on the nitrogen content in the droplets

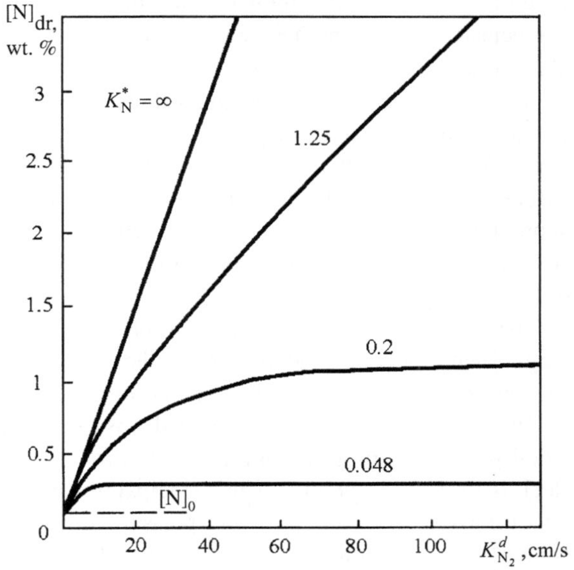

Fig. 5.8 Influence of the mass-transfer constant in the gas on the nitrogen content in the droplets

When nitrogen is absorbed from the atmosphere, rather than from a plasma, and $K_N = 0.048$, the diffusional hindrances in the gas have a considerably weaker influence (Fig. 5.9).

In addition, it was shown that the desorption of nitrogen, during which the gas is relatively stationary and $K_{N_2}^d \ll 5$ cm/s, is controlled by diffusion in the metal.

According to the calculations, the contribution of the pool stage to the absorption of nitrogen is approximately the same as that of the electrode stage

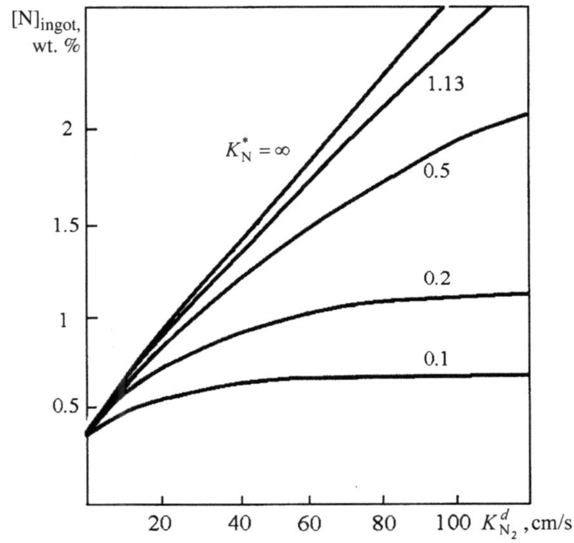

Fig. 5.9 Influence of the mass-transfer constant in the gas on the nitrogen content in the ingot

despite the unfavorable influence of the increase in temperature. The latter is compensated by the higher values of $K_{N_2}^d$ in the pool stage. Diffusion in the gas plays a decisive role (Fig. 5.9).

A fivefold increase in the mass-transfer constant in the metal (Table 5.10, variants 7 and 8) scarcely alters the value of $[N]_{ingot}$. An increase in the fraction of the surface that is not covered by the arc lowers $[N]_{ingot}$ significantly (Table 5.10, variants 4, 11–14). The influence of K_N^* (Table 5.10, variants 4, 9, and 10) is appreciable, but weaker than on the electrode. The region of values of K_N^* and f_N covers the temperature range from 1800 to 2800 K (larger values of K_N^* correspond to lower temperatures).

The calculated values of $[N]_{ingot}$ (Table 5.10) as a whole cover the experimental range, enabling proper adaptation of the model to the conditions of PAR for a specific technology by adjusting the corresponding constants.

The mathematical model developed was used to predict the composition of the metal [42, 43] during the multiple PAR processing of grade 06Kh21N11AN4 M steel in a U-365 facility. The principal parameters of the process were: the remelting rate, 4–5 mm/min; the pressure in the furnace, 1.5 atm; the nitrogen content in the gas mixture, 24–37%; the plasmatron current, 400 A; the concentration of nitrogen in the intermediate product, 0.1 wt.%.

The results of the calculations and experiments are presented in Table 5.11.

Thus, the calculated data demonstrated the possibility, in principle, of raising the concentration of nitrogen in the metal during multiple remelting processing. A comparison of the experimental and calculated values reveals sufficient convergence between them, providing a basis for using the mathematical model developed to predict the composition of steel with respect to the nitrogen content during PAR.

Table 5.11 Nitrogen content in grade 06Kh21N11AN4M steel during multiple PAR processing

Melt and remelt nos.	P_{N_2} (atm)	$[N]_{ingot}^{exp}$ (wt.%)	$[N]_{ingot}^{calc}$ (wt.%)
433 1st remelt	0.386	0.42–0.47	0.25–0.44
436 2nd remelt	0.480	0.65–0.74	0.61–0.90
439 3rd remelt	0.470	0.78–0.80	0.96–1.00
435 1st remelt	0.473	0.32–0.49	0.28–0.37
437 2nd remelt	0.430	0.46–0.70	0.38–0.43
440 3rd remelt	0.487	0.58–0.77	0.62–0.76
425 1st remelt	0.373	0.28–0.41	0.39–0.45
432 2nd remelt	0.480	0.41–0.60	0.58–0.64
445 3rd remelt	0.552	0.57–0.75	0.90

5.4.2 Mathematical Model of the Interaction of the Metal, Slag, and Gas During Vacuum Arc Remelting

Two variants of vacuum arc remelting (VAR) that use slag are known [44, 45]. In one of them, a slag pool with a mass up to 4% of the mass of the ingot to be melted is formed on the surface of the metal. In the other variant a large part of the surface of the metal is exposed, and the slag floats in the form of thin "lenses." This technique allows refinement of the metal by feeding small portions of a flux into the pool at a rate equal to 0.01–0.1% of the mass remelting rate. This variant of metal treatment during VAR is called microslag refining (MSR).

A diagram of the vacuum arc remelting process with microslag refining is presented in Fig. 5.10. According to this diagram, an interaction takes place on the following interfaces:

- pool-metal/slag-lens;
- pool-metal-surface/gas;
- remelted-electrode-metal-surface/gas;
- slag-surface/gas.

In addition, there are specific factors that are characteristic of VAR: the presence of an oxide film on the electrode $(E_iO_m)_{ox}$; the presence in the metal of nonmetallic inclusions (NI), which float up to the surface during the remelt and are assimilated by the slag, altering its mass and composition; the formation of a slag skin. The distribution coefficients of the elements between the liquid and solid slag (the liquation coefficients) κ_i are not equal to unity in the general case.

With consideration of the foregoing statements, the actual picture of the phase interaction during VAR with MSR can be depicted in the form of the block diagram presented in Fig. 5.11.

The general model of the processes involved in VAR with MSR is based on the following theoretical reasoning and assumptions:

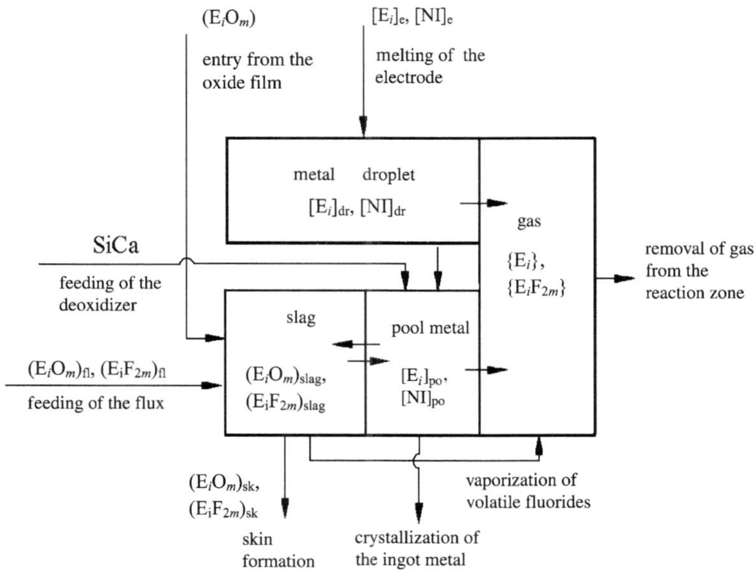

Fig. 5.10 Block diagram of the phase interactions that occur during vacuum arc remelting with microslag refining

- The composition of the metal and the slag is identical within each phase at the time τ.
- The phase interaction at any time in each stage can be assigned to a specific area with average values of the mass-transfer constant, temperature, and other factors that influence the character of the processes.
- There are no losses of the components of the metallic melt.
- The vaporization of the slag components significantly alters the composition of the melt when it has a small mass and a large specific surface.
- The vaporization of the slag components can be described by the following reactions:

$$(SiO_2) + 2(CaF_2) = 2(CaO) + \{SiF_4\}, \tag{5.74}$$

$$(TiO_2) + 2(CaF_2) = 2(CaO) + \{TiF_4\}, \tag{5.75}$$

$$(MgO) + (CaF_2) = (CaO) + \{MgF_2\}, \tag{5.76}$$

$$2(AlO_{1.5}) + 3(CaF_2) = 3(CaO) + 2\{AlF_3\}, \tag{5.77}$$

$$(CaF_2) = \{CaF_2\}. \tag{5.78}$$

- The following reaction occurs on the exposed surface of the metal:

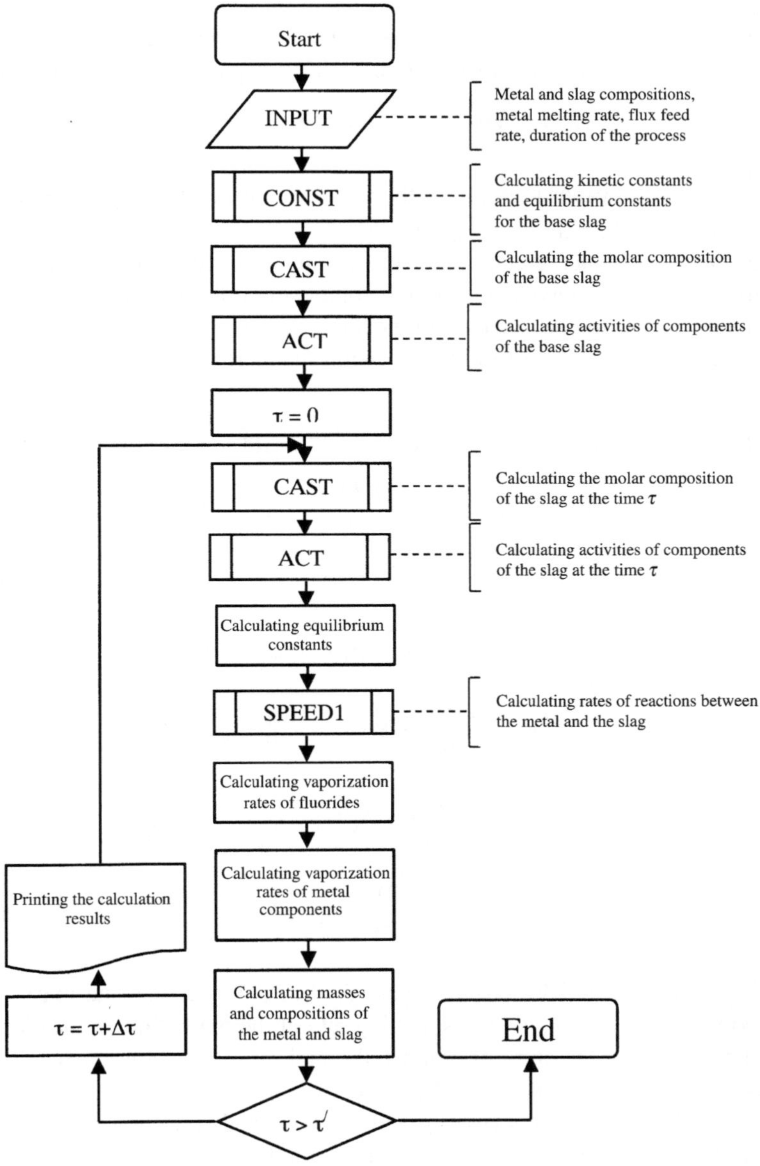

Fig. 5.11 Block diagram of the algorithm of a computational experiment for predicting the composition of the phases during VAR with MSR

$$[C] + [O] = \{CO\}. \tag{5.79}$$

- The mass of the melts varies as a consequence of the chemical reactions between the metal, the slag, and the gas.

The equations presented in Sect. 4.3.2 were used to calculate the rates of the interaction. In addition, the vaporization rates of the fluorides were taken into account under the assumption that the rate-limiting step is their diffusion in the gas:

$$J^{\mathrm{vap}}_{\mathrm{E}_i\mathrm{F}_{2m}} = \beta_{\mathrm{g}} D^{0.5}_{\mathrm{E}_i\mathrm{F}_{2m}} \frac{P_{\mathrm{E}_i\mathrm{F}_{2m}}}{RT}, \tag{5.80}$$

where $P_{\mathrm{E}_i\mathrm{F}_{2m}}$ is the vapor pressure of the fluoride (atm) and $D_{\mathrm{E}_i\mathrm{F}_{2m}}$ is the diffusion coefficient of the fluoride in the gas (m^2/s).

The material balance equations of the components of the metal and the slag were used to estimate the changes in the composition of the phases. In particular, the changes in the masses of the components of the slag phase during the time interval $\Delta\tau$ can be expressed in the following manner:

$$\Delta m^{\tau}_{\mathrm{E}_i\mathrm{O}_m} = \left(V_{\mathrm{fl}} \frac{(\mathrm{E}_i\mathrm{O}_m)_{\mathrm{fl}}}{100} + V_{\mathrm{ox}} \frac{(\mathrm{E}_i\mathrm{O}_m)_{\mathrm{ox}}}{100} + V_{\mathrm{NI}} \frac{(\mathrm{E}_i\mathrm{O}_m)_{\mathrm{NI}}}{100} \right.$$
$$\left. - J_{\mathrm{E}_i} S_{\mathrm{Me/slag}} M_{\mathrm{E}_i\mathrm{O}_m} - V_{\mathrm{sk}} \frac{(\mathrm{E}_i\mathrm{O}_m)^{\tau-1}}{100} \kappa_{\mathrm{E}i} - J^{\mathrm{vap}}_{\mathrm{E}_i\mathrm{F}_{2m}} S_{\mathrm{slag/gas}} M_{\mathrm{E}_i\mathrm{O}_m} \right) \Delta\tau, \tag{5.81}$$

$$\Delta m^{\tau}_{\mathrm{CaF}_2} = \left(V_{\mathrm{fl}} \frac{(\mathrm{CaF}_2)_{\mathrm{fl}}}{100} - \sum m J^{\mathrm{vap}}_{\mathrm{E}_i\mathrm{F}_{2m}} S_{\mathrm{slag/gas}} M_{\mathrm{CaF}_2} \right.$$
$$\left. - V_{\mathrm{sk}} \frac{(\mathrm{CaF}_2)^{\tau-1}}{100} \kappa_{\mathrm{CaF}_2} - J^{\mathrm{vap}}_{\mathrm{CaF}_2} S_{\mathrm{slag/gas}} M_{\mathrm{CaF}_2} \right) \Delta\tau, \tag{5.82}$$

$$\Delta m^{\tau}_{\mathrm{CaO}} = \left(V_{\mathrm{fl}} \frac{(\mathrm{CaO})_{\mathrm{fl}}}{100} + \sum m J^{\mathrm{vap}}_{\mathrm{E}_i\mathrm{F}_{2m}} S_{\mathrm{slag/gas}} M_{\mathrm{CaO}} \right.$$
$$\left. - V_{\mathrm{sk}} \frac{(\mathrm{CaO})^{\tau-1}}{100} \kappa_{\mathrm{CaO}} \right) \Delta\tau. \tag{5.83}$$

where V_{fl}, V_{sk}, V_{ox}, and V_{NI} are the mass feed rate of the flux, the mass rate of formation of the slag skin, the mass melting rate of the oxide film on the electrode, and the mass rate of the entry of nonmetallic inclusions as a consequence of their assimilation by the slag (kg/s), $(\mathrm{E}_i\mathrm{O}_m)_{\mathrm{fl}}$, $(\mathrm{CaF}_2)_{\mathrm{fl}}$, and $(\mathrm{CaO})_{\mathrm{fl}}$ are the concentrations of the flux components (wt.%), and $(\mathrm{E}_i\mathrm{O}_m)_{\mathrm{ox}}$ and $(\mathrm{E}_i\mathrm{O}_m)_{\mathrm{NI}}$ are the concentrations of the oxide of the element E_i in the oxide film on the electrode and in the nonmetallic inclusions (wt.%).

In Eqs. 5.81–5.83 the first terms characterize the increase in the mass of the components in the slag due to their entry when the flux is fed. The increase in the mass of the oxides in the slag due to melting of the oxide film on the electrode, assimilation of the nonmetallic inclusions, and chemical reactions between the metal and the slag are characterized by the second, third, and fourth terms in Eq. 5.81, respectively. The fifth term in Eq. 5.81 and the second term in Eq. 5.83 characterize the increase in the mass of CaO and the decrease in the mass of the other oxides as a result of the occurrence of reactions (5.74–5.78). The second and third terms in Eq. 5.82 describe the decline in the amount of calcium fluoride due to its vaporization, as well as due to the formation and vaporization of fluorides of other elements. The last terms in Eqs. 5.81–5.83 describe the decrease in the mass of the oxides and calcium fluoride in the liquid slag due to formation of the slag skin.

We present the equation for the assimilation rate of the nonmetallic inclusions in the example of aluminum oxide:

$$(V_{AlO_{1.5}})_{NI} = \frac{[O]_{NI}(AlO_{1.5})_{NI}}{(O)_{NI}100} S_{po}\rho_{Me}K_{NI}. \quad (5.84)$$

According to the data in [46, 47], $K_{NI.} = 0.046$, and $(AlO_{1.5})_{NI.}/(O)_{NI.} = 2.04$.

Dissolved oxygen arrives together with the metal of the melted electrode, passes into the metal of the crystallizing ingot, and is removed when carbon monoxide is formed as a result of reaction (5.79). Also, an additional decrease in the concentration of dissolved oxygen occurs when deoxidizers are introduced into the flux. Thus, the oxygen content in the metal at the time τ equals

$$[O]^{\tau} = [O]^{\tau-1} + \Big([O]_e V_e - [O]^{\tau-1} V_{cryst} - V_{deox} - 100 J_{CO} S_{Me/gas} M_O + 100 J_O S_{Me/gas} M_O\Big)\frac{\Delta\tau^{\tau}}{m_{Me}}, \quad (5.85)$$

where V_{cryst}, V_e, and V_{deox} are the mass crystallization rate of the metal, the mass melting rate of the electrode, and the mass deoxidization rate, respectively, kg/s. Under a steady-state remelting regime, under which the mass of the metal pool remains practically constant, it may be assumed that $V_{cryst} = V_e$.

As our investigations showed, the deoxidization rate is proportional to the concentration of dissolved oxygen in the bulk of the melt and the feed rate of the deoxidizer (V_{SiCa} for silicocalcium):

$$V_{deox} = kV_{SiCa}[O]. \quad (5.86)$$

The value of k for our conditions equals 58.78.

The oxygen content in the nonmetallic inclusions at the time τ was determined from the data in [45]:

$$[O]_{NI}^{\tau} = [O]_{NI}^{\tau-1} + \Big([O]_{NI_e} V_e - V_{cryst}[O]_{NI}^{\tau-1} - 0.667 k_{NI.} S_{Me/slag}[O]^{\tau-1}\Big)\frac{\Delta\tau}{m_{Me}^{\tau}}. \quad (5.87)$$

The rate of the entry of oxides from the oxide film on the electrode V_{ox} was determined from the diameter of the electrode, the thickness of the film δ_{ox}, and the linear melting rate of the electrode $\bar{V}_e$:

$$V_e = \bar{V}_e \pi d_e \delta_{ox} \rho_{ox}, \quad (5.88)$$

where ρ_{ox} is the density of the scale (kg/m^3).

Table 5.12 Liquation coefficients of the slag components

Component	CaF_2	CaO	SiO_2	Al_2O_3	MgO
κ_i	1.21	0.59	0.52	1	6.36

According to the literature data, during the remelting of low-alloy steels, reduction of the scale by solid-state carbon occurs in the high-temperature zone, and an electrode with a clean surface enters the melting zone as a result. This phenomenon is not observed during the remelting of alloyed steels. It was taken as an assumption that the oxide film for stainless steels consists solely of TiO_2.

To construct a quantitative mathematical model, we used several experimental parameters and dependences.

In particular, the rate of formation of the slag skin and the area of the metal/slag interface depend on the mass of the slag:

$$V_{sk} = 4.6 \times 10^{-3} M_{slag}, \tag{5.89}$$

$$S_{slag} = 0.357 M_{slag}. \tag{5.90}$$

Since the slag on the metal surface exists in the form of flat "lenses", it can be assumed with sufficient accuracy that the areas of the metal/slag and slag/gas interfaces are equal.

The distribution coefficients of the components between the liquid and solid phases are presented in Table 5.12.

To adapt the software developed, calculations of the remelting of grade 16KhN3MA steel with type FI-1 refining flux, which contains 20 wt.% SiCa, were performed.

The following technological parameters were used: flux feed rate, 3.3×10^{-4} kg/s; ingot melting rate, 0.166 kg/s; electrode diameter, 0.57 m; mold diameter, 0.67 m; furnace chamber pressure, 1.3×10^{-4} atm; temperature, 1873 K.

The compositions of the starting materials, as well as the experimental and calculated data, are presented in Table 5.13. A comparison of the experimental and calculated values shows that they are in fairly good agreement with one another. In particular, it is noteworthy that the calculated values of the slag mass differ from the experimental value by no more than 10%.

Thus, the model created describes the actual interaction processes of the metal, slag, and gas during vacuum arc remelting with microslag refining with sufficient accuracy and can be used to perfect the remelting technology for steels of various grades.

The model developed was used to analyze the evolution of the slag during the remelting of grade 12Kh18N10T stainless steels with unremoved scale. The remelting of these steels is accompanied by a significant change in the composition of the slag, as a result of which it becomes more viscous. This results in contamination of the subcutaneous layer of the ingot with slag inclusions. A calculation confirmed that the slag composition changes rapidly in this case as a result

Table 5.13 Composition and mass of the metal and slag during the vacuum arc remelting of 16KhN3MA steel with FI-1 flux

Material	Composition of metal and slag (wt.%)													Slag mass (g)
	Fe	C	Mn/MnO	Si/SiO_2	Cr	Al/$AlO_{1.5}$	P	O	Ni	Mo	Ca/CaO	MgO	CaF_2	
Original metal	base	0.16	0.58		0.49	0.010	0.011	0.006	3.06	0.31	–	–	–	–
Original flux	0.6	–	–	4.90	–	0.70	–	–	–	–	4.50	1.40	87.80	–
Silicocalcium	5.8	–	–	59.6	–	1.1	–	–	–	–	33.0	–	–	–
Slag[a]	3.93	–	0.43	12.25	–	5.11	–	–	–	–	12.54	0.43	65.23	28
	0.42		0.24	10.16		5.64					15.24	0.41	66.96	30
Ingot metal[a]	base	0.16	0.45	0.290	0.47	0.010	0.011	0.005	3.06	0.31	–	–	–	–
		0.16	0.56	0.291	0.49	0.010	0.010	0.007	3.06	0.31				

[a] The values above the bars are experimental, and the values below the bars are calculated

Table 5.14 Composition and mass of the slag during VAR of grade 12X18N10T steel with unremoved scale

Oxide film thickness (mm)	Content of components (wt.%)							Slag mass (g)
	MnO	SiO_2	TiO_2	$AlO_{1.5}$	CaF_2	CaO	MgO	
0	0.34	6.19	6.61	4.94	64.3	14.6	0.4	32
0.1	0.43	6.13	20.70	5.19	51.0	14.0	0.4	35
0.2	0.23	7.60	53.20	7.94	2.2	25.0	0.4	21

of the occurrence of reactions (5.74–5.78) and the entry of a large amount of TiO_2 from the oxide film. The extent of its regeneration depends very strongly on the thickness of the casting scale. The results of calculations for the actual range of oxide film thicknesses are presented in Table 5.14.

It follows from the data in Table 5.14 that when the thickness of the oxide film on the electrode is less than 0.1 mm, the composition of the slag varies only slightly with respect to the calcium fluoride content, ensuring its low melting point and maintenance of its refining properties. Increasing δ_{ox} to 0.2 mm results in significant increases in the concentrations of TiO_2 and CaO and in the nearly complete disappearance of calcium fluoride from the slag. The melting point of such a system is high, and the slag thickens and solidifies on the mold walls, creating defects in the surface part of the ingot.

During the development of the mathematical model of the denitrogenation process of a metal during VAR, the following theoretical reasoning and simplifying assumptions were adopted.

- The rate-limiting step of the denitrogenation process is the convective mass transport of atoms of dissolved nitrogen from the bulk of the metallic melt to the metal/gas interface.
- Two stages, one related to the liquid film on the electrode and one related to the pool, play significant roles in the gas removal process. In the pool stage, degassing occurs uniformly over the entire surface of the molten metal that is free of slag.
- In each stage the denitrogenation process can be characterized by average values of the interaction area, the convection constant, and the denitrogenation rates.
- The influence of alloying elements and impurities on the nitrogen removal process is manifested by the change in equilibrium solubility of nitrogen in the metal.

Under such an approach the rate of nitrogen removal in the electrode stage equals

$$V_{\mathrm{N}}^{\mathrm{e}} = \frac{[\mathrm{N}]_{\mathrm{e}} - [\mathrm{N}]_{\mathrm{eq}}}{\frac{100 M_{\mathrm{N}}}{\beta_{\mathrm{Me}} D_{\mathrm{N}}^{0.5} \rho_{\mathrm{Me}}} + \frac{100 S_{\mathrm{e}} M_{\mathrm{N}}}{m_{\mathrm{e}}}}, \tag{5.91}$$

where $[\mathrm{N}]_{\mathrm{e}}$ and $[\mathrm{N}]_{\mathrm{eq}}$ are the current and equilibrium concentrations of nitrogen in the liquid film on the end surface of the electrode (wt.%), β_{Me} is the convection constant for the metal pool ($s^{-0.5}$), D_{N} is the diffusion coefficient of nitrogen in the metal (m^2/s), and m_{e} is the melting rate of the electrode (kg/s).

Table 5.15 Calculated and experimental data [48] for grade 12Kh18N10T steel

Linear remelting rate (cm/s)	$[N]_e/[N]_{ingot}$, experiment	$[N]_e/[N]_{ingot}$, calculation
1.29	1.3–1.4	1.41
0.89	1.4–1.6	1.60
0.70	1.7–1.9	1.87
0.50	2.0–2.6	2.10

Table 5.16 Calculated and experimental data for grade 14KhN3MA steel

Linear remelting rate (kg/s)	$[N]_e$ (wt.%) experiment	$[N]_{ingot}$ (wt.%) experiment	$[N]_{ingot}$ (wt.%) calculation
0.123	0.0099	0.0061	0.0058
0.208	0.0099	0.0076	0.0069

For the pool stage, a non-steady-state process is considered. At each time τ the nitrogen removal rate equals

$$V^{\tau}_{po} = \frac{\beta_{Me} D_N^{0.5} \rho_{Me}}{100 M_N} \left([N]^{\tau}_{po} - [N]_{eq}\right). \tag{5.92}$$

Here $[N]^{\tau}_{po}$ is the current concentration of nitrogen in the pool, wt.%.

Accordingly, the nitrogen content in the droplets of the electrode metal entering the pool equals

$$[N]_{dr} = [N]_e - \frac{100 V^{e}_{N} S_e M_N}{m_e}, \tag{5.93}$$

and we find the nitrogen concentration in the liquid metal of the pool at the time τ from the equation

$$[N]^{\tau}_{dr} = \{([N]_{dr} - [N]^{\tau-1}_{po} \cdot \kappa_N) m_e - 100 V^{\tau}_{po} S_{po} M_N\} \Delta\tau / m_{po} + [N]^{\tau-1}_{po}, \tag{5.94}$$

where $\Delta\tau$ is the calculation time interval (s) and m_{po} is the current mass of the metal pool (kg).

To test the model and confirm its faithfulness, we compared calculated and experimental data that are available in the literature with regard to the behavior of nitrogen during VAR.

Table 5.15 presents the results of the calculations for grade 12Kh18N10T steel and the data in [48], which were obtained under the following melting conditions: electrode diameter, 0.57 m; mold diameter, 0.65 m; furnace chamber pressure, 1×10^{-4} atm; nitrogen pressure, 1.11×10^{-5} atm; convection constant for the electrode stage, $\beta_{Me} = 0.8\ s^{-0.5}$; convection constant for the pool stage, $\beta_{Me} = 0.5\ s^{-0.5}$; convection constant for the gas phase, $\beta_{Me} = 8\ s^{-0.5}$. As can be seen from Table 5.15, the calculation gives an extent of nitrogen removal close to the experimental value.

Table 5.16 presents calculated and experimental data for the remelting of grade 14KhN3MA low-alloy steel, which were obtained for the following mass-transfer parameters: for the electrode stage, $\beta_{Me} = 1\ s^{-0.5}$; for the pool stage, $\beta_{Me} = 0.4\ s^{-0.5}$; for the gas phase, $\beta_{Me} = 8\ s^{-0.5}$.

The models and computer calculation programs developed enable us to predict the changes in the composition of the ingot metal and can be used to design the VAR technology for structural steels [49].

5.4.3 *Mathematical Model of the Physicochemical Processes in the Electron-Beam Remelting of Alloys*

An analysis of the available data regarding the features of the heating, melting, vaporization, and refining operations for metals and alloys during electron-beam remelting (EBR) enables us to formulate the main aspects of the physical model and the assumptions and hypotheses which will be taken into account in developing a mathematical model that describes the behavior of components during EBR.

1. Hitherto there have not been any sufficiently rigorous methods that provide a way to calculate the dynamics of the melting of an intermediate product and the temperature fields on the surface of the melting intermediate product and the metal in the mold under the action of a scanning electron beam. Analytical solution of the heat-conduction equation is complicated by the influence of the vigorous convective transport. For this reason, the following scheme of the thermal processes was employed: the surface temperature of the liquid–metal film (with a thickness less than 0.1 mm) is about 50 K higher than the melting point of the alloy [50]; the surface temperature of the liquid metal in the mold outside of the focal spot T_0 exceeds the melting point of the metal T_m by 150 K; and in a rapid-scanning regime the temperature within the focal spot T_{fs} is [51]

$$T_{fs} = T_0 + W(2a)^{3/2}(v_{sc}\lambda_{eff}C_V)^{1/2}. \tag{5.95}$$

Here W is the power of the beam, kW; a is the radius of the focal spot, m; v_{sc} is the scanning rate of the electron beam, m/s; λ_{eff} is the effective thermal conductivity, W/(m K); C_V is the volumetric specific heat, J/(m^{3}K). According to data from several studies, the value of λ_{eff} with allowance for convective heat transfer is approximately three times greater than the thermal conductivity of solid metals. In some cases boiling of the metals within the focal spot can be expected. An estimate of the influence of vaporization on the value of $T_{fs} - T_0$ was given in [50], and it was shown that at small values of a, the energy loses related to vaporization significantly lower the superheating.

2. Vaporization of the base metal of an alloy occurs mainly from the focal spot [52]. In more exact calculations the parallel vaporization from the remaining surface of the liquid metal should be taken into account. Therefore, it is expedient to include the radius of the focal spot, the scanning rate, and the power of the electron beam among the input parameters of the model.

3. We will represent the individual stages of EBR in the form of ideally mixed reactors.

4. For the sake of generality in the analysis, the influence of all factors on the vaporization rate, including the diffusion in the metal and the gas and the slowness of the step involving adsorption followed by discharge and ionization, must be taken into account. Under the conditions of a fairly high vacuum, under which the mean free path of the vapor molecules is large, a molecular vaporization regime is realized, under which the vaporization rate is specified by the Langmuir formula [53]

$$v = \alpha P_{\mathrm{E}_i} \sqrt{\frac{1}{2\pi RTM_{\mathrm{E}_i}}}, \tag{5.96}$$

where α is the accommodation coefficient and P_{E_i} is the saturated vapor pressure of the element E_i.

The rate defined by Eq. 5.96 may be regarded as the rate of the adsorption followed by discharge and ionization.

5. Equation 5.96 enables us to calculate the maximum rate of loss of the metal upon vaporization

$$\Delta[\mathrm{E}_i] = \frac{vM_{\mathrm{E}_i}S}{\dot{m}}100, \tag{5.97}$$

where S is the area of the metal/gas interface and $\dot{m}$ is the melting rate of the metal, whose value can be defined in the following manner:

$$\dot{m} = \frac{\bar{V}\rho}{\tau_{\mathrm{av}}}. \tag{5.98}$$

Here $\bar{V}$ is the average volume of a poured portion of the metal; ρ is the density of the material; τ_{av} is the average refining (melting) time.

The calculated loss of nickel and chromium as a result of vaporization is large and is close to the experimental data [54–56]. The large calculated values of the loss of chromium, cobalt, and nickel at 2273 K, i.e., at the temperature of a focal spot, are associated with the large vaporization surface. At the same time, the vaporization rates of aluminum and yttrium are so small that even the approximate character of the calculations cannot account for the decreases in their concentrations during EBR. The reason for the loss of these elements is their oxidation by gaseous oxygen in the chamber atmosphere followed by removal of the oxide inclusions.

6. We assume that the residual pressure in the chamber ($P_{\mathrm{res}} = 0.006$–0.8 Pa) is related to air inleakage. Then the oxygen pressure in the chamber is $P_{\mathrm{O}_2} = 0.21P_{\mathrm{res}}$. Calculations based on Eq. 5.96 show that the presence of oxygen in such amounts in the chamber atmosphere can account for the loss of aluminum and yttrium on the exposed metal surface. In addition, we took into account that oxygen is distributed between aluminum and yttrium in accordance with their

oxygen affinities. We showed that nearly all the oxygen is bound in the oxide Y_2O_3. The oxidation of aluminum begins only after the thorough oxidation of yttrium. We note that this is possible, because more oxygen enters the metal than is needed for the oxidation and vaporization of yttrium. Thus, the oxygen pressure in the chamber must be included among the input parameters of the model.

7. It has been shown that the formation of nitrides during EBR can be disregarded.

8. The composition of the gas phase in the chamber varies over the course of EBR due to air inleakage, continuous operation of the vacuum pumps, and vaporization and condensation of the alloy components.

9. We divide the vaporization of elements during EBR into several stages, each of which can be treated separately:

- stage 1, which includes processes on the intermediate product end surface being melted, i.e., on the surface of the thin film of liquid metal draining into the intermediate vessel;
- stage 2, which includes processes in the pool of liquid metal in the intermediate vessel;
- stage 3, which includes processes in the pool of liquid metal in the mold from the time when the current portion was poured from the intermediate vessel until it crystallizes.

Stage 3 precedes the heating of the solid metal in the mold to temperatures close to the melting point, which can cause appreciable vaporization of elements from the solid phase (sublimation), especially elements that have a high volatility (chromium, manganese).

The material presented above enables us to develop a mathematical model of the physicochemical metal-refining processes that occur during EBR.

We introduce the notations for the parameters and regimes of the process: $\dot{m}_1^0$ is the melting rate of the intermediate product, kg/s: $\dot{m}_1$ is the rate of the entry of the metal into the intermediate vessel, kg/s; $\dot{m}_{\text{cryst}}$ is the crystallization (melting) rate of the metal in the intermediate vessel, kg/s; S_1, S_2, and S_3 are the surface areas of the tip of the intermediate product being melted, the liquid metal in the intermediate vessel, and the liquid metal in the EBR ingot pool, respectively, m^2; m_2 and m_3 are the masses of the liquid metal in the intermediate vessel and in the EBR ingot pool, respectively, kg; $[E_i]_0$, $[E_i]_1$, $[E_i]_2$, $[E_i]_3$, and $[E_i]_4$ are the concentrations of the respective element in the intermediate product, in the droplets entering the intermediate vessel, in the intermediate vessel, in the EBR ingot pool in the mold, and in the EBR ingot, respectively, wt.%; M_{E_i} is the molar weight, kg/mole; C_{E_i} is the concentration of the element in the alloy, mole/m^3; S_j^{fs} is the area of the focal spot in stage j, m^2; $(v_{E_i})_j$ and $\left(v_{E_i}^{\text{fs}}\right)_j$ are the rates of vaporization (degassing) of the element from the exposed surface and the focal spot in the j-th stage of the refining process, mole/(m^2 s); $v_{E_i}^{\text{cond}}$ is the condensation rate of the element on the chamber walls, mole/(m^2 s); V is the volume of the chamber, m^3; $\dot{V}$ is the capacity of the pumping system, m^3/s; τ is the time of the process, s; $\Delta m_{vap,2}$ is the loss of metal due to vaporization from the intermediate vessel, kg.

We formulate the fundamental equations of the mathematical model that are needed to determine the composition of the metal in each stage of the EBR process.

1. *Balance equations*

Stage 1. The melting regime is steady-state because of the small mass of the metal on the end surface of the intermediate product (the thickness of the film is 0.1–0.2 mm). The mass balance of the element E_i can be expressed in the following manner:

$$\dot{m}_1^0 \frac{[E_i]_0}{100} = \dot{m}_1 \frac{[E_i]_1}{100} + \nu_{E_i,1} S_1 M_{E_i} + \nu_{E_i,1}^{fs} S_1^{fs} M_{E_i}, \tag{5.99}$$

$$\dot{m}_1 = \dot{m}_1^0 - \sum_{i=1}^{l} \left(\nu_{E_i,1} S_1 M_{E_i} + \nu_{E_i,1}^{fs} S_1^{fs} M_{E_i} \right), \tag{5.100}$$

where l is the number of alloy components and removable impurities.

In Eq. 5.100 the vaporization rates of nitrogen and hydrogen can clearly be neglected. To solve system of Eqs. 5.99 and 5.100, we need to know the dependence of $\nu_{E_i,1}$ on $[E_i]$.

Stage 2. Metal accumulates as the intermediate product melts; therefore the regime is non-steady-state, and the calculation must be performed over an accumulation–pouring cycle.

The mass balance of the element E_i in the liquid metal in the intermediate vessel can be described by the equation

$$\frac{dm_{E_i}}{d\tau} = \dot{m}_1 - \dot{m}_{cryst2} - \left(\nu_{E_i,2} S_2 M_{E_i} + \nu_{E_i,2}^{fs} S_2^{fs} M_{E_i} \right). \tag{5.101}$$

In this equation the first term characterizes the entry of the element with the metal droplets, the second term characterizes its removal (entry) upon crystallization (melting) in the intermediate vessel, and the third term characterizes the losses to the gas.

Since $m_{E_i} = m_2 \frac{[E_i]_2}{100}$, we have

$$\frac{dm_{E_i}}{d\tau} = \frac{1}{100} \left(m_2 \frac{d[E_i]_2}{d\tau} + [E_i]_2 \frac{dm_2}{d\tau} \right), \tag{5.102}$$

where $\frac{dm_2}{d\tau} = -\dot{m}_{cryst,2} + \dot{m}_1$ (if we assume that $\dot{m}_{cryst,2}$ is a positive number).

Then

$$m_2 \frac{d[E_i]_2}{d\tau} = \dot{m}_1 \left([E_i]_1 - [E_i]_2 \right) - 100 M_{E_i} \left(\nu_{E_i,2} S_2 + \nu_{E_i,2}^{fs} S_2^{fs} \right). \tag{5.103}$$

Dividing the filling time of the intermediate vessel into k intervals, we obtain

$$\frac{d[\mathrm{E}_i]_2}{d\tau} = \frac{[\mathrm{E}_i]_2^{k+1} - [\mathrm{E}_i]_2^k}{\Delta\tau}, \tag{5.104}$$

hence for $\dot{m}_{\mathrm{cryst},\,2} = 0$ we have

$$[\mathrm{E}_i]_2^{k+1} = [\mathrm{E}_i]_2^k + \frac{\dot{m}_1}{m_2^k}\Delta\tau\Delta[\mathrm{E}_i] - \frac{100M_{\mathrm{E}_i}\Delta\tau\left(v_{\mathrm{E}_i,2}S_2 + v_{\mathrm{E}_i,2}^{\mathrm{fs}}S_2^{\mathrm{fs}}\right)}{m_2^k}, \tag{5.105}$$

where $\Delta[\mathrm{E}_i] = [\mathrm{E}_i]_1 - [\mathrm{E}_i]_2^k$.

We note that the losses of metal from the intermediate vessel due to vaporization are not taken into account in the determination of m_2 in expression (5.103). For a more exact determination of $m_2 = f(\tau)$, it is expedient to take into account the change in the mass of the metal due to vaporization

$$\Delta m_{\mathrm{vap},\,2} = \sum_{i=1}^{l} v_{\mathrm{E}_i,\,2}^k \Delta\tau S_2^k M_{\mathrm{E}_i}. \tag{5.106}$$

Stage 3. The process of pouring the metal from the intermediate vessel into the mold is quick and can be neglected. When the crystallization process has a steady-state regime, it is described by an equation similar to (5.101), but without the first term on the right-hand side. If the influence of the pouring time is considered, the first term on the right-hand side, in which $m_1(\tau)$ will play the role of the pouring rate, should be retained. For the steady-state regime, instead of Eq. 5.105, we have

$$[\mathrm{E}_i]_3^{k+1} = [\mathrm{E}_i]_3^k - \frac{100\Delta\tau v_{\mathrm{E}_i,3}^k S_3^k M_{\mathrm{E}_i}}{m_3^k}. \tag{5.107}$$

When the composition of the gas is calculated, the dynamics of its formation in the chamber during the complete smelting time of the ingot is taken into account by treating the gas phase as a separate ideally mixed reactor (Fig. 5.12).

The *initial state* is a vacuum with a residual pressure equal to $P_{\mathrm{res}} = P_{\mathrm{res}}^{\mathrm{O}_2} + P_{\mathrm{res}}^{\mathrm{N}_2}$.

Let us calculate the variation of the composition of the gas with time. The variation of the composition of the gas also alters the composition of the metal in each of the poured portions.

For vaporizable elements (Ni, Co, Cr, Al, and Y) the material balance equation has the form

$$\sum_{j=1}^{4}\left(v_{\mathrm{E}_i}^{\mathrm{fs}}\right)_j S^{\mathrm{fs}} M_{\mathrm{E}_i} + \sum_{j=1}^{4}\left(v_{\mathrm{E}_i}\right)_j\left(S^n - S^{\mathrm{fs}}\right)M_{\mathrm{E}_i} = v_{\mathrm{E}_i}^{\mathrm{cond}} S^{\mathrm{cond}} M_{\mathrm{E}_i} + \frac{d\left(m_{\mathrm{g}}\frac{\{\mathrm{E}_i\}}{100}\right)}{d\tau}, \tag{5.108}$$

where $\{\mathrm{E}_i\}$ is the concentration of element i in the vapor (wt.%) and m_{g} is the mass of the vapor phase (kg).

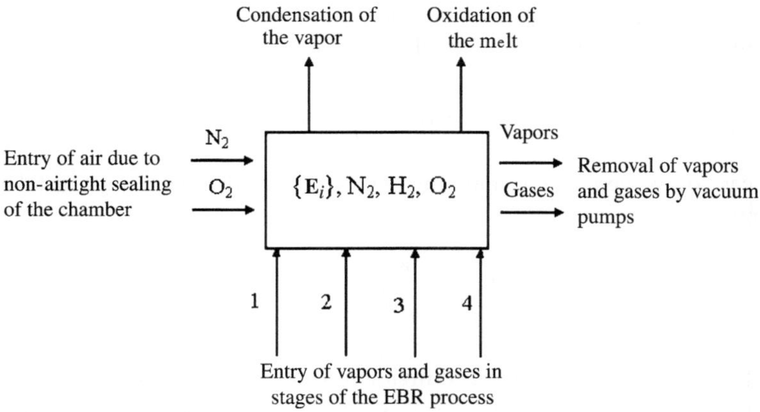

Fig. 5.12 Diagram of the formation of the vapor and gas phases during smelting of an EBR ingot

The terms on the left-hand side of Eq. 5.108 take into account the entry of element i from four reaction surfaces (the fourth stage is vaporization from the surface of the solid phase). The first term on the right-hand side of the equation represents the removal of the element upon condensation, and the second term represents the accumulation of the element in the gas. Going over to partial pressures and volumes and dividing the time of the process into small intervals, we obtained the following equation:

$$P_{\mathrm{E}_i}^{k+1} = P_{\mathrm{E}_i}^{k} + \left\{ \sum_{j=1}^{4} \left(v_{\mathrm{E}_i}^{k} \right)_j (S_j - S^{\mathrm{fs}}) + \sum_{j=1}^{4} \left(v_{\mathrm{E}_i}^{\mathrm{fs}_k} \right)_j S^{\mathrm{fs}} - v_{\mathrm{E}_i}^{\mathrm{cond}} S^{\mathrm{cond}} - \dot{V} \frac{P_{\mathrm{E}_i}^{k}}{RT_{\mathrm{g}}} \right\} \frac{RT}{V} \Delta\tau. \tag{5.109}$$

The condensation rates $v_{\mathrm{E}_i}^{\mathrm{cond}}$ can be determined only to an approximation on the basis of several assumptions. At low pressures, at which the mean free path of the particles is greater than the dimensions of the chamber, each vapor molecule condenses when it collides with the cold wall of the latter. The values of $v_{\mathrm{E}_i}^{\mathrm{cond}}$ can be calculated to an approximation using the Langmuir equation with an indefinite value for the accommodation coefficient, which can be adjusted empirically. We note that the temperature of the gas phase (a mixture of metal vapors and gaseous N_2, O_2, and H_2) is also indefinite. The inleakage of cold air and cooling by the walls should clearly lower the temperature of the vapors, possibly to the condensation point.

For N_2 and H_2, with allowance for the operation of the pumping system, we obtained

$$P_{\mathrm{E}_i}^{k+1} = P_{\mathrm{E}_i}^{k} + \left\{ \dot{V}_{\mathrm{inl}} \frac{P_{\mathrm{E}_i}^{0}}{RT_0} - \dot{V} \frac{P_{\mathrm{E}_i}^{k}}{RT_{\mathrm{g}}} + \sum_{j=1}^{4} \left(v_{\mathrm{E}_i}^{k} \right)_j (S_j - S^{\mathrm{fs}}) + \sum_{j=1}^{4} \left(v_{\mathrm{E}_i}^{\mathrm{fs}_k} \right)_j S^{\mathrm{fs}} \right\} \frac{RT}{V} \Delta\tau. \tag{5.110}$$

The gas inleakage rate $\dot{V}_{\text{inl}}$ can be determined from the value of the residual pressure in the chamber over the melt

$$\dot{V}_{\text{inl}} = \dot{V}\frac{P_{\text{res}}}{P_{\text{atm}}}. \tag{5.111}$$

In the case of oxygen, its consumption in oxidation of the metal must be taken into account:

$$P_{O_2}^{k+1} = P_{O_2}^{k} + \left\{ \dot{V}_{\text{inl}}\frac{P_{O_2}^0}{RT_0} - \dot{V}\frac{P_{O_2}^k}{RT_g} - \sum_{j=1}^{4}\left(v_{O_2}^k\right)_j S_j \right\} \frac{RT}{V}\Delta\tau. \tag{5.112}$$

Solid metal that has been heated to high temperatures (up to the melting point) is present in all stages of EBR. Vaporization of the components of the alloy from the solid surface can be taken into account by introducing the terms S_s and $v_{E_i,4}^{s}$ into the balance equations and taking a certain averaged temperature and surface area of the solid metal. Special difficulties arise in the calculation of the vaporization rate $v_{E_i,4}^{s}$, because the influence of diffusion in the solid phase and the heightened vaporization in the focal spot must be taken into account in the general case. In a first approximation we calculate the vaporization rate from Eq. 5.96 and take into account sublimation only in the stage of heating the ingot in the mold.

2. *Equations for the vaporization and degassing rates*

The process resulting in the passage of elements from the metal into the gas consists of at least three successive stages: diffusion in the metal, a chemical reaction on the interface, and diffusion in the gas.

The following kinetic equation of the process of refining a liquid metal in a vacuum, which takes into account all three stages, was obtained in [57] for condensed impurities:

$$v = \frac{\beta_g}{RT}\frac{\frac{\rho_{E_i}P_{E_i}^0 C_{E_i}}{v_l} - P_{E_i}}{1 + \frac{\rho_{E_i}P_{E_i}}{RTv_l}\left(\frac{\beta_g}{\beta_{\text{Me}}} + \frac{\beta_g}{k_x}\right)}, \tag{5.113}$$

where β_{Me} and β_g are the mass-transfer coefficients of the alloy component in the liquid metal and the vapor phase, respectively, $v_l = \sum_{i=1}^{l} C_{E_i}$ is the total number of moles of the alloy components and impurities in 1 m^3 of the melt (mole/m^3), and $k_x = \frac{\alpha P_{E_i}^0 \rho_{E_i}}{\sqrt{2\pi RTM}}$ is the rate constant of the chemical reaction on the interface according to the Langmuir equation.

For gases (nitrogen and hydrogen), a diffusion-controlled regime for their removal from the metal was assumed. In this case, under the condition of equilibrium on the boundary, for the known equilibrium constants K_N and K_H we obtained

$$[N]' = K_N\sqrt{P'_{N_2}}, \tag{5.114}$$

$$[H]' = K_H \sqrt{P'_{H_2}}, \tag{5.115}$$

where $[N]'$ and $[H]'$ are the concentrations of nitrogen and hydrogen in the surface layer of the melt (wt.%), and P'_{N_2} and P'_{H_2} are the partial pressures of nitrogen and hydrogen at the surface of the melt (Pa).

Then for nitrogen, in the stage of diffusion in the metal, we have

$$v_N^{Me} = \beta_{Me} \frac{\rho_{Me}}{100 M_{E_i}} \left([N] - [N]' \right), \tag{5.116}$$

and in the stage of diffusion in the gas, we have

$$v_N^g = \frac{\beta_g}{RT} \left(P'_{N_2} - P_{N_2} \right). \tag{5.117}$$

The last equations together with the condition $v_N^{Me} = v_N^g = v_N$ enables us to find the value of v_N:

$$v_N = \frac{\beta_{Me} \rho_{Me}}{100 M_{E_i}} \left\{ [N] + \frac{b}{2} - \sqrt{\frac{b^2}{4} + b[N] + K_N P_{N_2}} \right\}, \tag{5.118}$$

where $b = K_N^2 \frac{RT \rho_{Me}}{100 M_{E_i}} \frac{\beta_{Me}}{\beta_g}$.

Similar equations can be written for hydrogen.

Thus, after writing the material balance equations for all the alloy components and impurities on the end surface of the intermediate product being melted (Eqs. 5.99 and 5.100), in the intermediate vessel (5.105), in the mold (5.107), and in the vapor phase [(5.109), (5.110), and (5.112)] with consideration of Eqs. 5.113 and 5.118, we obtained a mathematical model of the physicochemical processes in EBR, which enables us to calculate the composition of the ingot metal. It should be kept in mind that the losses of aluminum and yttrium are due to their oxidation. Therefore, expression (5.96) must be used to determine the rate of the entry of oxygen into the metal.

3. *Adaptation of the Mathematical Model*

To adapt the mathematical model, calculations were performed for the conditions of EBR of the SDP-1A alloy in the ÉLU-194 facility in the experimental plant of the E. O. Paton Electric Welding Institute.

Figure 5.13 presents a block diagram of the algorithm of the program.

Implementation of the algorithm requires introduction of the chemical composition of the metal in the intermediate product, technological parameters, and several constants that are needed for the calculation.

The diffusion coefficients of the elements in the metal were calculated using the temperature dependence

$$D_{E_i} = D_0 \exp\left(-\frac{\Delta E_{E_i}}{RT} \right), \tag{5.119}$$

Fig. 5.13 Block diagram of the algorithm of the program for calculating the composition of the metal during EBR

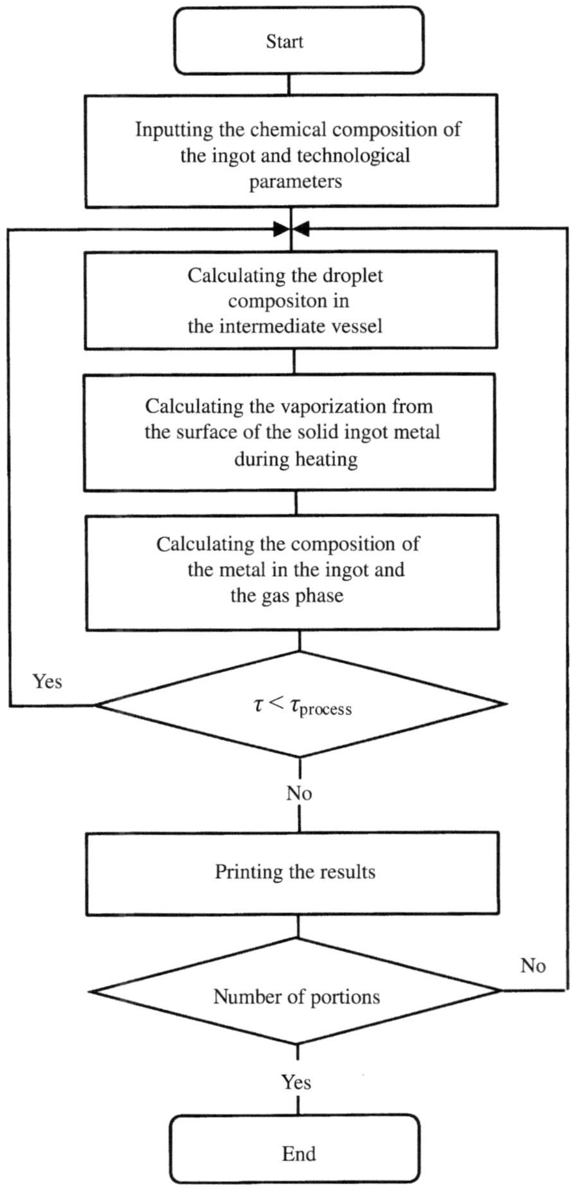

where ΔE_{E_i} is the activation energy for diffusion, J/(mole K).

The data for calculating D_{E_i} and the values of the activity coefficients of the elements are presented in Table 5.17.

The value $D^{g}_{E_i} = 10^{-2}$ m^2/s was taken for the gas phase.

Table 5.17 Data for calculating the physicochemical parameters

Parameter	Ni	Co	Cr	Al	Y	N	H
γ_{E_i}	1.0	1.0	0.6	2.5×10^{-4}	0.094	–	–
D_0	0.34×10^{-4}	0.2×10^{-4}	0.11×10^{-6}	0.16×10^{-5}	1.0×10^{-5}	1.0×10^{-4}	1.5×10^{-3}
ΔE_{E_i}	264200	259600	52000	78000	0	0	0

Table 5.18 Data from the calculation of the chemical composition of an EBR ingot obtained from the SDP-1A alloy

Object	Concentration (wt.%)							
	Ni	Co	Cr	Al	Y	N	H	Impurities
Requirements of Technical Specification 1-92-80-86 for the SDP-1A alloy	Base	18–22	18–22	11–13	0.2–0.6	–	–	–
Intermediate product	47.456	18.369	22.423	11.262	0.415	0.001	0.010	0.064
EBR ingot	47.456	18.375	22.421	11.293	0.414	0.0006	0.005	0.064

The equilibrium constants for Eqs. 5.114 and 5.115 were found from the data in [38]:

$$\log K_{\mathrm{N}} = -\frac{1970}{T} - 1.81, \tag{5.120}$$

$$\log K_{\mathrm{H}} = -\frac{924}{T} - 1.88. \tag{5.121}$$

Table 5.18 presents the data from the calculation of the chemical composition of an ingot of a refractory alloy before and after EBR.

A review of the literature data, adjustment of the parameters of the mathematical model, and experience from using the program allow us to recommend work items for improving the model for the purpose of raising the reliability of predictions of the composition of EBR ingots: enhancement of the accuracy of the determination of the surface temperature of the liquid metal, especially in the region of the focal spot; development of a method for describing the kinetics of the vaporization of components of the alloy from the solid phase with allowance for diffusion in the metal and the gas; accumulation of experimental data regarding the behavior of oxygen and oxide nonmetallic inclusions during EBR; experimental investigation of the vaporization of individual components of the alloy under steady-state conditions for the purpose of determining the kinetic constants; theoretical analysis and experimental study of the behavior of vapors in the vacuum chamber and their condensation; refinement of the thermodynamic properties of the alloy components; accumulation of experimental data regarding the dynamics of the melting of the intermediate product, formation of the pool of liquid metal in the intermediate vessel, and crystallization of the EBR ingot.

5.5 Conclusion

The preceding sections of this book presented information regarding a procedure for constructing mathematical models and described thermodynamic and kinetic methods for analyzing interaction processes in multicomponent systems.

This information was used in this chapter to develop mathematical models of physicochemical processes as applied to several technological processes for obtaining and refining metals.

Let us summarize the main results presented in this chapter.

1. A mathematical model of the physicochemical processes that occur during the smelting of steel in steelmaking arc furnaces, which can be used to predict the composition of the metal and the slag, was developed. It takes into account the mutual influence of all the reactions that occur on different interfaces of the liquid phases, the kinetic hindrances of the electrode processes, the diffusion of all the reactants in the metal and the slag, and the changes in the masses and compositions of the phases as a result of melting of the charge, chemical reactions, and additions of slag formers, oxidizing agents, alloying elements, and deoxidizers. The changes in the temperature and hydrodynamic regimes during oxidation of the metal were taken into account.
2. A mathematical model of the interaction processes that occur in a metal–slag–gas system during the ladle treatment of steel by a synthetic slag was developed.
3. Physical models of refining remelting processes, viz., plasma arc remelting, vacuum arc remelting, and electron-beam remelting, were proposed.
4. A mathematical model of the PAR of nitrogen-containing steels, which can be used to predict the composition of the metal, was developed. It takes into account the diffusional hindrances in the metal and the gas, the occurrence of the process in two successive stages, viz., an electrode stage and a pool stage, the desorption of nitrogen from the pool surface that is free from the arc, the liquation of nitrogen during crystallization of the metal, the appearance of "boiling" of the metal pool, and the influence of the PAR regimes. Calculations established that the main rate-limiting step in the adsorption of nitrogen from the plasma during the PAR of high-alloy steels and its desorption from the metal is diffusion in the gas and that the role of the reversibility of the process is significant.
5. A mathematical model of processes involving the VAR of structural steels with MSR, which takes into account the simultaneous occurrences of reactions, the vaporization of fluorides from the slag, and the formation of a slag skin, was developed.
6. A mathematical model of the physical and chemical processes that occur during the EBR of alloys was developed. The performance of computational experiments enabled us to formulate the principal work items for improving the model.

One of the modern methods for improving and optimizing metallurgical production is to create automated design and control systems for various technological processes. This provides a way to solve many important problems:

- to significantly shorten the time devoted to technological preparations for production;
- to improve the quality of the design by selecting more efficient and optimal variants;
- to improve the quality of the product produced;
- to ensure savings in material and other resources both by reducing the number of experimental melts and by selecting the optimal technological solutions;
- to implement operational control over the course of the process and to make the necessary corrections.

Then the composition of the reacting phases can be successfully predicted both over the course of a technological process and at its completion using the mathematical models created.

References

1. Nikol'skii, L.E., Zinurov, M.Yu.: Outfitting and Designing Electric Steelmaking Shops (in Russian), p. 272. Metallurgiya, Moscow (1993)
2. Enderal, F.P., Filippov, A.F.: Calculations for the Electrometallurgy of Steel and Ferroalloys (in Russian), p. 230. Nauka, Moscow (1962)
3. Boronenkov, V.N., et al.: Mathematical modeling of the interaction of metal and slag in a ladle. Izv Vyssh Uchebn Zaved Chern Metall **1**, 36–41 (1983)
4. Rozhkov, I.M., Travin, O.M., Turkevich, D.K.: Mathematical Modeling of Convector Processes (in Russian), p. 184. Nauka, Moscow (1978)
5. Yavoiskii, V.I.: Theory of Steel Production Processes (in Russian), p. 792. Metallurgizdat, Moscow (1967)
6. Esin, O.N., Gel'd, P.V.: Physical Chemistry of Pyrometallurgical Processes, Part 2 (in Russian), p. 704. Metallurgiya, Moscow (1966)
7. Yavoiskii, V.I., et al.: Metallurgy of Steel (in Russian), p. 816. Metallurgiya, Moscow (1973)
8. Ogurtsov, A.P., et al.: Numerical investigation of heat- and mass-transfer processes during the dissolution of lime in a slag melt. Izv Vyssh Uchebn Zaved Chern Metall **2**, 12–16 (1982)
9. Amdur, A.M., et al.: Melting of metalized pellets in a metal phase. Article deposited in the Ukrainian Scientific-Research Institute of Scientific and Technical Information, No. 327 Uk-85Dep, p. 23, 13 Feb 1985
10. Denisov, V.A., Denisov, A.V.: Method for calculating the solidification temperatures of steel. Litein Proizv **5**, 11 (1983)
11. Konakchiev, S.S.: Mathematical models of technological processes in the oxidation period of steel smelting in steelmaking arc furnaces for predicting characteristics of the melt. Candidate dissertation, Leningrad, p. 198 (1984)
12. Terskov, Yu.I., Khlynov, V.V., Babkin, V.G.: Investigation of the kinetics of the dissolution of ferrochromium in iron, pig iron, and steel. Izv Vyssh Uchebn Zaved Chern Metall **4**, 49–51 (1978)
13. Hansen, M., Anderko, K.: Constitution of Binary Alloys. McGraw-Hill, New York (1958)
14. Turkdogan, E.T.: Physical Chemistry of High-Temperature Processes (in Russian), p. 344. Metallurgiya, Moscow (1985)
15. Tokovoi, O.K., et al.: Optimization of steel deoxidation processes. Report 2. Izv Vyssh Uchebn Zaved Chern Metall, No. 2, pp.19–22 (1983)
16. Chuchmarev, S.K., Esin, O.A., Dobryden', A.A.: Kinetics of the oxidation of slag sulfur by gaseous oxygen. Izv Vyssh Uchebn Zaved Chern Metall **7**, 12–18 (1962)

17. Sokolov, G.A.: Ladle Refinement of Steel (in Russian), p. 208. Metallurgiya, Moscow (1977)
18. Voinov, S.G., Shalimov, A.G., Kosoi, L.F., Kalinnikov, E.S.: Refining of Steel by Synthetic Slags (in Russian). Metallurgiya, Moscow (1970)
19. Pozdnyakov, A.D.: Kinetics of the simultaneous oxidation of impurities dissolved in iron by a liquid slag. Candidate dissertation, Sverdlovsk, p. 233 (1980)
20. Boronenkov, V.N., Pozdnyakov, A.D.: Mathematical model of the kinetics of the combined oxidation of iron impurities by a molten slag. In: Physicochemical Investigations of Metallurgical Processes, No. 7 (in Russian), pp.75–83. Ural Polytechnic Institute Press, Sverdlovsk (1979)
21. Levich, V.G.: Physicochemical Hydrodynamics. Prentice-Hall, Englewood Cliffs (1962)
22. Voinov, S.G., Shalimov, A.G.: Ball-Bearing Steel (in Russian), p. 480. Metallurgiya, Moscow (1962)
23. Bakumenko, S.P., Gulyaev, V.B., Verkhovtsev, É.V.: Lowering the Waste from a Steel Ingot (in Russian), p. 219. Metallurgiya, Moscow (1967)
24. Shalimov, M.P., Boronenkov, V.N., Davydov, Yu.S.: Mathematical model for predicting the composition of the head part of an ingot during teeming under heat-generating mixtures. In: Modeling of Physicochemical Systems and Technological Processes in Metallurgy (in Russian), pp.148–149. Novokuznetsk Institute of Metal Press, Novokuznetsk (1991)
25. Grigoryan, V.A., Belyanchikov, L.P., Stomakhin, A.Ya.: Theoretical Principles of Electric Steelmaking Processes (in Russian), p. 272. Metallurgiya, Moscow (1987)
26. Lakomskii, V.I.: Plasma Arc Remelting (in Russian), p. 336. Technika, Kiev (1974)
27. Erokhin, A.A.: Laws of the Plasma Arc Alloying and Refining of Metals (in Russian), p. 185. Nauka, Moscow (1984)
28. Erokhin, A.A.: Plasma Arc Melting of Metals and Alloys. Physicochemical Processes (in Russian), p. 188. Nauka, Moscow (1975)
29. Klyuev, N.M.: Plasma Arc Remelting (in Russian), p. 256. Nauka, Moscow (1980)
30. Kalugin, A.S., Kalugina, K.V.: Efficiency of Refining in Remelting Processes (in Russian), p. 200. Metallurgiya, Moscow (1988)
31. Grigorenko, G.M., Pomarin, Yu.M.: Hydrogen and Nitrogen in Metals during Plasma Melting (in Russian), p. 200. Naukova Dumka, Kiev (1989)
32. Lakomskii, V.I., Torkhov, G.F.: Calculation of the solubility of nitrogen in liquid high-alloy steels. Fiz Khim Obrab Mater **4**, 43–49 (1971)
33. Grigoryan, V.A., et al.: Investigation of the interaction of nitrogen with liquid iron under the conditions of electric-arc heating. In: Interaction of Gases with Metals (in Russian), pp. 146–152. Nauka, Moscow (1973)
34. Lakomskii, V.I., Grigorenko, G.M., Torkhov, G.F.: Investigation of interaction processes of nitrogen with the metal during PAR. In: Refining Remelts, vol. 2 (in Russian), pp. 151–158. Naukova Dumka, Kiev (1975)
35. Bobrovskaya, R.S., et al.: Parameters of an open arc stabilized by an argon flow. Zh Prikl Mekh Tekh Fiz **1**, 66–74 (1973)
36. Ryzhonkov, D.I. et al.: Computer Calculations of Metallurgical Processes (in Russian), p. 231. Metallurgiya, Moscow (1987)
37. Pokhodnya, I.K.: Gases in Welded Joints (in Russian), p. 256. Mashinostroenie, Moscow (1972)
38. Linchevskii, B.V.: Thermodynamics and Kinetics of the Interaction of Gases with Liquid Metals (in Russian), p. 222. Metallurgiya, Moscow (1986)
39. Afanas'ev, N.D., Zhdanovich, K.K.: Alloying of steel with nitrogen in a melt boiling regime during plasma arc remelting. In: High-Nitrogen Steels: Proceedings of the 1st All-Union Conference (in Russian), pp. 251–257. Kiev (1990)
40. Shalimov, M.P., Boronenkov, V.N., Afanas'ev, N.D.: Modeling of the behavior of nitrogen during plasma arc remelting. In: High-Nitrogen Steels: Proceedings of the 1st All-Union Conference (in Russian), pp. 57–64. Kiev (1990)

41. Khokhlov, A.A.: Investigation of the interaction of nitrogen with liquid steel during plasma melting. In: High-Nitrogen Steels: Abstracts of Reports to the 2nd All-Union Conference (in Russian), pp. 90–91. Kiev (1992)
42. Shalimov, M.P., Davydov, Yu.S., Boronenkov, V.N.: Modernization of the mathematical model of the alloying of a metal with nitrogen during plasma arc remelting. In: High-Nitrogen Steels: Abstracts of Reports to the 2nd All-Union Conference (in Russian), pp. 108–109. Kiev (1992)
43. Boronenkov, V., et al.: Mathematical modeling of nitrogen behavior in plasma-arc and vacuum-arc remelting. In: Proceedings of 3rd International Conference on High-Nitrogen Steels, pp. 122–130. Kiev (1993)
44. Gerasimov, V.S.: Improving the quality of the subcutaneous zone in large vacuum arc remelting ingots. Candidate dissertation, p. 180, Chelyabinsk (1980)
45. Shved, F.I., Gerasimov, V.S., Soskov, R.A.: Microslag Refining of Steel during Vacuum Arc Remelting (in Russian), p. 16. Chermetinformazia, Moscow (1976)
46. Afanas'ev, N.D.: Improving the technology of vacuum arc remelting using mathematical modeling of the interaction of a metal with a slag and a gas. Candidate dissertation, p. 166, Sverdlovsk (1983)
47. Boronenkov, V.N., Afanas'ev, N.D., Bakumenko, S.P.: Optimization of vacuum arc remelting using mathematical modeling. Proizv Tekh Byull **10**, 19–21 (1983)
48. Shalimov, A.G., Gotin, V.N., Tulin, A.N.: Intensification of Special Electrometallurgy Processes (in Russian), p. 334. Metallurgiya, Moscow (1988)
49. Sergeev, A.B., Shved, F.I., Tulin, A.N.: Vacuum Arc Remelting of Structural Steels (in Russian), p. 192. Metallurgiya, Moscow (1974)
50. Movchan, B.A., Tikhonovskii, A.L., Kurapov, Yu.A.: Electron-Beam Melting and Refining of Metals, p. 239. Naukova Dumka, Kiev (1973)
51. Ladokhin, S.V., Kornyukhin, Yu.V.: Electron-Beam Slag-Skin Melting of Metals and Alloys (in Russian), p. 144. Naukova Dumka, Kiev (1988)
52. Paton, B.E., Movchan, B.A., Tikhonovskii, A.L.: Analysis of modern melting and refining methods in electron-beam furnaces. In: Refining Remelts (in Russian), vol. 2, pp. 179–195. Naukova Dumka, Kiev (1975)
53. Ryzhonkov, D.I., et al.: Theory of Metallurgical Processes (in Russian), p. 392. Metallurgiya, Moscow (1989)
54. Zhadkevich, M.L., Statnikov, B.Sh., Ishchenko, D.A.: Numerical analysis of heat-transfer processes during electron-beam treatment. In: Reports to the 3rd International Conference on Electron-Beam Technologies—EBT-91 (in Russian), pp. 312–317. Varna (1991)
55. Zhadkevich, M.L., Statnikov, B.Sh., Ishchenko, D.A.: Investigation of the thermal state of the ingot–mold system during electron-beam remelting. Fiz Khim Obrab Mater **3**, 53–57 (1991)
56. Boronenkov, V.N., et al.: Mathematical model of physicochemical processes during electron-beam remelting. In: Modeling of Physicochemical Systems and Technological Processes in Metallurgy: Abstracts of Reports to the All-Union Conference (in Russian), pp. 21–22. Novokuznetsk (1991)
57. Akhonin, S.V., Tikhonovskii, A.L.: Kinetics of the mass-transfer process in a real-solution/vapor-phase system. Probl Spets Élektrometall **2**, 61–64 (1992)

Chapter 6
Modeling and Simulation of Welding, Surfacing and Coating Processes

The composition and properties of a weld metal are shaped during welding and surfacing as a result of complex thermal, physicochemical, and thermal deformation processes. Their quantitative description is hampered by the transient nature of the welding process, the nonuniformity of the thermal fields, the presence of a large number of chemically interacting phases, and the complex hydrodynamic circumstances.

The creation of theoretically rigorous methods for predicting the composition of the metal during welding, surfacing, and the application of coatings based on the fundamental principles of thermodynamics and chemical kinetics is very attractive and would allow their use for a broad class of materials, different types of joints, etc.

The development of methods for predicting the chemical composition of the weld metal opens up the possibility of optimizing the welding process in the design stage by selecting welding materials and regime parameters that would ensure assigned service characteristics of the welded joint.

The creation of such methods is also important from the theoretical standpoint, since close agreement between the calculated and experimental data is a definite indication of the correspondence of the calculation scheme used to the physical picture of the phase interaction occurring during welding, surfacing, and the application of coatings.

6.1 Mathematical Modeling of Phase Interaction Processes in Electroslag Technologies

The interaction of the gas and the molten metal and slag during an electroslag process determines the chemical composition of the final metal. This interaction has special significance when the steels being treated contain active elements

M. Zinigrad et al., *Phase Interaction in the Metal–Oxide Melts–Gas System*, Engineering Materials, DOI: 10.1007/978-3-642-22377-8_6,

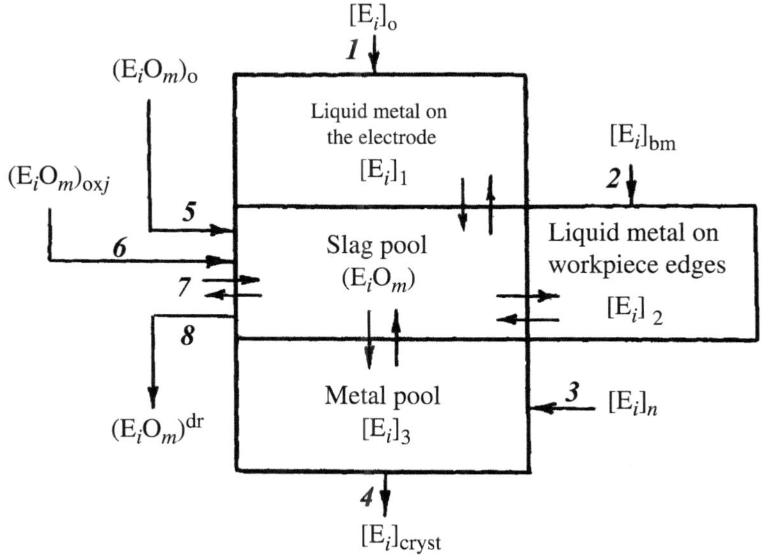

Fig. 6.1 Block diagram of the directions of transport of the element E_i and variation of the composition of the phases during an electroslag process: *1* melting of the electrode; *2* melting of the base metal; *3* introduction of metal additives; *4* crystallization of the metal; *5* feeding of the flux; *6* dissolution of the oxide films from the electrode or part surface in the slag; *7* interaction of the slag with the atmosphere; *8* solidification of the slag

(Ti, Al, etc.), which can be the reason for the appearance of chemical heterogeneity in the metal of the workpiece. Preliminary calculation of the chemical processes involved in electroslag technologies (EST) provides a way to foresee measures that would prevent deviation of the metal composition from the required.

This chapter is devoted to the development of mathematical models of the kinetics of the phase interaction in electroslag remelting (ESR) [1–8], electroslag welding (ESW) [9, 10], electroslag surfacing (ESS) [11], and centrifugal electroslag casting (CESC) [12] that enable prediction of the chemical composition of the ingot, welded joint, and surfaced metal.

6.1.1 Fundamentals of the Mathematical Model of the Phase Interaction in Electroslag Technologies

The diagram presented in Fig. 6.1 illustrates the directions of transport of the element E_i and the changes in the composition of the phases in an electroslag process. The mathematical model developed takes into account the following special features of an electroslag process [1–9]:

- simultaneous occurrence of all the reactions studied on the metal/slag and slag/gas interfaces and their mutual thermodynamic and kinetic influence;
- slowness of the diffusion of all the reactants and the steps involving adsorption followed by discharge and ionization in the reactions;
- interaction of the liquid metal and slag in the stages at the following locations: the electrode, the film of liquid metal on melted edges of the workpiece, the continuously melted tip or electrode of a different chemical composition, the metal pool;
- electrolytic effects;
- influence of the changes in the chemical composition and amount of the slag with time due to the interaction of the slag with the metal and the gas phase at the surface of the slag pool, deoxidization of the slag pool over the course of the process, feeding of the flux into the slag pool, and dissolution of the oxide films from the electrode surface and workpiece edges in the slag on the reaction rate;
- introduction deoxidizers and modifiers into the metal pool through the slag pool;
- ionic composition of the slag melt;
- variation of the mass of the metal pool.

The following simplifying assumptions were adopted.

- The liquid metal in each stage and the liquid slag are homogeneous with respect to their chemical composition.
- When an alternating current is used, the electrolysis phenomena are neglected.

We also note that the theory of the polymeric constitution of the multicomponent oxyfluoride slag systems that are employed in ESTs has not been adequately developed. In the preceding chapters it was shown that the rates of reactions involving the principal polymer-forming element, namely, silicon, are fairly close to those obtained using molecular theories. Therefore, in this work we used the mathematical machinery for describing the kinetics of reactions based on molecular notation, but with consideration of the ionic structural units in the slag [13].

Because the heterogeneity of the composition of ESR ingots is caused, first, by the chemical processes taking place and, second, by the variable mass of the metal pool, the model incorporates equations that enable calculating the profile of the crystallization front and the mass of the metal pool at any given time. Since the metal composition found refers to the crystallization front, rather than a specific transverse section (height) of the ingot, the set of data on the composition and profile of the pool enables us to construct the distribution of the concentrations of all the elements in any given cross section of the ingot.

The material balance (kg/s) of the element under consideration E_i in a given phase (for example, in the metal phase) can be written in the general case in the following form [1, 13]:

$$\sum_{k=0}^{k} m_k \frac{[E_i]_0^k}{100} = \sum_{l=0}^{l} m_l \frac{[E_i]^l}{100} + \sum_{m=0}^{m} \int_{S_m} I_{E_i}^m M_{E_i} dS_m + \int_{v} \frac{d([E_i]p_m)}{100 d\tau} dV. \qquad (6.1)$$

Here $\dot{m}_k$ is the rate of entry of the substance into the particular phase with one of the k incoming flows, kg/s; $\dot{m}_l$ is the rate of departure of the substance from the phase with one of the l outgoing flows; $[E_i]_0^k$ and $[E_i]^l$ are the concentrations of the element E_i in the incoming and outgoing flows, wt.%; $[E_i]$ is the concentration of the element E_i within the phase in a given element of its volume dV, wt.%; M_{E_i} is the atomic weight of the element, kg/mole; $I_{E_i}^m$ is the rate of passage of the element from the current phase into one of the m phases interacting with the current phase on an interfacial surface element dS_m, mole $E_i/(m^2/s)$.

The sum on the left-hand side of Eq. 6.1 is the total rate of entry of the element into the particular phase with the incoming flows. The first sum on the right-hand side is the total rate of departure of the element from the phase with the outgoing flows, the second sum is the total rate of departure or entry (when $I_{E_i}^m < 0$) of the element into the adjacent phases as a result of the interaction, and the third term is the rate of entry of the element within the phase as the concentration varies. Equation 6.1 contains integral dependences, because the rates of the chemical reactions can be different at different points on the interface due to the heterogeneity of the phase with respect to temperature, composition, and convection conditions, and the concentrations at different points in the phase can vary with time with unequal rates. The density of the phase can also be variable.

The evaluation of the integrals on the right-hand side of Eq. 6.1 is extremely difficult, because the integrands are usually not known *a priori* and must be determined by solving the entire system of equations of the mathematical model, including an equation of type (6.1) written for each phase. To avoid the complexities of a mathematical nature that would arise in this process, a phase that is heterogeneous with respect to its properties is treated as a chemical reactor of a simple type, that is, as an ideally mixed reactor. This allows replacement of the integral expressions in Eq. 6.1 by simpler algebraic equations. This method was used extensively in the development of mathematical models of chemical engineering processes by Kafarov [14].

6.1.2 Mathematical Model of the Chemical Processes Occurring in the Electroslag Remelting of Steels

Equation 6.1 is simplified significantly under the conditions of ESR. When only one incoming flow and one outgoing flow with identical flow rates exist, we have $\dot{m}_k = \dot{m}_l = \dot{m}$. In addition, we will assume that the electrode and pool stages under consideration are kinetically homogeneous and that the phases are compositionally homogeneous. Therefore, instead of Eq. 6.1, the material balance (kg/s) of the element under consideration E_i in the metal in the electrode stage (stage 1) can be described in the following form [1, 4, 5, 9]:

$$\frac{G_{\mathrm{Me},1}}{100}\frac{d[\mathrm{E}_i]_1}{d\tau}=\dot{m}\frac{[\mathrm{E}_i]_0}{100}+I_{\mathrm{E}_i,1}S_1M_{\mathrm{E}_i}-\dot{m}\frac{[\mathrm{E}_i]_1}{100}, \tag{6.2}$$

$$\begin{aligned}\frac{[\mathrm{E}_i]_2}{100}\frac{dG_{\mathrm{Me},2}}{d\tau}+\frac{G_{\mathrm{Me},2}}{100}\frac{d[\mathrm{E}_i]_2}{d\tau}=\dot{m}\frac{[\mathrm{E}_i]_1}{100}+V_{\mathrm{Me}}^n\frac{[\mathrm{E}_i]_n}{100}+I_{\mathrm{E}_i,2}S_2M_{\mathrm{E}_i}\\-\dot{m}_{\mathrm{cryst}}\frac{[\mathrm{E}_i]_{\mathrm{ingot}}}{100},\end{aligned} \tag{6.3}$$

where $\dot{m}$ is the melting rate ($\dot{m}_{\mathrm{cryst}}$ is the crystallization rate) of the metal (kg/s), $G_{\mathrm{Me},j}$ is the mass of the liquid metal in stage j (kg), $[\mathrm{E}_i]_j$ is the mass percentage of the element E_i in stage j (wt.%), $[\mathrm{E}_i]_n$ is the mass percentage of the element E_i in the metal additives at the time of their absorption by the metal pool (wt.%), and V_{Me}^n is the feed rate of the additives to the metal pool (kg/s).

The left-hand side of Eq. 6.2 is the rate of variation of the concentration of the element E_i in the metal film on the electrode tip. The first term on the right-hand side is the rate of entry of the element (with the molten metal) into the liquid metal on the electrode tip, the second term is the rate of passage of the element from the metal into the slag as a result of chemical reactions, and the third term is the rate of departure of the element from the metal film with metal droplets.

The metal-pool stage is characterized by a rate of variation of the concentration of the element in the metal pool (see Eq. 6.3) as a result of variation of the mass of the metal pool and variation of the content of the element in the metal due to the entry of the element into the pool with portions of the electrode metal and the introduction of the metal additives, as well as by its loss as a consequence of chemical reactions and crystallization of the weld (see Fig. 6.1).

In a steady-state regime the left-hand side of Eq. 6.2 can be neglected. An analysis of the non-steady-state nature of the ESR process showed that it should be taken into account only for the metal pool due to the large values of $G_{M,2}$ at the beginning of the process and when the alloying additives are introduced into the pool.

The material balance of the same element E_i in the form of its oxide in the slag can be expressed by the equation

$$\frac{G_{\mathrm{slag}}}{100}\frac{d(\mathrm{E}_i\mathrm{O}_m)}{d\tau}=\sum_{j=1}^{2}I_{\mathrm{E}_i,j}S_j\frac{M_{\mathrm{E}_i\mathrm{O}_m}}{M_{\mathrm{E}_i}}+V_{\mathrm{ox}}+V_{\mathrm{fl}}-V_{\mathrm{g}}-V_{\mathrm{sk}}, \tag{6.4}$$

where $(\mathrm{E}_i\mathrm{O}_m)$ is the concentration of the oxide in the slag (wt.%), G_{slag} is the mass of the slag (kg), and $M_{\mathrm{E}_i\mathrm{O}_m}$ is the molecular weight of the oxide (kg/mole).

The first term (the sum) on the right-hand side of Eq. 6.4 is the rate of entry of the element into the slag from the metal as a result of chemical reactions in the two stages, and V_{OK} is the rate of entry of the element upon dissolution of the oxide film from the electrode. The value of V_{ox} can easily be calculated from the composition of the oxide film $(\mathrm{E}_i\mathrm{O}_m)$, its density ρ_{ox}, its cross-sectional area S_{ox}, and the linear melting rate of the electrode V_{m} using the formula $V_{\mathrm{ox}}=V_{\mathrm{m}}S_{\mathrm{ox}}\rho_{\mathrm{ox}}(\mathrm{E}_i\mathrm{O}_m)_{\mathrm{ox}}/100$. The term $V_{\mathrm{fl}}=V_{\mathrm{add}}(\mathrm{E}_i\mathrm{O}_m)_{\mathrm{ox}}/100$ represents the rate of entry of the oxide into the slag due to the feeding of fresh flux at the rate V_{add}, kg/s. The rate of departure of the oxide into the slag skin is given by the formula $V_{sk} = V_{su}S_{sk}\rho_{sk}(\mathrm{E}_i\mathrm{O}_m)_g/100$, where V_{su} is the surfacing rate (m/s) and S_{sk} and ρ_{sk} are the cross-sectional area and density of the slag skin. The losses to the gas can be estimated using the expression $V_g = I^g_{\mathrm{E}_i}[P_{\mathrm{O}_2}, (\mathrm{E}_i)\ldots]S_g$, where S_g is the area of the gas/slag interface (m^2) and $I^g_{\mathrm{E}_i}[P_{\mathrm{O}_2}, (\mathrm{E}_i)\ldots]$ is the rate of the interaction of the element E_i with the gas, which depends on the composition of the slag and the gas (kg $\mathrm{E}_i\mathrm{O}_m/(\mathrm{m}^2\mathrm{s})$).

The compositions of the pools (wt.%) are related to the compositions of the ingot and the slag skin by the relations

$$[\mathrm{E}_i]_{\mathrm{ingot}}\big/[\mathrm{E}_i]_2 = \kappa_{\mathrm{E}_i}; \quad (\mathrm{E}_i\mathrm{O}_m)_g\big/(\mathrm{E}_i\mathrm{O}_m) = \kappa_{\mathrm{E}_i\mathrm{O}_m}, \tag{6.5}$$

where κ_{E_i} and $\kappa_{\mathrm{E}_i\mathrm{O}_m}$ are the liquation coefficients.

The value of κ_{E_i} depends on the ingot crystallization conditions and can be calculated using the formula [13]

$$\kappa_{\mathrm{E}_i} = \frac{\kappa'_{\mathrm{E}_i}}{\kappa'_{\mathrm{E}_i} + (1 - \kappa'_{\mathrm{E}_i})\exp(-\dot{m}_{\mathrm{cryst}}\delta/D_i)}, \tag{6.6}$$

where κ'_{E_i} is the equilibrium liquation coefficient of the ith element and δ is the thickness of the diffusion layer (m). The values of κ'_{E_i} for impurities in liquid iron are as follows [13]: sulfur, 0.05; phosphorus, 0.07; carbon, 0.13; silicon, 0.66; chromium, 0.95; nickel, 0.80.

Neglecting the non-steady-state nature of the process, from (6.2) we find expressions for calculating the concentration of the element in the electrode stage:

$$[\mathrm{E}_i]_1 = [\mathrm{E}_i]_0 - I_{\mathrm{E}_i,1}S_1M_{\mathrm{E}_i}100/\dot{m}. \tag{6.7}$$

To take into account the non-steady-state nature of the processes in the metal and slag pools, we divide the melting time (τ_m) into sufficiently short time intervals of length $\Delta\tau$ (usually into 100–200 intervals), during which the rates of the chemical reactions may be assumed constant. Then, for the time interval from $k\Delta\tau$ to $(k+1)\Delta\tau$, from Eqs. 6.3–6.5 we obtain

$$\begin{gathered}\Delta[\mathrm{E}_i]_2^{k+1,k} = \left\{\dot{m}[\mathrm{E}_i]_j + V^n_{\mathrm{Me}}[\mathrm{E}_i]_n - I^k_{\mathrm{E}_i,2}S_2M_{\mathrm{E}_i}100 - \dot{m}_{\mathrm{cryst}}\,\kappa_{\mathrm{E}_i}[\mathrm{E}_i]_2^k\right\}\frac{\Delta\tau}{G_{\mathrm{Me},2}} \\ [\mathrm{E}_i]_2^{k+1} = [\mathrm{E}_i]_2^k + \Delta[\mathrm{E}_i]_2^{k+1,k},\end{gathered} \tag{6.8}$$

$$\begin{gathered}\Delta(\mathrm{E}_i\mathrm{O}_m)^{k+1,k} = \left\{\sum_{j=1}^{2} I^k_{\mathrm{E}_i,j}S_j\frac{M_{\mathrm{E}_i\mathrm{O}_m}}{M_{\mathrm{E}_i}} + V_{\mathrm{ox}} + V_{\mathrm{fl}} - V_g - V_{sk}\right\}\frac{100\Delta\tau}{G_{\mathrm{slag}}} \\ (\mathrm{E}_i\mathrm{O}_m)^{k+1} = (\mathrm{E}_i\mathrm{O}_m)^k + \Delta(\mathrm{E}_i\mathrm{O}_m)^{k+1,k}.\end{gathered} \tag{6.9}$$

The choice of the integration step with respect to time $\Delta\tau$ can be made automatically on the basis of a condition that there is an assigned maximum value for the change in the composition of the metal and the slag, for example, 1 rel.%.

We note that Eq. 6.7 enables us to determine $[E_i]_1^k$ for any time interval k, if the value of $I_{E_i 1}$ is known.

The calculation of the ingot composition is performed in successive steps. We first find the composition of the metal after the interaction in the electrode stage for the first time interval, when the compositions of the phases are known, using Eq. 6.7. Then we find the composition of the metal in the pool stage at the beginning of the time interval at $(k+1)\Delta\tau$ using expression (6.8). The content of an element in the ingot is determined from Eq. 6.5. If the influence of the non-steady-state nature of the process is also insignificant for the metal pool, from (6.3) and (6.7) we have

$$[E_i]_2^k = \left(\dot{m}[E_i]_1 + V_{Me}^n [E_i]_n - I_{E_i 2}^k S_2 M_{E_i} 100\right) \Big/ \left(\dot{m}_{cryst} \kappa_{E_i}\right). \tag{6.10}$$

In this case the composition of the ingot varies with time only as a result of changes in the composition of the slag, which are reflected in the values of $I_{E_i}^k$, as well as variations in the feed rate of the alloying additives V_{Me}^n.

To find the reaction rates $I_{E_i,j}^k$ we use the procedure that we developed in [15], which enables us to take into account the mutual influence of the reactions, the hindrances to the diffusion of all the reactants in the metal and the slag, and the influence of electrolysis.

We describe the interaction of the components of the metal with the slag using the following reactions with the common reactant FeO:

$$1/m[E_i] + (FeO) = [Fe] + 1/m(E_iO_m), \tag{6.11}$$

$$[C] + (FeO) = [Fe] + \{CO\}, \tag{6.12}$$

$$(FeO) = [Fe] + [O], \tag{6.13}$$

$$\left(S^{2-}\right) + (FeO) = [Fe] + [S] + \left(O^{2-}\right), \tag{6.14}$$

$$2[H] + \left(O^{2-}\right) + (FeO) = [Fe] + 2(OH^-). \tag{6.15}$$

Calculation of the rates of reactions (6.11–6.15) is possible when the compositions of the metal and the slag are known for the time $k\Delta\tau$. We introduce the following notations into the expressions for the limiting diffusion fluxes [16]:

$$I_{E_i,j}^n = \beta_j D_{E_i,j}^{0.5} [E_i]_j \rho_{Me} (100 M_{E_i})^{-1} = [E_i]_j \Big/ a_{ij}, \tag{6.16}$$

$$I_{E_iO_m}^n = \beta_j D_{ox,j}^{0.5} (E_iO_m)_j \rho_S (100 M_{ox})^{-1} = (E_iO_m)_j \Big/ b_{ij}. \tag{6.17}$$

After several transformations, we obtain the following expression for calculating the rates of reactions of type (6.11) in the metal-pool stage [15]:

$$I_{E_i2} = \frac{(K_{i,2}x_2)^m [E_i]_1 - (E_iO_m)}{(K_{i,2}x_2)^m a_{i,2} + b_{i,2}}. \quad (6.18)$$

The slowness of the step involving adsorption followed by discharge and ionization on the metal/slag boundary must be taken into account for the reactions involving silicon and carbon [17]:

$$I_{Si,2} = \frac{(K_{Si,2}x_2)^2 [Si]_1 - (SiO_2)}{(K_{Si,2}x_2)^{1.6} / K_{Si,2}^{exch} + (K_{Si,2}x_2)^2 a_{Si,2} + b_{Si,2}}. \quad (6.19)$$

$$I_{C,2} = \frac{(K_{C,2}x_2)\ [C]_1 - P_{CO}}{(K_{C,2}x_2)^{0.5} / K_{C,2}^{exch} + K_{Si,2}x_2 a_{C,2}}. \quad (6.20)$$

where $K_{Si,2}^{exch}$ and $K_{C,2}^{exch}$ are the rate constants of the exchange of silicon and carbon atoms between the metal and the slag in the equilibrium state, which were found experimentally.

Following the method in [15], for reactions (6.13–6.15) we found

$$I_{O,2} = (K_{O,2}x_2 - [O]_1)/a_{O,2}, \quad (6.21)$$

$$I_{S,2} = \frac{K_{S,2}x_2(S) - [S]_1}{K_{S,2}x_2 b_{S,2} + a_{S,2}}, \quad (6.22)$$

$$I_{H,2} = \frac{\sqrt{K_{H,2}x_2}[H]_1 - (OH)}{\sqrt{K_{H,2}x_2}a_{H,2} + b_{H,2}}. \quad (6.23)$$

In Eqs. 6.22 and 6.23 $K_{S,j}$ and $K_{H,j}$ represent the equilibrium concentration ratios for reactions (6.14) and (6.15):

$$K_{S,j} = \frac{[S]_j}{(S)(FeO)}; \quad K_{H,j} = \frac{(OH)^2}{[H]_j^2\ (FeO)}.$$

The diffusion flux of FeO in stage j with allowance for Stefan flow equals [18]

$$I_{FeO,j} = \frac{(FeO) - x_j}{b_{FeO,j}(100 - x_j)/100}. \quad (6.24)$$

Equations 6.18–6.23 do not enable us to directly calculate the reaction rates in the electrode stage, because the concentrations of the elements in the liquid metal are unknown. The composition of the latter is shaped, as in the case of a flow reactor, in accordance with Eq. 6.2. Therefore, $[E_i]_1$ must be found preliminarily by the combined solution of Eq. 6.2 and one of the Eqs. 6.18–6.23. The solution is obtained most simply for a steady-state regime, for which the left-hand side of Eq. 6.2 can be neglected. Then

$$[\mathrm{E}_i]_1 = [\mathrm{E}_i]_0 - I_{\mathrm{E}_i,1} S_1 M_{\mathrm{E}_i,1} 100/\dot{m} = [\mathrm{E}_i]_0 - I_{\mathrm{E}_i,1} c_{i,1}. \tag{6.25}$$

Substituting $[\mathrm{E}_i]_1$ into Eqs. 6.18–6.23, we obtain [1, 9]

$$I_{\mathrm{E}_i,1} = \frac{(K_{i,1}x_1)^m [\mathrm{E}_i]_0 - (\mathrm{E}_i\mathrm{O}_m)}{(K_{i,1}x_1)^m (a_{i,1} + C_{i,1}) + b_{i,1}}, \tag{6.26}$$

$$I_{\mathrm{Si},1} = \frac{(K_{\mathrm{Si},1}x_1)^2 [\mathrm{Si}]_0 - (\mathrm{SiO}_2)}{(K_{\mathrm{Si},1}x_1)^{1.6} / K^0_{\mathrm{Si},1} + (K_{\mathrm{Si},1}x_1)^2 (a_{\mathrm{Si},1} + C_{\mathrm{Si},2}) + b_{\mathrm{Si},1}}, \tag{6.27}$$

$$I_{\mathrm{C},1} = \frac{K_{\mathrm{C},1}x_1 [\mathrm{C}]_0 - P_{\mathrm{CO}}}{(K_{\mathrm{C},1}x_1)^{0.5} / K^0_{\mathrm{C},1} + K_{\mathrm{C},1}x_1 (a_{\mathrm{C},1} + C_{\mathrm{C},1})}, \tag{6.28}$$

$$I_{\mathrm{O},1} = (K_{\mathrm{O},1}x_1 - [\mathrm{O}]_0) / (a_{\mathrm{O},1} + c_{\mathrm{O},1}), \tag{6.29}$$

$$I_{\mathrm{S},1} = \frac{K_{\mathrm{S},1}x_1 (\mathrm{S}) - [\mathrm{S}]_0}{K_{\mathrm{S},1}x_1 b_{\mathrm{S},1} + a_{\mathrm{S},1} + b_{\mathrm{S},1}}, \tag{6.30}$$

$$I_{\mathrm{H},1} = \frac{\sqrt{K_{\mathrm{H},1}x_1} [\mathrm{H}]_0 - (\mathrm{OH})}{\sqrt{K_{\mathrm{H},1}x_1} a_{\mathrm{H},1} + b_{\mathrm{H},1} + c_{\mathrm{H},1}}. \tag{6.31}$$

Expressions (6.18–6.24) and (6.26–6.31) contain the single unknown x_j, which can be found from the material balance equation when the electrolytic effects are taken into account. Since the total diffusion flux for FeO, that is, $I_{\mathrm{FeO},j}$, is the sum of the partial diffusion fluxes for the oxidation of all the elements, to find x_j we solve the equation [9, 15, 19]

$$\sum_{i=1}^{i} m I_{\mathrm{E}_i,j} = I_{\mathrm{FeO},j} \pm \frac{i_j}{2F}, \tag{6.32}$$

where i_j is the current density in the respective stage (A/m^2).

The influence of all the factors on the rate of each reaction is taken into account here by the corresponding change in x_j when Eq. 6.32 is solved.

According to the literature data, the composition of the gas phase has a great influence on the behavior of sulfur and hydrogen in the slag. The presence of oxygen results in the oxidation of sulfur according to the reaction

$$(\mathrm{S}^{2-}) + 3/2\{\mathrm{O}_2\} = \{\mathrm{SO}_2\} + (\mathrm{O}^{2-}). \tag{6.33}$$

At low gas flow rates this reaction is limited by the diffusion of oxygen in the gas, and at moderate flow rates it is limited by the diffusion of sulfur in the slag to the interface with the gas. In the latter case the rate of reaction (6.33) was described by the equation

$$I_{\mathrm{S}}^{n} = \beta D_{(\mathrm{S})}^{0.5}(\mathrm{S})\rho_{\mathrm{S}}/(100M_{\mathrm{S}}). \tag{6.34}$$

When water vapor is present in the atmosphere, it is actively absorbed by the slag according to the reaction

$$\{\mathrm{H_2O}\} + \mathrm{O}^{2-} = 2(\mathrm{OH}^-). \tag{6.35}$$

Assuming that this process is limited mainly by the diffusion of water vapor in the gas, its rate can be determined from the equation

$$I_{\mathrm{H}}^{\mathrm{g}} = K_{\mathrm{H_2O}}^{d}\left[P_{\mathrm{H_2O}} - (\mathrm{OH}^-)^2 \Big/ K_{\mathrm{eq}(6.35)}\right], \tag{6.36}$$

where $K_{\mathrm{H_2O}}^{d}$ is the rate constant of the diffusion of H_2O in the gas and $K_{\mathrm{eq}(6.35)}$ is the equilibrium constant of reaction (6.35).

The interaction of water vapor with slag was examined in [20] with allowance for the polymerization of a silicon-containing slag, the vaporization of CaF_2, and the formation of volatile fluorides.

As a result of the chemical reactions and the effects of other factors (Fig. 6.1), the mass of the slag varies with time. This variation can be taken into account using the equation

$$G_{\mathrm{slag}}^{k+1} = G_{\mathrm{slag}}^{k} + \sum_{i=1}^{i}\sum_{j=1}^{2} I_{\mathrm{E}_i,j}^{k} S_j (M_{\mathrm{E}_i} - mM_{\mathrm{Fe}})\Delta\tau + V_{\mathrm{m}} S_{\mathrm{ox}} \rho_{\mathrm{ox}} \Delta\tau + V_{\mathrm{fl}}\Delta\tau$$
$$- V_{\mathrm{dr}} S_{\mathrm{sk}} \rho_{\mathrm{sk}} \Delta\tau - \sum_{i=1}^{i} V_{\mathrm{g}}\Delta\tau. \tag{6.37}$$

The double sum on right-hand side of Eq. 6.37 describes the increase in the mass of the slag during a time interval of length $\Delta\tau$ due to chemical reactions with all the elements in both stages, and the remaining terms describe, in order of their appearance, the changes in the mass of the slag due to dissolution of the oxide film, feeding of the flux, formation of the slag skin, and departure of elements into the gas phase.

The metal pool is shaped as a result of the entry of molten electrode metal into it, crystallization of the ingot, the introduction of metal additives, and the rate of its interaction with the slag. The material balance condition can be described by the equation

$$G_{\mathrm{Me},2}^{k+1} = G_{\mathrm{Me},2}^{k} + \left(\dot{m} + V_{\mathrm{Me}}^{n} - \dot{m}_{\mathrm{cryst}} - V_x\right)\Delta\tau, \tag{6.38}$$

where V_x is the change in the mass of the pool caused by the chemical reactions. The term V_x can be neglected.

The distribution of the concentrations of the elements in a transverse section of the ingot was constructed using the method described in [6, 7, 21], which provides ways to calculate the dynamics of the mass and profile of the metal pool during ESR.

Thus, system of Eqs. 6.5–6.10, 6.16–6.32, 6.34, and 6.36–6.38 can be regarded as a general model of the chemical processes occurring during ESR.

The input parameters of the model are the compositions of the electrode metal and the flux, the remelting rate, the areas of the interaction surfaces, the thickness and composition of the oxide film on the electrode, the liquation coefficients, physicochemical parameters, and other quantities that appear in the equations.

The output parameters are the compositions of the metal in the stages considered, including the ingot, in both longitudinal and transverse sections, the composition and mass of the slag, the reaction rates, and the depth and shape of the metal pool.

6.1.3 Mathematical Modeling of the Chemical Processes Occurring in Electroslag Welding and Surfacing

The electroslag remelting process and the ESW and surfacing processes are similar in many respects. Therefore, the general approach considered above to describing the kinetics of the phase interaction in an electroslag process can be used to develop a mathematical model of the chemical processes occurring in ESW and ESS. In addition, the following special features of ESW and ESS were taken into account.

- A welded joint and a deposited layer are formed by the molten base and electrode metals.
- Molten metal from the electrode and the continuously melted tip, falling droplets of the electrode metal, the surface of the liquid metal pool, and the molten surface of the base metal interact with the liquid slag.
- The differences between the mass-transfer conditions in each stage of the processes, which characterize the rates of the chemical reactions, are taken into account using the corresponding values of β_j.
- Due to the high current density and the occurrence of processes at the electrode, the surface temperature of the metal droplets that form on the electrode during ESW and ESS is higher than the temperature of the liquid metal film on the electrode during ESR and is determined by the regime of the electroslag process [22–26]. The temperature of the metal in the liquid film on the melted edges of the workpiece is close to the melting point of the metal. The surface of the metal pool in ESW and ESS probably has the same temperature as in ESR.

The material balance of the elements in the metal in the electrode stage, the falling-droplet stage, and the stage at the melted surface of the base metal can be written in the form of Eq. 6.2, and the expression for calculating the concentration of an element in these stages is similar to Eq. 6.7.

When the material balance equation for the metal-pool stage is written, the percentage of the participation of the base metal in the metal of the welded joint and the deposited layer should be taken into account [22].

In the case under consideration, when the entry of the electrode metal (stage 1) and the base metal (stage 2), as well as the continuously melted tip (stage 3), into the pool (stage 4) is taken into account, Eq. 6.3 has the form

$$\frac{G_{\mathrm{Me},4}}{100}\frac{d[\mathrm{E}_i]_4}{d\tau} = \sum_{j=1}^{3} \dot{m}_j \frac{[\mathrm{E}_i]_j}{100} + V^n_{\mathrm{Me}} \frac{[\mathrm{E}_i]_n}{100} - I_{\mathrm{E}_i,4} S_1 M_{\mathrm{E}_i} - \dot{m}_{\mathrm{cryst}} \frac{[\mathrm{E}_i]_{\mathrm{w}}}{100}, \qquad (6.39)$$

where $[\mathrm{E}_i]_{\mathrm{w}}$ is the mass percentage of the ith element in the weld, wt.%.

Then for the time interval from $k\Delta\tau$ to $(k+1)\Delta\tau$, we obtain an expression for calculating the concentration of an element in the metal-pool stage:

$$\begin{aligned} \Delta[\mathrm{E}_i]_4^{k+1,k}, &= \left\{ \sum_{j=1}^{3} \dot{m}_j [\mathrm{E}_i]_j + V^n_{\mathrm{Me}} [\mathrm{E}_i]_n - I^k_{\mathrm{E}_i,4} S_4 M_{\mathrm{E}_i} 100 \right. \\ &\quad \left. - \dot{m}_{\mathrm{cryst}} \kappa_{\mathrm{E_i}} [\mathrm{E}_i]_4^k \right\} \frac{\Delta\tau}{G_{\mathrm{Me},4}}, \\ [\mathrm{E}_i]_4^{k+1} &= [\mathrm{E}_i]_4^k + \Delta[\mathrm{E}_i]_4^{k+1,k}. \end{aligned} \qquad (6.40)$$

Similar changes should also be made in expressions (6.9) and (6.37) for calculating the composition and mass of the slag pool by supplementing them with terms that take into account the interaction of the metal with the slag during the stages involving the base metal and the continuously melted tip, as well as the dissolution of the oxide films from the surfaces of the latter.

Since the mass of the slag is held constant during ESW and ESS, Eq. 6.37 enables us to calculate the required flux feed rate V_{fl}. To do this, it is sufficient to set $G^{k+1}_{\mathrm{slag}} = G^k_{\mathrm{slag}}$.

The calculation of the reaction rates in the stages enumerated was performed using Eqs. 6.18–6.24 and 6.26–6.31, and the reaction rates on the slag/gas interface were found using Eqs. 6.34 and 6.36.

The concentrations of elements that undergo practically no oxidation during the welding (surfacing) of steels, namely, nickel, tungsten, cobalt, etc., were calculated according to the mixing rule

$$[\mathrm{E}_i]_{\mathrm{w}} = (1-\gamma)[\mathrm{E}_i]_1 + \gamma[\mathrm{E}_i]_2, \qquad (6.41)$$

where $\gamma = [\dot{m}_2/(\dot{m}_1 + \dot{m}_2)]$ is the fraction of the base metal in the weld, $\dot{m}_1$ is the melting rate of the additive metal (kg/s), and $\dot{m}_2$ is the melting rate of the base metal (kg/s).

As in the case of electroslag remelting, the mathematical model of the chemical processes occurring in ESW and ESS offers a system of material balance equations and kinetic equations of the reactions for each stage.

6.1.4 Allowing for Deoxidization of the Slag Pool by Metallic Components in Electroslag Technologies

Several variations of the mathematical description of the processes under consideration have been developed.

The first variant of the mathematical machinery provides a way to take into account the interaction of slag oxides with a deoxidizer (aluminum, silicocalcium, etc.) that is added to the slag pool [27]. The calculation method provided for the fact that the deoxidizer reduces elements from their oxides in succession, in order of increasing oxygen affinity of the elements. Such an approach has a well defined theoretical and experimental foundation [13] and is fully acceptable if the interaction rates of the deoxidizer with the slag are fairly high, for example, in the case of a very small fraction of a powdered deoxidizer, in which the interaction surface is large. In this case the reactions are close to equilibrium, which can be described by the laws of thermodynamics. However, this calculation method leads to errors if the reaction does not go to completion. In that situation the unreacted deoxidizer enters the metal pool, dissolves in it, and reacts primarily with the slag oxides, but far less strongly due to the decreases in concentration and interaction surface (here the latter equals the surface of the pool). In addition, the method described in [27] does not take into account the possibility of the simultaneous reduction of different elements from the slag.

Another, more complete method treats the flow of droplets that continuously enters with the feed and exits into the pool as a separate parallel step of the process, which is similar to the electrode stage and has an interaction surface equal to the total surface of all the droplets found within the slag at a particular time. This enables us to take into account all the reactions and the fact that they do not go to completion. The approximate nature of such a calculation is due to the fact that the variation of the reaction rates during the lifetime of the droplets is ignored and average rates are used. We note that for fairly strong deoxidizers, for example, aluminum, the reduction reactions are limited by the diffusion of the oxides of these elements in the slag [13]. Therefore, in some cases this method will be practically exact.

Further development enabled us to take into account the kinetics of all possible, simultaneously occurring reactions on the slag/deoxidizer-droplet interface, the variation of the composition of a droplet during its residence in the slag, and the absorption of droplets by the metal pool.

A definite complexity is introduced here by the calculation of the average lifetime of a droplet τ_{dr}, i.e., the time segment from the time when the deoxidizer particles are completely melted until the time when they are absorbed by the slag pool. The value of τ_{dr} depends on the viscosity of the slag, the depth of the slag pool, and the size of the droplets. The value of τ_{dr} can be estimated only approximately, because it is practically impossible to take into account the influence of the convective flows in the slag, which can either increase τ_{dr} in the ascending part of a flow or decrease it if the flow is directed downward. In addition, the density of a droplet varies over the course of its interaction with the slag due to the entry of iron, silicon, and other elements into it and the oxidation of aluminum and titanium. To estimate τ_{dr} we used Stokes's formula:

$$V_{dr} = h_{slag}/\tau_{dr} = 2\left(\rho_{dr} - \rho_{slag}\right)gr_{dr}^2/\left(9\eta_{slag}\right), \tag{6.42}$$

where V_{dr} is the droplet velocity (m/s), h_{slag} is the depth of the slag pool (m), η_{slag} is the viscosity of the slag (P), ρ_{dr} is the density of the droplet metal (kg/m^3), r_{dr} is the droplet radius (m), and g is the acceleration of gravity (m^2/s).

The material balance of the element Ei in a droplet of mass m_{dr} that traverses the slag pool during the time τ_{dr} equals

$$m_{dr}\frac{d[\mathrm{E}_i]_{dr}}{d\tau_{dr}} = I_{\mathrm{E}_i,dr}S_{dr}M_{\mathrm{E}_i}100\tau_{dr}, \tag{6.43}$$

where S_{dr} is the surface area of the droplet (m^2), $[\mathrm{E}_i]_{dr}$ is the mass percentage of the element E_i in the droplet metal (wt.%), and $I_{\mathrm{E}_i,dr}$ is the reaction rate on the droplet/slag interface (mole/m^2 s).

We calculate the variation of the concentration $[\mathrm{E}_i]_{dr}$ using the previously considered technique. We divide the entire time τ_{dr} into sufficiently short time intervals of length $\Delta\tau$ (10–20 intervals), during which we assume that the rates of the reactions on the droplet/slag interface are constant.

After the transformation of Eq. 6.43 for the time interval from $k\Delta\tau$ to $(k+1)\Delta\tau$, the change in the concentration of an element in the droplet metal is given by the equation

$$[\mathrm{E}_i]_{dr}^{k+1} = [\mathrm{E}_i]_{dr}^{k} - I_{\mathrm{E}_i,dr}M_{\mathrm{E}_i}S_{dr}100\Delta\tau_{dr}/m_{dr}. \tag{6.44}$$

We calculate the droplet density ρ_{dr} for each time interval $k\Delta\tau$ using the equation

$$\rho_{dr} = \sum_{i=1}^{i} \rho_{\mathrm{E}_i}[\mathrm{E}_i]_{dr}/100, \tag{6.45}$$

where ρ_{E_i} is the density of a substance that consists entirely of the element E_i (kg/m).

The change in the composition of the metal pool during a time interval of length $\Delta\tau$ as a result of the interaction with the deoxidizer droplets was taken into account using an additional term in Eq. 6.9:

$$\Delta(E_iO_m)^{k+1,k}_{dr} = \left\{[E_i]_{dr,f} - [E_i]_{abs}\right\} \frac{M_{E_iO_m} V^n_M}{M_{E_i} G_{slag}} \Delta\tau, \tag{6.46}$$

where $[E_i]_{dr,f}$ and $[E_i]_{abs}$ are the mass percentages of the *i*th element in the deoxidizer droplets at the time of their formation and at the time of their absorption by the metal pool, wt.%.

The change in the composition of the metal pool as a result of absorption of the deoxidizer droplets is taken into account by the second term in Eq. 6.8.

6.1.5 Mathematical Model of the Chemical Processes Occurring in Centrifugal Electroslag Casting

The use of EST provides ways to obtain castings with properties that do not fall short of those of forgings and stampings and to raise the metal utilization ratio by shortening the metallurgical cycle for the production of intermediate products and by reducing the loss of metal as shavings during machining [28].

The mathematical model of the interaction processes in CESC enables us to predict the chemical composition of the metal across the thickness of a casting and to take into account a variety of factors of thermodynamic, kinetic, chemical, diffusive, and electrolytic nature that arise when the metal, slag, crucible lining, atmosphere, and deoxidizers react in the various stages of the manufacturing process from melting of the electrode to crystallization of the casting in the mold.

Figure 6.2 shows a diagram of the directions of transport of the element E_i and the variation of the composition of the phases during CESC. The material balance of the element E_i in the metal can be described in the following manner:

- in the electrode stage ($j = 1$)

$$\frac{G_{Me,1}}{100} \frac{d[E_i]_1}{d\tau} = V_{Me,1} \frac{[E_i]_0}{100} - I_1 S_1 A_{E_i} - V_{Me,1} \frac{[E_i]_1}{100}, \tag{6.47}$$

- in the pool stage ($j = 2$)

$$\frac{G_{Me,2}}{100} \frac{d[E_i]_2}{d\tau} + \frac{[E_i]_2}{100} \frac{dG_{Me,2}}{d\tau} = V_{Me,2} \frac{[E_i]_1}{100} + V_{deox} \frac{[E_i]_{deox}}{100} + V_{mod} \frac{[E_i]_{mod}}{100} - I_2 S_2 A_{E_i}, \tag{6.48}$$

- in the pouring stage ($j = 3$)

$$\frac{G_{Me,3}}{100} \frac{d[E_i]_3}{d\tau} = -I_3 S_3 A_{E_i}, \tag{6.49}$$

- in the crystallization stage ($j = 4$)

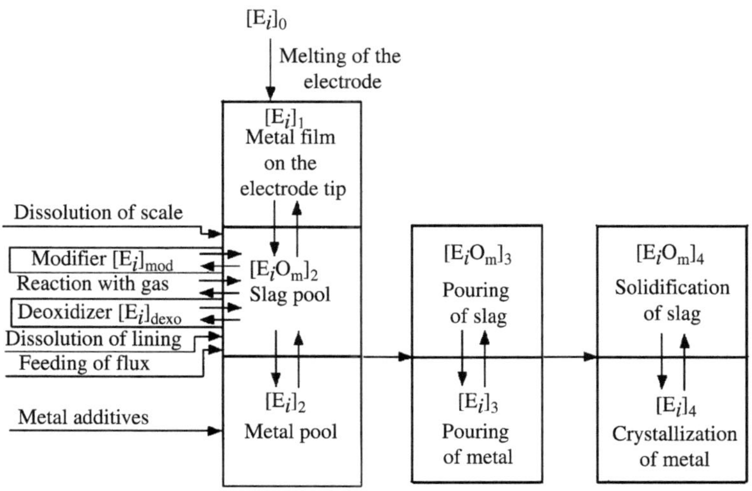

Fig. 6.2 Block diagram of the directions of transport of the element E_i and variation of the composition of the phases during CESC

$$\frac{G_{\mathrm{Me},4}}{100}\frac{d[\mathrm{E}_i]_4}{d\tau}+\frac{[\mathrm{E}_i]_4}{100}\frac{dG_{\mathrm{Me},4}}{d\tau}=V_{\mathrm{Me},4}\frac{[\mathrm{E}_i]_3}{100}-I_2S_2A_{\mathrm{E}_i}-V_{\mathrm{Me}_{\mathrm{cryst}}}\frac{[\mathrm{E}_i]_4}{100\kappa}, \qquad (6.50)$$

where $[\mathrm{E}_i]_0$, $[\mathrm{E}_i]_j$, $[\mathrm{E}_i]_{\mathrm{mod}}$, and $[\mathrm{E}_i]_{\mathrm{deox}}$ are the concentrations of the *i*th element in the electrode, in stage *j*, in the modifier, and in the deoxidizer, respectively (wt.%), I_j is the rate of passage of the element from the metal into the slag in stage *j* (mole/m^2 s), S_j is the area of the reaction surface in stage *j* (m^2), A_{E_i} is the atomic weight of the *i*th element (kg/mole), $V_{\mathrm{Me},1}$, $V_{\mathrm{Me},2}$, V_{mod}, V_{deox}, $V_{\mathrm{Me},4}$,and $V_{\mathrm{Me}_{\mathrm{cryst}}}$ are the melting rate of the electrode, the accumulation rate of the metal in the crucible, the feed rates of the modifier and the deoxidizer, the pouring rate of the metal, and its crystallization rate, respectively (kg/s), τ is the time of the process (s), $G_{\mathrm{Me},2}$ and $G_{\mathrm{Me},4}$ are the masses of the metal in the crucible and in the mold (kg), and κ is the liquation coefficient.

Equation 6.47 describes the rate of variation of the content of the *i*th element in the metal in the first stage, where the first term on the right-hand side describes the rate of entry of the element from the electrode into the liquid metal, the second term describes the rate of passage of the element through the metal/slag interface, and the third term describes the rate of departure of the element from the metal with the droplets.

The left-hand side of Eq. 6.48 describes the rate of variation of the content of element *i* in the metal pool in the second stage, when both the mass of the metal and the concentrations of the components vary with time. The first term on the right-hand side describes the rate of entry of the element with portions of the electrode metal, the second and third terms describe the rates of entry of the element with the deoxidizer and the modifier, and the fourth term describes the rate of passage of the element from the metal into the slag as a result of chemical reactions.

The pouring stage (6.49) is characterized by variation of the content of the *i*th element in the metal due to the interaction with the slag.

In the crystallization stage (6.50) the left-hand side of the equation is analogous in the mathematical respect to the left-hand side of Eq. 6.48. The terms on the right-hand side describe, in their order of appearance, the decrease in the amount of the element due to the occurrence of reactions between the metal and the slag and crystallization of the metal, as well as its entry with portions of the metal when the latter is poured into the mold.

The material balance for the element under consideration in the slag is described by the following equations:

- in the pool stage ($j = 2$)

$$\frac{G_{\text{slag},2}}{100}\frac{d(\text{E}_i\text{O}_m)_2}{d\tau} = V_{\text{fl}}\frac{(\text{E}_i\text{O}_m)_0}{100} + I_1 S_1 M_{\text{ox}} + I_2 S_2 M_{\text{ox}} + V_{\text{ox}_{\text{dis}}}\frac{(\ni_i\,\text{O}_m)_{\text{lin}}}{100} + V^{\text{e}}_{\text{ox}_{\text{dis}}}\frac{(\text{FeO})_{\text{e}}}{100}, \tag{6.51}$$

- in the pouring stage ($j = 3$)

$$\frac{G_{\text{slag},3}}{100}\frac{d(\text{E}_i\text{O}_m)_3}{d\tau} = I_3 S_3 M_{\text{ox}}, \tag{6.52}$$

- in the crystallization stage ($j = 4$)

$$\frac{(\text{E}_i\text{O}_m)_4}{100}\frac{dG_{\text{slag},4}}{d\tau} + \frac{G_{\text{slag},4}}{100}\frac{d(\text{E}_i\text{O}_m)_4}{d\tau} = V_{\text{slag},4}\frac{(\text{E}_i\text{O}_m)_3}{100} + I_4 S_4 M_{\text{ox}} - V^{\text{cryst}}_{\text{slag}}\frac{(\text{E}_i\text{O}_m)_4}{100}, \tag{6.53}$$

where $(\text{E}_i\text{O}_m)_0$, $(\text{E}_i\text{O}_m)_2$, $(\text{E}_i\text{O}_m)_3$, $(\text{E}_i\text{O}_m)_4$, and $(\text{E}_i\text{O}_m)_{\text{lin}}$ are the concentrations of the oxide of element *i* in the original flux, in the slag pool, in the pouring and crystallization stages, and in the crucible lining, respectively (wt.%), V_{fl}, $V_{\text{lin}_{\text{dis}}}$, $V^{\text{e}}_{\text{ox}_{\text{dis}}}$, $V^{\text{cryst}}_{\text{slag}}$, and $V_{\text{slag},4}$ are the flux feed rate, the rate of solution of the crucible lining, the rate of solution of the oxide films on the electrode, the solidification rate, and the slag pouring rate, respectively (kg/s), $M_{\text{E}_i\text{O}_m}$ is the molecular weight of the oxide (kg/mole), $G_{\text{slag},2}$ and $G_{\text{slag},4}$ are the masses of the slag in the crucible and the mold (kg), and (FeO) is the concentration of the oxide film on the electrode (wt.%).

The left-hand side of Eq. 6.51 describes the variation of the content of the oxide of element *i* in the slag in the pool stage, and the right-hand side describes the entry of the oxide upon feeding of the flux, chemical reactions on the electrode and in the metal pool, dissolution of the crucible lining, and dissolution of the oxide films.

In the pouring stage (6.52) the change in the content of the oxide in the slag is due to chemical reactions.

In the crystallization stage (6.53) the change in the amount of the oxide of the element in the slag pool is due to its entry with the liquid slag during pouring into the mold, reactions with the metal, and solidification of the slag.

On the basis of the data in [7, 9], it was shown that the expression for the concentration of the element in the metal in the electrode stage will be as follows:

$$[\mathrm{E}_i]_1 = [\mathrm{E}_i]_0 - \frac{100 I_1 S_1 A_{\mathrm{E}_i}}{V_{\mathrm{slag},1}}. \tag{6.54}$$

For the calculation of the concentrations of any element in the metal pool and of its oxide in the slag pool, we divide the time of the process τ into sufficiently short time intervals of length $\Delta\tau$, during which the rates of the chemical reactions may be assumed constant. Then, after performing some transformations on Eqs. 6.48 and 6.51, the expressions for calculating the changes in the concentrations of the elements in the metal and of their oxides in the slag for the time interval from $k \cdot \Delta\tau$ to $(k+1) \cdot \Delta\tau$ will be, respectively, as follows:

$$[\mathrm{E}_i]_2^{k+1} = [\mathrm{E}_i]_2^k \frac{G_{\mathrm{Me},2}^k}{G_{\mathrm{Me},2}^{k+1}} + \left(V_{\mathrm{Me},1}[\mathrm{E}_i]_1 + V_{\mathrm{mod}}[\mathrm{E}_i]_{\mathrm{mod}} + V_{\mathrm{deox}}[\mathrm{E}_i]_{\mathrm{deox}} - 100 I_2 S_2 A_{\mathrm{E}_i}\right) \frac{\Delta\tau}{G_{\mathrm{Me},2}^{k+1}}, \tag{6.55}$$

$$(\mathrm{E}_i\mathrm{O}_m)_2^{k+1} = (\mathrm{E}_i\mathrm{O}_m)_2^k \Big\{ V_{\mathrm{slag},1}(\mathrm{E}_i\mathrm{O}_m)_0 + 100 M_{\mathrm{ox}}(I_1 S_1 + I_2 S_2) + V_{\mathrm{lin}_{\mathrm{dis}}}.(\mathrm{E}_i\mathrm{O}_m)_{\mathrm{lin}} + V_{\mathrm{ox}_{\mathrm{dis}}}^{\mathrm{e}}(\mathrm{FeO})_{\mathrm{e}} \Big\} \frac{\Delta\tau}{G_{\mathrm{slag},2}} + \Delta\tau(\mathrm{E}_i\mathrm{O}_m)_{\mathrm{dr}}^{k+1}. \tag{6.56}$$

To calculate the changes in the concentrations of an element in the metal and of its oxide in the slag during the pouring and crystallization of the casting, we used a similar approach. For the time interval from $k \cdot \Delta\tau$ to $(k+1) \cdot \Delta\tau$, from Eqs. 6.49, 6.50, 6.52, and 6.53 we found

- in the pouring stage

$$[\mathrm{E}_i]_3^{k+1} = [\mathrm{E}_i]_3^k - \frac{400 I_3 A_{\mathrm{E}_i} \Delta\tau}{d \rho_{\mathrm{Me}}}, \tag{6.57}$$

$$(\mathrm{E}_i\mathrm{O}_m)_3^{k+1} = (\mathrm{E}_i\mathrm{O}_m)_3^k + \frac{400 I_3 M_{\mathrm{ox}} d \Delta\tau}{\left(d_1^2 - d^2\right)\rho_{\mathrm{slag}}} + \Delta(\mathrm{E}_i\mathrm{O}_m), \tag{6.58}$$

where d is the diameter of the metal stream (m), d_1 is the outer diameter of the liquid slag stream (m), ρ_{Me} and ρ_{slag} are the densities of the liquid metal and liquid slag, respectively (kg/m^3), and $\Delta(E_iO_m)$ is the amount of the oxide of element i that enters the slag when the deoxidizer (or modifier) interacts with the slag;

- in the crystallization stage

$$[E_i]_4^{k+1} = [E_i]_4^k \left\{ 1 - \left(V_{\mathrm{Me},4} - V_{\mathrm{Me_{cryst}}}(1-\kappa) \right) \frac{\Delta\tau}{G_{\mathrm{Me},4}^{k+1}} \right\} + \left(V_{\mathrm{Me},4}[E_i]_3 - 100 I_4 S_4 A_{E_i} \right) \frac{\Delta\tau}{G_{\mathrm{Me},4}^{k+1}}, \quad (6.59)$$

$$(E_iO_m)_4^{k+1} = (E_iO_m)_4^k \left\{ 1 - \left(V_{\mathrm{slag}}^{\mathrm{cryst}}(1-\kappa) - V_{\mathrm{slag},4} \right) \right\} \frac{\Delta\tau}{G_{\mathrm{slag},4}^{k+1}} + \left(V_{\mathrm{slag},4}(E_iO_m) + I_4 S_4 M_{\mathrm{ox}} \right) \frac{\Delta\tau}{G_{\mathrm{slag},4}^{k+1}}. \quad (6.60)$$

The composition of the metal and the slag was calculated successively for each of the stages.

When modifiers and deoxidizers are introduced, they melt in the slag pool to form droplets, which pass through the slag layer and react with it. In this case the material balance of the element E_i in a droplet of mass m_{dr} that has passed through the slag pool during the time τ_{dr} has the form

$$m_{\mathrm{dr}} \frac{d[E_i]_{\mathrm{dr}}}{d\tau_{\mathrm{dr}}} = 100 I_{E_i}^{\mathrm{dr}} S_{\mathrm{dr}} \tau_{\mathrm{dr}}, \quad (6.61)$$

where S_{dr} is the surface area of the droplet (m^2) and $[E_i]_{\mathrm{dr}}$ is the concentration of the element in the droplet metal (wt.%).

To calculate the variation of the concentration of an element $[E_i]_{\mathrm{dr}}$, we also divided the entire process into short time intervals of length $\Delta\tau$, during which the rates of all the reactions on the metal-droplet/slag interface may be assumed constant. For the time interval from $k \cdot \Delta\tau$ to $(k+1) \cdot \Delta\tau$, expression (6.61) takes the form

$$[E_i]_{\mathrm{dr}}^{k+1} = [E_i]_{\mathrm{dr}}^k - \frac{100 I_{E_i}^k A_{E_i} S_{\mathrm{dr}} \Delta\tau_{\mathrm{dr}}}{m_{\mathrm{dr}}}. \quad (6.62)$$

The value of τ_{dr} can be estimated only approximately, because it is impossible to take into account the influence of the convective flows in the slag, which can either increase τ_{dr} in the case of ascending flows or decrease it if the flow is directed downward. In addition, the density of a droplet varies over the course of its interaction with the slag due to the entry of iron, silicon, and other elements into it and the oxidation of aluminum and titanium. To estimate τ_{dr} we used Stokes's formula (6.42).

From Eq. 6.42 we found

$$\tau_{dr} = \frac{4.5\eta_{slag}h_{slag}}{(\rho_{dr} - \rho_{slag})gr_{dr}^2}. \tag{6.63}$$

The droplet density was calculated for each time interval using additive approximation (6.45).

The change in the composition of the slag pool as a result of the interaction with deoxidizer (or modifier) droplets was taken into account in the following manner:

$$\Delta(E_iO_m)_{dr}^{k+1} = \left\{[E_i]_{dr}^k - [E_i]_{dr}^{k+1}\right\}\frac{M_{ox}V_{deox}\Delta\tau}{A_{E_i}G_{slag}}. \tag{6.64}$$

To find the reaction rates I_i^k, we used the method for the kinetic analysis of simultaneously occurring reactions with a common reactant of the type

$$1/m[E_i] + (FeO) = [Fe] + 1/m(E_iO_m), \tag{6.11}$$

which was considered in detail in Chap. 4

As a result of the chemical reactions, the interaction with the crucible lining, and other factors, the mass of the slag pool varies. Its variation can be taken into account using the expression

$$G_{slag}^{k+1} = G_{slag}^k + \left\{V_{ox_{dis}}^e S_{ox}\rho_{ox} + V_{fl} - V_g + \sum_{i=1}^{i}\sum_{j=1}^{j} I_j^k S_j\left(A_{E_i} - \frac{m}{n}A_{E_i}\right)\right\}\Delta\tau. \tag{6.65}$$

In Eq. 6.65 the double sum on the right-hand side describes the variation of the mass of the slag due to the occurrence of reactions of type (6.11) in the electrode, pool, and deoxidizer (or modifier) droplet stages, and the remaining terms, in their order of appearance, describe the changes in the mass of the slag due to dissolution of the oxide film from the electrode, feeding of the flux, and the passage of components into the gas phase.

System of Eqs. 6.54–6.65 is a general mathematical model of the chemical processes occurring during CESC and can be used to predict the composition of the phases. Figure 6.3 presents a simplified block diagram of the algorithm of this mathematical model.

The input parameters of the model are the compositions of the electrode metal, flux, crucible lining, deoxidizers, and modifiers, their melting and feed rates, the areas of the interaction surfaces, the thickness of the oxide film on the electrode surface, the liquation coefficients, and other parameters. In addition, information regarding some physicochemical parameters, namely, the equilibrium constants, diffusion coefficients, and convection constants, is needed for calculating the reaction rates [1, 3].

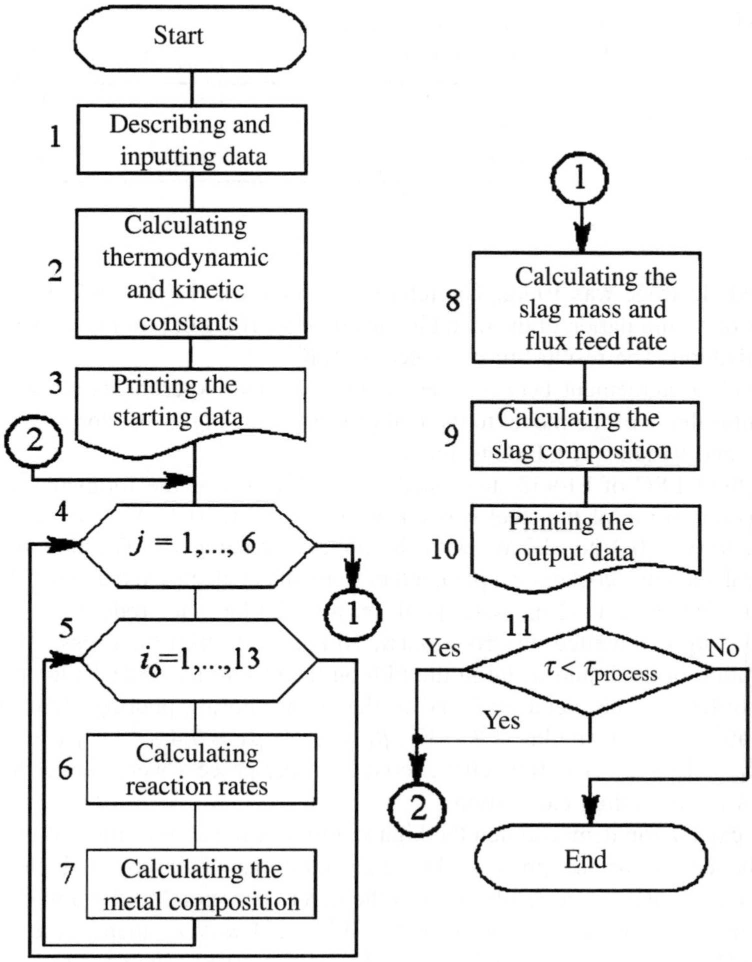

Fig. 6.3 Flowchart of a mathematical model of the chemical processes occurring in CESC

The output parameters are the composition of the metal in all of the stages of the CESC process, the composition and mass of the slag, and the rates of the chemical reactions.

In order to test the faithfulness of the mathematical model to a real process, trial melts of R6M5 steel with the addition of titanium to the metal pool over the course of the remelting of the electrode and of EL-958 (40Kh5V2FS) steel were carried out.

According to the experimental data, the composition of the metal in a casting of R6M5 steel with respect to the principal alloying components scarcely differs from the composition of the remelted electrode. Therefore, to evaluate the faithfulness of the model, titanium was introduced over the course of the process, and its concentration was determined experimentally and mathematically. The mass of the

Table 6.1 Titanium content in the ingot metal during remelting

Time from the beginning of the melt (s)	[Ti], wt.%	
	Calculation	Experiment
855	0.80	0.79
975	0.32	0.30
1,095	0.085	0.080

remelted electrode was 10 kg. The total CESC time was 1,410 s, which included 1,140 s of accumulation of the metal in the crucible, 100 s of pouring, and 180 s of crystallization. The results are presented in Table 6.1.

The close agreement between the calculations and experiments attests to the high faithfulness of the model to the real process, allowing its use for other grades of steel and variants of CESC technologies.

For the CESC of EI-958 steel under type ANF-6 flux, the following average grade composition of the steel was taken, wt.%: 0.4 C, 0.35 Mn, 6.0 Cr, 1.0 V, 0.03 P, 0.03 S, 0.3 Ni, 2.0 W, with the remainder being Fe. The flowing technological and physicochemical parameters were taken: electrode diameter, 0.06 m; crucible diameter, 0.32 m; slag pool mass, 12.8 kg; electrode melting rate, 8×10^{-3} kg/s; remelted electrode mass, 10 kg; oxide film thickness, 0.01 mm; temperature of the liquid metal in the film on the electrode, 1,823 K; temperature at the surface of the metal pool and in the stream during pouring, 1,863 K [8], temperature in the mold, 1,813 K; $\beta_1 = \beta_2 = \beta_3 = 7{,}4\ \mathrm{s}^{-0.5}$; $\beta_4 = 2\ \mathrm{s}^{-0.5}$; $\beta_5 = \beta_6 = 10\ \mathrm{s}^{-0.6}$. The time characteristics of the process were similar to those taken in the preceding calculation.

The calculation demonstrated the high stability of the composition of the metal over the course of the process. The concentrations of carbon and chromium scarcely vary, and the concentrations in the ingot correspond to the initial values. Only insignificant losses of silicon (0.12–0.14 wt.%), manganese (0.05–0.06 wt.%), and vanadium (0.05–0.06 wt.%) are observed. The desulfurization of the steel proceeds fairly rapidly in two stages of the process: in the liquid film on the electrode tip, it decrease from an initial value to 0.018 wt.%, in the pool it decreases from 0.018 to 0.004 wt.%. The high stability of the composition of the metal is attributed to the stability of the composition of the slag pool.

6.2 Calculated Evaluation of the Kinetics of the Interaction of a Multicomponent Metal and a Slag During Submerged-Arc Welding

It has been reported that equilibrium between the metal and the slag is not achieved during submerged-arc welding in most cases [29, 30]. This conclusion was additionally confirmed as applied to multicomponent metals on the basis of improved calculation procedures and refined equilibrium constants in [31]. Thus, it may be stated that it is impossible to obtain reliable data regarding the composition of the phases using the thoroughly developed thermodynamic methods that were described, for example, in [31, 32].

The kinetic hindrances to the reactions that occur during welding were systematically taken into account for the first time by Erokhin [30] in the example of the passage of the element E_i from the metal into the slag

$$[E_i] = (E_i). \tag{6.66}$$

Unfortunately, the mathematical machinery used in [30] is not suitable for multicomponent melts, because the process indicated is usually described by a reaction involving three or four substances, rather than by Eq. 6.66. In addition, the presence of common reactants causes a mutual influence of the reactions on the rates and direction of their occurrence.

In this section we will consider a mathematical model that reflects the kinetics of the chemical processes occurring during submerged-arc welding and enables us to predict the composition of the weld metal with allowance for the following special features of the process:

- occurrence of the interaction of the metal with the slag in two successive stages, namely, an electrode (droplet) stage and a pool stage, which have different temperature, geometric, and hydrodynamic conditions;
- continuous renewal of the interacting metal and slag in each stage as a result of the melting and crystallization of the additive and base metals and of the flux;
- the occurrence of electrolysis when part of the arc current is shunted by the liquid slag;
- the parallel occurrence of all the reactions in each stage and their thermodynamic and kinetic influence on one another;
- the influence of the hindrances to the diffusion of all the reactions in the metal and the slag on the rate of each reaction and the slowness of the chemical act of ion discharge on the metal/slag interface.

Several simplifying assumptions were used in the model.

- The chemical composition of the metal and the slag is identical throughout the phase at a given time in each stage (ideal mixing).
- The interaction surface between the metal and the slag is kinetically homogeneous in each stage, i.e., it can be described by time-averaged contact areas and a mass-transfer constant.
- In accordance with the data from [33–35], the influence of the reactions with the gas phase can be neglected in a first approximation. Reactions involving hydrogen and nitrogen are exceptions [36].

The structural diagram of the phase interaction drawn is presented in Fig. 6.4. The arrows depict the directions of transport of the substance. It was taken into account that in the general case definite portions of the metal (α) from the

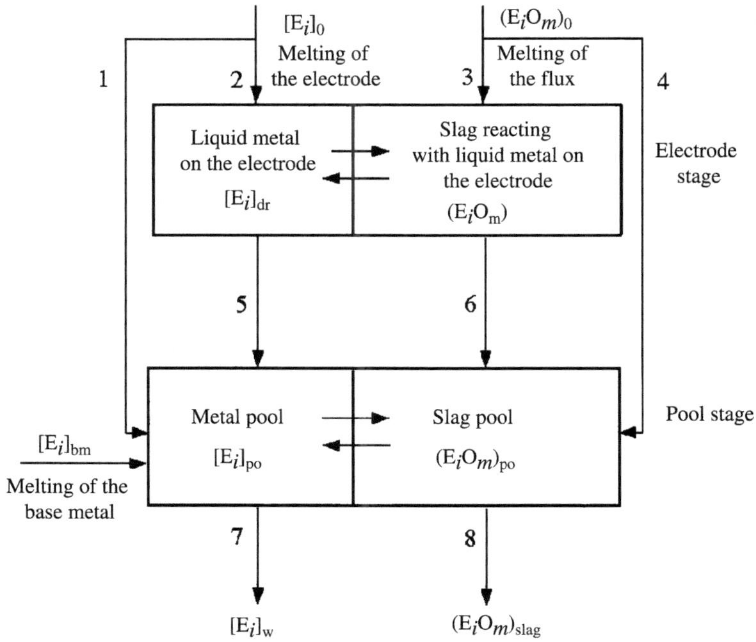

Fig. 6.4 Block diagram of the phase interaction during submerged-arc welding: *1* transport of electrode droplets into the pool without coming into contact with the slag; *2* formation of liquid metal on the electrode surface; *3* melting of the flux in the electrode zone; *4* melting of flux which reacts only with the metal pool; *5* transport of metal droplets that have reacted on the electrode into the pool; *6* mixing of the slag that has reacted with electrode droplets with the slag pool; *7* crystallization of the metal pool; *8* solidification of the slag pool

continuously melted electrode and of the flux (α') enter the metal and slag pools without interacting in the electrode stage.

The concentrations of the elements in the pools (wt.%) are related to the composition of the weld metal and the slag crust by the relations

$$\frac{[\mathrm{E}_i]_{\mathrm{w}}}{[\mathrm{E}_i]_{\mathrm{po}}} = \kappa; \quad \frac{(\mathrm{E}_i\mathrm{O}_m)_{\mathrm{slag}}}{(\mathrm{E}_i\mathrm{O}_m)_{\mathrm{po}}} = \kappa_{\mathrm{ox}}, \tag{6.67}$$

where κ and κ_{ox} are the liquation coefficients of the element E_i in the metal and the slag, respectively.

The material balance of the element E_i (kg/s) in a given volume of a phase (for example, the metal phase) can be written in the following form:

$$\sum V_k \frac{[\mathrm{E}_i]_k}{100} = \sum V_l \frac{[\mathrm{E}_i]_l}{100} + ISM + \frac{G}{100}\frac{d[\mathrm{E}_i]}{d\tau}, \tag{6.68}$$

where V_k and V_l are, respectively, the rate of entry of the metal into the particular phase and its rate of departure with one of the incoming or outgoing flows (kg/s),

$[E_i]_k$ and $[E_i]_l$ are the concentrations of the element E_i in these flows (wt.%), I is the rate of the reaction in which the element E_i passes from the metal into the slag (mole/m^2 s), S is the area of the reaction surface (m^2), M is the molecular (atomic) weight of the element E_i (kg/mole), and G is the mass of the phase (kg).

The left-hand side of Eq. 6.68 expresses the rate of entry of the element E_i into the particular phase, and the terms on the right-hand side represent, respectively, its rate of departure with the outgoing flows, its rate of departure due to chemical reactions, and its rate of departure due to accumulation. In a steady-state regime $(d[E_i]/d\tau) = 0$.

Using Eq. 6.68 for a steady-state regime we find:

- for the electrode (droplet) stage

$$V_e(1-\alpha)[E_i]_0 = V_e(1-\alpha)[E_i]_{dr} + 100 I_{dr} S_{dr} M, \tag{6.69}$$

$$V_{fl}(1-\alpha')(E_iO_m)_0 = V_{fl}(1-\alpha')(E_iO_m)_{dr} - 100 I_{dr} S_{dr} M_{ox}, \tag{6.70}$$

- for the pool stage

$$\begin{aligned} &V_e(1-\alpha)[E_i]_{dr} + V_e\alpha[E_i]_0 + V_{bm}[E_i]_{bm} \\ &= (V_e + V_{bm})\kappa[E_i]_{po} + 100 I_{po} S_{po} M, \end{aligned} \tag{6.71}$$

$$V_{fl}(1-\alpha')(E_iO_m)_{dr} + V_{fl}\alpha'(E_iO_m)_{dr} = V_{fl}\kappa'(E_iO_m)_{po} - 100 I_{po} S_{po} M_{ox}, \tag{6.72}$$

where V_e, V_{bm}, and V_{fl} are the melting rates of the electrode, the base metal, and the flux (kg/s) and M_{ox} is the molecular weight of the oxide E_iO_m (kg/mole).

The calculation is performed in successive steps. The compositions of the metal and the slag in the droplet stage was first found from Eqs. 6.69 and 6.70, and then the compositions in the pool stage were found using expressions (6.71) and (6.72).

The reaction rates in the droplet and pool stages were found using the method described in Sect. 4.3.2.

We describe the interaction of the metal with the slag by the following set of reactions with a common reactant (FeO), through which their mutual influence is taken into account:

$$1/m[E_i] + (FeO) = [Fe] + 1/m(E_iO_m), \tag{6.11}$$

$$[C] + (FeO) = [Fe] + \{CO\}, \tag{6.12}$$

$$(FeO) = [Fe] + [O], \tag{6.13}$$

where E_i denotes silicon, manganese, titanium, chromium, aluminum, phosphorus, etc.

System of Eqs. 6.67–6.72 along with the expressions for the rates of all the reactions comprises a mathematical model of the chemical processes occurring during submerged-arc welding and surfacing. The input parameters of the model are the initial compositions of the base metal, the electrode metal, and the flux, the technological parameters, and the temperatures of the reaction zones, which are needed to calculate the physicochemical parameters. The output parameters are the compositions of the electrode droplets, the pool, and the slag crust and the rates of all the reactions in both stages of the process.

The concentrations of the elements which are scarcely oxidized during welding of the steels, namely, cobalt, copper, and nickel, were calculated according to the mixing rule

$$[E_i]_{mix} = (1 - -\gamma)[E_i]_e + \gamma[E_i]_{bm}, \tag{6.73}$$

where γ is the fraction of the base metal in the welded joint.

The temperatures of the reaction zones in the droplet and pool stages were taken from literature data [36]: $T_{dr} = 2{,}500$ K (for alternating current and direct current of reverse polarity), $T_{dr} = 2{,}300$ K (for direct current of forward polarity), and $T_{po} = 2{,}070$ K. We note that variation of the temperature in the ± 100 K range has little effect on the final results of the calculations if the other parameters remain unchanged.

To evaluate the faithfulness of the model developed, computational experiments were performed for 10 variants of submerged-arc welding and surfacing.

Table 6.2 presents the calculated and experimental data.

The calculation error for the alloying elements studied does not exceed 18% and is usually within the range of accuracy of the chemical analysis. The large discrepancies between the experimental and calculated data regarding the concentration of oxygen in the weld metal (up to 44%) are most probably attributable to the presence of nonmetallic inclusions and deoxidization products that do not manage to leave the volume of the weld pool [30], while the calculation gives the value of the concentration of oxygen dissolved in the pool according to reaction (6.13).

The practical use of the model is greatly simplified if dependences of the input parameters on the parameters of the welding regime are known. Statistical treatment of the data from the experiments yielded several dependences:

$$\gamma = 5.3 \cdot 10^{-2} I_w^{0.42} d_e^{-0.21} k, \tag{6.74}$$

$$S_{po} = 1.2 \cdot 10^{-5} I_w^{1.62} U_{arc}^{1.42} V_w^{-0.51} k, \tag{6.75}$$

$$S_{dr} = 8.85 \cdot 10^{-4} I_w^{1.3} d_e^{0.5}, \tag{6.76}$$

$$K_{fl} = 17.8 I_w^{-1.43} U_{arc}^{1.28} d_e^{0.72}, \tag{6.77}$$

$$K_{slag} = 8.1 I_w^{-0.5} V_w^{-0.68} d. \tag{6.78}$$

Here d_e is the diameter of the electrode wire, mm; U_{arc} is the arc voltage, V; V_w is the welding speed, m/h; k is a coefficient that takes into account the degree of alloying of the weld metal ($k = 1$ for low-carbon steels, $k = 1.1$ for steels with a total concentration of alloying additives ≥ 4 wt.%).

Table 6.2 Comparative evaluation of the calculated and experimental data for submerged-arc welding and surfacing

Base metal	Flux	Wire	Experiment No.	Concentration of elements in the weld metal, wt. % (the values above the bars are calculated, and the values below the bars are experimental)					
				C	Si	Ti	Mn	Cr	O
Grade 3 steel	AN–348A	Sv-25Kh10G10T	1	0.24	0.55	0.14	4.21	3.42	0.037
				0.23	0.63	0.14	4.30	3.35	0.065
			2	0.24	0.48	0.13	4.10	3.24	0.029
				0.24	0.71	0.15	4.40	3.20	0.060
	AN-20		3	0.24	0.62	0.17	3.70	3.40	0.030
				0.24	0.50	0.22	3.90	3.40	0.035
110G13L	AN-348A		4	0.66	1.07	0.24	11.13	6.01	0.036
				0.72	0.84	0.20	11.47	5.71	0.070
		Sv-08	5	0.68	0.66	0	7.22	0.20	0.042
				0.64	0.66	0	6.70	0.19	0.078
Grade 3 steel			6	0.10	0.32	0	1.01	0.05	0.06
				0.08	0.27	0	1.19	−−	0.15
		Sv-08GA	7	0.12	0.21	0	1.40	0.02	0.05
				0.10	0.20	0	1.20	0.02	−−
		Sv-08G2S	8	0.11	0.51	0	2.20	0.06	0.04
				0.10	0.48	0	1.90	0.05	−−
	AN-20		9	0.11	0.43	0	1.10	0.06	0.03
				0.10	0.40	0	1.10	0.06	−−
09G2S	AN-22	Sv-10NMA	10	0.07	0.59	0.026	1.19	0.06	0.046
				0.07	0.46	0.020	1.20	0.06	−−

Note Reverse polarity was used in the second experiment, and forward polarity was used in the remaining experiments

According to the Student and Fisher criteria, the confidence level of Eqs. 6.74–6.78 is 95–99%, and the regression coefficient is 99% for the welding of low- and medium-alloy steels under silicon- and manganese-containing fluxes in the following regime: $I_w = 200$ to 400 A, $U_{arc} = 20$ to 30 V, $V_w = 10$ to 50 m/h, $d_e = 2$ to 5 mm.

In view of the small volume of liquid metal on the electrode [30], the rapid stabilization of the droplet composition enables us to assume that the process in the droplet stage is steady-state.

In the pool stage, in which the process has a non-steady-state regime, the last term in Eq. 6.68, which characterizes the variation of the mass of the phase and the concentration of the element E_i in it, must be taken into account.

After dividing the process into sufficiently small time intervals of length $\Delta\tau$, we obtained the following material balance equations for the components of the metal and the slag:

$$\frac{dm_{E_i}}{d\tau} = V_e \frac{[E_i]_{dr}}{100} + V_{bm} \frac{[E_i]_{bm}}{100} - V_{cryst} \frac{[E_i]}{100} \kappa_{E_i} - I_{E_i} S M_{E_i}, \qquad (6.79)$$

$$\frac{dm_{\mathrm{E}_i\mathrm{O}_m}}{d\tau} = V_{\mathrm{fl}}^{\mathrm{dr}}\frac{(\mathrm{E}_i\mathrm{O}_m)_{\mathrm{dr}}}{100} + V_{\mathrm{fl}}\frac{(\mathrm{E}_i\mathrm{O}_m)_{\mathrm{fl}}}{100} - V_{\mathrm{solid}}\frac{(\mathrm{E}_i\mathrm{O}_m)}{100}\kappa_{\mathrm{E}_i\mathrm{O}_m} - I_{\mathrm{E}_i}SM_{\mathrm{E}_i\mathrm{O}_m}, \quad (6.80)$$

$$\frac{dm_{\mathrm{CaF}_2}}{d\tau} = V_{\mathrm{fl}}^{\mathrm{dr}}\frac{(\mathrm{CaF}_2)_{\mathrm{dr}}}{100} + V_{\mathrm{fl}}\frac{(\mathrm{CaF}_2)_{\mathrm{fl}}}{100} - V_{\mathrm{solid}}\frac{(\mathrm{CaF}_2)}{100}\kappa_{\mathrm{CaF}_2}. \quad (6.81)$$

Here $[\mathrm{E}_i]_{\mathrm{dr}}$, $[\mathrm{E}_i]_{\mathrm{bm}}$, and $[\mathrm{E}_i]$ are the concentrations of the element E_i in the droplet metal, the base metal, and the pool metal, wt.%; $(\mathrm{E}_i\mathrm{O}_m)_{\mathrm{dr}}$, $(\mathrm{E}_i\mathrm{O}_m)_{\mathrm{fl}}$, and $(\mathrm{E}_i\mathrm{O}_m)$ are the concentrations of the oxide of the element E_i in the slag that reacted with the metal in the droplet stage, in the original flux, and in the slag pool, wt.%; (CaF_2) is the concentration of calcium fluoride, wt.%; V_{e}, V_{bm}, and V_{cryst} are the melting rates of the electrode and base metals and the crystallization rate of the weld metal, kg/s; $V_{\mathrm{fl}}^{\mathrm{dr}}$, V_{fl}, and V_{solid} are the melting rates of the flux in the droplet and pool stages and the solidification rate of the slag crust, kg/s; $\kappa_{\mathrm{E}_i}$is the liquation coefficient of the element E_i between the solid and liquid phases. The concentrations of CaO and MgO in the slag pool are determined from equations that are analogous to expression (6.81).

The mass of a phase, for example, the slag phase, at the time τ equals

$$m_{\mathrm{slag}}^{\tau} = \sum_i m_{\mathrm{E}_i\mathrm{O}_m}^{\tau} + m_{\mathrm{CaF}_2}^{\tau} + m_{\mathrm{CaO}}^{\tau} + m_{\mathrm{MgO}}^{\tau}, \quad (6.82)$$

and the concentration (wt.%) of each component, for example, calcium fluoride, is

$$(\mathrm{CaF}_2)^{\tau} = \frac{m_{\mathrm{CaF}_2}^{\tau}}{m_{\mathrm{slag}}^{\tau}}100. \quad (6.83)$$

The model constructed enables us to predict the composition of the deposited metal, including its variation along the length of the deposited layer in response to various perturbing factors: switching to a base metal of a different grade in the welded joint, intersection of previously formed welded joints, fluctuations of the technological parameters of the process, etc.

A series of calculations was performed to verify that the non-steady-state nature of the process is taken into account.

The results of the calculations and experimental data related to the variation of the composition of the weld pool at the beginning of the process are presented in Fig. 6.5. The surfacing was carried out on 12Kh18N10T steel by a three-phase arc with Sv-08A wire under AN-26 flux.

The intersection of a previously formed welded joint was simulated in the following manner: an insert of 09G2S steel was placed in a plate of 08Kh22N6T steel, and three-phase arc surfacing by Sv-08A wire under AN-26 flux was carried out in a direction perpendicular to the insert. Figure 6.6 presents the calculated and experimental data on the distribution of nickel, titanium, and silicon along the length of the deposited layer in the plane of a longitudinal central section at a

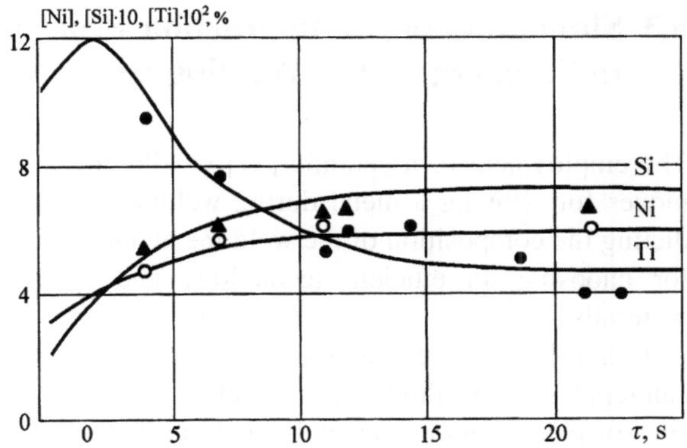

Fig. 6.5 Variation of the concentration of nickel (○), silicon (▲), and titanium (●) in the weld pool at the beginning of the surfacing process: *curves* calculation; *points* experiment

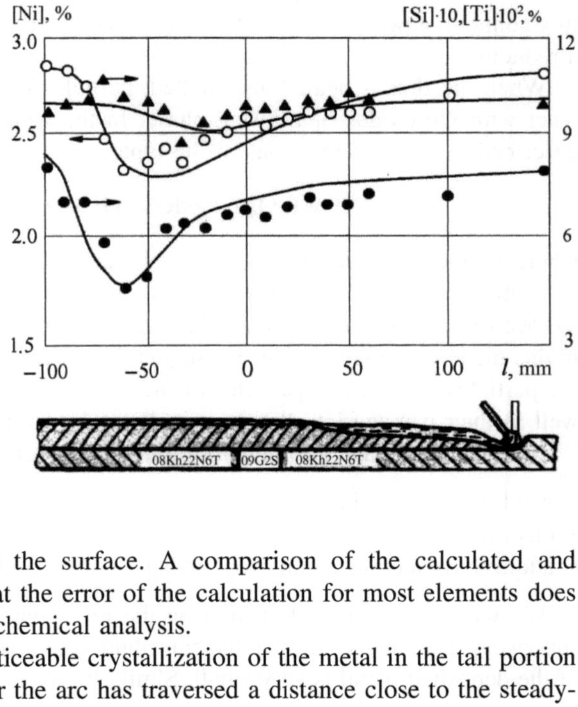

Fig. 6.6 Distribution of nickel (○), silicon (▲), and titanium (●) during the surfacing of an insert on a composite sample (the *curves* show calculated results, and the *points* indicate experimental results) and diagram of the arrangement of the plates

depth of 0.8–1.0 mm from the surface. A comparison of the calculated and experimental data shows that the error of the calculation for most elements does not exceed the error of the chemical analysis.

It was also found that noticeable crystallization of the metal in the tail portion of the pool begins only after the arc has traversed a distance close to the steady-state length of the weld pool.

The results obtained confirm the promising nature of the use of kinetic analysis of the reactions between the metal and the slag during submerged-arc welding. The model developed is applicable to single- and multiple-electrode welding (surfacing) processes with a wire or ribbon and can be used to select welding materials and to determine numerous regime parameters for developing welding and surfacing technologies, especially in cases in which welded joints of variable composition must be obtained.

6.3 Simulation of the Interaction Processes in Submerged-Arc Welding with a Ceramic Flux

The employment of a ceramic [37] or alloying flux is one of the effective techniques for alloying a metal during welding or surfacing. The possibility of predicting the composition of the weld metal and the ability to control it are the basis for choosing an efficient technology and the composition of the surfacing materials.

Using the cybernetic approach that is well-known in chemical engineering, we can represent a welding zone as a set of continuous reactors, whose boundaries are sites of phase interactions and which exchange matter with one another [14].

Figure 6.7 presents a block diagram of the phase interaction during submerged-arc welding with an alloying or ceramic flux.

Let us examine of the scheme of the phase interaction during surfacing under a flux containing not only a metal additive, but also oxides of alloying elements and a reducing agent.

When the flux is heated and melted, particles of the reducing agent (carbon) react with the oxides of the slag phase. In the general case this process can be described by a set of reactions of the type

$$y(\mathrm{E}_i\mathrm{O}_m) + m\mathrm{R}_j = y\mathrm{E}_i + m\left(\mathrm{R}_j\mathrm{O}_y\right), \tag{6.84}$$

where R_j and $\mathrm{R}_j\mathrm{O}_y$ are the reducing agent and its oxide.

A melt of the metals whose oxides are contained in the slag forms on the surface of the reducing agent. At the same time, dissolution of the reducing agent in this metal phase is possible. Subsequently, processes involving an interaction of the particles formed and particles of the metal additive with slag components, as well as their passage into the deposited metal, take place.

In the general case the following interaction surfaces can be identified:

- weld-pool-metal/slag;
- electrode-metal-droplet/slag;
- additive (reducing)-metal-droplet/slag.

We will neglect the interaction on the metal/gas boundary, since it has been shown experimentally [33] that the contribution of this stage to the alloying process of the deposited metal is very small. Similarly, we will disregard the losses of the alloying elements due to vaporization [38]. Alloying of the weld metal occurs mainly as a result of the assimilation of particles of the metal additive by the weld pool. These particles are conveyed to the pool surface by convective flows.

To develop a mathematical model based on the physical picture of the process considered, we make the following assumptions.

- The particles of the reducing agent, whose role was played by carbon, or the particles of the metal additive have identical dimensions and are uniformly distributed throughout the volume of the weld pool.

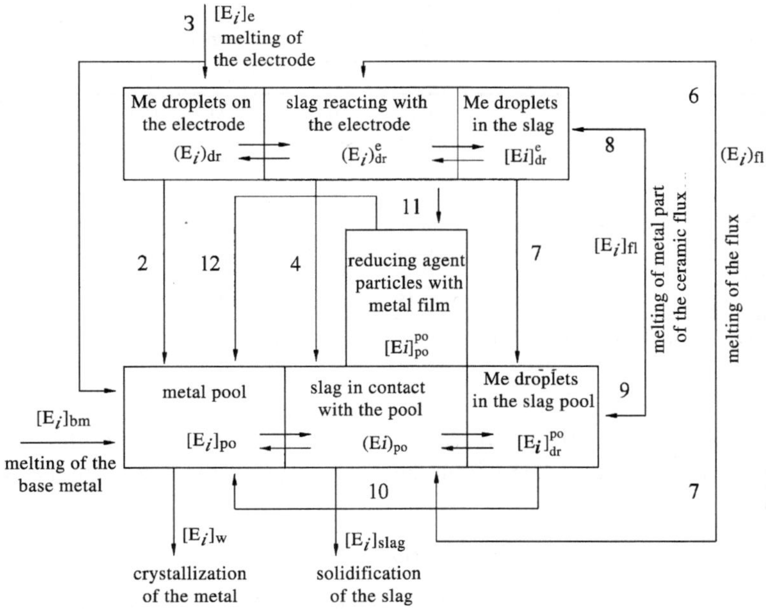

Fig. 6.7 Block diagram of the phase interaction during submerged-arc welding with a ceramic or alloying flux: *1* transport of electrode droplets without contact with the slag into the pool; *2* transport of electrode droplets after interaction with the slag; *3* formation of metal droplets on the electrode and of metal droplets that interact with the slag; *4* passage of the slag that has reacted with the electrode metal into the slag pool; *5* passage of metal droplets from the slag that has reacted with the electrode into the slag pool; *6, 7* melting of the flux in the electrode and pool zones, respectively; *8, 9* formation of metal droplets in the slag from the ceramic flux in the electrode and pool zones, respectively; *10* settling of metal droplets from the slag pool into the metal pool; *11* passage of reducing-agent particles surrounded by a metal film from the electrode zone into the slag pool; *12* settling of these particles into the metal pool

- Initially, the reduction of iron with the formation of a metal phase on the surface of a solid particle occurs in the slag according to the reaction

$$C_s + FeO = Fe + CO. \tag{6.85}$$

- The dissolution of carbon in iron occurs until the saturation concentration is reached. Taking into account that as the concentration of carbon in iron increases, the contact angle between the Fe–C melt and the solid carbon decreases sharply and the spreading rate, conversely, increases [39], we presumed that the particles formed are solid spheres of carbon surrounded by a film of the Fe–C_{sat} melt of uniform thickness.
- The rate of entry of particles of the additive and droplets of the electrode metal into the weld pool is determined by their sedimentation rate in the slag, which depends on the size of the particles, the parameters of the welding regime, and the hydrodynamic conditions in the welding zone.

Table 6.3 Composition of the metal deposited under boron-containing ceramic fluxes

Slag base	Boron content,wt.%	Concentration of elements, wt.%					
		C	B	Mn	Ti	Si	Cr
Flux AN-28	1.0	0.25	0.35	9.71	0.38	0.71	7.55
		0.28	0.30	9.98	--	0.89	8.35
	2.0	0.25	1.00	9.95	0.41	0.81	7.56
		0.25	0.96	10.43	0.21	1.03	7.90
Flux AN-20	2.0	0.25	0.62	9.28	0.36	1.00	7.20
		0.17	0.68	9.00	0.16	1.41	7.98
Flux AN-26	2.0	0.24	0.48	8.78	0.32	1.14	7.17
		0.18	0.45	8.34	--	1.53	7.58

The values above the bars are calculated, and the values below the bars are experimental

- The interaction with the gas phase was taken into account only for the oxidation of particles of solid carbon within the composition of the flux when the latter is heated:

$$C_s + {}^1\!/_2\{O_2\} = \{CO\}. \tag{6.86}$$

- The interaction of the metal and the slag was described by a set of reactions with a common reactant that occur on all the interfaces of the type

$$1/m[E_i] + (FeO) = 1/m(E_iO_m) + [Fe]. \tag{6.11}$$

- The metal and slag pools are ideally mixed reactors.

To perform the calculation, we divide the surfacing process into short time intervals of length $\Delta\tau$, during which the rates of all the reactions of type (6.11) may be assumed constant. During this time, N_i droplets of the electrode metal pass into the weld pool. We will also assume that this time is sufficient for particles of the additive (reducing) metal that are at a level close to the weld pool to settle into it. The calculation is performed in successive steps. The reaction rates for the initial compositions of the phases are first determined, and then the compositions of the phases at the beginning of the next calculation step are found with allowance for all the processes that lead to changes in the compositions of the metal and the slag. After this, the calculation cycle is repeated. Such a scheme enables us to calculate the composition of the metal and the slag not only when the welding or surfacing process has reach a steady state, but also in its initial and final stages, when the metal and slag pools form or they decrease in size as a result of crystallization, respectively.

In order to test and adapt the model, calculations were performed for available experimental variants of developed ceramic fluxes with the use of ferroboron as the additive [40] and for alloying fluxes with the use of boron-containing materials and reducing agents as necessary components [41].

The results of the calculations and a comparison of the results with experimental data [31, 42] (Tables 6.3, 6.4) demonstrate the possibility of predicting the

Table 6.4 Chemical composition of the deposited metal

Wire	Flux[a], wt.%	Concentration of elements, wt.%					
		C	B	Cr	Ni	Si	Mn
Sv-0Kh13N9 M	95 SHB + 5 C	0.72	0.74	11.70	7.90	–	–
		0.70	0.80	12.10	8.30		
	90 SHB + 10 C	0.86	0.90	11.80	8.00	–	–
		0.82	0.96	12.20	8.20		
	80 SHB + 20 C	0.92	0.94	11.90	8.05	–	–
		1.05	1.12	12.30	8.30		
Np–25Kh10G10T	90 SHB + 10 C	0.84	0.95	8.30	–	0.74	8.20
		0.90	1.00	8.70		0.80	8.70
Sv-08A	40 CRO, 40 SHB, 10 C, 10 Al	1.26	1.07	6.50	–	–	–
		0.90	1.00	8.70			
	30 CRO, 30 SHB, 20 C, 20 Al	1.20	1.05	7.12	–	–	–
		1.50	1.28	6.86			

The values above the bars are calculated, and the values below the bars are experimental
[a] *SHB* sodium hexaborate, *CRO* chromium ore

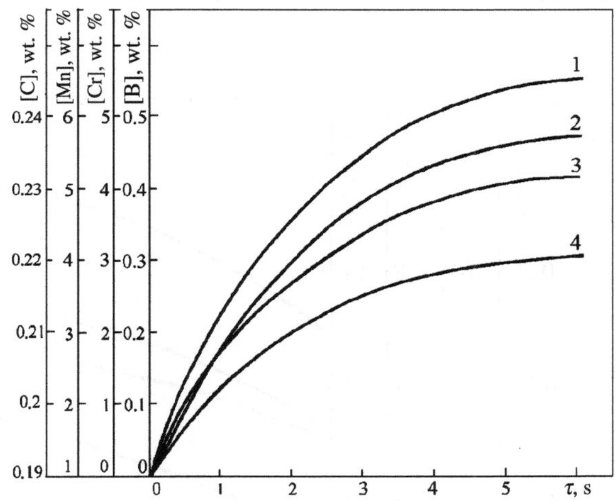

Fig. 6.8 Variation of the concentrations of the elements in the first layer of the deposited metal with time: *1* boron; *2* manganese; *3* chromium; *4* carbon

composition of the phases during submerged-arc welding or surfacing with a ceramic flux.

Figures 6.8, 6.9, 6.10 present the results of calculations of the variation of the concentration of the alloying elements with time in the initial (transitional) period of the surfacing process. The composition of the first and second layers of the deposited metal is stabilized within a period of 4–5 s after the beginning of the process, which corresponds to ~2 lengths of the weld pool. The time for the third layer is ~3 s. In the subsequent layers the concentrations of the elements in the metal remain practically constant. These data were also confirmed experimentally [42]. A similar type of variation of the composition of a metal deposited under a

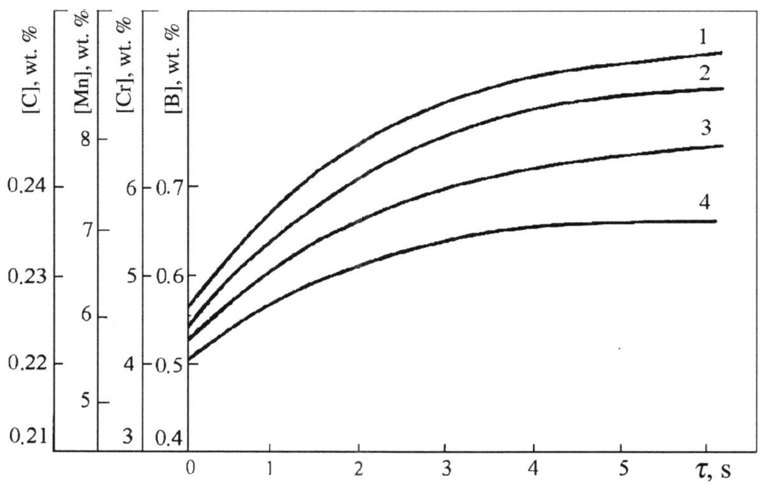

Fig. 6.9 Variation of the concentrations of the elements in the second layer of the deposited metal with time: *1* boron; *2* manganese; *3* chromium; *4* carbon

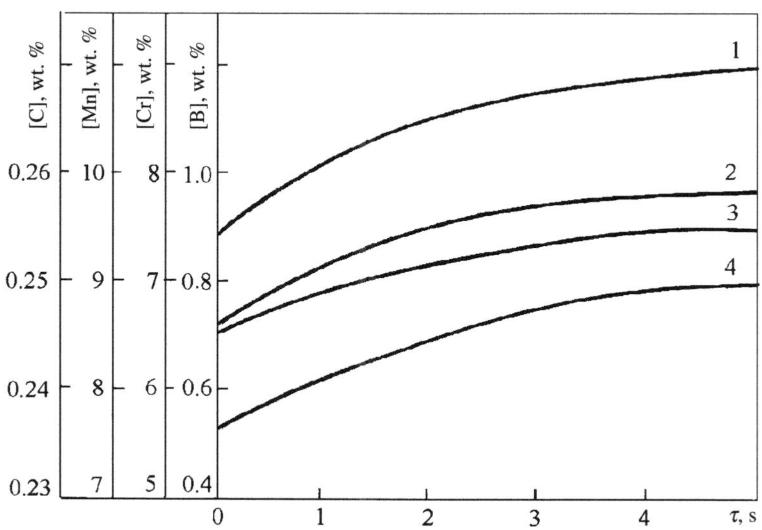

Fig. 6.10 Variation of the concentrations of the elements in the third layer of the deposited metal with time: *1* boron; *2* manganese; *3* chromium; *4* carbon

flux with time should be taken into account in the case of the surfacing of critical parts, for which stricter requirements are placed on the stability of the composition.

6.4 Mathematical Model of the Kinetics of the Chemical Processes Occurring in Gas Welding

Arc welding in oxygen-containing shielding gases (hereinafter, gas welding) is one of the most common types of fusion welding. It is accompanied by several physicochemical processes that determine the composition and properties of the weld metal. Simulation of these processes based on calculation methods makes it possible to predict the chemical composition of the weld metal, on which the mechanical and other properties of the welded joint depend.

The creation of theoretically rigorous prediction methods based on the fundamental principles of thermodynamics and chemical kinetics is very attractive.

A thermodynamic model of the interaction processes occurring in gas welding was first developed in [43]. Comparison of the experimental and calculated data obtained using the thermodynamic model showed that equilibrium is not achieved for reactions involving elements with a high oxygen affinity (C, Al, and Ti), since the calculated values of their concentrations are much lower than the experimental. Therefore, it was concluded that the kinetic hindrances to the chemical reactions that occur in gas welding must be taken into account.

For this purpose, a method that describes the kinetics of the phase interaction and provides a way to calculate the chemical compositions of the metal and the slag formed as a result of oxidation of the alloying elements was developed.

Separation of the interaction of the metal with the gas and with the slag in the welding zone into two successive stages, namely, an electrode-droplet stage and a pool stage, is generally accepted. These stages differ strongly with respect to the temperature, convection conditions, and interaction surface [44]. In the opinion of several authors [45], the interaction of the superheated metal of the electrode droplets in the head portion of the pool at the base of the arc (the "hot" zone of the pool) should be assigned to the first stage.

It is very difficult to experimentally separate the roles of the individual stages in the overall interaction. It particular, the mechanism of the passage of oxygen from the arc atmosphere into the metal is unclear. It may occur by dissolution in the metal followed by interaction with elements in the bulk or on the metal/slag interface or by the formation of FeO or oxides. The role of the vaporization of the metal and the formation of the volatile oxides SiO, AlO, etc. is unknown. For this reason, several variants of the physical model of gas welding processes were considered. Allowance for the kinetic restrictions in several probable phase interaction schemes enabled us to reveal the most probable mechanism of the processes and to predict the concentrations of most alloying elements in the weld metal with sufficient accuracy.

Variant 1. This variant is similar to the model for submerged-arc welding [44], in which the droplet and pool stages are treated as ideally mixed reactors operating in a steady-state regime. The quantity of the flux needed for the calculations was estimated on the basis of the assumption that all the oxygen absorbed by the metal forms ferrous oxide (FeO), which then reacts with the metal.

Variant 2. This variant corresponds to the two-stage scheme, but neglects the processes occurring in the head portion of the pool. The continuous absorption of oxygen on the free surface of the droplet and the interaction of the metal with the slag were taken into account in the droplet stage as the droplet grows and in the pool.

Variant 3. In addition to variant 2, the interaction in the head portion of the pool, where the droplets entering the pool crater form a certain superheated volume, which is continuously renewed due to the entry of droplets and the departure of metal as a result of mixing with the cooler melt in the tail portion of the pool, was taken into account.

Variant 4. In addition to variant 3, the deoxidization in the bulk of the droplet metal and in the head portion of the pool until an equilibrium state is achieved was taken into account. The deoxidization products formed rise to the surface and are mixed with the remaining slag.

Variant 5. In addition to variant 4, the deoxidization in the "cold" tail portion of the pool was also taken into account.

In all the variants the same temperature, geometric, and hydrodynamic parameters were taken when the calculations were performed.

The interaction of the metal with the slag was described by a set of reactions of types (6.11–6.13).

The mathematical models have the form of a system of material balance equations for the phases and equations for the rates of the reactions. For example, for the metal phase the concentrations of the alloying elements and oxygen in any time interval were found using the equations

$$[\mathrm{E}_i]^{k+1} = [\mathrm{E}_i]^k + \left(V_{\mathrm{in}}^k[\mathrm{E}_i]_0 - V_{\mathrm{out}}^k[\mathrm{E}_i]^k + 100 I_{\mathrm{E}_i}^k S_{\mathrm{Me-slag}}^k M_{\mathrm{E}_i}\right)\frac{\Delta\tau}{m_{\mathrm{Me}}^k}, \tag{6.87}$$

$$[\mathrm{O}]^{k+1} = [\mathrm{O}]^k + \left(V_{\mathrm{in}}^k[\mathrm{O}]_0 - V_{\mathrm{out}}^k[\mathrm{O}]^k + 100 M_{\mathrm{O}}\left(I_{\mathrm{O}}^k S_{\mathrm{Me-slag}}^k + V_{\mathrm{O}}^k S_{\mathrm{Me-g}}^k\right)\right)\frac{\Delta\tau}{m_{\mathrm{Me}}^k}. \tag{6.88}$$

Here V_{in} and V_{out} are the incoming and outgoing mass flow rates, kg/s; $I_{\mathrm{E}_i}^k$ and I_{O}^k are the rates of reactions (6.11–6.13), mole/(m^2 s); V_{O}^k is the rate of absorption of oxygen from the gas phase, mole/(m^2 s); $S_{\mathrm{Me-slag}}^k$ and $S_{\mathrm{Me-g}}^k$ are the areas of the interaction surfaces of the metal with the slag and the gas, m^2; m_{Me}^k is the mass of the metal, kg; M_{E_i} and M_{O} are the atomic weights of the elements and oxygen, kg/mole.

The absorption of oxygen by the metal was taken into account in all the variants through the relative mass of the absorbed oxygen φ, whose experimental values were presented in [44].

Table 6.5 Composition of the metal during CO_2 welding

Calculation variant	In the droplet stage/in the head portion of the pool						In the pool stage					
	C	Mn	Si	Ti	Cr	O	C	Mn	Si	Ti	Cr	O
Thermodynamic model	1×10^{-4}	0.37	0.32	0.22	18.2	0.070	1×10^{-4}	0.40	0.30	0.04	18.7	0.006
1	0.086	0.45	0.48	0.59	17.9	0.010	0.070	0.33	0.63	0.31	16.0	0.013
2	0.088	0.46	0.47	0.55	18.8	0.160	0.070	0.40	0.57	0.35	18.3	0.125
3	0.086/0.049	0.46/0.42	0.47/0.43	0.54/0.41	18.6/18.6	0.150/0.650	0.066	0.46	0.55	0.34	18.7	0.340
4	0.086/0.049	0.45/0.34	0.45/0.32	0.49/0.20	18.6/18.1	0.066/0.066	0.066	0.41	0.49	0.25	18.7	0.078
5	0.086/0.049	0.45/0.34	0.45/0.32	0.49/0.20	18.6/18.1	0.066/0.066	0.066	0.42	0.48	0.10	18.5	0.008
Experiment [2]	0.09	0.28	0.43	0.16	17.2	–	0.1	0.4	0.45	0.24	17.7	–

The input parameters of the model are the initial compositions of the base metal and the electrode metal, the values of φ, the fractions of participation of the base metal in the weld metal, and the equilibrium constants of the reactions in each stage. The output parameters are the compositions and relative changes in the mass of the metal and the slag in each stage.

Table 6.5 presents the results of the calculations and experiments [44] for the CO_2 welding of 12Kh18N9T steel by Sv-08Kh18N9T wire. The welding parameters were: current, 256 A; voltage, 29 V; wire diameter, 2 mm; fraction of the participation of the base metal in the welded joint, 0.43.

It follows from the table that the thermodynamic model incorrectly predicts low concentrations of carbon and titanium in the pool. The model of submerged-arc welding predicts only little oxidation of the principal alloying elements. The same result was also obtained in the calculation based on variant 2. Only the oxygen content in the metal rose significantly. This indicates relatively weak development of the processes on the metal/slag interface and their vigorous occurrence on the metal/gas interface, where the absorption of oxygen occurs. Variant 3, which takes into account the processes in the head portion of the pool is distinguished from variant 2 practically solely by the large area of the metal/gas interface. This ensured greater inflow of oxygen into the metal and of FeO into the slag and an increase in the loss of the alloying elements. However, the concentration of oxygen increased to the unlikely value of 0.65 wt.%. Allowance for the deoxidization reactions in the bulk of the droplets falling into the head portion of the pool (variant 4) significantly lowered the oxygen content, and the compositions of the metal and the slag approximated the experimental values. Additional allowance for deoxidization of the metal in the tail portion of the pool (variant 5) resulted in excessive oxidation of titanium, although the concentration of oxygen dropped to more realistic values. Apparently, the deoxidization reactions do no manage to reach equilibrium completely in the cooler metal.

Thus, the most probable physical model of the chemical processes occurring in gas welding includes the step of deoxidization of the metal in the head portion of the pool. We stress that all the results were obtained without calling upon empirical correction factors. The parameters of the model have explicit physical meanings, and refinement of their values will enable us to raise the accuracy of predictions of the composition of the phases in gas welding.

6.5 Calculation of the Composition of the Metal and the Slag During the Melting of Coated Electrodes and Flux-Cored Wires

During welding by coated (stick) electrodes and flux-cored wires, metal and slag droplets form on the tip of the electrode (wire). These droplets determine the composition of the deposited metal and the welding-technological properties of the materials: the shape of the weld surface, the detachability of the slag crust, the possibility of welding in different spatial positions, etc. The droplet composition is shaped during the melting of the sheath of a flux-cored wire or the powdered components of the flux mixture (core or coating of the electrode) and the interaction of the phases formed with one another and with the gas.

The calculation method was based on the complete material balance for all elements in each of the phases. The influence of the chemical reactions was taken into account using the following transfer coefficients:

$K_1^{E_i}$—the fraction of the mass of the component E_i in the metallic part of the flux mixture that is oxidized by the gas;

$K_2^{E_i}$—the fraction of the mass of the component E_i in the sheath (core wire) metal that is oxidized by the gas;

$K_3^{E_i}$—the fraction of the mass of the component E_iO_m in the nonmetallic part of the flux mixture (coating) that passes into the metal as a result of reduction reactions in the droplet stage.

The expression for calculating the mass of the oxide component (E_iO_m) for welding with coated electrodes is as follows:

$$m_{E_iO_m} = \frac{m_e}{1 + K_{m_{coat}}} \left\{ \frac{[E_i]_{core}}{100} \frac{M_{E_iO_m}}{M_{E_i}} K_2^{E_i} \cdot \right.$$
$$+ \frac{K_{m_{coat}}}{1 + \frac{\alpha\beta}{100}} \left(\sum_{k=1}^{k} \frac{\%coat_k}{100} \frac{[E_i]_k}{100} \frac{M_{E_iO_m}}{M_{E_i}} K_1^{E_i} \right.$$
$$\left. \left. + \sum_{j=1}^{j} \frac{\%coat_j}{100} \frac{(E_iO_m)_j}{100} \left(1 - K_3^{E_i}\right) + \frac{\alpha\beta}{100} \frac{(E_iO_m)_b}{100} \right) \right\}, \tag{6.89}$$

where m_e is the mass of the electrode (kg), j and k are the indices of the nonmetallic and metallic components of the coating, $[E_i]_k$ and $[E_i]_{core}$ are the concentrations of the element E_i in the kth component of the metallic part of the coating and in the core wire of the electrode, respectively (wt.%), $(E_iO_m)_j$ and $(E_iO_m)_b$ are the concentrations of the oxide of the element E_i in the jth component of the nonmetallic part of the coating and in the binder, respectively (wt.%), α and Me_k are the concentrations of the binder and the kth component of the metallic part of the coating, respectively (wt.%), β is the mass fraction of the dry residue of the binder, M_{E_i} and $M_{E_iO_m}$ are the molecular weights of the element E_i and its oxide (kg/mole), and K_{mcoat} is the mass ratio of the coating

The value of β is calculated from the formula

$$\beta = 0.59\rho + 0.028m - 0.535, \tag{6.90}$$

where ρ is the density of the binder (water glass) and m is the modulus of the binder.

The summed terms in expression (6.89) represent the mass of the oxide entering the slag from the mineral part of the coating, from the binder, and, when the component E_i is oxidized, from the metallic part of the coating and the core wire.

We specify the ratio between the mass of the coating and the mass of the electrode core using the mass ratio of the coating (K_{mcoat}):

$$K_{m\,\text{coat}} = \frac{m_{\text{coat}}}{m_{\text{core}}} = \frac{m_{\text{al}} + m_{\text{slag}} + m_{\text{dry}}}{m_{\text{core}}}, \tag{6.91}$$

where m_{coat}, m_{core}, m_{al}, m_{slag}, and m_{dry} are the masses of the coating and core wire of the electrode, the alloying components of the coating, the slag-forming components of the coating, and the dry residue of the binder (water glass), respectively (kg).

The mass of the slag and the concentration of the element E_i are determined by the following trivial equations:

$$\begin{aligned} m_{\text{slag}} &= \sum m_{E_iO_m}, \\ (E_iO_m) &= \frac{m_{E_iO_m}}{m_{\text{slag}}} 100 = \frac{m_{E_iO_m}}{\sum m_{E_iO_m}} 100. \end{aligned} \tag{6.92}$$

The composition of the metal is specified by structurally similar equations.

The calculations enable us to determine the initial compositions of the metal and the slag upon melting. Using the equations of the mathematical model for submerged-arc welding, we can estimate the composition of the weld metal and the slag crust. By assigning the composition of the weld metal required by the service conditions, we can essentially solve the inverse problem, that is, select the required welding material (flux-cored wire or stick electrode).

In addition, the calculation program provides a relatively simple way to find the transfer coefficients of the elements from the experimental data.

Table 6.6 Composition of the metal and the slag

Metal components	C	Si	Mn	Cr	Ni	S	P	Fe
Calculation	0.10	0.03	0.54	0.12	0.25	0.03	0.03	98.90
Experiment	0.08	0.05	0.50	0.10	0.25	0.03	0.03	98.90
Slag components	TiO_2	SiO_2	MnO	CaO	MgO	Al_2O_3	Na_2O	FeO
Calculation	46.34	17.24	16.50	9.71	3.63	1.62	1.70	1.65
Experiment	40.60	22.42	14.75	10.71	2.15	–	–	–

When a flux-cored wire was designed, the following technological parameters for its production were also used: the diameter of the wire and strand, the width and thickness of the ribbon, the tapering coefficient of the ribbon, the density of the flux mixture, the packing factor of the flux mixture, and the ratio of slag formers.

To evaluate the faithfulness of the method developed, we performed calculations that simulate the melting of MR-3 electrodes. The following starting data were used:

- coating composition, wt.%: rutile, 50.40; marble, 17.35; talc, 11.13; mica, 4.64; cellulose, 1.50; ferromanganese, 14.98;
- binder—soda glass, 25% of the coating mass;
- water content in the binder—60 wt.%;
- mass ratio of the coating—0.4.

Table 6.6 presents the experimental and calculated data.

6.6 Mathematical Model of the Motion, Heating and Oxidation Processes Occurring in Electric-Arc Spraying

The electric-arc spraying (EAS) process has found application in the restoration of worn parts and the deposition of layers with assigned service properties owing to such properties as the atomization of metal that is guaranteed to be molten and the high velocities of the transporting gas and particles.

EAS is 3–10 times less costly than other coating methods, but several features of the process have prevented its widespread use for practical purposes. The difficulties in forming a concentrated flow of the material being atomized result in lowering of the utilization ratio of the material and the quality of the coating. In addition, the quality of coatings produced by EAS depends on the amounts of both the oxygen dissolved in the metal particles and the oxygen that is present in the form of oxides.

In EAS the coating is formed from droplets of liquid metal moving in a stream of a transporting gas. The metal being atomized is heated and melted by the heat of an electric arc burning between consumable wires, that is, the electrodes from

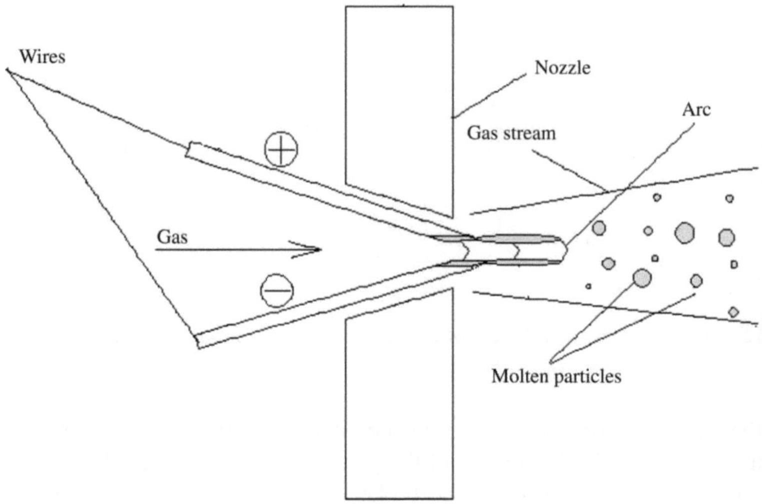

Fig. 6.11 Diagram of the EAS process

which the material being atomized is formed (Fig. 6.11). The liquid metal is swept off the electrode tips and broken up under the action of gas-dynamic and electromagnetic forces, and then it travels in the form of droplets in the direction in which these forces act.

In order to optimize the technology and improve the equipment, a model of the processes that occur in EAS was developed. A version of the process that is called high-velocity arc spraying (HVAS), in which the metal is atomized by the combustion products of a propane–air mixture, was considered as the object of application.

To construct the physical and mathematical models, we divided the process into three successive stages, which differ with respect to features of the temperature and gas-dynamic regimes, as well as the interaction of the metal with components of the gas phase.

Stage 1—Metal on the Electrode Tip

The metal is continuously fed into the interaction zone and is carried away from it in droplets. The concentration of the reactant (oxygen) in the gas phase is constant. The temperature of the metal on the tip is of the order of 2,400–2,700 K, according to the data in [36]. In the calculations we assume that $T = 2{,}400$ K. According to the available estimates [30, 46], in this range the mass-transfer coefficient of oxygen in the metal is an order of magnitude higher than in ordinary metallurgical practice with contact between the clean surface of the liquid metal and air.

Stage 2—Metal in the Arcing Zone

On the basis of an analysis of oscillographic data [47, 48], it may be stated for steel that droplets with a midsection diameter equal to 550–900 μm are detached

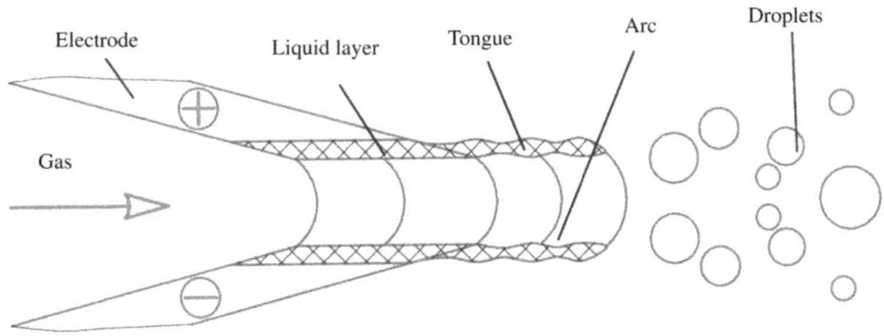

Fig. 6.12 Diagram of the detachment of droplets from the electrode tip during EAS

from the electrode tips with a period equal to 5×10^{-4} to 11×10^{-4} s. Our calculation of the critical diameter of the particles shows that metal droplets in the range of diameters indicated can be broken up near the nozzle exit section into droplets with a minimum diameter equal to 165 μm. As they move further along the spraying distance, the stability of these droplets increases. This is at variance with both our experimental data and the literature data [49], according to which the percentage of droplets with a diameter less than the critical value is 20–80%, although the surface tension forces should protect droplets with diameters in this range from fragmentation by the gas stream. In particular, the data from a sieve analysis of samples collected during spraying into water show that the diameter of the steel droplets lies mainly in the range from 20 to 150 μm.

Elimination of this contradiction is possible if it is assumed that detachment of a droplet with $d = 690$ μm from the electrode tip due to the dynamic head of the gas stream and its fragmentation into smaller droplets occur simultaneously in the arc. The following scheme can be proposed for this process (Fig. 6.12).

Under the action of the dynamic head of the gas, metal is carried away from the electrode tip and stretched out into the stream in the form of a "tongue." However, due to the surface tension forces, it is not yet detached from the tip. An electric current passes through the metal droplet and the electric arc, which is also stretched out under the action of the dynamic head of the gas. According to the classical scheme, fragmentation of the stream of liquid metal by the gas flow occurs at the constrictions formed under the effect of the difference between the surface tension forces at sites of local perturbations [50].

The presence of a flowing current supplements the constriction mechanism with another, more powerful mechanism involving disintegration of the molten metal bridges, which was considered in [51]. The evolution of Joule heat is more intense in the constrictions than in other parts of a conductor due to the increased ohmic resistance. Therefore, the metal and gas near molten metal bridges are heated to higher temperatures. Further disintegration of the stream develops due to the combined processes of ebullition of the liquid metal and shock expansion of the gas at the constrictions. The length of this zone is comparable to the length of the electrode tip according to data from high-speed motion-picture photography [47].

The time during which the metal is found in this zone corresponds to the frequency at which droplets are detached from the electrode tip. This time is of the same order as the time of flight of a droplet over the spraying distance. The specific surface of the droplets increases significantly compared with the electrode tip, and a further increase in temperature occurs due to the additional evolution of heat in the constrictions. The value of the mass-transfer coefficient of oxygen in the metal was assumed to be the same as in the first stage.

In addition, electromagnetic forces appear in the arcing zone due to the interaction of the flowing current and the induced magnetic fluxes. We note some special features of EAS compared with arc welding, for which investigations of the liquid-metal–electric-arc interactions have been conducted: the velocities of the gas flow is 50–100 times higher in EAS, and the inclined relative orientations of the electrodes causes an unusual interaction between the electromagnetic fields.

The calculation showed that when a transverse gas flow acts on the arc, the column maintains a cylindrical shape and the plasma flows are attached to reference spots and are carried away under the combined action of Ampère's force law and the dynamic head of the gas.

Let us estimate the magnitude and direction of the forces acting on an individual droplet in the "tongue" when the droplet is detached from the electrode tip. If it is assumed that the relative pressures of the metal vapors from the two tips are equilibrated, the following forces act on the droplet:

- the force of the aerodynamic drag;
- the force due to the pressure of the shock wave from the explosive expansion of the metal vapor and the gas at the molten metal bridge;
- Ampère forces: F_1 (the interaction of the magnetic field of the electrode gap and the current in the molten metal bridge), F_2 (the pinch effect), F_3 (the interaction of the magnetic field of the arc column and the current in the molten metal bridge).

The direction of flow of the currents and the arrangement of the magnetic fields and the Ampère forces are presented in Fig. 6.13. We will calculate these forces when the combustion products of a propane–air mixture serves as the transporting gas [52].

The force of aerodynamic drag (F_{ad}) is given by the expression

$$F_{\mathrm{ad}} = \pi \cdot C_d \cdot \left(V_{\mathrm{g}} - V_{\mathrm{p}}\right)^2 \cdot \rho_{\mathrm{g}} \cdot d_{\mathrm{dr}}^2 \big/ 8, \tag{6.93}$$

where C_d is the aerodynamic drag coefficient, V_{g} and V_{p} are the velocities of the gas and the particles (m/s), ρ_{g} is the density of the gas (kg/m^3), and d_{dr} is the diameter of the droplets detached (m).

The magnitude of the force F_{f} from the shock wave on the surface of a constriction of radius r_{con} and length Δh is given by the expression

$$F_{\mathrm{f}} = P_{\mathrm{f}} \cdot 2\pi \cdot r_{\mathrm{con}} \cdot \Delta h, \tag{6.94}$$

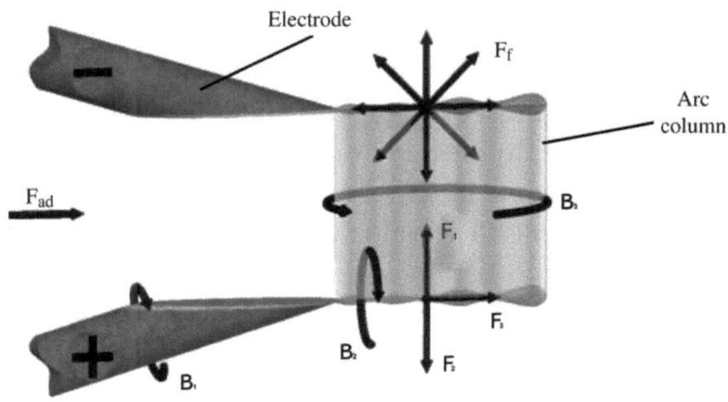

Fig. 6.13 Diagram of the forces in the arcing zone

where P_f is the pressure on the shock-wave front (MPa).

According to our proposed mechanism of "tongue" fragmentation, the current flowing in the constrictions sharply intensifies the evolution of heat in them due to the increased ohmic resistance. This initiates fragmentation of the tongue at the constrictions. Micro-arcs appear at the sites of fragmentation, resulting in explosive expansion of the metal vapors and gas. When the gas expands, a sharp increase in the volume of the heated gas and propagation of a shock wave accompanied by a release of energy occur. The pressure on the shock wave front can be expressed by the dependence [53]

$$P_f = 2 \cdot \rho_f^0 \cdot V_f^2 \Big/ (\gamma + 1) \tag{6.95}$$

where ρ_g^0 is the initial density of the gas (kg/m^3), γ is the adiabatic constant, V_f is the velocity of the front of the expanding gas (m/s).

The velocity (V_f) can be estimated from the equality between the forces of the dynamic head of the gas and the capillary pressure in a molten metal bridge, which prevents its disintegration [51]:

$$V_f = \sqrt{(16 \cdot C_d \cdot \sigma) \big/ (d_{br} \cdot \rho_g)} = 2857 \text{ m/s}. \tag{6.96}$$

The diameter of the molten metal bridge d_{br} is assumed to be equal to the thickness of the layer carried away from the electrode tip, which is calculated from the mass of metal carried away from the electrode tip during the detachment period.

Apart from the gas-dynamic forces, the electromagnetic forces created by the flow of current in the "tongue" and in the electrodes act on the molten metal bridge. Three principal directions of current flow relative to the axis of the gas flow can be distinguished (Fig. 6.14):

- along a segment of the electrode gap of length L at the angle α;
- over the layer of liquid metal;
- along the arc column of length x.

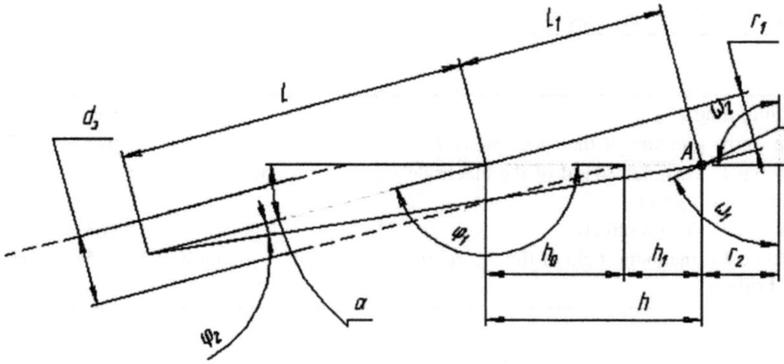

Fig. 6.14 Diagram of the electromagnetic fields in the arcing zone

The magnetic induction B_1 of the electrode gap at the center of a constriction (point A) is calculated from the formula for the magnetic field of a rectilinear conductor [54]:

$$B_1 = \mu \cdot I \cdot (\cos \phi_2 - \cos \phi_1)/(4 \cdot \pi \cdot r_1). \quad (6.97)$$

According to the diagram (Fig. 6.14), r_1 is the shortest distance from point A to the electrode; φ_1 and φ_2 are the angles between the current density vector in the conductor and the radii, that is, the vectors drawn from the beginning and end of the conductor to point A; μ is the permeability of a vacuum. The angles φ_1 and φ_2 are determined by the angle of inclination of the electrode (α):

$$\cos \phi_1 = \cos(180 - \alpha) = -\cos \alpha, \quad (6.98)$$

$$r_1 = h \cdot \sin \alpha, \quad (6.99)$$

$$\cos \phi_2 = L + h \cdot \cos \alpha \Big/ \sqrt{(h \cdot \sin \alpha)^2 + (L + h \cdot \cos \alpha)^2}, \quad (6.100)$$

$$h = d_e/2 \cdot \sin \alpha + h_1. \quad (6.101)$$

The force F_1 acting on a constriction of length Δh due to the interaction of the magnetic field (the magnetic induction B_1) and the current flowing through the constriction is given by the expression

$$F_1 = I \cdot B_1 \cdot \Delta h. \quad (6.102)$$

Due to the local narrowing of the conductor, the force F_2 of the pinch effect caused by the interaction of the flowing current and the magnetic field (the magnetic induction B_2) acts in the constriction:

$$F_2 = \mu \cdot I^2 \cdot \ln(r_{col}/r_{con})/(4\pi). \quad (6.103)$$

Table 6.7 Calculated estimates of the forces acting on the liquid metal in the arcing zone

Forces	Equation	F_i 10^5, N
Aerodynamic drag F_{ad}	(5.93)	16.9
Force due to the pressure of the shock wave F_{f}	(5.94)	2.7
Ampère force F_1 (magnetic field of the electrode gap and current in the molten metal bridge)	(5.102)	6.2
Ampère force F_2 (pinch effect)	(5.103)	9.4
Ampère force F_3 (magnetic field of the arc column and current in the molten metal bridge)	(5.105)	0.03

In analogy to Eq. 6.97, the numerical value of the magnetic induction B_3 at point A is

$$B_3 = \mu \cdot I \cdot (\cos \omega_2 - \cos \omega_1)/(4\pi \cdot r_2) = \mu \cdot I \cdot (1 - x/h_2)/4\pi \cdot h_2 \qquad (6.104)$$

Here ω_1 and ω_2 are the angles between the current density vector in the arc column and the radii, that is, the vectors drawn from the beginning and end of the column to point A (Fig. 6.14), and $\omega_2 = 90°$ according to the assumption made regarding the vertical orientation of the arc column relative to the axis of the gas flow.

The interaction of the magnetic field of the arc column with the current in the constriction produces the force F_3:

$$F_3 = I \cdot B_3 \cdot \Delta h. \qquad (6.105)$$

The results of the calculation of the forces that detach droplets from the electrode tips are presented in Table 6.7. It was assumed in the calculations that the "tongue" splits into identical droplets with a diameter equal to 1×10^{-5} m, the thickness of the "tongue" equals 7×10^{-6} m, the magnetic permeability of the medium is $\mu = 4 \times 10^{-7}$ H/m, the velocity of the gas stream is $V_g = 800$ m/s, $C_{\text{d}} = 0.4$, and the length of the constriction is $\Delta h = 1 \times 10^{-6}$ m.

The calculation results presented in Table 6.7 show that the forces of an electromagnetic nature are of the same order of magnitude as the force of aerodynamic drag, but differ in direction from the desirable direction for motion of the droplets. Thus, our assumption regarding the possibility of fragmentation of the droplets in the arcing zone under the action of not only aerodynamic forces, but also electrodynamic forces is fairly sound.

Stage 3—Particle Flight over the Spraying Distance

For particles emerging at an angle α to the stream axis, the velocity in the axial direction (along the x axis) is calculated for the parameters assigned to the point with the y coordinate determined from the equation

$$y_i = x_j \cdot \tan \alpha_i. \qquad (6.106)$$

Such a scheme presumes that the gas does not have any dynamic effect on the particle along the y axis and that the initial momentum, which determined its displacement at an angle to the x axis, was imparted to it at the initial point due to the electrodynamic action of the arc and the head of the gas stream.

The calculation was performed for different particle diameters according to a sieve analysis of samples.

It was assumed that the throughput of the particles G_p at the distance x from the nozzle in an annular cross section with the coordinates y_i and y_{i+1} obeys a normal distribution law:

$$G_p = G_p^0 \left(e^{-k_c \cdot y_i} - e^{-k_c \cdot y_{i+1}}\right), \tag{6.107}$$

where the concentration coefficient with allowance for the experimental condition that $\approx$ 10% of the particles escape from the confines of a stream of radius δ is $k_c = -\ln 0.1/\delta^2$. The temperature of the particles decreases to $T_m = 1{,}809$ K and then remains unchanged to the end of the distance owing to the release of the enthalpy of fusion.

The extent of oxidation ψ of the particles in the gas was determined from the condition that oxidation is limited by the diffusion of oxygen in the gas, and it was calculated for the diffusive flow of oxygen onto a particle with a density ρ_p and a diameter d_p traveling with a velocity W_p. The concentration of the components C_{ox} was assumed to be the total concentration with allowance for the fact that not only oxygen, but also the oxygen-containing gases CO_2 and H_2O can be the oxidizing agent:

$$\psi = \frac{600}{\rho_p \cdot d_p^2} \cdot \int_0^l \frac{\mathrm{Nu} \cdot D \cdot C_{ox}}{W_p} \cdot dl. \tag{6.108}$$

The equilibrium composition of the combustion products was calculated on the basis of the equilibrium constants of the reactions

$$\{CO_2\} = \{CO\} + 0.5\{O_2\}, \tag{6.109}$$

$$\{H_2O\} = \{H_2\} + 0.5\{O_2\} \tag{6.110}$$

with allowance for the material balance of the elements in the gas [55].

The addition of air from the atmosphere to the stream was calculated with allowance for the proportional relationships of the concentrations and temperatures of a gas to its velocity. Expressions for calculating the concentrations of nitrogen and oxygen added to the stream were obtained for each interval Δx and the cross section Δy:

$$n_{N_2}^{add} = 0.79 \cdot \sum n_i \cdot \left(1 - \gamma_x \cdot \gamma_y\right) / \left(\gamma_x \cdot \gamma_y\right), \tag{6.111}$$

$$n_{O_2}^{add} = 0.21 \cdot \sum n_i \cdot \left(1 - \gamma_x \cdot \gamma_y\right) / \left(\gamma_x \cdot \gamma_y\right). \tag{6.112}$$

Here $\sum n_i$ is the number of moles of gas in the particular cross section, and γ is the proportionality factor.

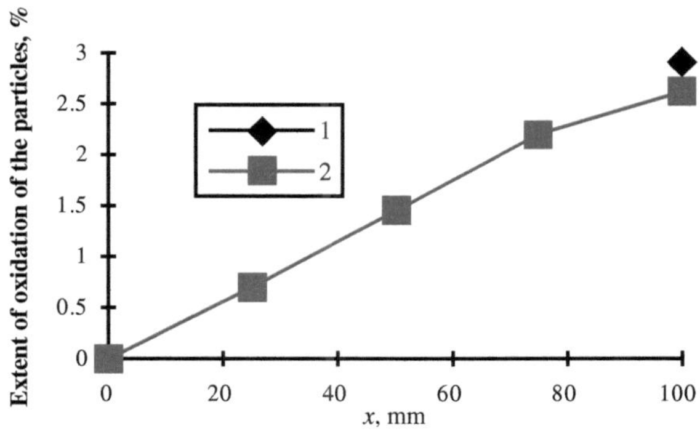

Fig. 6.15 Diagram of the extent of oxidation of the particles, %: *1* experimental data; *2* weighted-mean calculated results in a cross section

The kinetic properties of the gas mixture, namely, the thermal conductivity, viscosity, and diffusion coefficients, were calculated non-additively with allowance for the mutual influence of the components in the mixture [56].

The temperature dependences of the thermal conductivity, heat capacity, density, and viscosity of the components in the gas mixture were approximated by polynomials using table data [57].

The calculated data were compared with experimental data for the velocity of the particles and the extent of oxidation of the coating. The velocity of the luminescent particles was measured using the rotating mirror technique on an ISSO-1 instrument, and the extent of oxidation of the coating was determined by the reductive melting method in a stream of an inert gas and in a vacuum on RO-116 and EAN-220 instruments (Fig. 6.15). The disparity between the results of the calculations and the experiments is 10–16 rel.%.

The calculation of the velocity of the particles in fractions of different diameters at a working distance of 100 mm shows that the velocities differ by more than a factor of 2. This results in variation of the interaction time of a particle with components of the gas phase (oxygen) and the mass-transfer coefficient in the gas. The temperature of the gas drops to 600–800 K. As a consequence of the intense mixing of the gas stream with the atmosphere, it consists nearly entirely of air after the conclusion of the initial segment.

Let us consider a specific example. As the original metal, we take Sv-08G2S wire with the following composition (wt.%): 0.11 C, 1.92 Mn, 0.79 Si, 0.020 P, and 0.021 S.

We adopt the following scheme for the oxidation of a particle. Oxygen dissolves in the liquid metal of the particle up to the saturation limit $[\mathrm{O}]_{\mathrm{Fe}}^{\mathrm{sat}}$. When the oxygen concentration on the metal surface reaches $[\mathrm{O}]' = [\mathrm{O}]_{\mathrm{Fe}}^{\mathrm{sat}}$, the formation of slag around the droplet begins.

According to [58],

$$\log[\mathrm{O}]_{\mathrm{Fe}}^{\mathrm{sat}} = -6320/T + 2.734. \tag{6.113}$$

In the 1,809–3,000 K temperature range $[O]_{Fe}^{sat} = 0.17$to 4.24 wt.%.

The total amount of oxygen m_O interacting with the metal will be equal to the sum of the concentrations of the oxygen dissolved in the liquid metal and the oxygen found in the slag:

$$m_O = m_{Me}\frac{[O]}{100} + m_{slag}\left(\frac{(FeO)M_O}{100M_{FeO}} + \frac{(MnO)M_O}{100M_{MnO}} + \frac{(SiO_2)M_O}{100M_{SiO_2}}\right). \quad (6.114)$$

The extent of oxidation of the particles in a given fraction r_i is given by the expression

$$r_i = m_O / (m_{Me} + m_{slag}). \quad (6.115)$$

The total extent of oxidation R is summed according to the distribution of the particles with respect to their share of the surface:

$$R = \sum r_i \cdot \frac{f_i}{F}. \quad (6.116)$$

Here m_{Me} is the mass of the liquid metal; m_{slag} is the mass of the slag; f_i is the surface area of the particles of the ith fraction; F is the total surface area of the particles; M is the molecular weight of the component.

As in [13, 30], we assume that the oxygen absorption process takes place in a diffusion-controlled regime. To select the rate-limiting step, we estimate the possible limiting diffusion fluxes in the gas (I_g^{lim}) and in the metal (I_{Me}^{lim}):

$$I_g^{lim} = \beta_g \cdot c_g = \beta_i \cdot P_{O_2}/RT, \text{ mole/cm}^2\text{s}, \quad (6.117)$$

$$I_{Me}^{lim} = \beta_{Me} \cdot c_{Me} = \beta_{Me} \cdot \rho_{Me} \cdot [O]_{Fe}^{sat}/100M_O, \text{ mole/cm}^2\text{s}, \quad (6.118)$$

Here β_g and β_{Me} are the mass-transfer coefficients of oxygen in the gas and in the metal; c_i is the concentration; P_{O_2} is the partial pressure of oxygen; T is the temperature.

The calculation results for particles with a diameter equal to 100 μm under various conditions regarding the spraying distance S are presented in Table 6.8. The value of β_{Me} for the arcing zone was taken from an estimate [30, 46], and the value for the flight stage was assumed to be the same as for a melt during hydrogen plasma deoxidization [46].

The mass-transfer coefficient of oxygen in the gas at the electrode tip was calculated for the case in which the flow impinges on the surface of a flat plate with a velocity V_0 [58]. The distance from the impingement point (x) was assumed to be equal to the length of the electrode tip. For particles in the arcing zone and in flight, the value of the mass-transfer coefficient in the gas was estimated from the Nusselt number as in the case of external flow around a body:

$$\beta_g = \text{Nu}\frac{D}{d}, \quad (6.119)$$

Table 6.8 Estimation of he limiting diffusion fluxes of oxygen in the gas and in the metal

Stage	S (mm)	T_g (K)	P (atm)	Re	Nu	D (cm^2/s)	v (cm^2/s)	β_{Me} (cm/s)	β_g (cm/s)	I_g^{lim} (mole/cm^2s)	I_{Me}^{lim} (mole/cm^2s)
Stage 1	0	2,400	0.014			6.6	4.8	0.250	181	1.3×10^{-5}	1.4×10^{-3}
Stage 2	0	2,700	0.014	206	9.8	8.1	5.6	0.250	7,916	5.0×10^{-4}	2.8×10^{-3}
Stage 3	0	2,076	0.21	288	11.4	5.2	3.9	0.025	5,931	7.3×10^{-3}	5.6×10^{-5}
	10	1,714	0.21	287	11.4	3.8	2.9	0.025	4,301	6.4×10^{-3}	2.0×10^{-5}
	20	1,423	0.21	289	11.4	2.8	2.1	0.025	3,144	5.7×10^{-3}	2.0×10^{-5}
	50	926	0.21	250	10.7	1.3	0.5	0.025	1,428	4.0×10^{-3}	2.0×10^{-5}
	100	635	0.21	136	8.2	0.7	0.6	0.025	576	2.3×10^{-3}	2.0×10^{-5}

where $\mathrm{Nu} = 0.459 + \mathrm{Re}^{0.55} \cdot \mathrm{Pr}^{0.75}$ [59].

The calculations were performed for the atomization of a metal by the combustion products of a propane–air mixture at the nozzle exit section, $P_{O_2} = 0.014$ atm, and $V_0 = 950$ m/s.

The calculation results presented show that on the electrode tip and in the arcing zone (stages 1 and 2) the rate-limiting step of the oxidation process is external diffusion and that along the spraying distance it is diffusion in the metal. The very high value of the mass-transfer coefficient in the gas, which is higher by an order of magnitude or more than in welding or plasma arc remelting, is noteworthy. It is due to the small size of the droplets being transported compared with their size in the processes just cited.

We note some specific features of the characteristics of droplet transport in EAS in comparison to the data for arc welding [36] with similar values of the current and the electrode diameters. In EAS the specific surface of the droplets and the mass-transfer coefficient in the gas are 1–2 orders of magnitude higher. On the other hand, the time for the interaction of a droplet of liquid metal with the air is 100–1,000 times shorter.

Let us consider the kinetics of the interaction of the molten metal with the ambient atmosphere in the stages indicated above.

As in [30], but with allowance for diffusion in the metal and in the gas, we describe the process for an open system with the supply and removal of reactants using the equation of a first-order reaction:

$$\frac{dm_O}{d\tau} = \frac{g_m \cdot [O]_0}{100} - \frac{g_{rem} \cdot [O]}{100} + \frac{\beta_{Me} \cdot \rho \cdot F}{100} \cdot \left([O]' - [O]\right). \tag{6.120}$$

Here $m_O = \rho \cdot V \cdot [O]$ is the amount of oxygen in the molten metal, g; ρ and V are the density and volume of the molten metal; g_m and g_{rem} are the mass melting and removal rates of the metal; $[O]_0$, $[O]$, and $[O]'$ are the initial, current, and interfacial concentration of oxygen, wt.%.

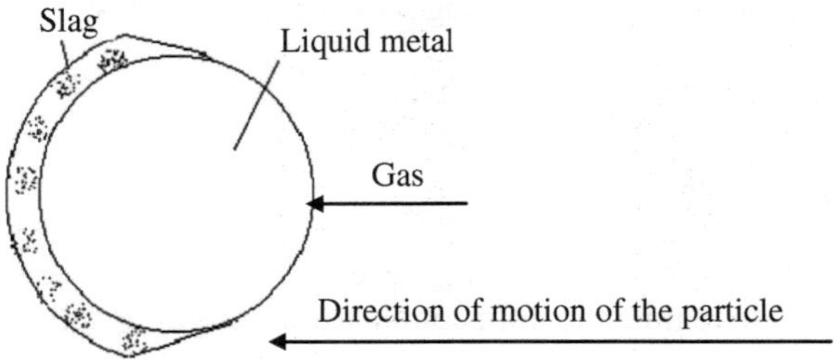

Fig. 6.16 Physical model of the interaction of the droplet metal and the slag over the spraying distance

Table 6.9 Diffusion coefficients of the components [13, 45]

Component	(FeO)	[C]	[O]	(MnO)	[Mn]	(SiO_2)	[Si]
$D{\cdot}10^5$ (cm^2/s)	1.0	6.4	0.12	64	4.9	64	8.9

We will adopt the following physical model of the interaction between a droplet and the slag (Fig. 6.16). A droplet having a spherical shape develops a velocity that is lower than that of the gas. Its aft part is a clean metal surface, on which the entry of oxygen occurs. The fraction of the surface area in the aft part is $L \leq 0.6$. All the slag formed is forced into the fore part by the gas flowing around the particle.

It was not ruled that a gas in an activated state can dissolve in the liquid metal in concentrations above the saturation limit, as in the case of nitrogen [46]. However, the formation of a supersaturated solution is less likely for oxygen because of its reaction with iron to form FeO.

To calculate the rates of the interaction between the metal and the slag, we use the method described in Sect. 4.3.2.

We take the value $\beta = 1.2$ from the range of most probable values from $\beta = 0.5$ to 2.0 [13]. The values of the diffusion coefficients of the components in the melt are presented in Table 6.9.

The concentrations of the components containing the element E_i in the metal and the slag were determined from the following equations:

$$(E_iO_m)^{k+1} = (E_iO_m)^k + I_{E_i}\frac{M_{E_iO_m}\cdot f\cdot(1-L)\cdot\Delta\tau\cdot 100}{k\cdot m_{slag}}, \tag{6.121}$$

$$[E_i]^{k+1} = [E_i]^k - I_{E_i}\frac{M_{E_i}\cdot f\cdot(1-L)\cdot\Delta\tau\cdot 100}{m_{Me}}. \tag{6.122}$$

The concentration of oxygen in the metal was determined from the equation

$$[O]^{k+1} = [O]^k + I_O\frac{M_O\cdot f\cdot(1-L)\cdot\Delta\tau}{m_{Me}}. \tag{6.123}$$

The concentration of FeO in the slag was determined from the equation

$$(\mathrm{FeO})^{k+1} = (\mathrm{FeO})^{k} - I_{\mathrm{FeO}} \frac{M_{\mathrm{FeO}} \cdot f \cdot (1-L) \cdot \Delta\tau \cdot 100}{m_{\mathrm{slag}}}. \tag{6.124}$$

In the first stage, a steady state, in which $\frac{dm_{\mathrm{O}}}{d\tau} = 0$, is established on the electrode tip after a definite time interval. We will assume that $g_{\mathrm{m}} = g_{\mathrm{rem}} = g$.

In the steady-state regime the diffusion fluxes in the gas and the metal are equal:

$$1\Big/2\beta_{\mathrm{g}}\left(C_{\mathrm{O_2}} - C'_{\mathrm{O_2}}\right) = \beta_{\mathrm{Me}}\left(C'_{\mathrm{O}} - C_{\mathrm{O}}\right) \tag{6.125}$$

or

$$\frac{\beta_{\mathrm{g}}}{2RT}\left(P_{\mathrm{O_2}} - P'_{\mathrm{O_2}}\right) = \frac{\beta_{\mathrm{Me}}\rho}{100M_{\mathrm{O}}}\left([\mathrm{O}]' - [\mathrm{O}]\right). \tag{6.126}$$

Then we can write

$$aP_{\mathrm{O_2}} - a\left([\mathrm{O}]'\right)^2\Big/k^2 = [\mathrm{O}]' - [\mathrm{O}], \tag{6.127}$$

where $P'_{\mathrm{O_2}} = \left([\mathrm{O}]'\right)^2\Big/k^2$, $\log k = 6,120/T + 0.15$ [8], and $a = \frac{100\beta_{\mathrm{g}}M_{\mathrm{O}}}{2\beta_{\mathrm{Me}}RT_{\rho}}$.

We find the surface concentration of oxygen on the interface $[\mathrm{O}]'$ from mass-transfer Eq. 6.120:

$$\frac{g_{\mathrm{rem}} \cdot [\mathrm{O}]_0}{100} - \frac{g_{\mathrm{m}} \cdot [\mathrm{O}]}{100} + \frac{\beta_{\mathrm{g}} \cdot M_{\mathrm{O_2}} \cdot F}{2RT} \cdot \left(P_{\mathrm{O_2}} - \left([\mathrm{O}]'\right)^2\Big/k^2\right) = 0, \tag{6.128}$$

$$A_2\left([\mathrm{O}] - [\mathrm{O}]_0\right) = P_{\mathrm{O_2}} - \left([\mathrm{O}]'/k\right)^2, \tag{6.129}$$

where $A_2 = \frac{2RTg}{100\beta_{\mathrm{g}} \cdot M_{\mathrm{O_2}} \cdot F}$.

From the combined solution of Eqs. 6.127–6.129 we have

$$[\mathrm{O}] = \frac{P_{\mathrm{O_2}} - B^2 + A_2[\mathrm{O}]_0}{A_2}, \tag{6.130}$$

$$[\mathrm{O}]' = kB, \tag{6.131}$$

where $B = \dfrac{-\frac{A_2k}{2} + \sqrt{\left(\frac{A_2k}{2}\right)^2 + (1+aA_2)\left(P_{\mathrm{O_2}} + aA_2P_{\mathrm{O_2}} + A_2[\mathrm{O}]_0\right)}}{(1+aA_2)}$.

For the values of the parameters taken, on the electrode tip $[\mathrm{O}]_0 = 0.012$ wt.%, $T = 2{,}400$ K, $\beta_{\mathrm{g}} = 181$ cm/s, $\beta_{\mathrm{Me}}=0.25$ cm/s, the area of the tip is 0.13 cm^2, and$P_{\mathrm{O_2}} = 0.014$ atm. The calculations give $[\mathrm{O}] = 0.0126$ wt.% and $[\mathrm{O}]' = 0.0183$ wt.%. On the electrode tip $[\mathrm{O}]' < [\mathrm{O}]^{\mathrm{sat}}_{\mathrm{Fe}}$, and all the oxygen entering the metal in this stage dissolves in it.

In stage 2 the temperature of the metal increases, in accordance with the assumption regarding the nature of the fragmentation of the droplets in this stage. In addition, the total area f_S of the droplets of a particular fraction that are obtained in each occurrence of detachment is

$$f_S = f_i \frac{g_i}{m_i}, \tag{6.132}$$

where f_i is the surface area of a single droplet (cm^2), g_i is the consumption of the metal per second (g/s), and m_i is the mass of a single droplet (g).

When the surface concentration of oxygen is less than $[\mathrm{O}]_{\mathrm{Fe}}^{\mathrm{sat}}$, we have

$$\frac{d(\rho V[\mathrm{O}])}{100 d\tau} = \frac{\beta_{\mathrm{Me}} \rho F}{100} \left([\mathrm{O}]' - [\mathrm{O}]\right). \tag{6.133}$$

If we divide the duration of the process into small time intervals which are such that $\frac{d[\mathrm{O}]}{d\tau} \approx \frac{\Delta[\mathrm{O}]}{\Delta\tau}$, we can write the following equality for the $(k + 1)$th time interval:

$$\frac{[\mathrm{O}]_{k+1} - [\mathrm{O}]_k}{\left([\mathrm{O}]' - [\mathrm{O}]_k\right)} = A\Delta\tau, \tag{6.134}$$

where $A = \frac{\beta_{\mathrm{Me}} F}{V}$.

From the condition of equality between the diffusion fluxes in the gas and the metal $I_{\mathrm{g}} = I_{\mathrm{Me}}$, Eq. 6.127 takes the following form for the $(k + 1)$th step (time interval):

$$a(P_{\mathrm{O}_2} - [\mathrm{O}]')^2/k^2 = [\mathrm{O}]' - [\mathrm{O}]_k, \tag{6.135}$$

where $\beta_{\mathrm{g}} = 7,916\,\mathrm{cm/s}$.

The oxygen concentrations $[\mathrm{O}]_{k+1}$ and $[\mathrm{O}]'$ are determined from the combined solution of Eqs. 6.134 and 6.135:

$$[\mathrm{O}]' = \frac{-\frac{1}{2} + \sqrt{\left(\frac{1}{2}\right)^2 + \frac{a}{k^2}\left([\mathrm{O}]_n + aP_{\mathrm{O}_2}\right)}}{\frac{a}{k^2}}, \tag{6.136}$$

$$[\mathrm{O}]_{k+1} = [\mathrm{O}]_k + A\Delta\tau\left([\mathrm{O}]' - [\mathrm{O}]_k\right). \tag{6.137}$$

If $[\mathrm{O}]' > [\mathrm{O}]_{\mathrm{Fe}}^{\mathrm{sat}}$, slag formation occurs. The calculations showed that the saturation limit of the metal surface with oxygen is exceeded in the second stage for particles in all the fractions at the possible temperatures of the metal in the range from 2,400 to 3,000 K. In this case the oxygen content in the metal was determined from Eq. 6.134 when $[\mathrm{O}]' > [\mathrm{O}]_{\mathrm{Fe}}^{\mathrm{sat}}$.

The diffusion flux of oxygen from the gas into the metal I_{g} equals

$$I_{\mathrm{g}} = \frac{\beta_{\mathrm{g}}\left(P_{\mathrm{O}_2} - \left([\mathrm{O}]_{\mathrm{sat}}/k\right)^2\right)}{2RT}. \tag{6.138}$$

It is distributed in such a manner that one part of it (I_{Me}) provides for the dissolution of oxygen in the metal, while the other part (I_{slag}) provides for slag formation. The values of the fluxes are given by the formulas

$$I_{Me} = \frac{\beta_{Me}\rho([O]_{sat} - [O]_k)}{100M_O}, \tag{6.139}$$

$$I_{slag} = I_g - I_{Me}. \tag{6.140}$$

The quantity of oxygen that can dissolve in each step is

$$\Delta m_O = I_{Me} \cdot F \cdot \Delta\tau \cdot L \cdot M_O. \tag{6.141}$$

The quantity of slag that moves into the fore part in each step is

$$\Delta m_{slag} = \sum_i \Delta m_{slag_i} = I_{slag} \cdot F \cdot \Delta\tau \cdot L \cdot \sum_i (M_i \cdot (\%)_i/100), \tag{6.142}$$

Here M_i and $(\%)_i$ are the molecular weight and mass concentration of the ith component of the slag formed on the surface.

In accordance with [60], we take the following typical composition of the slag on droplets of Sv08G2S steel during welding, wt.%: $(MnO)_O$, = 45, $(FeO)_O = 3.4$, $(SiO_2)_O = 51.6$.

The mass of the liquid metal decreases in comparison to the previous step due to the passage of components into the slag:

$$m_{Me}^{k+1} = m_{Me}^k - \sum_i (\Delta m_{slag_i} \cdot M_{E_i}/M_i). \tag{6.143}$$

The final concentrations of the components in the $(k + 1)$th step are as follows:

$$[O]_S^{k+1} = [O]^{k+1} + \Delta m_O \cdot 100/m_{Me}^{k+1}, \tag{6.144}$$

$$(E_iO_m)_S^{k+1} = (E_iO_m)^{k+1} + \Delta m_{slag_i} \cdot 100/m_{slag}, \tag{6.145}$$

$$FeO_S^{k+1} = FeO^{k+1} + \Delta m_{slag_i} \cdot 100/m_{slag}. \tag{6.146}$$

Stage 3, in which the particles travel over the spraying distance, is distinguished from stage 2 by the fact that the temperatures of the gas and the metal drop, the partial pressure of oxygen increases, and the mass-transfer coefficient in the gas decreases (Table 6.8).

The calculations showed the following.

The final total extent of oxidation R depends strongly on the size of the clean metal surface. In the calculations we took the value $L = 0.15$, which corresponds to the closest approach of R to the experimental data in [60]. The diameter of the particles also has a considerable influence on the extent of oxidation.

A decrease in the difference between the velocities of the particles and the gas and a decrease in the gas temperature result in lowering of the rate of oxidation over the spraying distance.

The oxygen content in the liquid metal reaches the saturation limit at a distance of 100 mm only in the case of particles of small diameter. The rate of solution of oxygen in the liquid metal corresponds to the total rate of entry of oxygen into a particle.

The value for the second stage increases markedly if measures are not taken to protect the arcing zone from oxidation, for example, when compressed air is employed as the atomizing gas. In this case 30–40% of all the oxygen reaching a particle during its flight enters it. This is due to the extremely high value of the mass-transfer coefficient in the gas and the large specific surface of the particles.

These features must be taken into account both in the development of coating technologies and in designing the equipment.

The mathematical model developed for the movement, heating, and oxidation of particles during EAS enables us to analyze the parameters of the two-phase flow over the cross section of the turbulent stream and the spraying distance. The model takes into account the nonuniformity of the size distribution of the particles, the cross-sectional distribution of the flow rates of the gas and particles, the variation of the composition of the gas in each microscopic volume due the occurrence of chemical reactions and the addition of air from the ambient atmosphere, the non-additivity of the kinetic properties of gases, and the influence of the wire electrode on the characteristics of the flow.

This model of the processes was employed, in particular, for developing a series of HVAS units [52].

6.7 Conclusions

The material presented in this chapter allows us to say that the models of metallurgical processes created in reference to various welding technologies provide ways not only to obtain a formalized description of their fundamental laws, but also to effectively control them and to optimize the conditions under which they occur. Some results obtained using the tools of mathematical modeling enabled us to formulate technical solutions on the level of inventions or proprietary know-how, which are protected by corresponding patent documents (see Chap. 8).

Conclusions regarding the faithfulness of the models developed have been drawn, and they corroborate the qualitative and quantitative correspondence of the simulation results to real welding processes.

First of all, the models do not contradict any fundamental physical laws or principles of conservation and are consistent with the influence of the most important laws of chemical thermodynamics and kinetics, and the simulation results do not exceed the statistical spread of the experimental data or the accuracy of the chemical analysis.

The high intellectual and practical value of the mathematical models, which enable us to predict previously unknown results, to recommend areas for improving both the mathematical model itself and the process being simulated, and to give optimal combinations of the parameters of the processes, may be noted.

An important result of modeling is the creation of regression equations, which facilitate the employment of models and are the foundation for creating engineering variants of computational experiments.

References

1. Boronenkov, V.N., Shanchurov, S.M.: Mathematical model of the interaction processes of a multicomponent metal and a multicomponent slag during electroslag remelting, 38 pp. Ural Polytechnic Institute, Sverdlovsk, Article deposited in Chermetinformatsiya, No. 700 (1979)
2. Shanchurov, S.M., Boronenkov, V.N.: Mathematical modeling of chemical processes in electroslag remelting. In: The 8th All-Union Conference on the Physicochemical Foundations of Steel Production: Abstracts of Reports, Part 1 (in Russian), pp. 81–82, Moscow (1977)
3. Boronenkov, V.N., et al.: Mathematical modeling of the interaction processes of the metal and the slag during the electroslag remelting of an alloy steel. In: Physicochemical Investigations of Metallurgical Processes, vol. 9 (in Russian), pp. 104–111. UPI, Sverdlovsk (1981)
4. Boronenkov, V.N., et al.: Prediction of the composition of electroslag remelting ingots by mathematical modeling. Spets. Élektrometall. **63**, 13–19 (1987)
5. Shanchurov, S.M., Boronenkov, V.N.: Estimation of the chemical heterogeneity of a metal during electroslag remelting. Spets. Élektrometall. **41**, 41–50 (1979)
6. Boronenkov, V.N., Shanchurov, S.M., Zalomov, N.I.: Mathematical model of prediction of electroslag ingot composition. In: The 10th International Conference on Vacuum Metallurgy, pp. 62–63, Beijing (1990)
7. Shanchurov, S.M., Zalomov, N.I., Boronenkov, V.N.: Mathematical model for predicting the composition of electroslag remelting ingots. Probl. Spets. Élektrometall. **1**, 26–33 (1991)
8. Shanchurov, S.M., Boronenkov, V.N., Bakumenko, S.P.: Behavior of sulfur and oxygen during the electroslag remelting of electrodes with scale. Peredovoi Opyt: Ezhemesyachnyi Proizvodstvenno-Tekhnicheskii Sbornik **5**, 38–41 (1990)
9. Boronenkov, V.N., Shanchurov, S.M.: Mathematical model of the kinetics of the chemical processes occurring in the electroslag welding of steel. Avt. Svarka **6**, 22–27 (1985)
10. Shanchurov, S.M., Boronenkov, V.N.,Yakovlev, V.V.: Predicting the composition of the weld metal during the electroslag welding of press columns. In: Theory and Practice in Industrial Welding (in Russian), pp. 9–14. Ural Polytechnic Institute Press, Sverdlovsk (1986)
11. Shanchurov, S.M., Trofimov, I.F., Babaev, G.M.: Predicting the composition of the cladding layer during electroslag surfacing. In: Mathematical Methods and CAD in Industrial Welding: Abstracts of Reports to the Scientific–Technical Conference (in Russian), p. 41, Sverdlovsk (1990)
12. Shanchurov, S.M., Boronenkov, V.N., Yanishevskaya, A.G.: Mathematical model of the chemical processes occurring in electroslag casting. In: Physicochemical Foundations of Metallurigical Processes: Scientific Reports to the 10th All-Union Conference, Part 3 (in Russian), pp. 194–195, Moscow (1991)
13. Popel', S.I., Sotnikov, A.I., Boronenkov, V.N.: Theory of Metallurgical Processes (in Russian), 463 pp. Metallurgiya, Moscow (1986)
14. Kafarov, V.: Cybernetic Methods in Chemistry and Chemical Engineering. Mir, Moscow (1976)

15. Boronenkov, V.N., Shanchurov, S.M., Zinigrad, M.I.: Kinetics of the interaction of a multicomponent metal with slag in a diffusion-controlled regime. Izv. Akad. Nauk. SSSR. Met. **6**, 21–27 (1979)
16. Boronenkov, V.N., Esin, O.A.: Kinetics of the simultaneous occurrence of several heterogeneous reactions with common reactants. Zh. Fiz. Khim. **8**, 2022–2027 (1970)
17. Salamatov, A.M.: Mathematical model of the interaction processes of the metal and slag during submerged-arc welding, 199 pp. Candidate Dissertation, Sverdlovsk (1982)
18. Boronenkov, V.N., Salamatov, A.M.: Prediction of the chemical composition of the weld metal during arc welding by mathematical modeling of the interaction processes of the metal, slag, and gas, No. 1591, 50 pp. S.M. Kirov Ural Polytechnic Institute, Sverdlovsk, 1982, Article deposited in Chermetinformatsiya (1983)
19. Boronenkov, V.N., Shanchurov, S.M., Zinigrad, M.I.: Kinetic analysis of the processes occurring in the combined oxidation of metal impurities by slag. In: Current Problems in the Electrometallurgy of Steel: Proceedings of the 3rd All-Union Scientific Conference (in Russian), vol. 206, pp. 83–89, Chelyabinsk (1978)
20. Zalomov, N.I., Boronenkov, V.N.: Mathematical modeling of the kinetics of the interaction of water vapor with molten fluxes. In: Application of Mathematical Methods in Welding: A Collection of Scientific Papers (in Russian), pp. 34–40. Naukova Dumka, Kiev (1988)
21. Zalomov, N.I.: Development of a mathematical model for predicting the composition of the metal during electroslag remelting and welding, 233 pp. Candidate Dissertation, Sverdlovsk (1988)
22. Paton, B.E. (ed.): Electroslag Welding (in Russian), 410 pp. Mashinostroenie, Moscow (1959)
23. D'yakonov, V.M., Vachugov, G.A., Chumanov, V.I.: Temperature field in a liquid metal film and a droplet during ESR. In: Scientific Papers of the Chelyabinsk Polytechnic Institute (in Russian), vol. 271, pp. 3–7 (1982)
24. Mitchell, A.: Mechanism of heat generation and transfer during ESR. In: Electroslag Remelting, pp. 148–161. Naukova Dumka, Kiev (1971)
25. Dewsnap, P., Schlatter, R.: Direct-current electroslag remelting of alloy steels. In: Electroslag Remelting, vol. 4, pp. 76–96. Naukova Dumka, Kiev (1978)
26. Medovar, V.I., Marinskii, G.S., Shevtsov, V.L.: Centrifugal Electroslag Casting (in Russian), 48 pp. Znanie, Kiev (1983)
27. Shanchurov, S.M.: Mathematical Modeling of Metal-Slag-Gas Interaction for Electroslag Technology. Doctor Dissertation, p. 440. Ural State Technical University Press, Yekaterinburg (1992)
28. Paton, B.E., Medovar, B.I. (eds.): Elecroslag Metal (in Russian), 679 pp. Naukova Dumka, Kiev (1981)
29. Podgaetskii, V.V., Galinich, V.I.: Achievability of thermodynamic equilibrium during arc welding. Avt. Svarka **1**, 1–6 (1961)
30. Erokhin, A.A.: Principles of Fusion Welding (in Russian), p. 448. Mashinostroenie, Moscow (1973)
31. Boronenkov, V.N., Salamatov, A.M.: Calculation of the equilibrium between a multicompoent metal and a multicomponent slag during submerged-arc welding. Avt. Svarka **7**, 19–23 (1984)
32. Kokh, B.A.: Thermodynamic calculation of alloying from the slag and deoxidization of the metal during welding. Avt. Svarka **7**, 20–25 (1977)
33. Bagryanskii, K.B., Lavrik, P.F.: Mechanism of the alloying of the droplet metal during submerged-arc welding with a ceramic flux. Svarochn. Proizv. **10**, 5–7 (1963)
34. Petrov, G.L.: Welding Materials (in Russian), 280 pp. Mashinostroenie, Leningrad (1972)
35. Frumin, I.I.: Automated Arc Surfacing (in Russian), 422 pp. Metallurgizdat, Moscow (1961)
36. Pokhodnya, I.K.: Gases in Welded Joints (in Russian), 256 pp. Mashinostroenie, Moscow (1972)
37. Bagryanskii, K.V.: Submerged-Arc Welding and Surfacing with Ceramic Fluxes (in Russian), 184 pp. Tekhnika, Kiev (1976)

38. Beskhlebnyi, V.A., et al.: Alloying of the metal during submerged-arc welding with a ceramic flux. Svarochn. Proizv. **6**, 11–13 (1968)
39. Belyaeva, E.V., Nikitin, Y.P.: Contact interaction of iron–carbon alloys with graphite. In: Physicochemical Investigations of Metallurgical Processes, vol. 6 (in Russian), pp. 15–21. Ural Polytechnic Institute Press, Sverdlovsk (1977)
40. Shalimov, M.P., et al.: Calculation of the composition of the metal deposited during surfacing with a ceramic flux. Avt. Svarka **1**, 33–36 (1984)
41. Fefelov, A.S., et al.: Mathematical modeling of surfacing processes for selecting the composition of the alloying flux. In: Prospective Surfacing Methods with Corrosion- and Wear-Resistant Materials (in Russian), pp. 20–24. Mashinostroenie, Leningrad (1987)
42. Balin, A.N.: Alloying of the deposited metal with chromium and boron by reducing them from oxide compounds, 177 pp. Candidate Dissertation, Sverdlovsk (1987)
43. Boronenkov, V.N.: Applicability of thermodynamic methods for calculating the composition of the metal during arc welding in gases. In: Current Problems in Welding Science and Technology: Abstracts of Reports to the International Scientific-Technical Conference (in Russian), Rostov-on-Don (1993)
44. Novozhilov, N.M.: Foundations of the Metallurgy of Arc Welding in Gases (in Russian), 231 pp. Mashinostroenie, Moscow (1979)
45. Boronenkov, V.N., Salamatov, A.M.: Calculated estimation of the kinetics of the interaction of a multicomponent metal and a multicomponent slag during submerged-arc welding. Avt. Svarka **8**, 19–24 (1985)
46. Lakomskii, V.I.: Plasma Arc Remelting (in Russian), 336 pp. Technica, Kiev (1974)
47. Vakhalin, V.A., et al.: The process of melting and atomization of the electrode material during electric-arc spraying. Fiz. Khim. Obrat. Mater. **3**, 58–63 (1981)
48. Kretzschmar, E.: Metall-, Keramik- und Plastspritzen, 392 pp. Verlag Technik, Berlin (1963)
49. Gubenko, B.G., et al.: Structural heterogeneity of arc-sprayed grade 65G steel coatings. Avt. Svarka **6**, 19–23 (1990)
50. Levich, V.G.: Physicochemical Hydrodynamics. Prentice–Hall, Englewood Cliffs (1962)
51. Sheikhaliev, S.M.: Investigation of the breakup of metal jets by an electric current, 235 pp. Candidate Dissertation, Sverdlovsk (1977)
52. Korobov, Y.S., et al.: Restoration of parts by high-velocity arc spraying. Avtomob. Prom-st. **1**, 23–24 (2000)
53. Marshak, I.S.: Pulsed Light Sources, 461 pp. Consultants Bureau, New York, (1984)
54. Yavorsky, B., Detlaf, A.: A Modern Handbook of Physics, p. 712. Mir, Moscow (1982)
55. Esin, O.A., Gel'd, P.V.: Physical Chemistry of Pyrometallurgical Processes, Part 1 (in Russian), 671 pp. Metallurgizdat, Moscow (1962)
56. Reid, R.C., Prausnitz, J.M., Sherwood, T.K.: The Properties of Gases and Liquids, 688 pp. McGraw–Hill, New York (1977)
57. Vargaftik, N.B.: Handbook of Physical Properties of Liquids and Gases: Pure Substances and Mixtures, 758 pp. Hemisphere, Washington (1983)
58. Ryzhonkov, D.I. (ed.): Theory of Metallurgical Processes (in Russian), 392 pp. Metallurgiya, Moscow (1989)
59. Tsvetkov, Y.V., Panfilov, S.A.: Low-Temperature Plasma in Reduction Processes (in Russian), 358 pp. Nauka, Moscow (1980)
60. Ivashko, V.S., Belotserkovskii, M.A., Buikus, K.V.: Restoration of friction assembly parts by means of high-velocity arc spraying. Avt. Svarka **4**, 47–49 (1999)

Chapter 7
Predicting the Structure, Phase Composition and Properties of the Metal During Welding and Surfacing

This chapter proposes a model for predicting the phase and structural composition of a weld metal. This model takes into account the primary and secondary crystallization of the weld metal, as well as the results of thermodynamic and kinetic analyses of these processes. Let us formulate the problems whose solution will enable us to create such a model.

1. To develop a phenomenological description of the carbide formation processes in the primary and secondary crystallization stages.
2. To modify the Schaeffler constitution diagram for the purpose of predicting the phase and structural composition of the matrix of the deposited metal. For this purpose, it is necessary:
   - to derive an equation for calculating the chromium and nickel equivalents that takes into account the carbide formation process and the nonequilibrium character of the crystallization process;
   - to develop a procedure for calculating the quantities of the phases in multiphase regions of the diagram;
   - to present a mathematical description of the diagram that will provide for automation of the calculation of the phase composition of the weld metal.

After solving these problems, we will be able to develop analytical methods for calculating the phase and structural composition of the weld metal and, on their basis, to create an efficient closed algorithm and, consequently, a program package for calculating both the direct and inverse problems.

M. Zinigrad et al., *Phase Interaction in the Metal–Oxide Melts–Gas System*, Engineering Materials, DOI: 10.1007/978-3-642-22377-8_7,

7.1 Predicting the Quantitative and Qualitative Composition of Hardening Phases

Let us examine the mechanisms of the primary and secondary crystallization processes of a weld metal.

The distribution of the impurities (alloying elements and carbon) between the liquid and solid phases is specified by the expression [1, 2]

$$E_i^{(s)} = E_i^0 \left[1 - (1 - K_{\mathrm{eff}_i}) \exp\left(-\frac{L^t V_{\mathrm{cryst}}}{D_i^{(l)}} \right) \right], \tag{7.1}$$

where $E_i^{(s)}$ is the concentration of the ith component in the solid phase at the crystallization time t; E_i^0 is the initial average concentration of the ith component in the melt, L^t is the distance from the crystallization starting point (the length of the crystallite at the crystallization time t); V_{cryst} is the crystallization rate, $D_i^{(l)}$ is the diffusion coefficient of the ith component in the melt, and K_{eff} is the effective heterogeneity factor, which is defined in the following manner [1, 2]:

$$K_{\mathrm{eff}} = \frac{K_0}{K_0 + (1 - K_0)\exp(-v_s \delta / D_l)}, \tag{7.2}$$

where δ is the depth of the impurity-enriched layer of the melt adjacent to the crystallization front (for real welding processes $\delta = 10^{-2}$ to 10^{-3} cm) and K_0 is the equilibrium distribution coefficient.

In accordance with the assumptions made, diffusionless decomposition of the supersaturated solid solution into austenite and carbide phases occurs at the time of crystallization. The distribution of carbon and the alloying elements between these solid solutions must be ascertained. Based on the principles of carbide formation in alloys, we will assume that the carbide-forming elements tend to bind all the carbon in carbides in accordance with their carbide-forming ability. An analysis of the available data enables us to assert that the amount of carbon that is used to form the carbide of the ith metal is proportional to the atomic radius of that metal (R_i) and is inversely proportional to the number of electrons in the d subshell of that metal. We introduce the concept of the absolute carbide-forming ability of the ith d metal and represent it as

$$\Theta_i = \frac{R_i}{d_i}. \tag{7.3}$$

It follows from an analysis of expression (7.3) that the carbide-forming ability increases along the series consisting of Fe, Mn, Cr, Mo, W, Nb, V, Ta, Ti, Zr, and Hf, in good agreement with several studies [3–8] and with the calculations of the temperature dependence of the chemical affinity of these elements toward carbon that were performed within the present research for individually taken carbide formation reactions (Fig. 7.1). At the same time, this series specifies the thermal stability of carbides in alloys.

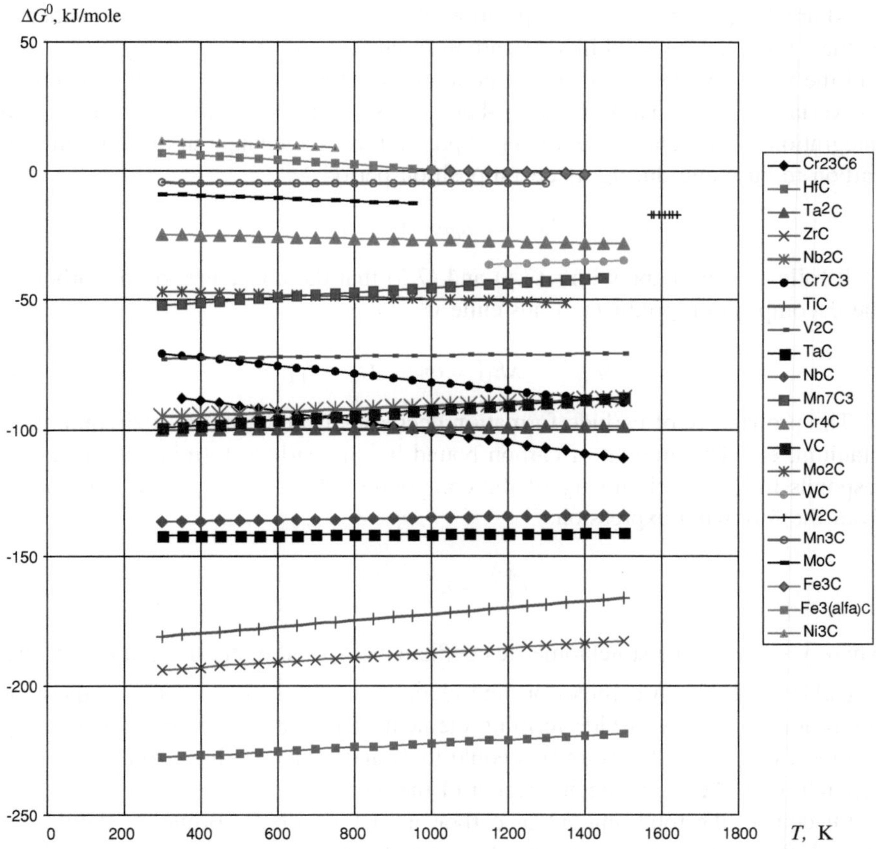

Fig. 7.1 Temperature dependence of the change in the Gibbs free energy for carbide formation reactions

The distribution coefficient of carbon between austenite and its decomposition products is specified by the ratio

$$K_d = \frac{E_{\mathrm{C}}^{(d)}}{E_{\mathrm{C}}^{(a)\,\mathrm{lim}}} = \frac{\int_{t_1}^{t_2} E_{\mathrm{C}}(t)\,dt}{E_{\mathrm{C}}^{(a)} - \int_{t_1}^{t_2} E_{\mathrm{C}}(t)\,dt} = \frac{\Delta E_{\mathrm{C}}}{E_{\mathrm{C}}^{(a)} - \Delta E_{\mathrm{C}}}, \tag{7.4}$$

where $E_{\mathrm{C}}^{(a)\,\mathrm{lim}}$ is the solubility limit of carbon in austenite at a given temperature, $E_{\mathrm{C}}^{(d)}$ is the concentration of carbon in the decomposition products of austenite, $E_{\mathrm{C}}^{(a)}$ is the concentration of carbon in austenite before it begins to decompose, and ΔE_{C} is the change in the solubility of carbon in austenite after the decomposition period.

At any time during the decomposition of austenite, the temperature dependence of the change in the solubility of carbon in austenite is defined as $\omega_T = dE_C/dT$, and the rate of variation of the temperature is defined as $\omega_d = dT/dt$. Separating the variables and equating the right-hand sides, after several transformations and integration, we obtain the following expression for the change in the solubility of carbon in austenite during its decomposition period:

$$\Delta E_C = \omega_d \omega_T (t_2 - t_1). \tag{7.5}$$

It follows from expressions (7.4) and (7.5) that the concentration of carbon in the decomposition products of austenite is

$$E_C^{(d)} = \Delta E_C = \omega_d \omega_T (t_2 - t_1). \tag{7.6}$$

The probability of carbide formation decreases from hafnium to zirconium, to titanium, etc. The amount of carbon bound by any carbide-forming element corresponds to the stoichiometry of the compound (Me_xC_y) and can be determined from the following expression:

$$E_{Ci}^{(c)} = E_i^{(c)} \frac{yA_C}{xA_i}, \tag{7.7}$$

where x and y are the stoichiometric coefficients, A_C and A_i are the atomic weights of carbon and the ith carbide-forming element, respectively, and $E_i^{(c)}$ is the concentration of the ith carbide-forming element in the carbide phase. For primary carbides, the value of x is always equal to 1, and y takes values from 0.4 to 1.0, depending on the homogeneity region of the particular carbide.

Of course, the total amount of carbon used for carbide formation should not exceed the carbon concentration determined for the crystallization time from Eq. 7.1. Moreover, it is known [3, 8] that not all the carbon and carbide-forming elements are consumed in carbide formation and that a portion of these elements remains dissolved in the austenite. It has been established [3, 6–8] that all the carbide-forming elements except chromium, manganese, and iron (which are completely dissolved in the austenite because their atomic radii are very close) participate in carbide formation during primary crystallization and form primary carbides (of the type MeC) with a carbon deficiency. As was postulated above, only the portion of the alloying elements and carbon that cannot be dissolved in the austenite at the particular temperature will be used for carbide formation, and precisely as much of the alloying elements as can be bound in carbides by the carbon that is not dissolved in the austenite will be used for carbide formation:

$$E_C^{(c)t_k} = E_C^{(s)t_k} - E_C^{(\lim)t_k}, \tag{7.8}$$

where $E_C^{(c)t_k}$ is the concentration of the carbon that is not dissolved in the austenite, $E_C^{(s)t_k}$ is the carbon concentration given by expression (7.1) at the crystallization time t_k, and $E_{\mathrm{C}}^{(\lim)t_k}$ is the solubility limit of carbon in austenite at the particular crystallization temperature at the time t_k.

According to the principles governing carbide formation, the distribution of carbon between the carbide phases in an alloy is proportional to the relative carbide-forming ability of the respective transition element $\frac{\Theta_i}{\sum_{i=1}^{n}\Theta_i}$ and its atomic percentage in the alloy a_i. It is, therefore, clear that the proportionality factor for the ith carbide-forming element is

$$\eta_i = \frac{1}{2}\left(\frac{a_i}{100} + \frac{\Theta_i}{\sum\limits_{i=1}^{n}\Theta_i}\right). \tag{7.9}$$

Then the concentration of the ith carbide-forming element dissolved in a particular carbide phase at the time t_k can be defined as

$$E_i^{(c)t_k} = \eta_i E_{\mathrm{C}}^{(c)t_k}\frac{xA_i}{yA_{\mathrm{C}}}, \tag{7.10}$$

Therefore, the concentration of the ith carbide-forming element dissolved in austenite at the time t_k equals

$$E_i^{(a)t_k} = E_i^{(s)t_k} - E_i^{(c)t_k}, \tag{7.11}$$

where $E_i^{(s)t_k}$ is the concentration of the carbide-forming element (wt.%) at the crystallization time t_k and is defined by expression (7.1).

The concentration of carbides formed at the time t_k is the sum of the amount of carbon and the total amount of the carbide-forming elements that are used for carbide formation:

$$Q_k^{t_k} = E_{\mathrm{C}}^{(c)t_k} + \sum_{i=1}^{n} E_i^{(c)t_k}. \tag{7.12}$$

In the next time interval t_{k+1}, the compositions of the austenite and carbide phases will be different. The calculation is repeated z times ($z = t_{rc}/t_k$, where t_{rc} is the cooling time, which is determined by the parameters of the thermal-straining cycle during welding). The total carbide content at the end of primary crystallization (t_{rc}) is defined as

$$Q^I = \sum_{k=1}^{z} Q_k^{t_k}. \tag{7.13}$$

Then the austenite content equals

$$Q^{(a)} = 100\% - Q^I. \tag{7.14}$$

The average concentrations of carbon and the alloying elements in the austenite phase are defined, respectively, as

$$E_{\mathrm{C}}^{(a)} = \frac{\sum_{k=1}^{z} E_{\mathrm{C}}^{t_k}}{zQ^{(a)}} 100\%; \quad E_i^{(a)} = \frac{\sum_{k=1}^{z} E_i^{t_k}}{zQ^{(a)}} 100\%. \tag{7.15}$$

Thus, at the end of primary crystallization, we know the average chemical composition of the austenite, as well as the quantitative and qualitative composition of the carbide phases in different zones of the welded joint. We note that as a result of primary crystallization, we obtained a chemically inhomogeneous welded joint.

The secondary crystallization that takes place during cooling is accompanied by diffusion-controlled evening of its composition to the composition specified by expression (7.15) and by partial coagulation of the primary carbides along their grain boundaries [3, 6–11]. When the temperature for the limiting solubility of carbon and the alloying elements in austenite is reached, the spinodal decomposition of austenite occurs, and the distribution of the impurities between the austenite and its decomposition products becomes extremely complex. If we take into account the assumptions made, we can determine this distribution and the concentrations of the alloying elements and carbon in the various phases of the deposited metal.

The distribution of carbon between the carbide phases is proportional to the relative carbide-forming ability of the respective transition element $\frac{\Theta_i}{\sum_{i=1}^{n} \Theta_i}$ and its atomic concentration in the alloy a_i, in analogy to the assumptions that we made when we considered primary crystallization (in accordance with the principles of carbide formation in alloys). In analogy to (7.9), we can write

$$E_i^{(c)} = \sum_{j=1}^{k} w_j \eta_i E_{\mathrm{C}}^{(d)} \frac{xA_i}{yA_{\mathrm{C}}}. \tag{7.16}$$

Here w_j is the fraction of carbon in the jth carbide phase relative to the total amount of carbon used to form carbides of the ith alloying element; $E_{\mathrm{C}}^{(d)}$ is the concentration of carbon in the austenite decomposition products;η_i is the coefficient defined by expression (7.9). Then the concentrations of the carbide-forming element and carbon dissolved in the matrix can be found from the expressions

$$E_i^{(b)} = E_i^{(a)} - E_i^{(c)}, \tag{7.17}$$

$$E_{\mathrm{C}}^{(b)} = E_{\mathrm{C}}^{(a)} - \sum_{i=1}^{n} \sum_{j=1}^{k} w_j \eta_i E_i^{(c)} \frac{yA_{\mathrm{C}}}{xA_i}. \tag{7.18}$$

The content of the carbide phases formed as a result of secondary crystallization is

$$Q^{II} = E_{\mathrm{C}}^{(a)} - E_{\mathrm{C}}^{(b)} + \sum_{i=1}^{n} E_i^{(c)}, \tag{7.19}$$

and the total content of the carbide phases (with consideration of (7.13)) is

$$Q_{Hd} = Q^{I} + Q^{II}. \tag{7.20}$$

Then the content of the matrix in the deposited metal is determined as

$$Q^{(b)} = 100\% - Q_{Hd}. \tag{7.21}$$

We find the concentration of carbon E_{C}^{b} and the concentrations of the alloying elements E_i^b in the matrix from the following expressions, respectively:

$$E_{\mathrm{C}}^{b} = \frac{E_{\mathrm{C}}^{(b)}}{Q^{(b)}}100\%; \quad E_i^b = \frac{E_i^{(b)}}{Q^{(b)}}100\%. \tag{7.22}$$

The equations obtained (Eqs. 7.1, 7.3–7.22) comprise a phenomenological model of the process of the nonequilibrium crystallization of a weld pool and the formation of a weld metal (or a deposited metal). Ultimately, at the end of the secondary crystallization, we have a deposited metal with a complex phase and structural composition that consists of primary and secondary carbides, on the one hand, and of the residual austenite and other austenite decomposition products (the matrix of the deposited metal), on the other hand. Equations 7.1 and 7.3–7.22 provide a way to calculate the concentrations of carbon and the alloying elements in the matrix, the amounts of carbon and the alloying components that were used to form the carbides in the deposited metal during its secondary crystallization, and, consequently, the amounts of the carbides themselves. The qualitative composition of the carbide phases is determined in parallel.

7.2 Predicting the Phase and Structural Composition of the Matrix of a Deposited Metal

The foregoing analysis shows that the problem of predicting the phase and structural composition calls for modification of the Schaeffler constitution diagram (Fig. 7.2) [12]. Numerous attempts have been undertaken to modify the Schaeffler diagram for the purpose of eliminating its shortcomings, which were examined in the preceding chapter. One of the first endeavors in this area was made by Schneider [13], who took into account the influence of alloying elements, such as N, Co, Cu, Al, and W, on the values of the Cr and Ni equivalents. Subsequently, Espy modified the Schaeffler diagram for nitrogen-containing austenite steels [14]. A formally different, but essentially similar type of diagram was devised by Potak and Sagalevich [15]. The mutual influence of alloying components was partially taken into account, and it became possible to predict the phase and structural composition of the deposited metal when the carbon content in it is <0.3 wt.%. Further modification of the Potak–Sagalevich diagram [16] significantly expanded the region for predicting deposited metals with different degrees of alloying, but

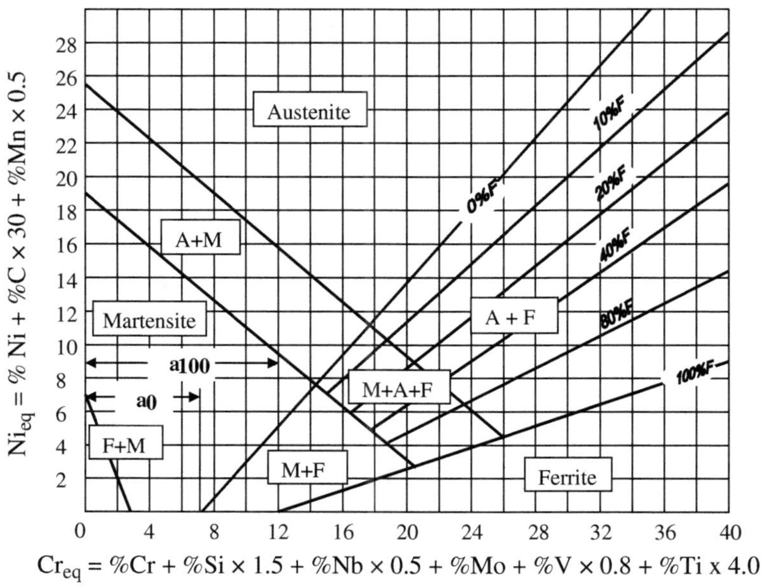

Fig. 7.2 The Schaeffler constitution diagram

did not eliminate the problems in predicting the phase and structural composition of a deposited metal with a carbon content >1 wt.% and a high concentration of alloying elements.

Questions related to the more exact prediction of the phase composition in two-phase regions were also considered in several publications and were solved by creating special diagrams [15, 17–19] or modifying the Schaeffler diagram [20]. The work of Kotecki [17, 21–23], which was devoted to more exact prediction of the phase and structural composition and establishing the boundaries for the formation of martensite in the steels investigated at different levels of alloying with manganese, merits special attention. David, Babu, and other investigators [24–28] carried out important studies in the area of predicting the phase and structural composition of austenite steels. However, none of these studies solved the general problem of determining the phase and structural composition of the matrix of a deposited metal over a broad alloying range, and, as a rule, they all focused on individual groups of steels or special cases of alloying.

In the first stage of the development of a new constitution diagram, parametric equations for calculating the chromium and nickel equivalents that take into account the nonequilibrium nature of the crystallization of a weld metal and the binding of a portion of the alloying components and a portion of the carbon in the formation of primary (special) and secondary (eutectic) carbides must be derived in order to predict the phase composition of the matrix. An analysis of the structures of more than one hundred deposited materials led to the establishment of empirical dependences of the changes in the coefficients in the formulas for calculating the equivalent values of chromium (Cr_{eq}) and nickel (Ni_{eq}) for the Schaeffler diagram [12]. The modified forms of these formulas are as follows:

$$\mathrm{Cr_{eq}} = \sum_{i=1}^{l} f_i(\omega) E_{m_a(Frt)_i}, \tag{7.23}$$

$$\mathrm{Ni_{eq}} = \sum_{i=1}^{z} \varphi_i(\lambda) E_{m_a(Ast)_i} + \varphi_c(\lambda) C_{m_a}, \tag{7.24}$$

where l is the number of ferrite-forming elements, $E_{m_a(Frt)_i}$ is the concentration of the ith ferrite-forming element (wt.%), $f_i(\omega)$ is the value of the coefficient in front of the concentration of the ith ferrite-forming element as a function of the complex factor ω, z is the number of austenite-forming elements, $E_{m_a(Ast)_i}$ is the concentration of the ith austenite-forming element (wt.%), $\varphi_i(\lambda)$ is the value of the coefficient in front of the concentration of the ith austenite-forming element as a function of the complex factor λ, and ω and λ are complex factors that are determined for the ferrite- and austenite-forming elements on the basis of an analysis of the phase and structural composition of deposited layers and a weld metal with a broad alloying range.

After this we can proceed to determining the phase ratio in multiphase regions of the diagram. The phase ratios in multiphase regions of the diagram were determined with allowance for carbide formation according to the structural analysis that was performed to determine the coefficients in Eq. 7.19 and 7.20, and percentage grids for the concentration of the second (and third) phase were drawn accordingly on the corresponding regions of the diagram. For the three-phase regions, algorithms were devised to determine the content of each phase (the austenite + ferrite + martensite region, where a grid for determining the ferrite phases was given by Schaeffler, is such a region). As a result of such transformations, a fundamentally new constitution diagram of the matrix of a deposited metal with an unrestricted alloying range that is obtained during nonequilibrium crystallization of the molten metal in the weld pool was constructed. The new diagram takes into account the use of part of the alloying d elements and carbon for carbide formation, the mutual influence of the elements on shaping the phase composition of the matrix, and the parameters of the thermal-straining cycle during welding, which is especially important. Unlike the Schaeffler diagram, which specifies the qualitative content of austenite and martensite and the quantitative content of the ferrite phase, the new constitution diagram specifies both the qualitative and the quantitative content of all the phase and structural components of the matrix. As is clear from the foregoing, before calculating the chromium and nickel equivalents using Eq. 7.23 and 7.24, it is first necessary to calculate the quantitative and qualitative content of the hardening phases and to determine the amounts of alloying d elements and carbon that are used for carbide formation. Afterwards we can calculate the chemical composition of the matrix of the deposited metal and, on its basis, the chromium and nickel equivalents. Knowing the values of the chromium and nickel equivalents, we determine the coordinates of the point on the new diagram which characterizes the qualitative phase and structural composition of the matrix for us. To quantitatively determine the phase components of the matrix of a deposited metal, it is necessary to use the percentage content lines of the phase components for each specific region of the diagram and, when necessary, to resort to interpolation. As we see, direct use of the diagram in its graphical form is a fairly tedious process, and the accuracy of the calculations is lost. Moreover, it is impossible to create a complete system for the computerized calculation of the quantitative and qualitative composition of the hardening phases and the phase and structural composition of the matrix of the deposited metal obtained during the nonequilibrium crystallization of the molten metal in a weld pool. This problem can be solved on the basis of a mathematical description of the new constitution diagram.

7.3 Mathematical Description of the New Constitution Diagram of the Matrix of a Deposited Metal

The problem under consideration was solved in several stages [29]. First, the boundary conditions for regions of the diagram (that is, the conditions, outside of which a given phase does not exist or exists only in contact with other phases) were determined. In the second stage, regression analysis of the experimental data for the structural components of different groups in the deposited metal was performed. Regression equations that provide a way to calculate the phase and structural composition for a specific region of the diagram (with specific values of the equivalents Cr_{eq} and Ni_{eq}) and the conditions that determine the applicability of specific regression equations as a function of the alloying level of the alloy were obtained.

Determination of the boundary conditions for regions of the diagram. The boundary conditions for regions of the diagram can be assigned using the mathematical equations for the separating lines that mark the boundaries of the structural regions. They describe the dependence of the Ni equivalent (Ni_{eq}) on the Cr equivalent (Cr_{eq}):

$$Ni_{eq} = f(Cr_{eq}) \tag{7.25}$$

or, since they are linear dependences,

$$Ni_{eq} = aCr_{eq} + b. \tag{7.26}$$

Thus, the position of a specific line on the diagram is specified by the coefficients a and b. Figure 7.3 presents the general form of the new constitution diagram with the notations of the separating lines. We note that the concentration of the ferrite phase is specified on it by the ferrite number (*FN*) lines [30].

The values of the coefficients a and b for each specific separating line are presented below (Table 7.1).

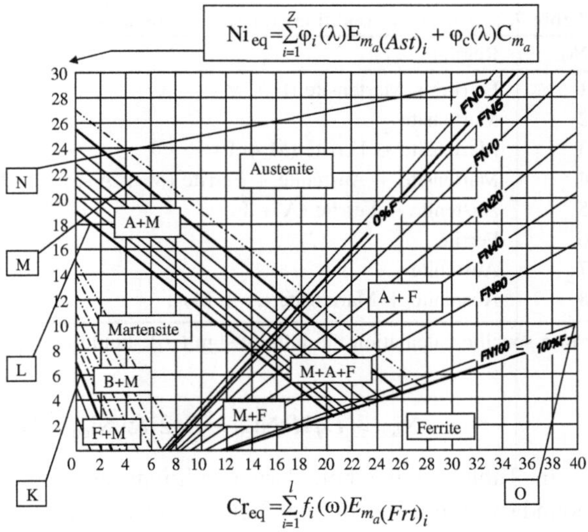

Fig. 7.3 New constitution diagram that takes into account the carbide formation processes in a deposited metal

Table 7.1 Values of the coefficients *a* and *b* for the equations of the separating lines

Line	*a*	*B*
K	−2.45	+7.2
L	−0.8	+19
M	−0.8	+25.4
N	1.13	−8.1
Q	0.35	−4.0

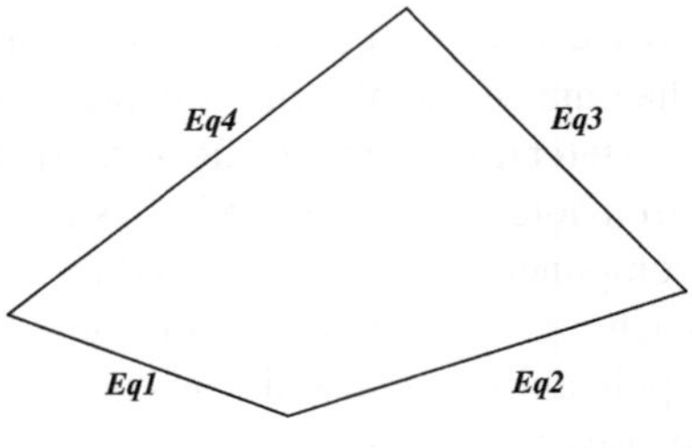

Fig. 7.4 Schematic representation of the *i*th region of the diagram

Now, having the values for the boundary conditions, we can devise an algorithm that determines the region in which the point with the coordinates (Cr_{eq}, Ni_{eq}) is located. Any region on the diagram can be represented in the form of irregular quadrangle (Fig. 7.4), which degenerates in some cases into a triangle (the F + M and ferrite regions).

In Fig. 7.4 the boundary conditions are designated as *Eq*1, *Eq*2, *Eq*3, and *Eq*4. Then the condition that satisfies finding a point in a specific region when Cr_{eq} and Ni_{eq} have specific values can be written in the following form:

Table 7.2 Boundary conditions for specific regions of the diagram

No.	Region	*Eq*1	*Eq*2	*Eq*3	*Eq*4
1	Ferrite + Martensite (F + M)	0.0	0.0	K	7.2
2	Martensite	K	N	L	19.0
3	Martensite + Ferrite (M + F)	0	Q	L	N
4	Martensite + Austenite + Ferrite (M + A+F)	L	Q	M	N
5	Austenite + Ferrite (A + F)	M	Q	37.5	N
6	Ferrite	0.0	0.0	10	Q
7	Austenite	M	N	∞*	∞*
8	Austenite + Martensite (A + M)	L	N	M	25.4

*In a specific calculation the infinity sign (∞) should be replaced by a large real number

$$(\mathrm{Ni_{eq}} \geq Eq1 \& \mathrm{Ni_{eq}} \geq Eq2) \& (\mathrm{Ni_{eq}} < Eq3 \& \mathrm{Ni_{eq}} < Eq4). \quad (7.27)$$

In Table 7.2 the boundary conditions are represented by the names of the boundary lines, and the coefficients corresponding to them in Eq. 7.25 are presented in Table 7.1. If boundaries of regions coincide with a boundary of the diagram (i.e., lines that coincide with or are parallel to the Cr_{eq} and Ni_{eq} axes), the equations for the boundary conditions degenerate into constants.

The algorithm for solving the problem of determining the regions where a specific point with the coordinates (Cr_{eq}, Ni_{eq}) is located has the form of a cycle, in which values of Cr_{eq} and Ni_{eq} corresponding to the coordinates of the point sought (these values do not vary during the cycle) are successively introduced into condition (7.27) along with the set of data from Table 7.2 corresponding to the cycle counter. This set of data contains the name of the region and the values of its boundary conditions. If condition (7.27) is satisfied in any step of the cycle, the cycle is terminated, and the row number with the code of the region sought is provided as output. Otherwise, the next iteration of the cycle begins. A block diagram of the algorithm is presented in Fig. 7.5.

Determination of the structural components within each region. As has been noted above, dependences (regression equations) for calculating the structural components were determined by regression analysis of the experimental data for each of the regions of the new constitution diagram, and the conditions for applicability of specific dependences for calculating the amounts of the individual phases in the multiphase regions of the diagram were determined. These conditions and the dependences for each of the eight regions are presented below. It should be noted that these conditions and dependences can be refined as experimental data are accumulated. The results of calculations performed using these dependences with allowance for their applicability conditions are in good agreement with the experimental data.

The ferrite + martensite (F + M) region. This region is located in the lower left-hand corner of the diagram and the structures specified by it are characteristic of a low-alloy deposited metal. If the condition

$$\mathrm{Cr_{eq}} < 2.8 \& (0.8 < \mathrm{Mn} < 2.0) \& SmLg < 3.5, \quad (7.28)$$

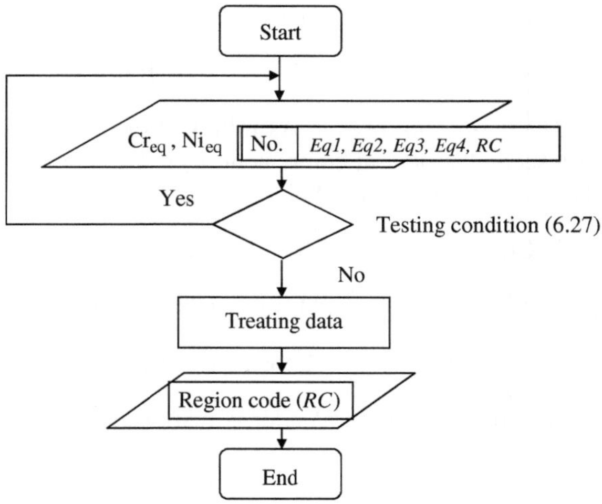

Fig. 7.5 Block diagram of the algorithm for determining the region where a point with the coordinates (Cr_{eq}, Ni_{eq}) is located

holds, the structure of this region will consist of two structural components, namely, pearlite and bainite. In expression (7.28) *SmLg* is the sum of the ferrite-forming alloying elements, with the exception of Cr. The content of the bainite structural component (wt.%) is given by the following dependence:

$$Q_B = 100.75\ln(Ni_{eq} + 2.45Cr_{eq}) - 99.324. \tag{7.29}$$

The remainder is the content of the pearlite structural component (wt.%):

$$Q_P = 100 - Q_B. \tag{7.30}$$

As has been noted, to determine the complete phase and structural composition, the previously calculated concentrations of carbides and other hardening phases must be taken into account. Then the complete phase and structural composition of the deposited metal is (wt.%)

$$Q_{Me} = Q_{Hd} + k_f Q_B + k_f Q_P = 100, \tag{7.31}$$

where Q_{Hd} is the total concentration of carbides and other hardening phases (wt.%) and $k_f$is a coefficient that takes into account the phase ratio:

$$k_f = \frac{100 - Q_{Hd}}{100}. \tag{7.32}$$

Dependences (7.31) and (7.32) are characteristic of all the regions, and we will not dwell further on the determination of the complete phase and structural composition of a deposited metal, but we will examine the structure of the matrix. If condition (7.28) does not hold, the constitution of the region consists of ferrite and martensite. The concentration of the martensite phase (wt.%) is determined from a dependence that is analogous to (7.29), and, and the concentration of ferrite (wt.%) is determined from a dependence that is analogous to (7.30):

$$Q_M = 100.75\ln(\mathrm{Ni_{eq}} + 2.45\mathrm{Cr_{eq}}) - 99.324, \quad (7.33)$$

$$Q_F = 100 - Q_B, \quad (7.34)$$

The martensite region. As follows from the experimental data, the appearance of a second structural component, namely, bainite, is possible in this region when the following condition holds:

$$\mathrm{Cr_{eq}} < 5.6 \& (0.8 < \mathrm{Mn} < 2.0) \& SmLg < 3.5 \& (\mathrm{Ni_{eq}} - 2.05) < 15.0. \quad (7.35)$$

The bainite content (wt.%) can be calculated from the following dependence:

$$Q_B = 2.0802(\mathrm{Ni_{eq}} + 2.45\mathrm{Cr_{eq}})^2 - 59.012(\mathrm{Ni_{eq}} + 2.45\mathrm{Cr_{eq}}) + 418.1. \quad (7.36)$$

Accordingly, the remainder is martensite (wt.%):

$$Q_M = 100 - Q_B. \quad (7.37)$$

If condition (7.35) does not hold, the matrix consists entirely of the martensite phase.

The martensite + ferrite (M + F) region. This region is located in the lower central part of the diagram and defines a region of a two-phase deposited metal. Under the condition that the ratio has a value

$$R_q = \frac{\mathrm{Ni_{eq}} + 0.4455\mathrm{Cr_{eq}}}{1 - 0.1908\mathrm{Cr_{eq}}} \leq -6.4, \quad (7.38)$$

the content of the ferrite phase (wt.%) is

$$Q_F = 10.142R_q + 85.071. \quad (7.39)$$

Otherwise, the content of the ferrite phase (wt.%) can be determined from the dependence

$$Q_F = -13.972R_q^2 - 112.22R_q - 125.38. \quad (7.40)$$

The other component of the matrix is martensite (wt.%):

$$Q_M = 100 - Q_F. \quad (7.41)$$

The martensite + austenite + ferrite (M + A + F) region. This region is located in the center of the diagram and represents an area with a three-phase constitution. The condition and dependences for calculating the content of the ferrite phase are analogous to Eqs 7.38–7.40. The content of the martensite phase (wt.%) is specified by the following dependence:

$$Q_M = k_z(0.313Q_Z^3 - 18.934Q_Z^2 + 359.25Q_Z - 2037.3), \quad (7.42)$$

where $k_z = \frac{100-Q_F}{100} = \frac{Q_Z}{100}$ is a coefficient that takes into account the phase ratio. Then the content of the austenite phase (wt.%) in the matrix is

$$Q_A = 100 - Q_F - Q_M. \tag{7.43}$$

The austenite + ferrite (A + F) region. This region embraces a broad spectrum of compositions of austenite–ferrite deposited metals and is extremely important for determining the content of the ferrite phase. For this region we use condition and dependences (7.28–7.40). The austenite content in the matrix (wt.%) is:

$$Q_A = 100 - Q_F. \tag{7.44}$$

The ferrite region. This region is located in the lower right-hand part of the diagram and characterizes a single-phase area. Thus, if it is determined in accordance with condition (7.28) that a point with the coordinates $(\mathrm{Cr_{eq}}, \mathrm{Ni_{eq}})$ is located in this region, the matrix consists entirely of ferrite.

The austenite region. This region is located in the upper part of the diagram and embraces a broad spectrum of single-phase austenite steels. At a specific ratio between the alloying components and specific cooling rates, a pearlite component can appear in the deposited metal due to the decomposition of austenite. Its content generally does not exceed 15%. It was established on the basis of a regression analysis of experimental data that when the condition

$$\mathrm{Cr_{eq}} < 14\&\mathrm{Ni_{eq}} > (-0.8\mathrm{Cr_{eq}} + 25.48)\&\mathrm{Cr} > 13.8\&(\mathrm{Ni_{eq}} + 0.8\mathrm{Cr_{eq}}) < 36.0 \tag{7.45}$$

holds, the content of the pearlite component (wt.%) can be determined from the dependence

$$Q_P = 0.0987Q_R^2 - 7.0412Q_R + 125.91, \tag{7.46}$$

where Q_P is the pearlite content (wt.%) and $Q_R = \mathrm{Ni_{eq}} + 0.8\mathrm{Cr_{eq}}$.

The remaining content of the matrix is austenite (wt.%):

$$Q_A = 100 - Q_P. \tag{7.47}$$

If condition (7.45) does not hold, the matrix of the deposited metal consists entirely of austenite.

The austenite + martensite (A + M) region. This region is located in the central left part of the diagram and embraces an extremely important and interesting spectrum of compositions of deposited metals, which is characteristic of many surfacing electrodes and wires intended for surfacing layers that operate under the conditions of impact-abrasive loading, hydro-abrasive wear, cavitation effects, etc. The martensite content for these types of deposited metals is given by a dependence of type (7.30). Then the austenite content (wt.%) is

$$Q_A = 100 - Q_F. \tag{7.48}$$

Now for each region of the new constitution diagram we have mathematical dependences that enable us to calculate the percentages of the structural components (phases) in a deposited metal with a definite chemical composition, which is characterized by a point with the coordinates $(\mathrm{Cr_{eq}}, \mathrm{Ni_{eq}})$ in a specific region (Table 7.3).

Thus, we have obtained a phenomenological representation of the nonequilibrium crystallization processes of a weld metal with a broad alloying range, which enables us to predict the phase and structural composition of the deposited metal. Equations 7.1 and 7.3–7.22 and the formalized description of the new constitution diagram (Eqs. 7.23–7.48) offer an analytical method for calculating the phase and structural composition of a weld metal that follows from a phenomenological description of its nonequilibrium crystallization. The set of expressions obtained provide for the calculation of the qualitative and quantitative compositions of the hardening phases and the chemical composition of the matrix of a deposited metal with allowance for carbide formation processes. Now, knowing the chemical composition of the matrix, we can determine the values of the chromium and nickel equivalents, and from them we can determine the phase and structural composition of the matrix of the deposited metal using the formalized description of the diagram.

To illustrate the possibilities of the proposed approach, we will examine an example of a highly simplified calculation of the phase and structural composition of a weld metal. We will determine the phase and structural composition of the weld metal without and with allowance for the formation of carbides when the metal has the following chemical composition (wt.%): C, 0.09; Si, 0.3; Mn, 1.4; Cr, 20.1; Ni, 9.85; Nb, 0.8. This chemical composition corresponds to grade 09Kh20N10 GB steel.

1. Calculation of the phase and structural composition of the weld metal without allowance for the formation of carbides.
   According to the formula [12]

$$\mathrm{Cr_{eq}} = \%\mathrm{Cr} + \%\mathrm{Si} \cdot 1.5 + \%\mathrm{Nb} \cdot 0.5 + \%\mathrm{Mo} + \%\mathrm{V} \cdot 0.8 + \%\mathrm{Ti} \cdot 4.0, \tag{7.49}$$

$$\mathrm{Ni_{eq}} = \%\mathrm{Ni} + \%\mathrm{C} \cdot 30 + \%\mathrm{Mn} \cdot 0.5, \tag{7.50}$$

without allowance for the formation of niobium carbides (NbC), the chromium and nickel equivalents are (wt.%):

$$\mathrm{Cr_{eq}} = 20.1 + 0.3 \cdot 1.5 + 0.8 \cdot 0.5 = 20.95,$$

$$\mathrm{Ni_{eq}} = 9.85 + 0.09 \cdot 30 + 1.4 \cdot 0.5 = 13.25.$$

The values determined for the coordinates of the point on the Schaeffler diagram correspond to an austenite–ferrite structure with a ferrite content equal to 4.9 wt.% (Fig. 7.2).

2. Calculation of the phase and structural composition of the weld metal with allowance for the formation of carbides.

Table 7.3 Equations for calculating the structural components (phases) in a deposited metal

Specification of the region		Calculation of the phase and structural composition of the deposited metal	Remarks
Condition	Name		
$0 \le Ni_{eq} < -2.45Cr_{eq} + 7.42$	F + M	$Q_M = 100.75 \ln (Ni_{eq} + 2.45Cr_{eq}) - 99.324$; $Q_F = 100 - Q_M$	
$Ni_{eq} \ge -2.45Cr_{eq} + 7.42$; $Ni_{eq} \ge 1.13Cr_{eq} - 8.1$; $Ni_{eq} < -0.8Cr_{eq} + 19$	Martensite	$Q_M = 100$	
$Ni_{eq} \ge 0$; $Ni_{eq} \ge 0.35Cr_{eq} - 4.0$; $Ni_{eq} < 1.13Cr_{eq} - 8.1$; $Ni_{eq} < -0.8Cr_{eq} + 19$	M + F	If $R_q \le -6.4$, $Q_F = 10.142R_q + 85.071$; otherwise, $Q_F = -13.972R_q^2 - 112.22R_q - 125.38$; $Q_M = 100 - Q_F$	$R_q = \frac{(Ni_{eq} + 0.4455Cr_{eq})}{(1 - 0.1908Cr_{eq})}$
$Ni_{eq} \ge -0.8Cr_{eq} + 19$; $Ni_{eq} \ge 0.35Cr_{eq} - 4.0$; $Ni_{eq} < -0.8Cr_{eq} + 25.4$; $Ni_{eq} < 1.13Cr_{eq} - 8.1$	M + A + F	If $R_q \le -6.4$, $Q_F = 10.142R_q + 85.071$; otherwise, $Q_F = -13.972R_q^2 - 112.22R_q - 125.38$; $Q_M = k_z(0.313Q_Z^3 - 18.3934Q_Z^2 + 359.25Q_Z - 2037.3)$; $Q_A = 100 - Q_F - Q_M$	$R_q = \frac{(Ni_{eq} + 0.4455Cr_{eq})}{(1 - 0.1908Cr_{eq})}$; $k_z = (100 - Q_F)/100$; $Q_Z = Ni_{eq} + 0.8Cr_{eq}$
$0 \le Ni_{eq} < 0.35Cr_{eq} - 4.0$	Ferrite	$Q_F = 100$	
$Ni_{eq} \ge -0.8Cr_{eq} + 25.4$; $Ni_{eq} \ge 1.13Cr_{eq} - 8.1$	Austenite	$Q_A = 100$	
$Ni_{eq} \ge -0.8Cr_{eq} + 19$; $Ni_{eq} \ge 1.13Cr_{eq} - 8.1$; $Ni_{eq} < -0.8Cr_{eq} + 25.4$	A + M	$Q_M = 0.313Q_Z^3 - 18.3934Q_Z^2 + +359.25Q_Z - 2037.3$; $Q_A = 100 - Q_M$	$Q_Z = Ni_{eq} + 0.8Cr_{eq}$

The weld metal is doped with niobium; therefore, carbides (NbC) will form in it with the following stoichiometric ratio:

$$\mathrm{Nb/C} = 92.9/12 \cong 8.$$

Assuming that the solubility limit of carbon in the solid solution is 0.03 wt.%, we determine the amount of excess carbon that is capable of forming niobium carbides (wt.%):

$$C_{\mathrm{exc}} = 0.09 - 0.03 = 0.06.$$

The quantity of free (excess) carbon obtained is capable of being bound in niobium carbides (wt.%):

$$\mathrm{Nb}_C = 0.06 \cdot 8 = 0.48.$$

The quantity of free ferrite-forming niobium is (wt.%):

$$\mathrm{Nb}_F = 9.8 - 0.48 = 0.32.$$

When the formation of niobium carbides is taken into account, the chromium and nickel equivalents given by formulas (7.49) and (7.50) become:

$$\mathrm{Cr_{eq}} = 20.1 + 0.3 \cdot 1.5 + 0.32 \cdot 0.5 = 20.71;$$

$$\mathrm{Ni_{eq}} = 9.85 + 0.03 \cdot 30 + 1.4 \cdot 0.5 = 11.45.$$

The values determined for the coordinates of the point on the Schaeffler diagram correspond to an austenite–ferrite structure with a ferrite concentration equal to 10% (Fig. 7.2).

The calculation presented reveals a significant difference between the results obtained. The increase in the ferrite concentration observed when the influence of carbide-forming elements is taken into account is associated with the abrupt decrease in the austenite-forming ability of carbon due to its binding in carbides. Already in this simple example, which takes into account only the formation of niobium carbides, the novelty and effectiveness of the proposed approach are apparent.

Figure 7.6 presents the complete algorithm for calculating the phase and structural composition of a weld metal.

7.4 Experimental Testing of the Methods Developed

The phenomenological model of carbide formation in an iron–carbon deposited metal and the mathematical description of the new constitution diagram both require thorough testing to determine their capability and their faithfulness to real processes. Such testing was implemented by conducting a set of experimental studies under the conditions of the advanced materials laboratory of the Materials Science Center of the Ariel University Center of Samaria (Israel).

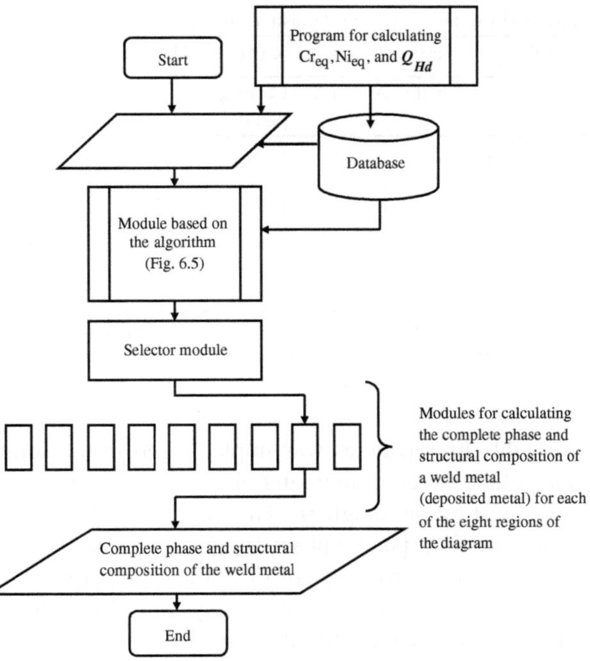

Fig. 7.6 Algorithm for calculating the phase and structural composition of a weld metal (deposited metal)

Table 7.4 Chemical composition of the deposited metal in the samples investigated

No.	Code number of the sample	Concentration (wt.%)					Remarks
		C	Si	Mn	Cr	Ti	
1	RI 1/01.7	1.58 / ≈1.40	0.68 / 0.554	0.347 / 0.427	8.54 / 8.50	3.61 / 3.35	The remainder is Fe.
2	RI 1/11.2	1.62 / ≈1.80	0.70 / 0.29	0.378 / 0.447	6.89 / 6.63	3.60 / 2.63	The remainder is Fe.
3	RI 1/08.3	1.306 / ≈1.50	0.53 / 0.60	0.385 / 0.362	7.87 / 8.15	2.57 / 2.10	The remainder is Fe.
4	RI 1/10.1	1.319 / ≈1.40	0.70 / 0.54	1.12 / 0.60	7.93 / 5.62	2.62 / 2.28	The remainder is Fe.

Note The values above the bars correspond to the expected composition, and the values below the bars correspond to the actual composition

The summary results of the investigations are presented in Tables 7.4 and 7.5.

The faithfulness of the model was evaluated on the basis of a comparison of the results of the design solution and the results obtained during the set of tests. As follows from the data presented, for the most part, the discrepancies do not exceed 20%, indicating the high faithfulness of the phenomenological model developed to the real process of nonequilibrium crystallization of the molten metal in a weld pool. In individual cases (especially in the evaluation of austenite) the discrepancies revealed exceed 20%. Such discrepancies are significant and indicate a need for adjustment of the model.

Table 7.5 Structure and properties of the deposited metal

No.	Code number of the sample	Structural components, wt.%			Hardness, HRC	Increase in hardness after loading, ΔHRC
		Martensite	Austenite	Carbide phase		
1	RI 1/01.7	55.52 / base	33.97 / 36.00	10.47 / 12.00	56.16 / 56–57	+(3–4) / ≈(+4)
2	RI 1/11.2	44.49 / 43.00	45.29 / 47.00	9.74 / 10.00	46.19 / 46–50	+(4–5) / ≈(+6)
3	RI 1/08.3	77.12 / base	13.65 / 11.00	9.20 / 8.00	56.39 / 54–56	+(2–3) / ≈(+3)
4	RI 1/10.1	81.16 / base	9.76 / 8.00	9.05 / 8.00	57.23 / 55–58	+(2–3) / ≈(+2)

Note The values above the bars correspond to the expected structure and properties, and the values below the bars correspond to the actual structure and properties

The analysis of the discrepancies was implemented in the form of correction functions for calculating the chromium and nickel equivalents using Eq. 7.23 and 7.24 in the new constitution diagram. The functions should ensure minimal deviation of the amount of a specific phase (in our case austenite) from the design solution, that is, the partial derivatives of (7.23) and (7.24) with respect to these functions should be equal to zero. Solving the system of equations obtained, we determined correcting coefficients for the functions $f_i(w)$, $\varphi_i(\lambda)$, and $\varphi_c(\lambda)$ in Eqs. 7.23 and 7.24, which ensured convergence of the calculation results with the actual data.

7.5 Calculation of the Composition and Amounts of the Phases in Equilibrium Multicomponent Alloys Based on Iron

Mathematical modeling and computational experiments have provided a way to solve problems involving the prediction of the composition of metals for diverse production and treatment technologies. The use of computer models reduces the duration and laboriousness of investigations, as well as enables the performance of computer experiments in cases in which the performance of full-scale trials is difficult or impossible. One of the important problems is the prediction of the composition and amounts of the phases formed when the components of a metal interact with one another both in the liquid state and during cooling of the solid alloy. Such phases can be carbides, oxides, nitrides, etc. The solution of this problem for multicomponent alloys is very complex. This is due, first, to the desire to provide for the formation of all the phases which may exist under particular conditions. Second, the kinetic hindrances, which actually determine the appearance of specific phases, as well as their amount, morphology, and composition under assigned crystallization conditions and a specified cooling rate, must be taken into account. To solve this problem, it is expedient to use the thermodynamic approach, which provides a way to predict the composition and amounts of the phases that are formed in the equilibrium state. The advantages of the thermodynamic approach include the relatively small amount of data needed for the calculation and the high reliability of the results, which depends solely on the accuracy of the thermodynamic characteristics used. In addition, in some cases, in which the rates of the phase formation reactions are fairly high, a thermodynamic calculation can give results that are close to the experimental values. Such conditions can be expected when finely dispersed primary phases form in a molten metal as a consequence of the high temperatures, large specific surface of the particles, and rapid diffusion, which significantly accelerate the interaction. For the same reasons, the actual phase composition of the alloy can be close to the equilibrium composition even in a certain temperature range below the solidus line when the cooling rate is not excessively high. The solution method described in this section also opens up a phenomenological possibility for taking into account the kinetic conditions for the formation and growth of new phases when there are kinetic data for the reactions under consideration.

Numerous studies devoted to the calculation of the phase equilibrium in multicomponent and multiphase systems, mainly involving gases and pure condensed phases, have been published [31, 32]. Techniques and computer programs that can be used to predict the appearance of scores of phases in systems containing condensed, generally ideal solutions were described in these investigations [32]. For metallic systems, which usually form non-ideal solutions, similar methods were considered in [33–39], particularly with regard to the calculation of phase diagrams. However, the solution methods used [33–36], for example, the Newton–Raphson method, require a satisfactory initial approximation and are extremely laborious already when three or four phase form. Nevertheless, impressive advances in the analysis of diagrams of state of metals [33–39], in taking into account the formation of carbides and nitrides in Fe–Ti–V–Cr alloys [34], and in calculations of the amount and composition of oxide inclusions in iron alloys [40] can be noted.

The proposed method provides a way to solve the problem described for a significantly larger and, in principle, arbitrary number of phases using a relative simple computer program.

The method is based on successive calculations of an approximation of the system to equilibrium using expressions for conditional reaction rates based on the law of mass action. Therefore, the method can be called a kinetic or relaxation method. Its important advantages include absolute convergence and the existence of only a unique solution [41]. In the case of multicomponent solutions, there is no need to solve systems of equations, and there is no requirement for a "good" initial approximation. The problem of determining the list of phases whose formation should be taken into account, which is a complex problem in other methods, is solved automatically. For the multiphase systems under consideration, the law of mass action is expressed in terms of mass concentrations, and in each calculation cycle the activity coefficients of the components of the solutions are refined, and the masses of all the phases, their composition, the list of possible reactions, etc. are determined.

We will describe the general approach in the example of multicomponent alloys based on iron.

The chemical and phase composition of such an alloy in the equilibrium state is calculated using the complete material balance with respect to all 16 elements that appear in the system: Fe, C, Si, Mn, Cr, Ti, V, Mo, W, Ni, Nb, Zr, Al, B, N, and O. The program takes into account the possibility of the formation of 60 individual phases of carbides, nitrides, borides, and oxides. In addition, the formation of solutions (for example, mixed carbides and oxide inclusions) is taken into account.

The input parameters of the model are the composition of the alloy and the temperature.

From the mathematical standpoint, the problem consists of finding 76 unknowns: the concentrations of the 16 elements in the metal matrix and the content (wt.%) of all 60 phases. This requires solving a system with a minimum of 76 equations: 16 material balance equations for the alloy elements and 60 equations for the equilibrium constants of all the phase formation reactions. If the compounds appearing are mutually soluble, the compositions of the solutions are also subject to determination.

The equilibrium state of the system is completely described by a set of independent reactions of the form

$$x[\mathrm{E}_i] + [\mathrm{C}] = (\mathrm{E}_{ix}\mathrm{C}), \tag{7.51}$$

$$y[\mathrm{E}_i] + [\mathrm{O}] = (\mathrm{E}_{iy}\mathrm{O}), \tag{7.52}$$

$$z[\mathrm{E}_i] + [\mathrm{N}] = (\mathrm{E}_{iz}\mathrm{N}), \tag{7.53}$$

$$n[\mathrm{E}_i] + [\mathrm{B}] = (\mathrm{E}_{in}\mathrm{B}), \tag{7.54}$$

where E_i, C, O, N, B are elements dissolved in the metal.

The number of elements and phases formed that are considered can easily be increased without introducing fundamental additions into the calculation procedure. For example, taking into account the formation of sulfides, silicides, etc. may be of practical interest.

The formation of several types of phases of one element (for example, the carbides V_2C and VC) is taken into account by altering the corresponding values of the stoichiometric coefficients x, y, z, and n, which can also have fractional values. The mutual influence of reactions is automatically taken into account through common reactants.

We write the expressions for the rates of the reactions resulting in the formation of carbides, oxides, nitrides, and borides in a unit mass of the alloy in the following form:

$$V_{\mathrm{C},i} = K^0_{\mathrm{C},i}\left([\mathrm{E}_i]^x[\mathrm{C}] - K_{\mathrm{C},i}a_{\mathrm{E}_{ix}\mathrm{C}}\right), \tag{7.55}$$

$$V_{\mathrm{O},i} = K^0_{\mathrm{O},i}\left([\mathrm{E}_i]^y[\mathrm{O}] - K_{\mathrm{O},i}a_{\mathrm{E}_{iy}\mathrm{O}}\right), \tag{7.56}$$

$$V_{N,i} = K^0_{N,i}\left([E_i]^z[N] - K_{N,i}a_{E_{iz}N}\right), \tag{7.57}$$

$$V_{B,i} = K^0_{B,i}\left([E_i]^n[B] - K_{B,i}a_{E_{in}B}\right). \tag{7.58}$$

Here the $V_{J,i}$ are the rates of reactions (7.51–7.54), mole *J*/kg s; *J* is a an index for denoting C, O, N, and B; the $K^0_{J,i}$ are the formal rate constants, whose values are chosen for reasons related to the stability and speed of the solution; the$K_{J,i}$ are the equilibrium concentration products for reactions (7.51–7.54), which depend on the composition of the matrix and can be refined in each calculation cycle; $[E_i]$ and $[J]$ are the concentrations of the elements dissolved in the metal at a given time, wt.%; and a_{E_iJ} is the activity of the compound E_iJ in the phases formed at the same time. When pure phases or solutions saturated with a particular compound form, we have $a_{E_iJ} = 1$.

We stress that Eqs. 7.55–7.58 are provisional and do not reflect the actual dependences of the rates of reactions (7.51–7.54) on the concentrations of the reactants. However, they correctly describe the tendency of the rates of reactions (7.51–7.54) to decrease to zero as these reactions approach equilibrium. This allows us to use these dependences in the analysis of the changes in a system that is relaxing toward equilibrium.

The values of the$K_{J,i}$ in Eqs. 7.55–7.58 equal the following:

$$K_{C,i} = \frac{[E_i]^x[C]}{a_{E_{ix}C}}, \tag{7.59}$$

$$K_{O,i} = \frac{[E_i]^y[O]}{a_{E_{iy}O}}, \tag{7.60}$$

$$K_{N,i} = \frac{[E_i]^z[N]}{a_{E_{iz}N}}, \tag{7.61}$$

$$K_{B,i} = \frac{[E_i]^n[B]}{a_{E_{in}B}}. \tag{7.62}$$

The concentrations and activities in Eqs. 7.59–7.62 refer to the equilibrium state, i.e., are the quantities sought, unlike the corresponding values in expressions (7.55–7.58), which refer to the current time.

As the system relaxes toward equilibrium, the concentrations of the dissolved elements decrease, in agreement with Eqs. 7.55–7.58, approaching the equilibrium values corresponding to expressions (7.59–7.62). A concentration difference no greater than 0.1 rel. % for all the reactions was taken as the criterion for achieving equilibrium. The accuracy of the solution depends first of all on the accuracy of the determination of the $K_{J,i}$ in the region of the equilibrium compositions. For this reason, this question is examined in detail below.

The rates in Eqs. 7.55–7.58 equal

$$V_{J,i} = -\frac{dm_{J,i}}{mM_J d\tau} = -\frac{dm_{\mathrm{E}_i}}{mM_{\mathrm{E}_i} d\tau} = +\frac{dm_{\mathrm{E}_i J}}{mM_{\mathrm{E}_i J} d\tau}, \tag{7.63}$$

where m_i and m are the mass of component i and of the alloy as a whole (kg) and M_i is the molar weight of component i (kg/mole).

We divide the time of the process into k time intervals of length $\Delta\tau$. Then, for each reactant, for example, the element J, instead of (7.63) we can write

$$V_{J,i}^k \approx \frac{m_J^{k+1} - m_J^k}{m^k M_J \Delta\tau} = \frac{m^{k+1}[J]^{k+1} - m^k [J]^k}{100 m^k M_J \Delta\tau}. \tag{7.64}$$

Equations 7.63 and 7.64 enable us to find the mass of each reactant in $k+1$ time intervals. For example, for each reactant J, if we take into account all the reactions involving it, we obtain

$$m_J^{k+1} = m_J^k - m^k M_J \Delta\tau \sum V_{J,i}^k. \tag{7.65}$$

Using (7.63) and (7.64) we can also express the concentrations of all the substances:

$$[J]^{k+1} = \frac{[J]^k - 100 M_J \Delta\tau \sum V_{J,i}^k}{S^{k+1}}, \tag{7.66}$$

$$[\mathrm{E}_i]^{k+1} = \frac{[\mathrm{E}_i]^k - 100 M_{\mathrm{E}_i} \Delta\tau \left(\sum x V_{\mathrm{C},i}^k + \sum y V_{\mathrm{O},i}^k + \sum z V_{\mathrm{N},i}^k + \sum n V_{\mathrm{B},i}^k\right)}{S^{k+1}}, \tag{7.67}$$

$$[\mathrm{E}_i J]^{k+1} = [\mathrm{E}_i J]^k + 100 M_{\mathrm{E}_i J} \Delta\tau V_{J,i}^k S_0^k, \tag{7.68}$$

where $S^{k+1} = m^{k+1}/m^k$ is the relative change in the mass of the alloy matrix during the current time interval and $S_0^k = m^k/m^0$ is the change in the mass of the alloy matrix at the time $k\Delta\tau$ relative to the original mass.

These changes can be calculated from the following dependences:

$$\begin{aligned} S^{k+1} = 1 - \Big(\Big(& M_{\mathrm{C}} \sum x V_{\mathrm{C},i}^k + M_{\mathrm{O}} \sum y V_{\mathrm{O},i}^k + M_{\mathrm{N}} \sum z V_{\mathrm{N},i}^k + M_{\mathrm{B}} \sum n V_{\mathrm{B},i}^k \Big) \\ & + M_{\mathrm{E}_i} \Big(\sum x V_{\mathrm{C},i}^k + \sum y V_{\mathrm{O},i}^k + \sum z V_{\mathrm{N},i}^k + \sum n V_{\mathrm{B},i}^k \Big) \Big) \Delta\tau, \end{aligned} \tag{7.69}$$

$$S_0^k = S^0 \cdot S^1 \cdot S^2 \cdot \ldots \cdot S^k. \tag{7.70}$$

When there are no compounds in the original system, we have $S^0 = 1$.

Equations 7.66 and 7.67 specify the concentrations of all the elements in the matrix, i.e., its composition, and Eq. 7.68 specifies the content of each of the compounds $\mathrm{E}_i J$ in the form of concentrations relative to the original mass of the alloy.

In the first step of the calculations, the thermodynamic possibility of the formation of all the compounds $\mathrm{E}_i J$ is determined for the initial phase composition and temperature. This corresponds to the condition $V_{J,i} > 0$. Therefore, the concentrations of all the compounds which can form under the particular conditions will increase in the first calculation cycles. As the concentrations of C, O, N, and B decline due to the their binding by elements with greater chemical affinity for the other elements, conditions for thermodynamic instability of their compounds are created, and they begin to dissolve (this corresponds to $V_{J,i} < 0$) until they vanish completely. In the concluding stages of the calculation near equilibrium, only the compounds that are thermodynamically stable under the assigned conditions remain in the system. Thus, in the proposed method there is no need to preliminarily determine the list of phases which can form in the particular system.

Let us examine the methods for determining the $K_{J,i}$ in greater detail.

For this purpose we will use expressions for the equilibrium constants of reactions (7.59–7.52). The pure condensed substance can be taken as the standard state for all the reactants in the determination of $\left(\Delta G_i^0\right)$.when carbides and borides form:

$$K_{\mathrm{C},i}^{\mathrm{eq}} = \frac{a_{\mathrm{E}_{ix}\mathrm{C}}}{a_{\mathrm{E}_i}^x a_{\mathrm{C}}}, \tag{7.71}$$

$$K_{\mathrm{B},i}^{\mathrm{eq}} = \frac{a_{E_{in}\mathrm{B}}}{a_{\mathrm{E}_i}^n a_{\mathrm{B}}}, \tag{7.72}$$

$$\Delta G_{\mathrm{E}_{ix}\mathrm{C}}^0 = -RT \ln K_{\mathrm{C},i}^{\mathrm{eq}}, \tag{7.73}$$

$$\Delta G_{\mathrm{E}_{in}\mathrm{B}}^0 = -RT \ln K_{\mathrm{B},i}^{\mathrm{eq}}, \tag{7.74}$$

where

$$a_i = \gamma_i N_i = \gamma_i \frac{\frac{[i]}{M_i}}{\sum \frac{[i]}{M_i}}, \tag{7.75}$$

and a_i, γ_i, N_i, and M_i are the activity, activity coefficient, mole fraction, and molar weight of component i, respectively.

From expressions (7.59), (7.52), (7.71), (7.72) and (7.75) we find

$$K_{\mathrm{C},i} = \frac{M_{\mathrm{E}_{ix}\mathrm{C}}^x M_{\mathrm{C}} S^{1+x}}{K_{\mathrm{C},i}^{\mathrm{eq}} \gamma_{\mathrm{E}_i}^x \gamma_{\mathrm{C}}}, \tag{7.76}$$

$$K_{\mathrm{B},i} = \frac{M_{\mathrm{E}_{in}\mathrm{B}}^n M_{\mathrm{B}} S^{1+n}}{K_{\mathrm{B},i}^{\mathrm{eq}} \gamma_{\mathrm{E}_i}^n \gamma_{\mathrm{B}}}, \tag{7.77}$$

where

$$S = \sum \frac{[E_i]}{M_i} + \frac{[\mathrm{C}]}{M_{\mathrm{C}}} + \frac{[\mathrm{O}]}{M_{\mathrm{O}}} + \frac{[\mathrm{N}]}{M_{\mathrm{N}}} + \frac{[\mathrm{B}]}{M_{\mathrm{B}}}. \tag{7.78}$$

Equations 7.73, 7.74, 7.76, and 7.77 cannot be used directly for the reactions resulting in the formation of oxides and nitrides, because only the values of the Gibbs free energy of the reactions for their formation by gaseous oxygen and nitrogen are known [42, 43]:

for $yE_i + 1/2O_2 = E_{iy}O$,

$$K_{O,i}^{eq1} = \frac{a_{E_{iy}O}}{a_{E_i}^{y} P_{O_2}^{0.5}}, \tag{7.79}$$

and for $zE_i + 1/2N_2 = E_{iz}N$,

$$K_{N,i}^{eq1} = \frac{a_{E_{iz}O}}{a_{E_i}^{z} P_{N_2}^{0.5}}. \tag{7.80}$$

Using the data on the equilibrium of the reactions
$1/2O_2 = [O]$,

$$K_{O}^{eq} = \frac{[O]f_{[O]}}{P_{O_2}^{0,5}}, \tag{7.81}$$

and $1/2N_2 = [N]$,

$$K_{N}^{eq} = \frac{[N]f_{[N]}}{P_{N_2}^{0,5}}, \tag{7.82}$$

from (7.60), (7.61), and (7.79–7.82) we obtained

$$K_{O,i} = \frac{K_{O}^{eq} M_{E_i}^{y} S^{y}}{K_{O,i}^{eq} \gamma_{E_i}^{y} f_{[O]}}, \tag{7.83}$$

$$K_{N,i} = \frac{K_{N}^{eq} M_{E_i}^{z} S^{z}}{K_{N,i}^{eq} \gamma_{E_i}^{z} f_{[N]}}. \tag{7.84}$$

Here $f_{[O]}$ and $f_{[N]}$ are the activity coefficients of oxygen and nitrogen dissolved in the metal, normalized to an infinitely dilute solution. The temperature dependences of K_{O}^{eq} and K_{N}^{eq} for liquid iron can be calculated from known equations [42].

Since oxide formation reactions (7.52) are the deoxidization reactions of the metal, the expressions for the deoxidization constants can also be used directly to calculate $K_{O,i}$:

$$K_{O,i} = K_{O,i}^{0} f_{[E_i]}^{y} f_{[O]}, \tag{7.85}$$

The temperature dependences of $K_{O,i}^{0}$ for most elements are known [42].

The accuracy of the determination of the activity coefficients γ_i and f_i for multicomponent systems has a great influence on the results of the calculations. The available experimental of such a kind are insufficient; therefore, calculation

Fig. 7.7 Variation of the concentrations of phases

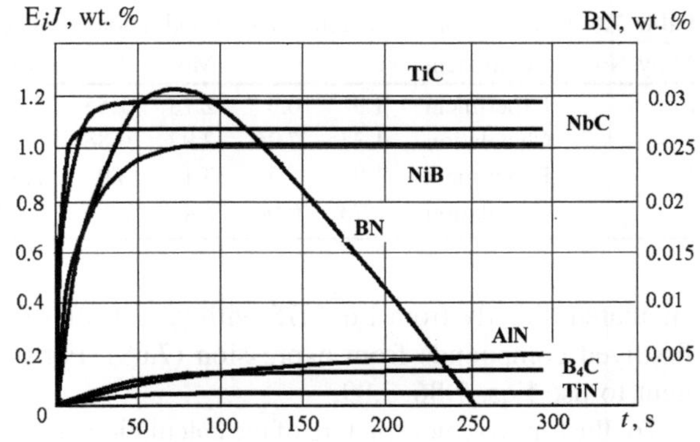

methods were used. At relatively low concentrations a good approximation is provided by the Wagner method:

$$\ln \gamma_i = \ln \gamma_i^0 + \sum_{j=2}^{j} \varepsilon_i^j N_j, \tag{7.86}$$

$$\log f_i = \sum_{j=2}^{j} e_i^j [j]. \tag{7.87}$$

The values of the interaction parameters ε_i^j and e_i^j, as well as their temperature dependences are known [42]. The values of the activity coefficients for dilute solutions (γ_i^0) at 1873 K are available in [42]. For other temperatures the values of γ_i^0 were recalculated from the equation:

$$\ln(\gamma_i^0)_T = \frac{T}{1873} \ln(\gamma_i^0)_{1873}. \tag{7.88}$$

To estimate the activities of the components in concentrated alloys it is preferable to use specific models of solutions (regular solutions, subregular solutions, etc.) supplemented by empirically selected parameters. For alloys based on iron, we can usually confine ourselves to the model of regular solutions. For an alloy consisting of k components, for component i we have

$$RT \ln \gamma_i = \sum_{j=1, i \neq j}^{k} N_j Q_{ij} + \sum_{i=1}^{k-1} \sum_{i=j+1}^{k} N_i N_j Q_{ij}, \tag{7.89}$$

where Q_{ij} is the mutual-exchange energy [44].

If the compounds appearing form solutions, $a_{E_iJ} \neq 1$. In this case information regarding the mutual solubility of the compounds and the types of these solutions is needed. In the simplest case, in which an ideal solution with unrestricted solubility of the components forms, the current values of the activities can be

Table 7.6 Composition of the alloy matrix and amounts of primary carbides (wt.%)

Alloy No.	Component	C	Cr	Mo	Ti	Nb	Ni	Amounts of carbides
1	Experiment	1.9	6.9	2.78	1.55	3.9	0.5	8 (TiC + NbC)
	Calculation	1.33	7.26	2.92	0.88	0.35	0.52	2.88 TiC + 3.75 NbC
2	Experiment	1.9	5.5	3.1	3.9	6.0	–	15 (TiC + NbC)
	Calculation	0.73	6.04	3.4	2.06	0.39	–	2.53 TiC + 6.37 NbC

calculated directly from Eq. 7.75 with $\gamma_i = 1$ after determining the content of the dissolved compounds from expression (7.68). For non-ideal solutions it is expedient to use Eqs. 7.86–7.89.

To illustrate a typical picture of the calculation procedure, Fig. 7.7 presents plots of the variation of the concentrations of compounds with time for an iron alloy containing (wt.%) 3 C, 10 Cr, 1 Si, 1 Ti, 1 V, 1 Nb, 1 Mo, 1 Ni, 1 B, 0.1 N, and 0.03 O at 1773 K. In accordance with the foregoing, the less stable compound BN, which appears at first, dissolves with time, since the concentrations of B and N in the alloy decrease due to the formation of more stable compounds (TiC, NbC, NiB, B_4C, AlN, and TiN).

Table 7.6 presents our calculated and experimental data [45] regarding the composition of the metal and the amounts of primary carbides for surfacing with abrasion-resistant alloys. The closeness of the calculated and experimental data attests to the applicability of the thermodynamic method for the purposes indicated.

Table 7.7 presents the results of calculation of the amount and composition of the nonmetallic oxide inclusions for welding by electrodes with a coating of the rutile type. The data on the composition of a weld metal from [46] were used for the calculation. The results obtained were compared with the experimental results from [47]. In the range of temperatures of the metal in the weld pool—from 2000 K to 1773 K (the crystallization temperature)—the formation of pure oxides of iron and manganese is possible. The calculation in the ideal solution approximation gives overestimated percentages of FeO and underestimated percentages of SiO_2. The calculation for high temperatures gives percentages of iron oxide that greatly exceed the experimental values. The results of the calculation for the crystallization temperature of steel using the theory of regular solutions to estimate the activities of the slag components are closest to the experimental data (Table 7.7, variant 3).

To conclude this section, we describe some additional possibilities and prospects for using the proposed method.

1. The model provides for calculation of the equilibrium state even when the original system contains an assigned amount of the compounds $[E_iJ]^0$. The calculation scheme is completely retained, but the value of S^0 in Eq. 7.70 in this case equals

Table 7.7 Composition of the nonmetallic inclusions in a weld metal (wt.%)

Calculation No.	T, K	FeO	SiO_2	MnO	TiO_2	Al_2O_3	Amount of inclusions	Slag model
1	1773	0	82.14	0	11.67	6.19	0.149	Pure oxides
2		17.79	46.17	9.28	14.72	5.22	0.181	Ideal solution
3		3.27	57.7	13.8	15.96	5.65	0.167	Regular solution
4	2000	0	84.9	0	8.9	6.1	0.140	Pure oxides
5		43.2	28.9	6.1	12.2	4.5	0.210	Ideal solution
6		15.22	49.9	11.78	13.0	4.66	0.205	Regular solution
Experiment [47]	–	0.9–5.3	55.5–73.1	6.6–19.0	5.3–11.5	7.2–13.1	0.109–0.131	–

$$S^0 = 1 - \sum_{i=1}^{16}\sum_{j=1}^{4}\frac{[E_i J]^0}{100}. \tag{7.90}$$

2. Some relatively simple additions enable us to solve a problem of practical importance, namely, the calculation of the equilibrium compositions and masses of the metal and slag for an assigned initial ratio between their masses $K_f^0 = m_f^0/m_{\mathrm{Me}}^0$. In this case it is convenient to express the amount of the oxides $E_{i_y}O$in terms of their concentration in the slag, rather than according to Eq. 7.68. From the material balance we have

$$(E_{i_y}O)^k = \frac{m}{m_f^k}[E_{i_y}O]^k = \frac{[E_{i_y}O]^k}{\mathrm{Sum} + \sum\limits_{i=1}^{i}\frac{[E_{i_y}O]^k}{100}}, \tag{7.91}$$

$$S^0 = 1 - \mathrm{Sum} - \frac{K_f^0}{100\left(1 + K_f^0\right)}\sum_{i=1}^{i}(E_{i_y}O)^0, \tag{7.92}$$

$$K_f^0 = \frac{\mathrm{Sum}}{S_0^k} + \frac{1}{100 S_0^k}\sum_{i=1}^{i}[E_{i_y}O]^k, \tag{7.93}$$

where

$$\mathrm{Sum} = \frac{K_f^0}{100\left(1 + K_f^0\right)}\sum_{j=1}^{j}(M_{j_m}O)^0. \tag{7.94}$$

In Eq. 7.94 $(M_{j_m}O)^0$ is the initial concentration of the oxides and salts that are not included among the compounds considered in the slag.

3. The method can be used to construct diagrams of state of metallic alloys and other systems. It should be kept in mind that this method does not require any new data compared with the known methods [36–39]. However, the requirements placed on the accuracy of the thermodynamic data are significantly higher in this case; therefore, the use of Eqs. 7.86–7.89 for calculating the activities of the components should be approached cautiously. It can only be noted that the use of Eqs. 7.86–7.89 is fully justified for highly stable compounds, which form already at low concentrations of the reactants. This allows us to expect a fairly exact prediction of the phase composition of multicomponent alloys based on iron in these cases.
4. The use of the method described here to analyze the homogeneous equilibrium in polymeric oxide–salt melts containing a large number of complex ions is promising. For example, in MeO–SiO_2 systems the formation reactions of all the polymeric silicon–oxygen anions can be described by a single equation [48]:

$$iSiO_4 = Si_iO_{3i+1-c}^{2(i+1-c)} + (i+1-c)O^{2-}, \tag{7.95}$$

where i and c are integers. This enables us to express the rates of all the reactions in the form of one equation and to write a simple program for calculating the concentrations of all the complexes.

5. It can be expected that the proposed approach will also be effective for analyzing the equilibrium state in other cases, in which a large number of reactions occur and there are many phases in the system studied.

7.6 Conclusion

1. A fundamentally new variant of the constitution diagram of the matrix of deposited metals, which takes into account the use of a portion of the carbon and alloying elements for carbide formation and the mutual influence of the elements on the formation of the structural components, has been developed.
2. A mathematical model for a formalized description of the Schaeffler diagram has been created. The mathematical dependences obtained enabled us to calculate and predict the phase and structural composition of a deposited metal over a broad alloying range for the first time.
3. A closed algorithm for calculating the phase and structural composition of a deposited metal, the quantitative and qualitative composition of the hardening phases, and their properties has been developed on the basis of a phenome-

nological model of the nonequilibrium crystallization of a weld pool and a mathematical description of the proposed constitution diagram.

4. A method for calculating the composition and amounts of the phases in equilibrium multicomponent alloys based on iron has been developed. Its application to several practical problems has been demonstrated. The calculated and experimental data are in fairly good agreement with one another, allowing us to conclude that the method is applicable under the conditions of welding and surfacing.
5. The prospects of using the method to predict the composition of the phases in different multicomponent systems have been demonstrated.

References

1. Volchenko, V.N., Yampolskii, V.M., Vinokurov, V.A.: Theory of Welding Processes (in Russian), p. 559. Vysshaya Shkola, Moscow (1988)
2. Ryzhonkov, D.I., et al.: Theory of Metallurgical Processes (in Russian), Moscow, pp.392. (1989)
3. Gulyaev, A.P.: Physical Metallurgy (in Russian), p. 648. Metallurgiya, Moscow (1977)
4. Maslenkov, S.B., Maslenkova, E.A.: Steels and Alloys for High Temperatures: A Handbook in Two Volumes (in Russian). vol. 1, p. 383. Metallurgiya, Moscow (1991)
5. Maslenkov, S.B., Maslenkova, E.A.: Steels and Alloys for High Temperatures: A Handbook in Two Volumes, Vol. 2 (in Russian), p. 832. Metallurgiya, Moscow (1991)
6. Oyama, S.T.: The Chemistry of Transition Metal Carbides and Nitrides, p. 1. Blackie Academic and Professional, London (1996)
7. Hugosson Håkan, W.: Theoretical studies of phase stabilities and electronic structure in molybdenum carbide. Licentiate Thesis in Physics, Department of Physics, Chapter 7, pp. 43–46. Uppsala University, Uppsala (1999)
8. Grossman, J.C., et al.: Transition metals and their carbides and nitrides: Trends in electronic and structural properties. Phys Rev B **60**(9), 6343–6347 (1999)
9. Ol'shanskii N.A. (ed).:Welding in Engineering. A Handbook (in Russian). vol 1, 501 pp. Moscow (1978)
10. Paton, B.E., Lebedev, V.K.: Electrical Equipment for Arc and Slag Welding (in Russian), p. 359. Mashinostroenie, Moscow (1966)
11. van Vlack, L.H.: Textbook of Materials Science. Addison –Wesley, Reading (1973)
12. Schaeffler, A.L.: Constitution diagram for stainless steel weld metal. Met Prog **56**(5), 680–680b (1949)
13. Schneider, H.: Investment casting of high-hot strength 12% chrome steel. Foundry Trade J **108**, 562–563 (1960)
14. Espy, R.H.: Weldability of nitrogen-strengthened stainless steels. Weld J **61**(5), 149–156 (1982)
15. Potak, Ya.M., Sagalevich E.A.: Constitution diagram of a low-carbon non-rusting cast deposited metal. Avtom. Svarka. (5):10–13 (1972)
16. Korolev, N.V., Boronenkov, V.N., Pimenova O.V.: Mathematical model of prediction of phase composition, structure and properties of weld metal. In: Proceedings of the 7th International Conference of Computer Technology in Welding, San Francisco, pp. 310–319 (1997)
17. Kotecki, D.J., Siewert, T.A.: Constitution diagram for stainless steel weld metals: a modification of the WRC diagram. Weld J **71**(5), 171–178 (1992)
18. Delong, W.T.: Ferrite in austenitic stainless steel weld metal. Weld J **53**(7), 273–286 (1974)

19. Balmforth, M.C., Lippold, J.C.: A new ferritic-martensitic stainless steel constitution diagram. Weld J **79**(12), 339–345 (2000)
20. Beres, L.: Proposed modification to Schaeffler diagram for chrome equivalents and carbon for more accurate prediction of martensite content. Weld J **77**(7), 273–276 (1998)
21. Kotecki, D.J.: A martensite boundary on the WRC-1992 diagram. Weld J **78**(5), 180–192 (1999)
22. Kotecki, D.J.: Forecasting weld microstructure. Adv Mater Processes **157**(6), 74–77 (2000)
23. Kotecki, D.J.: A martensite boundary on the WRC-1992 diagram Part 2: The effect of manganese. Weld J **79**(12), 346–354 (2000)
24. David, S.A., Babu, S.S.: Modeling microstructure development in weld metals. Trans Ind Inst Met **50,** 591 (1998)
25. Babu, S.S., et al.: High resolution analysis of elemental partitioning in nickel-base superalloy welds using atom probe field ion microscopy. Microsc. Microanal. **4**, 94 (1998)
26. Vitek, J.M., et al.: Neural net model for predicting ferrite number in austenitic stainless steel welds. In: Proceedings of the International Conference on Trends in Welding Re-search, Georgia. pp. 82–87 1998
27. Babu, S.S., David, S.A., Quintana, M.A.: Modeling microstructure evolution in self-shielded flux-cored arc welds. Weld J **80**(4), 98–105 (2001)
28. Quintana, M., et al.: Inclusion formation in self-shielded flux-cored arc welds Weld J **80**(4), 91–97 (2001)
29. Mazurovsky, V., Zinigrad, A., Zinigrad, M.: Predicting weld structure using modified Schaeffler constitution diagram. In: Proceedings of the International Conference Mathematical Modeling and Simulation of Metal Technologies, pp. 540–545. Ariel, Israel (2000)
30. Merinov, P.E., et al.: Method for the quantitative determination of the ferrite content in welded joints of chromium–nickel steels. Svarochn Proizv (3),49–50 (1977)
31. Alemasov, B.E., et al.: Thermodynamic and Thermophysical Properties of Combustion Products, edited by V. P. Glushko. vol. 1: Computation Methods, Israel Program for Scientific Translations, Jerusalem, 1974; Vol. 3: Oxygen and Air-Based Propellants, Israel Program for Scientific Translations, Jerusalem (1975)
32. Vatolin, N.A., Moiseev, G.K., Trusov ,B.G.: Thermodynamic Modeling in High-Temperature Inorganic Systems (in Russian). pp. 352, Metallurgiya, Moscow (1994)
33. Gol'dshtein, M.I., Popov, V.V.: Solubility of Interstitial Phases during the Thermal Treatment of Steel (in Russian). p. 200, Metallurgiya, Moscow (1989)
34. Mogutnov, V.M., Tomilin, I.A., Shvartsman, L.A.: Thermodynamics of Iron Alloys (in Russian). p. 208, Metallurgiya, Moscow (1984)
35. Kaufman, L., Bernstein, H.: Computer Calculation of Phase Diagrams. Academic, New York (1970)
36. Ansara, I.: Comparison of methods for thermodynamic calculation of phase diagrams. Int Met Rev **24**(l), 20–53 (1979)
37. Lukas, H.L., Weiss, J., Henig, E.Th.: Strategies for the calculation of phase diagrams. CALPHAD Comput. Coupling. Phase Diagr. Thermochem. 6(3):229–251
38. Udovskii, A.L.: Computer simulation of phase diagrams, thermodynamic properties, and structure of multicomponent systems. Metally (2):136–157 (1990)
39. Hillert, M.: Some viewpoints on the use of a computer for calculating phase diagrams. Physica **103B**(1), 31–40 (1981)
40. Mikhailov, G.G., Tyurin, A.T.: Calculation of the deoxidizing ability of calcium and aluminum in liquid steel. Izv Akad Nauk SSSR Met. (5):24–31 (1978)
41. Zel'dovich, Ya.B.: Proof of the uniqueness of solutions of mass action equations. Zh Fiz Khim **11**(5), 685–689 (1938)
42. Turkdogan, E.T.: Physical Chemistry of High Temperature Technology, p. 344. Academic, New York (1980)
43. Kulikov, I.S.: Thermodynamics of Carbides and Nitrides (in Russian) p. 320, Chelyabinsk (1988)

44. Burylev, B.P.: Thermodynamics of Metallic Interstitial Solutions (in Russian). p. 161, Rostov-on-Don (1984)
45. Mozok, V.M. et al.: Influencing of ferrum-based multialloying C–Cr–Nb alloys composition and structure on impact toughness. Avt. Svarka. (12):43–45, (1991)
46. De Rissone, N.M.R., et al.: ANSI/AWS A5–91 E6013 rutile electrodes: The effect of wollastonite. Weld J **76**(11), 498–507 (1997)
47. Gloor, K., et al.: Nonmetallic inclusions in weld metal. Weld World **4**(2), 70–87 (1966)
48. Baes, C.F.: A polymer model for BeF_2 and SiO_2 melts. J Solid State Chem (1):159–169 (1970)

Chapter 8
Using Modeling and Simulation to Improve High-Temperature Technologies

8.1 Improving the Technological Scheme of the Oxidative Remelting of Tungsten-Containing Scrap

The mathematical model of the process of smelting steel in an electric arc furnace (EAF) described in Chap. 4 was used to analyze the technology for obtaining tungsten-containing slags in [1].

The essence of the technological scheme for the remelting of tungsten-containing scrap that was developed and adopted in the Izhstal Corporation is as follows. A 30-ton electric arc furnace is charged with metal scrap, whose composition is presented in Table 8.1, and then, as it melts, lime and iron ore are added. In the time intervals between the additions of the slag formers, oxygen is blown through the metal.

At the conclusion of the process, the metal and slag, whose chemical compositions are presented in Table 8.2, are poured and air-cooled.

The materials obtained as a result of remelting are used as alloying additives during the smelting of some grades of high-speed steels. This technology provides a way to process metal scrap containing tungsten in any form, does not require complex equipment, and eliminates the need for subsequent treatment of the slag, for example, to recover tungsten for ferroalloys.

However, along with the obvious advantages, it also has some significant shortcomings, the main one being the low degree of recovery of tungsten in the oxide in the stage of the oxidative remelting of tungsten-containing scrap (see Table 8.2).

Therefore, the problem of developing recommendations for improving this technology for the purpose of increasing the degree of recovery of tungsten in the slag on the basis of a computational experiment was formulated.

M. Zinigrad et al., *Phase Interaction in the Metal–Oxide Melts–Gas System*, Engineering Materials, DOI: 10.1007/978-3-642-22377-8_8,

Table 8.1 Characteristics of the tungsten-containing scrap processed by oxidative remelting

Remelted parts	Material	Percentage of the total mass (%)
Drill bit legs	14KhNZMA steel	60
Drill bit cutters	14KhNZMA steel, 20KhNZA steel	35
Metal-ceramic inserts	Hard WC alloy	5

Table 8.2 Content of the principal components in the metal and slag before and after oxidative remelting of tungsten-containing scrap

Metal	Chemical composition of the metal (wt.%)						
	C	S	Mn	Si	W	P	Cr
Upon melting	0.4–0.7	0.01–0.02	0.1–0.3	0.04–0.44	4.5–6.0	0.004–0.090	0.4–0.7
Final product	0.04	0.01–0.02	0.01–0.04	0.01–0.06	1.2–0.8	0.002–0.005	0.09–0.20
Slag	Chemical composition of the slag (wt.%)						
	MnO	SiO_2	WO_3	P_2O_5	Cr_2O_5	FeO	CaO
Upon melting	3–8	6–25	0.2–14.0	<0.2	3–8	20–30	11–33
Final product	1–2	5–18	13–18	0.10	2–4	30–40	25–30

Table 8.3 Technology of the oxidative remelting of tungsten-containing scrap according to data from the Izhstal Corporation

Technological operation	Time after the moment of melting for melt No. (min)	
	8,742	8,739
1. Adding 1.5–2.0 ton of lime	9	9
2. Blowing oxygen through the metal	20–43	30–49
3. Adding of 1.0 ton of iron ore	48	55
4. Pouring the metal into a ladle	58	62

To adapt the model, the changes in the compositions of the phases for two melts carried out in the Izhstal Corporation were calculated. The time line of the industrial melts is presented in Table 8.3.

At first, the composition and masses of the metal and the slag upon melting, which were taken as the initial values, were determined according to the procedure described in Chap. 4. Then the processes occurring between the metal and the slag were analyzed. The following principal chemical reactions, which describe the interaction for the phase compositions considered (Table 8.2), were selected:

$$\tfrac{1}{3}[\mathrm{W}] + (\mathrm{FeO}) = \tfrac{1}{3}(\mathrm{WO_3}) + [\mathrm{Fe}], \tag{8.1}$$

$$[\mathrm{C}] + (\mathrm{FeO}) = \{\mathrm{CO}\} + [\mathrm{Fe}], \tag{8.2}$$

$$(FeO) = [O] + [Fe], \tag{8.3}$$

$$(CaS) + (FeO) = (CaO) + [S] + [Fe], \tag{8.4}$$

$$[Mn] + (FeO) = (MnO) + [Fe], \tag{8.5}$$

$$\tfrac{1}{2}[Si] + (FeO) = \tfrac{1}{2}[Si] + [Fe], \tag{8.6}$$

$$2/5[P] + (FeO) = 2/5(PO_{2.5}) + [Fe], \tag{8.7}$$

$$\tfrac{2}{3}[Cr] + (FeO) = \tfrac{2}{3}(CrO_{1.5}) + [Fe]. \tag{8.8}$$

The rates of the reactions just listed were calculated from the following equations:

$$i_W^f = \frac{(WO_3)_1^f - [W]_1^f x^6 K_{eq_W}^3 F^3}{\dfrac{(WO_3)^{1-\alpha-\beta} x^{6\alpha} K_{eq_W}^{3\alpha} F^{3\alpha}}{K_W^{exch}} + B_{[W]} x^6 K_{eq_W}^3 F^3 + B_{(WO_3)}}, \tag{8.9}$$

$$i_{Mn}^f = \frac{(MnO)_1^f - [Mn]_1^f x^2 K_{eq_{Mn}} F}{\dfrac{(MnO)^{1-\alpha-\beta} x^{2\alpha} K_{eq_{Mn}}^{\alpha} F^{\alpha}}{K_{Mn}^{exch}} + B_{[Mn]} x^2 K_{eq_{Mn}} F + B_{(MnO)}}, \tag{8.10}$$

$$i_{Si}^f = \frac{(SiO_2)_1^f - [Si]_1^f x^4 K_{eq_{Si}}^2 F^2}{\dfrac{(SiO_2)^{1-\alpha-\beta} x^{4\alpha} K_{eq_{Si}}^{2\alpha} F^{2\alpha}}{K_{Si}^{exch}} + B_{[Si]} x^4 K_{eq_{Si}}^2 F^2 + B_{(SiO_2)}}, \tag{8.11}$$

$$i_P^f = \frac{(PO_{2.5})_1^f - [P]_1^f x^5 K_{eq_P}^{2.5} F^{2.5}}{\dfrac{(PO_{2.5})^{1-\alpha-\beta} x^{5\alpha} K_{eq_P}^{2.5\alpha} F^{2.5\alpha}}{K_P^{exch}} + B_{[P]} x^5 K_{eq_P}^{2.5} F^{2.5} + B_{(PO_{2.5})}}, \tag{8.12}$$

$$i_{Cr}^f = \frac{(CrO_{1.5})_1^f - [Cr]_1^f x^3 K_{eq_{Cr}}^{1.5} F^{1.5}}{\dfrac{(CrO_{1.5})^{1-\alpha-\beta} x^{3\alpha} K_{eq_{Cr}}^{1.5\alpha} F^{1.5\alpha}}{K_{Cr}^{exch}} + B_{[Cr]} x^3 K_{eq_{Cr}}^{1.5} F^{1.5} + B_{(CrO_{1.5})}}, \tag{8.13}$$

$$i_S^f = \frac{(S)_1^f - [S]_1^f x^2 K_{eq_S} F}{\dfrac{(S)^{1-\alpha-\beta} x^{2\alpha} K_{eq_S}^{\alpha} F^{\alpha}}{K_S^{exch}} + B_{[S]} x^2 K_{eq_S} F + B_{(S)}}, \tag{8.14}$$

$$i_C^f = \frac{P_{CO} - [C]_1^f x^2 K_{eq_C} F}{\dfrac{P_{CO}^{1-\alpha-\beta} x^{2\alpha} K_{eq_C}^{\alpha} F^{\alpha}}{K_C^{exch}} + B_{[C]} x^2 K_{eq_C} + F}, \tag{8.15}$$

Table 8.4 Experimental (above the bar) and calculated (below the bar) values of the concentrations of the components for the calculation periods of industrial melts

Melt No.	τ (min)	Metal, wt.%						Slag (wt.%)					
		C	S	Mn	Si	W	P	MnO	SiO_2	WO_3	P_2O_5	$\sum FeO$	CaO
8,742	0	0.70	0.013	0.23	0.06	4.46	0.017	5.5	24.4	0.2	–	24.0	15.1
	22	0.41	0.009	0.24	0.08	3.33	0.013	1.9	7.4	8.7	0.10	20.2	41.5
		0.41	0.011	0.31	0.14	3.96	0.017	1.8	9.9	4.5	––	13.6	47.1
	43	0.04	0.009	0.08	0.02	1.67	0.007	2.4	7.1	19.4	0.10	29.I	30.I
		0.04	0.008	0.12	0.07	1.8	0.004	2.9	7.5	22.2	0.24	34.7	31.0
	58	0.04	0.011	0.04	0.06	1.52	0.005	1.6	7.3	16.4	0.04	34.4	30.0
		0.04	0.008	0.11	0.07	1.6	0.004	2.3	6.6	18.2	0.18	40.5	31.0
8,739	0	0.45	0.013	0.18	0.04	3.76	0.009	3.1	6.0	13.9	0.20	26.0	33.0
	30	0.38	0.013	0.19	0.04	2.15	0.010	1.7	3.8	15.9	0.20	24.0	41.8
		0.17	0.011	0.16	0.04	1.63	0.014	1.7	3.5	7.9	0.01	15.6	48.6
	49	0.04	0.008	0.17	0.07	1.27	0.006	1.7	3.5	15.1	0.20	28.4	51.8
		0.04	0.011	0.11	0.03	1.12	0.004	2.3	3.5	12.8	0.19	29.5	48.6
	62	0.04	0.009	0.01	0.01	0.90	0.003	1.4	5.8	12.9	0.10	39.3	51.8
		0.04	0.010	0.096	0.02	1.30	0.003	1.7	3.3	8.7	0.14	41.9	48.6

$$i_O^f = \frac{[\mathrm{O}]_1^f - x^2 K_{eq_O} F_1^f}{\dfrac{x^{2\alpha} K_{eq_O}^{\alpha} F^{\alpha}}{K_O^{exch}} + B_{(FeO)} x^2 K_{eq_O} + B_{[O]}}, \tag{8.16}$$

$$i_{Fe}^f = \frac{F_1^f - x^2 F}{\dfrac{x^{2\alpha} F^{2\alpha}}{K_{Fe}^{exch} [\mathrm{Fe}]^{\alpha}} + B_{(FeO)}}. \tag{8.17}$$

From the condition of electroneutrality we have

$$i_{Fe}^f + i_O^f + i_C^f + i_S^f + i_W^f + i_{Mn}^f + i_{Si}^f + i_P^f + i_{Cr}^f = 0. \tag{8.18}$$

The calculations were performed using the temperature dependences of the equilibrium constants of all the reactions considered, the diffusion coefficients of the reactants in the metal and the slag, and the exchange currents presented in preceding chapters of this book. The densities of the metal and the slag were assumed to be equal to 7.1×10^{-3} and 3.2×10^{-3} kg/m^3, respectively [2]. The results of the calculations are presented in Table 8.4, which compares them with industrial data. They support the possibility of predicting the composition of the metal and the slag during the oxidative remelting of tungsten-containing scrap.

It follows from an analysis of the calculated and experimental data (Table 8.4) that the highest oxidation rates of tungsten are observed in the oxygen blowing period when oxygen is blown through the metal.

This is clearly attributable to the presence of a constant large concentration of iron oxides, the abrupt lowering of the carbon content in the metal, and the vigorous stirring of the melts.

Table 8.5 Starting data used in the calculations for selecting an efficient technology for the process of the oxidative remelting of tungsten-containing scrap

Trial No.	Blowing time (min)	Mass of lime (kg)	Mass of ore (kg)	$[C]_{init}$ (wt.%)	$[W]_{init}$ (wt.%)	βS_{eff} in the blowing period
1	25	2,000	1,000	0.70	4.46	$6S_{geom}$
2	25	250	2,750	0.70	4.46	$6S_{geom}$
3	30	250	2,750	0.70	4.46	$6S_{geom}$
4	30	250	2,750	0.70	4.46	$6S_{geom}$
5	30	250	2,750	0.70	4.46	$6S_{geom}$
6	30	250	2,750	0.20	4.46	$6S_{geom}$
7	30	250	2,750	0.20	3.00	$6S_{geom}$
8	30	250	2,750	0.70	3.00	$6S_{geom}$
9	30	250	2,750	0.70	3.00	$9S_{geom}$

In the other periods, that is, after the addition of lime or iron ore, the recovery of tungsten in the slag is insignificant, even after prolonged holding (for 20–25 min) of the melts in the furnace.

To select an efficient technology for the process of the oxidative remelting of tungsten-containing scrap, we performed calculations of the composition of the phases under various interaction conditions (Table 8.5, variants 1–9).

The initial composition of the metal and the slag in variants 1–5 corresponded to the data from industrial melt No. 8,742 (Table 8.4).

The results of the calculations are presented in Table 8.6. It follows from these results that the variation of the masses of the lime and iron ore introduced into the slag, as well as of the initial concentration of carbon in the metal (within the ranges considered), does not have an appreciably influence on the final composition of the remelting products (Table 8.6, variants 1 and 2 and variants 5 and 6). Conversely, an increase in the time during which oxygen is blown through the metal pool from 25 to 30 min and an increase in the phase interaction temperature from 2,023 to 2,223 K result in lowering of the final tungsten content in the metal (Table 8.6, variants 2 and 3 and variants 3 and 5). The initial tungsten content in the metal and the stirring intensity of the melts have significant effects on the final composition of the phases. A decrease in $[W]_{init}$ from 4.46 to 3.00 wt.% and a 1.5-fold increase in the convection rate result in lowering of the final tungsten concentration by nearly an order of magnitude (Table 8.6, variants 5 and 9). We note that the composition of the slag obtained in all the cases considered corresponds to the technical requirements adopted in the Izhstal Corporation.

As a result of the analysis of the different variants for carrying out the process, the following recommendations were developed.

1. For the purpose of shortening the duration of the process of remelting tungsten-containing scrap, the blowing of oxygen through the metal should be started as soon as the feedstock melts. The blowing time should be increased from 20–25

Table 8.6 Calculated composition of the metal and the slag (wt.%) over the course of remelting

Trial No.	τ. min	[C]	[S]	[Mn]/(MnO)	[Si]/(SiO_2)	[W]/(WO_3)	[P]/(P_2O_5)	[Cr]/(Cr_2O_3)
1	15	0.038	0.0113	0.139/2.46	0.054/9.08	2.61/11.75	0.004/0.157	0.32/3.67
	30	0.031	0.0112	0.089/2.78	0.050/9.72	1.67/17.62	0.003/0.159	0.178/4.73
2	15	0.038	0.0116	0.150/2.48	0.054/10.63	2.60/11.78	0.004/0.159	0.335/3.58
	30	0.030	0.0112	0.098/2.81	0.050/10.67	1.67/17.62	0.004/0.159	0.180/4.69
3	30	0.030	0.0108	0.084/2.89	0.048/10.69	1.42/19.22	0.0038/0.168	0.154/4.91
4	30	0.027	0.0106	0.096/2.81	0.048/10.69	1.19/20.63	0.0038/0.168	0.150/4.93
5	30	0.026	0.0106	0.104/2.76	0.049/10.69	1.09/21.27	0.0036/0.170	0.150/4.93
6	30	0.011	0.0106	0.104/2.76	0.049/10.69	1.09/21.27	0.0037/0.170	0.153/4.91
7	30	0.011	0.0106	0.099/2.79	0.048/10.69	0.73/14.34	0.0036/0.170	0.144/4.98
8	35	0.025	0.0108	0.092/2.84	0.048/10.71	0.58/15.25	0.0036/0.170	0.118/5.18
9	15	0.039	0.0105	0.148/2.47	0.065/10.52	0.93/13.05	0.0037/0.170	0.23/4.29
	30	0.021	0.0103	0.075/2.95	0.052/10.65	0.24/17.43	0.0036/0.170	0.083/5.42
	35	0.010	0.0103	0.066/3.01	0.048/10.69	0.15/15.97	0.0035/0.171	0.070/5.53

to 30–35 min. At the conclusion of the blowing, the process should be stopped, and the metal and slag should be poured into a ladle.

2. The initial concentration of tungsten in the metal should be in the range from 3.0 to 3.5 wt.%, which ensures a fairly high (up to 95 rel.%) level of recovery of tungsten in the slag. A higher initial concentration of tungsten results in increases in its losses and in the duration of the process.
3. To increase the interaction temperature of the melts, as well as the stirring intensity, an increased initial concentration of carbon must be ensured, for example, by adding coke to the feedstock.
4. To prevent excessive overheating of the metallic melt and to maintain a constantly high oxidizing power of the slag, portions of iron ore should be added over the entire course of the period of blowing oxygen through the metal.

The computational experiment performed enabled us to reduce the expenditures on experimental technological investigations and industrial test melts, and implementation of the recommendations developed ensured the maximum recovery of tungsten in the slag.

8.2 Development of Compositions of Synthetic Slags for the Ladle Refinement of Steel

One of the most important contributors to the steady rise in the quality of metal that have recently been developed is the ladle refinement of liquid steel [3]. The numerous technological schemes for ladle refinement can be reduced on the basis of the characteristics of the refining media to three main types: the treatment of steel in a vacuum, a neutral gas, and liquid slags. As long-term practice has shown [3], iron-free slags based on the four-component $CaO–SiO_2–Al_2O_3–MgO$ system are most promising for ladle treatment. The refinement effect includes both the extraction of sulfur and oxygen dissolved in the metal and the removal of non-metallic inclusions from the steel. In addition, the slag refinement of steel that was smelted in arc furnaces leads to simplification and reduction of the cost of the electric smelting technology, as well as increases in the output of metallurgical plants. These objectives are achieved by shortening the reducing period of melts by transferring the final refinement, desulfurization, and deoxidization to a ladle.

A synthetic lime-alumina with the following composition is widely used for refining steel, wt.%: 52–55 CaO, 38–42 Al_2O_3, up to 5 SiO_2, up to 2 MgO, and less than 1 FeO. A slag with such a composition effectively removes sulfur from the metal (the degree of desulfurization is 55–75%, depending on the grade of steel) and deoxidizes the metal. The low viscosity of the slag and the high interfacial tension on its boundary with the metal enable the effective elimination of deoxidization products from the metal and promote the removal of slag droplets from the liquid metal. All of this is responsible for the high quality of metal treated by this slag. However, as was shown in [3], a synthetic slag of a single composition cannot ensure an equally high refinement effect in all cases. In principle, each group of grades of steel should have its own optimal composition for the refining slag. In addition, the high cost of the slag charge is a serious technical and economic deficiency.

Therefore, the problem of selecting compositions of synthetic slag with a reduced content of aluminum oxide for treating steel by using relatively inexpensive materials for smelting it and of testing them under industrial conditions was posed. The slag was selected for treating grade 10 steel, which represents a group of low-carbon steels, and 35KhGSA steel, which is a typical representative of medium-carbon steels.

Selecting a new slag composition by experimental methods under production conditions is associated with large expenditures of time and resources. The use of a mathematical model of the refining of a metal by a slag enables us to replace some of the industrial trials by computational experiments. During such a computational experiment, by assigning the chemical compositions of the metal and the slag before the refining, as well as the values of the technological parameters, we can obtain the compositions of the melts at the conclusion of the refinement or at any time during the treatment of the steel. A mathematical model also enables us to take into account the influence of various technological parameters and operations (the temperature, composition, and amounts of the metal and the slag, the metal

Table 8.7 Chemical composition of the steels

Steel grade	Technology variant	Content of elements, wt.%								
		C	Mn	Si	P	S	Cr	Ni	Cu	Al
35KhGSA	Experimental	0.34	1.04	1.20	0.030	0.011	1.24	0.27	0.20	0.01
	Reference	0.34	0.94	1.16	0.035	0.015	1.19	0.26	0.24	0.01
10	Experimental	0.14	0.63	0.25	0.015	0.014	0.15	0.26	0.19	0.01
	Reference	0.13	0.53	0.25	0.016	0.018	0.13	0.27	0.18	0.01

tapping rate, the amounts of ferroalloys and deoxidizers and their time of introduction, etc.) on the composition of the finished steel. By disregarding the different technological variants of the refining, we can solve the problem posed, that is, find the optimal composition of the slag, its amount, and the duration of the refining. In this case industrial testing is performed for the optimal variant of the refining technology, and the risk of producing an unacceptable product is thereby reduced to a considerable extent.

In developing the compositions of the synthetic slags, attention was focused on maintaining good deoxidizing and desulfurizing powers.

An analysis of the influence of various components on the chemical and technological properties of the refining slags that we performed enabled us to determine the following concentration ranges of the components, wt.%: 25–30 Al_2O_3, 45–55 CaO, 5–10 SiO_2, 5–10 MgO, and 0–10 CaF_2.

During the development of the optimal composition of the synthetic slag by mathematical modeling [4], the following technological features of the smelting of steels of different grades were taken into account.

- Grade 10 steel. The content of the principal components in the metal before treatment by the slag are presented in Table 8.7. The lacking manganese in the form of ferromanganese (MN-6) and all the silicon in the form of ferrosilicon (FS-45) are poured onto the bottom of the ladle. The metal is deoxidized by aluminum in the metal stream. The onset of deoxidization occurs when 1/4 of the height of the ladle is filled, and the end occurs when 2/3 of the ladle is filled. The amount of aluminum added is 1 kg/ton metal.
- Grade 35KhGSA steel. The composition of the metal before treatment by the slag is presented in Table 8.7. The lacking manganese is poured in the form of ferrosilicium (FS-75) onto the bottom of the ladle. The metal is not deoxidized by aluminum.

The calculations were performed for a temperature of 1,873 K, and the mass of the metal was assumed to be equal to 120 tons.

During the computational experiment, the final composition of the metal was calculated from the assigned compositions of the phases with variation of the composition of the slag in the ranges given above. After the optimal slag composition was selected, attention was turned to matching the composition of the refined metal to the ranges of the respective grade under the assumption that the sulfur content in the finished metal should not be above 0.015 wt.% and the CaF_2

content in the slag should be minimal. It was shown during the calculations for 35KhGSA steel that decreasing the concentration of Al_2O_3 to 30 wt.% has practically no influence on the desulfurizing power of the slag and that an SiO_2 content above 10 wt.% results in the additional entry of silicon into the metal due to its reduction from the slag. The introduction of MgO to 6 wt.% and of CaF_2 to 4 wt.% provides a way to lower the viscosity to values close to the viscosity of a basic lime–alumina alloy: 0.16–0.2 $N{\cdot}s/m^2$ at 1,843–1,873 K.

Therefore, a slag with the following composition was recommended for the treatment of 35KhGSA steel, wt.%: 50 CaO, 30 Al_2O_3, 10 SiO_2, 6 MgO, 4 CaF_2.

Calculations were also performed to reveal the influence of the relative mass of the slag on the composition of the final metal. It was shown that when the amount of slag is greater than 5% of the mass of the metal, the losses of the alloying elements increase considerably, particularly in the case of manganese, whose losses increase by 10 rel.%. If the slag is taken in an amount equal to less than 3% of the mass of the metal, the impurity-removal reactions do not develop to an appreciable extent. Therefore, the following range of amounts of slag was recommended for the ladle treatment of steel: 3.5–4.2% of the mass of the metal.

The influence of the tapping time on the composition of the steel was also analyzed. It was shown that increasing the time until the metal is poured into the ladle to more than 7 min results in a significant furnace loss of manganese. Moreover, there is a danger of secondary oxidation of the metal during prolonged contact with the air [2, 3]. It should also be noted that in the case of deoxidization by aluminum, the processes that remove the deoxidization products from the steel take place over a time period greater than 4 min. Restricting the tapping time to 4–6 min was recommended.

The slag for treating grade 10 steel should have a stronger desulfurizing power than the slag for treating grade 35KhGSA steel. The computational experiment showed that a slag with the following composition (wt.%) can be used for this steel: 50 CaO, 10 SiO_2, 25 Al_2O_3, 5 MgO, 10 CaF_2. The calculation showed that the use of such a slag provides a way to obtain a steel composition that satisfies the requirements of the standards documentation and lowers the concentration of sulfur to 0.013–0.015 wt.%.

Testing of new synthetic slags for treating steel, whose composition was found with the aid of a computational experiment, was conducted in unit 21 of the Izhstal Corporation.

The influence of the treatment of grade 35KhGSA steel and of grade 10 steel with the experimental liquid synthetic slags on their quality was investigated. The steels were smelted according to the standards documentation of the Izhstal Corporation in basic 120-ton furnaces and then treated with the experimental liquid synthetic slags.

The amount of slag added was equal to about 4% of the mass of the liquid metal, and the temperature of the slag before tapping was 1,993 K.

The experimental and reference melts were poured by the siphon method under the 50SK500 exothermic mixture into ingots weighing 3.5 tons. The ingots were rolled on blooming and section rolling mills.

Table 8.8 Degree of desulfurization of the steels

Steel grade	Technology variant	Sulfur content (wt.%)		Degree of desulfurization (%)
		Before tapping	In the finished metal	
35KhGSA	Experimental	0.035	0.011	68.6
	Reference	0.035	0.012	65.8
10	Experimental	0.033	0.014	57.6
	Reference	0.038	0.018	52.6

Table 8.9 Hydrogen content in the steels

Steel grade	Technology variant	Change in hydrogen content (cm^3/100 g)	
		Before tapping	In the finished metal
35KhGSA	Experimental	5.01	5.04
	Reference	5.53	6.00
10	Experimental	5.84	6.01
	Reference	4.34	5.80

The quality of the metal was investigated in accordance with the requirements of State Standard 4543-71 for grade 35KhGSA steel and State Standard 10-50-74 for grade 10 steel with additional monitoring of the nonmetallic inclusions and flakes. During the smelting and pouring of the experimental and reference melts, samples of the liquid metal were collected for determination of the chemical composition, including the content of gases, before tapping from the furnace and in the pouring stage.

Table 8.7 presents data on the chemical compositions of the experimental and reference steels after ladle treatment by the synthetic slags.

It can be seen from the data presented that the chemical composition of the experimental steels fully conforms to the requirements of State Standard 4543-71 and State Standard 10-50-74.

Data on the degree of desulfurization are presented in Table 8.8. It follows from the table that a somewhat higher degree of desulfurization is observed when the steels are treated with the experimental slags.

The changes in the hydrogen content in the steels smelted with treatment by the respective experimental slag are presented in Table 8.9. It follows from this table that the hydrogen content increases significantly less in the case of treatment of the metal with the experimental slags than it does in the case of refinement by an ordinary lime–alumina slag.

The content of nonmetallic inclusions in the experimental and ordinary metal is at approximately the same level.

No macrostructure defects or flakes were found in the steels treated with the experimental slags. The mechanical properties of the steels treated with the experimental slags satisfy the requirements of State Standard 4543-71 and State Standard 10-50-74.

Thus, it may be concluded that the experimental slags are suitable for refining steels.

The use of dolomite, which contains ~36 wt.% MgO and has a low (<0.1 wt.%) sulfur content in its roasted form, as the main supplier of magnesium oxide was proposed. The use of unroasted dolomite is more efficient, because the carbon dioxide bubbles released when it decomposes in the furnace during smelting of the slag pass through the slag and promote more complete removal of hydrogen from the slag. To reduce the use of synthetic corundum, the employment of crushed fire clay containing about 33 wt.% Al_2O_3 was proposed. A spent ESR flux was used to introduce CaF_2 into the slag. This material (depending on the grade) contains the following, wt.%: 50–60 CaF_2, 20–50 Al_2O_3, and 0.025 S. The presence of aluminum oxide in it enables further reduction of the use of synthetic corundum. Employment of the proposed slag resulted in a savings of 6.5 kg of synthetic corundum per ton of steel and lowering of its input rate by about 40%.

The mathematical model described in Sect. 4.2 was also used to optimize the technology for obtaining ShKh-15 steel by the mixing method [5]. As a result of an analysis of different variants of the technology, the following steel smelting conditions were recommended: rapid pouring of the alloying elements over the course of 20–30 s; simultaneous starting of the pouring of the alloying elements and the intermediate product; a tapping rate of the intermediate product equal to 15–18 ton/min. Adoption of the recommendations enabled us to decrease the expenditures for conducting industrial experiments at the Izhstal Corporation, to smelt ShKh-15 steel with the required composition, and to produce an economic benefit.

8.3 Improvement of Electroslag Technologies Based on Mathematical Modeling of the Interaction Processes of the Metal, Slag, and Gas

This section presents data from calculations of the chemical composition of the phases that appear in electroslag technologies for different grades of steel using previously presented mathematical models of the interaction of the metal, slag, and gas. The results of the calculations are compared with experimental data. Recommendations for improving the technologies of the processes indicated are developed on the basis of the calculations.

8.3.1 Influence of the Composition of the Weld Metal on Pore Formation during the Electroslag Welding of Press Columns

Electroslag welding (ESW) is used during the manufacture of 15,000-ton press columns at the Uralmash Machine-Building Corporation. Unfortunately, the

Table 8.10 Values of the input parameters of the model

Stage	β_j, $s^{-1/2}$	$S_{dr}\beta_{dr}$, $m^2 s^{-1/2}$	S_j, m^2	V_j, g/s	δ_{ox_j}, m
Electrode	–	0.029	–	20.8	1×10^{-6}
Base metal	2.0	–	0.02	19.7	5×10^{-6}
Consumable tip	2.0	–	0.0056	4.0	5×10^{-6}
Pool	0.4	–	0.057	44.5	–

welded joints are often found to contain defects (pores, cracks, etc.), whose appearance is associated with the heterogeneity of the chemical composition of the welds. In order to eliminate these defects, information regarding the influence of the technological parameters of the welding process and the composition of the welding material on the properties of the weld metal is needed. Since obtaining such data experimentally is associated with large material expenditures, the possibility of theoretically predicting the composition of welds takes on special significance.

In one study a mathematical model of the chemical processes that occur during ESW [6] was used to investigate the influence of the welding speed, the weld penetration depth in the grade 35 steel parts being welded, and the oxidation level of Sv-10G2 welding wire on the composition of the weld metal during the welding of half-columns of a press with a 1.05 × 1.05 m cross section [7]. The welding was performed by a grade 30 steel consumable tip under AN-8 flux with the feeding of twelve electrode wires with a diameter of 3×10^{-3} m.

The physicochemical constants of the model (equilibrium constants, diffusion coefficients, etc.) were taken from [8]. The temperatures of the liquid metal on the droplet surface (2,073 K), on the edges of the workpiece and the consumable tip (1,873 K), and of the metal pool (1,923 K) were taken according to [9].

Table 8.10 presents the values of the convection constants (β), as well as the mass-transfer constants in the droplet state in the form of the product $S_{dr}\beta_{dr}$, according to [10]. The table also presents the values of the thicknesses of the oxide films (δ_{ox_j}) and the other input parameters of the model.

The mass of the slag pool was $G_{slag} = 6.6$ kg. The chemical composition of the flux was as follows (wt.%): 6.12 FeO, 31.9 SiO_2 30.2 MnO, 10.4 Al_2O_3, with the remainder consisting of CaF_2, CaO, and MgO. The compositions of the starting materials, as well as the weld (the calculated and experimental values are given above and below the bars, respectively), including the composition obtained when the base and electrode metals were mixed and no chemical reactions occurred, are presented in Tables 8.11 and 8.12.

The calculations were performed for different welding speeds (variants 1 and 2), different penetration depths of the base metal (variants 1 and 5), and different oxidation levels of the electrode wire δ_{ox_j} (variants 1, 3, and 4), as well as for the case in which six of the twelve grade Sv-10G2 electrode wires were replaced by Sv-08G2S wires (variant 6). As a whole, the calculated composition of the weld metal corresponds to the experimental composition with the exception of the upper part of the weld.

Table 8.11 Chemical composition of the starting materials (wt.%)

Starting materials	C	Si	Mn	P	S	Cr	O
Grade 35 steel	0.35	0.36	0.78	0.03	0.03	0.38	0.005
Grade 30 steel	0.30	0.28	0.73	0.03	0.03	0.22	0.005
Sv-10G2	0.07	0.04	1.80	0.016	0.023	0.06	0.005
Sv-08G2S	0.07	0.80	2.00	0.03	0.025	0.20	0.005

Table 8.12 Chemical composition of the weld metal (wt.%)

Variant	Seam length (m)	C	Si	Mn	P	S	Cr	O	Remarks
1		0.20	0.17	1.23	0.024	0.027	0.16	0.0005	With no chemical reactions
	0.008	0.18/0.17	0.18/0.08	1.01/0.87	0.012/0.024	0.014/0.022	0.16/0.14	0.016/0.015	$V_w = 0.36$ m/h; $\delta_{pen} = 12 \times 10^{-3}$ m
	0.24	0.18/0.19	0.16/0.09	1.03/0.85	0.013/0.020	0.022/0.022	0.16/0.15	0.017/0.013	$\delta_{ox1} = 10^{-6}$ m
	0.48	0.18/0.17	0.16/0.17	1.06/0.92	0.014/0.020	0.025/0.021	0.16/0.14	0.017/0.011	
	0.72	0.18/0.17	0.15/0.23	1.08/1.17	0.014/– –	0.026/0.021	0.16/0.15	0.017/0.010	
	0.96	0.18/0.12	0.15/0.55	1.09/1.13	0.014/0.017	0.026/0.023	0.16/0.12	0.017/0.018	
	1.20	0.18/0.11	0.15/0.07	1.10/1.06	0.015/0.019	0.026/0.026	0.16/0.11	0.017/0.016	
2	0.24	0.18	0.16	1.04	0.014	0.022	0.16	0.017	$V_w = 0.4$ m/h
	0.72	0.18	0.15	1.08	0.015	0.026	0.16	0.017	
	1.20	0.18	0.14	1.10	0.015	0.026	0.16	0.017	
3	0.24	0.18	0.22	1.16	0.015	0.021	0.16	0.014	$\delta_{ox1} = 0$
	0.72	0.18	0.21	1.21	0.017	0.026	0.16	0.014	
	1.20	0.18	0.21	1.23	0.018	0.026	0.16	0.014	
4	0.24	0.17	0.11	0.91	0.012	0.023	0.16	0.021	$\delta_{ox1} = 2 \times 10^{-6}$ m
	0.72	0.17	0.10	0.93	0.012	0.026	0.16	0.022	
	1.20	0.17	0.10	0.95	0.013	0.026	0.16	0.023	
5		0.17	0.15	1.38	0.026	0.026	0.14	0.005	With no chemical reactions
	0.24	0.15	0.14	1.06	0.012	0.021	0.14	0.020	$\delta_{pen} = 6 \times 10^{-9}$ m
	0.72	0.15	0.12	1.11	0.013	0.025	0.14	0.020	
	1.20	0.15	0.12	1.14	0.013	0.025	0.14	0.021	
6		0.20	0.35	1.28	0.027	0.027	0.19	0.005	With no chemical reactions
	0.24	0.18	0.31	1.16	0.017	0.021	0.19	0.013	Sv-10G2 + Sv-08G2S
	0.72	0.18	0.31	1.18	0.018	0.026	0.19	0.014	
	1.20	0.18	0.31	1.19	0.019	0.027	0.19	0.014	

The values above the bars are calculated, and the values below the bars are experimental

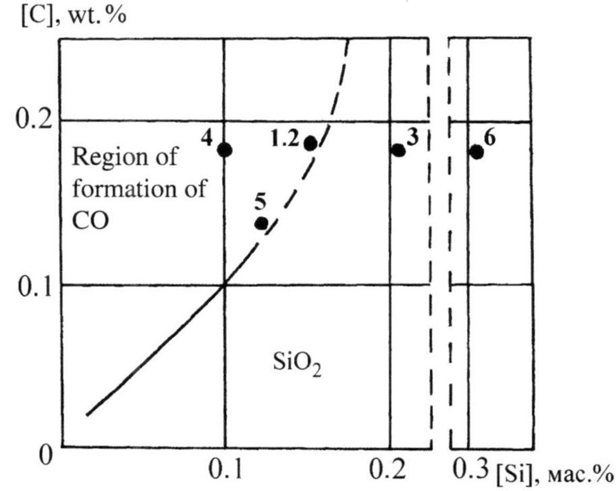

Fig. 8.1 Influence of silicon on the critical concentration of carbon (the numbers of the points correspond to the calculation variants in Table 8.12 for a seam length equal to 0.72 m)

As follows from Table 8.12, the welding speed does not have an appreciable influence on the composition of the weld in the assigned range. The experiment exhibits an abrupt rise in the loss of carbon, silicon, and chromium in upper part of the weld with a simultaneous increase in the oxygen content in the metal. Apparently, this is a result of a decrease in the concentration of silicon in the metal to a level below the critical concentration, at which the interaction of oxygen and carbon is suppressed [11]. This situation, which is not taken into account in the model, results in the formation of gas bubbles during crystallization of the metal. The presence of gas pores in the upper part of the weld confirms this.

Figure 8.1 presents a plot of the critical concentrations of carbon and silicon in the temperature range for crystallization of the metal based on the data in [11], which enables us to determine the composition region in which ebullition of the metal with the formation of CO is possible. The concentration of oxygen in the metal should then exceed its solubility in δ-iron [11]:

$$\log[\mathrm{O}]_{\delta-\mathrm{Fe}} = -\frac{2500}{T} - 0.86. \tag{8.19}$$

At 1,700–1,800 K, $[\mathrm{O}]_{\delta\text{-Fe}} = 0.0047 - 0.0056$ wt.%. This range is below the experimental values (0.01–0.016 wt.%).

It follows from Fig. 8.1 that when the thickness of the oxide film on the electrode wire is greater than 1×10^{-3} mm (variants 1, 2, 4, and 5), the entry of FeO into the slag results in lowering of the silicon concentration to the critical value 0.15 wt.% Si in the presence of 0.18 wt.% C, which also causes the appearance of gas pores. The stirring of the phases by CO bubbles intensifies mass transfer and accelerates the oxidation of the alloying elements, as well as the entry of oxygen into the metal (see Table 8.12).

The calculations show that the silicon content in the weld is lowered. Consequently, the probability of the formation of pores increases as the penetration depth of the workpiece edges decreases, since the concentration of silicon is higher in the base metal (0.36 wt.%) than in the electrode wire (0.04 wt.%).

Inadequate cleaning of the surface of the electrode wire also promotes the appearance of pores in the weld, because it results in an increase in the loss of silicon due to the accumulation of ferrous oxide in the slag pool. In fact, when the oxide film is completely absent (variant 3), the silicon content is sufficient for suppressing ebullition of the pool (see Fig. 8.1). One of the measures for preventing pore formation may be partial replacement of the Sv-10G2 electrode wire by a wire with an increased silicon content, for example, Sv-08G2S. Table 8.12 presents the calculated composition of the weld metal when six of the Sv-10G2 wires (every other wire) are replaced by Sv-08G2S wires (variant 6). It is seen from Fig. 8.1 that the resultant Si content rules out the formation of CO. Sufficiently thorough stirring of the metal pool during ESW ensures the homogeneity of its chemical composition.

The use of computational experiments enabled us to lower the number of test welds and to develop recommendations for improving the technology of the ESW of various steels, which were submitted to the Uralmash Machine-Building Corporation.

8.3.2 Influence of the Flux Composition and the Oxidation Level of the Electrode on the Chemical Heterogeneity of the Ingot during the Electroslag Remelting of 38Kh2MYuA Steel

On instruction from the Izhstal Corporation, the influence of the composition of the flux and the oxidation level of the 38Kh2MYuA steel electrode on the heterogeneity of the chemical composition of the ingot was analyzed using a quantitative model of the chemical processes that occur during electroslag remelting (ESR) [12, 13].

The starting data taken in the calculations were as follows: cross-sectional dimensions of the electrode, 0.37 × 0.37 m; cross-sectional area of the mold, 0.485 × 0.485 m; ingot mass, 3.5 tons; ingot length, 1.9 m; melting time, 5 h; mass of the slag pool (ANF-6 flux), 100 kg; melting rate, 700 kg/h; height of the consumable electrode cone, 0.09 m. The optimal values of the slag pool mass, the melting rate, and the electrode cone height for the industrial ingot smelting technology were taken.

The calculations were performed for various concentrations of SiO_2 (0–6 wt.%) and FeO (0–1 wt.%) in the slag, as well as for different thicknesses of the oxide film on the electrode in the range $\delta_{ox} = 0–2.8 \times 10^{-3}$ m and $\Delta\tau = 2$ min.

Using the starting data, we found the following: the area of the contact surface between the metal and the slag in the electrode stage $S_e = 0.152$ m^2, the surface area of the metal pool $S_{po} = 0.235$ m^2. The temperature of the metal in the liquid film on the electrode (1,823 K) and the temperature on the surface of the metal pool (1,863 K) were taken from the data in [14].

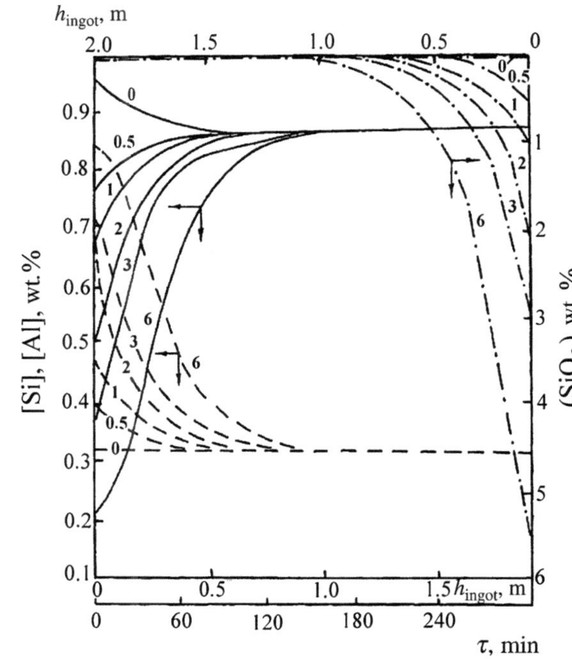

Fig. 8.2 Influence of the initial silicon dioxide content (the numbers on the curves are the initial concentration of SiO_2, wt.%) in the slag on the distribution of Al (*solid curves*), Si (*dashed curves*), and SiO_2 (*dot-dashed curves*) along the ingot height for $\delta_{ox} = 0.7$ mm

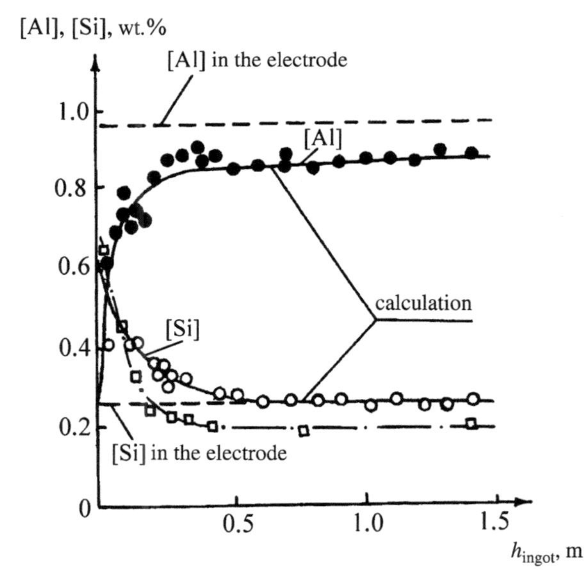

Fig. 8.3 Distribution of aluminum and silicon along the height of an ESR ingot of 38Kh2MYuA steel for $(FeO)_0 = 0.5$ wt.%, $(SiO)_0 = 2$ wt.%, and $\delta_{ox} = 0.7$ mm. The flux was deoxidized by ferraluminum

The initial composition of 38Kh2MYuA steel used in the calculations was as follows (wt.%): 1.0 Al, 0.38 C, 0.45 Mn, 1.5 Cr, 0.2 Mo, 0.325 Si, 0.02 S, 0.005 O, with the remainder being Fe.

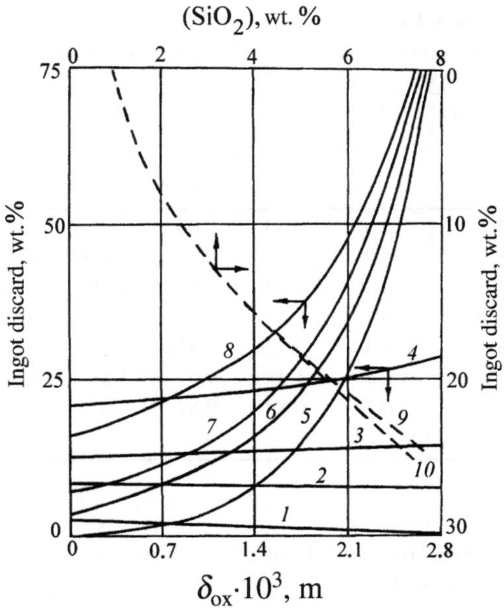

Fig. 8.4 Influence of the thickness of the oxide film on the electrode and the SiO_2 content in the flux on the amount of ingot discard for the ESR of 38Kh2MYuA steel: *1–4*—determined by the off-grade silicon content in the metal; *5–8*—determined by the off-grade aluminum content in the metal; *1*, *5*—$(SiO_2)_0 = 1$ wt.%; 2, 6—$(SiO_2)_0 = 2$ wt.%; 3, 7—$(SiO_2)_0 = 3$ wt.%; 4, 8—$(SiO_2)_0 = 6$ wt.%; 9—$\delta_{ox} = 0$; $\delta_{ox} = 0.7$ mm

The results of the calculations are presented in Figs. 8.2, 8.3, 8.4, and in Table 8.13. According to the calculations, the concentrations of C, Cr, Mo, and Mn in the metal scarcely vary and remain equal to the initial concentrations in the electrode at all the values of (SiO_2), (FeO), and δ_{ox}. This was confirmed by the data from the trial melts.

The concentrations of sulfur and oxygen decrease from 0.02 and 0.005 wt.%, respectively, to 0.006 and 0.001 wt.% and remain practically constant over the length of the ingot. The reaction rates are significantly higher on the electrode than in the pool. This is attributable to the higher concentrations of the elements and rates of mass transfer in the electrode stage. Accordingly, the concentration of (FeO) near the metal/slag interface in the electrode stage (0.003–0.002 wt.%) is lower than the concentration in the pool (0.022–0.016 wt.%).

The percentages of aluminum and silicon undergo the greatest changes, but the calculated dependences are close to the experimental (Fig. 8.3). In the lower part of the ingot the silicon concentration exceeds the permissible value (0.45 wt.%), and the corresponding part of the ingot is bottom discard. Therefore, the influence of the composition of the flux and the oxidation level of the electrode (δ_{ox}) on the distribution of Al and Si along the height of the ingot was examined.

The calculations showed that the presence of up to 1 wt.% FeO in the original flux does not have an appreciable influence on the content of [Al] and [Si], i.e., on the amount of ingot discard. The main causes of the heterogeneity of the chemical composition are the presence of silica in the slag and the oxide film on the electrode. Their influence on the amount of ingot discard is presented in Fig. 8.4. It can be seen that up to $\delta_{ox} = 0.85$ mm the amount of discard is determined by the off-grade silicon content and that at $\delta_{ox} > 0.85$ mm it is determined by the diminished aluminum content compared with the grade composition range.

Table 8.13 Concentrations of elements in the metal and of oxides in the slag (wt.%) during the ESR of 38Kh2MYuA steel: (FeO) = 1 wt.%, (SiO_2) = 1 wt.%, and $\delta_{ox} = 0$ (no oxidation of the electrode)

Element	$[E_i]_e$			$[E_i]_{ingot}$			(E_iO_m)		
	Melting time (min)								
	2	150	300	2	150	300	2	150	300
O	0.002	0.002	0.002	0.001	0.001	0.001	–	–	–
S	0.014	0.014	0.014	0.006	0.006	0.006	0.004	0.03	0.07
C	0.38	0.38	0.38	0.38	0.38	0.38	–	–	–
Mn	0.45	0.45	0.45	0.45	0.45	0.45	0.001	0.012	0.012
Si	0.38	0.326	0.325	0.469	0.326	0.325	0.93	0.013	0.009
Cr	1.5	1.5	1.5	1.5	1.5	1.5	0	0.002	0.002
Al	0.79	0.89	0.89	0.62	0.88	0.89	27.1	29.1	29.6
FeO	–	–	–	–	–	–	0.93	0.014	0.011

The results obtained enabled us to formulate measures for reducing the chemical heterogeneity of the ingot and increasing the yield of acceptable metal:

1. Lowering the content of (SiO_2) in the original flux to 1 wt.%, which provides a way to decrease the amount of discard from 12 to 3–5 rel.%. This can be achieved, in particular, by deoxidizing the flux with aluminum. For faster settling of the Al–Si alloy droplets formed, it is expedient to use ferroaluminum with a high density (Fig. 8.3).
2. The use of electrodes with an oxide film thickness no greater than 0.7 mm for remelting, which ensures a percentage of aluminum in the metal within the grade limits. This can be achieved by cleaning the electrodes or by creating conditions that ensure minimal oxidation by the heated lower part of the electrode (argon protection).
3. Repeated use of a flux with a lowered (SiO_2) content, which provides a way to decrease the amount of ingot discard associated with the increased silicon content in the metal.

The data obtained and the recommendations developed were submitted to the Izhstal Corporation for improving the ESR technology for 38Kh2MYuA steel. Adoption of these recommendations enabled the achievement of an increase in the yield of acceptable metal by lowering the chemical heterogeneity of the ingot.

8.3.3 Optimization of the Technology for Alloying 45 Steel with Lead from a Flux

Alloying steel with 0.1–0.5 wt.% lead significantly improves its machinability and friction properties. Because of its low solubility in iron, lead exists in an ingot in the form of inclusions, which act as a lubricant during mechanical treatment.

Lead can be introduced into the metal during ESR by different techniques. One of them is to feed lead shot into the metal pool. Due to the high density, as the lead particles melt, probably still in the slag layer, they sink to the bottom of the metal pool and are "trapped" by crystallites that are growing perpendicularly to the solidification front. In such a technology it is obviously difficult to ensure uniformity of the alloying over the height and cross section of an ingot. The accumulation of liquid lead in the central portion of the pool is possible. For greater homogeneity, the size of the lead particles introduced can be decreased. However, when tiny particles are introduced, their entrainment by the slag and metal flows and accumulation on the periphery of the ingot are possible. Since lead droplets can "slide down," the accumulation of liquid lead at the center of the ingot is also possible here.

Lead is known to have a high volatility [15], because its boiling point is $T_b^{Pb} = 2024\,K$. Since lead is toxic, such a technology is complicated by the need for additional safety measures.

In this section we will examine the possibility of alloying a metal with lead during ESR by reducing it from the oxide, which is introduced into the slag by periodic or continuous feeding. Because of its strong affinity toward oxygen, iron acts as the reducing agent here:

$$(\mathrm{PbO}) + [\mathrm{Fe}] = (\mathrm{FeO}) + [\mathrm{Pb}]. \tag{8.20}$$

The equilibrium constant equals

$$K_{eq} = \frac{a_{FeO}a_{Pb}}{a_{PbO}a_{Fe}} = \frac{\gamma_{FeO}N_{FeO}}{\gamma_{PbO}N_{PbO}} = \frac{\gamma_{FeO}(\mathrm{FeO})/M_{FeO}}{\gamma_{PbO}(\mathrm{PbO})/M_{PbO}}. \tag{8.21}$$

Due to the low mutual solubility of iron and lead, it is assumed here that $a_{Pb} = a_{Fe} = 1$.

We can obtain a thermodynamic estimate of the extent of completion of reaction (8.20) under the conditions of ESR. For reaction (8.20) at 2,000 K, we have

$$\Delta G^0 = \Delta G^0_{FeO} - \Delta G^0_{PbO},$$

whence we obtain

$$\ln K_{eq} = -\Delta G^0/(RT) = 3.66 \text{ and } K_{eq} = 38.9.$$

Using the value found for K_{eq}, as well as the known value for the PbO–SiO_2 system $\gamma_{FeO} \approx 0.2$, from Eq. 8.21 we can estimate the ratio:

$$\frac{(\mathrm{FeO})}{(\mathrm{PbO})} = K_{eq}\frac{\gamma_{Pb}M_{FeO}}{\gamma_{FeO}M_{PbO}} = 2.51. \tag{8.22}$$

The concentration of (FeO) during the ESR of structural steels is ≈0.5 wt.%. According to expression (8.22), this value corresponds to a concentration of (PbO) equal to 0.2 wt.%. Thus, when the slag contains more than 2 wt.% PbO, a regime of reaction (8.20) that is close to the limiting diffusion-controlled regime with respect to (PbO) is realized, i.e., the expression for the rate of this reaction can be written in the form

$$J_{\mathrm{Pb}} = K^{\mathrm{d}}_{\mathrm{PbO}}(C_{\mathrm{PbO}} - C'_{\mathrm{PbO}}) \approx K^{\mathrm{d}}_{\mathrm{PbO}} C_{\mathrm{PbO}}. \tag{8.23}$$

It is known [16] that lead oxides are also volatile ($T_{\mathrm{b}}^{\mathrm{PbO}} = 1928\,\mathrm{K}$); therefore, the use of PbO in pure form is not expedient. It is more practically feasible to use materials that contain PbO in a bound form, for example, glass ("crystal glass"), which contains 20–50 wt.% PbO. When broken glass of a definite fraction enters the slag pool, it melts and dissolves in it. This significantly lowers the concentration, that is, the activity, of PbO. Since $P_{\mathrm{PbO}} = P^0_{\mathrm{PbO}} N_{\mathrm{PbO}}$, this also lowers the vapor pressure of PbO.

Reaction (8.20) occurs on all contact surfaces of the molten metal with the slag. The result of the reaction is the formation of films and then tiny droplets, which are uniformly distributed in the metal flowing from the tip of the electrode, as well as on the surface of the metal pool, and settle in a uniform "shower" onto the bottom of the metal pool. The rate of entry of lead is regulated by the PbO content in the slag. It can be assumed that this ensures a more uniform lead content in the ingot both along its height and over a transverse section. Here the slag and reaction (8.20) act as a damper, which reduces the fluctuations of the lead content in the steel in the case of nonuniform feeding of the lead-containing components into the slag pool.

There are two possible ways to analyze the behavior of lead in the alloying of steel during ESR:

- by including lead in the list of elements in the overall mathematical model of the chemical processes that occur during ESR [8];
- by employing an approximate, but more graphic analytical method.

We will first consider the analytical method for calculating the concentration of Pb in steel. The material balance of lead in a slag in a steady-state regime can be written in the form of the following equation:

$$\dot{m}_{\mathrm{fl}} \frac{(\mathrm{PbO})_{\mathrm{fl}}}{100} = J, \ \mathrm{kg/s}, \tag{8.24}$$

where $\dot{m}_{\mathrm{fl}}$ is the feeding rate of the glass into the slag pool (kg/s) and $(\mathrm{PbO})_{\mathrm{fl}}$ is the mass concentration of PbO in the glass (wt.%).

The left-hand side of Eq. 8.24 represents the rate of entry of PbO into the slag with the glass containing PbO, and the right-hand side (J) represents the rate of the passage of lead into the steel through reaction (8.20). Here

$$J = K^{\mathrm{d}}_{\mathrm{PbO}} S\, C_{\mathrm{PbO}} M_{\mathrm{PbO}} = K^{\mathrm{d}}_{\mathrm{PbO}} S (\mathrm{PbO}) \rho_{\mathrm{slag}}/100, \ \mathrm{kg/s}, \tag{8.25}$$

where $K^{\mathrm{d}}_{\mathrm{PbO}} = \beta D^{0.5}$ (m/s), S is the total surface area of the metal/slag interface (m^2), and D is the diffusion coefficient of PbO in the slag (m/s).

Under the assumption that lead passes completely into the metal pool and is distributed uniformly over the surface of the crystallizing ingot (the bottom of the metal pool), the concentration of lead in the steel equals:

$$[\mathrm{Pb}]\frac{\dot{m}_{\mathrm{Pb}}}{(\dot{m}+\dot{m}_{\mathrm{Pb}})}100\% = \frac{\dot{m}_{\mathrm{fl}}(\mathrm{PbO})_{\mathrm{fl}}M_{\mathrm{Pb}}100\%}{M_{\mathrm{PbO}}\left[100\dot{m}_{\mathrm{fl}}+\dot{m}_{\mathrm{fl}}(\mathrm{PbO})_{\mathrm{fl}}M_{\mathrm{Pb}}/M_{\mathrm{PbO}}\right]}, \tag{8.26}$$

where $\dot{m}_{\mathrm{Pb}} = J\frac{M_{\mathrm{Pb}}}{M_{\mathrm{PbO}}}$ is the rate of entry of Pb into the metal undergoing remelting due to chemical reaction (8.20) (kg Pb/s) and $\dot{m}$ is the remelting rate (kg/s).

Neglecting the second term in square brackets in the denominator of Eq. 8.26, we obtain

$$[\mathrm{Pb}] = \frac{\dot{m}_{\mathrm{fl}}(\mathrm{PbO})_{\mathrm{fl}}M_{\mathrm{Pb}}}{M_{\mathrm{PbO}}\dot{m}}. \tag{8.27}$$

Substituting the values $M_{\mathrm{Pb}} = 207$ g/mole and $M_{\mathrm{PbO}} = 223$ g/mole into Eq. 8.27, we obtain an expression for the technological parameter, namely, the feeding rate of the glass ($\dot{m}_{\mathrm{fl}}$) that ensures the assigned concentration of lead in the metal [Pb]:

$$\dot{m}_{\mathrm{fl}} = 1.077\,\frac{\dot{m}[\mathrm{Pb}]}{(\mathrm{PbO})_{\mathrm{fl}}}. \tag{8.28}$$

Clearly, a steady-state alloying regime is realized for a definite concentration of (PbO) in the slag. Solving the combined system of Eqs. 8.24, 8.25, and 8.27, we obtain an expression for calculating the value of (PbO) that ensures constancy of the value of [Pb]:

$$(\mathrm{PbO}) = \frac{\dot{m}_{\mathrm{fl}}(\mathrm{PbO})_{\mathrm{fl}}}{K^{\mathrm{d}}_{\mathrm{PbO}}S\rho_{\mathrm{slag}}} = \frac{[\mathrm{Pb}]M_{\mathrm{PbO}}\dot{m}}{M_{\mathrm{Pb}}K^{\mathrm{d}}_{\mathrm{PbO}}S\rho_{\mathrm{slag}}}. \tag{8.29}$$

After substituting the value $\rho_{\mathrm{slag}} = 3000\mathrm{kg/m^3}$ into this expression, we found

$$(\mathrm{PbO}) = 3.6\cdot 10^{-4}\,\frac{[\mathrm{Pb}]\dot{m}}{K^{\mathrm{d}}_{\mathrm{PbO}}S}. \tag{8.30}$$

where $K^{\mathrm{d}}_{\mathrm{PbO}}S = \beta_{\mathrm{e}}S_{\mathrm{e}}D^{0.5}_{\mathrm{e}} + \beta_{\mathrm{po}}S_{\mathrm{po}}\,D^{0.5}_{\mathrm{po}}$.

A more exact solution can be obtained when the variation of the concentration of PbO in the slag during the remelting is taken into account. In this case the material balance equation for (PbO) can be written in the form

$$\dot{m}_{\mathrm{fl}}\frac{(\mathrm{PbO})_{\mathrm{fl}}}{100} = K^{\mathrm{d}}_{\mathrm{PbO}}S\,\frac{(\mathrm{PbO})\rho_{\mathrm{slag}}}{100} + \frac{m_{\mathrm{slag}}}{100}\frac{d(\mathrm{PbO})}{d\tau}, \tag{8.31}$$

where m_{slag} is the mass of the slag pool (kg).

The left-hand side of Eq. 8.31 represents the rate of entry of PbO into the slag due to the feeding of the glass, the first term on the right-hand side represents the rate of the loss of (PbO) as a result of the chemical reaction on the metal/slag interface, and the second term represents the rate of variation of the concentration of (PbO) in the slag pool.

Integrating Eq. 8.31, we found

$$(\mathrm{PbO}) = \frac{\dot{m}_{\mathrm{fl}}(\mathrm{PbO})_{\mathrm{fl}}}{K^{\mathrm{d}}_{\mathrm{PbO}} S \rho_{\mathrm{slag}}}\left[1 - \exp\left(-K^{\mathrm{d}}_{\mathrm{PbO}} S \rho_{\mathrm{slag}} \tau / m_{\mathrm{slag}}\right)\right]. \tag{8.32}$$

Substituting (8.32) into (8.31) and performing some transformations, we obtain

$$[\mathrm{Pb}] = \frac{0.928 \dot{m}_{\mathrm{fl}}(\mathrm{PbO})_{\mathrm{fl}}}{\dot{m}_{\mathrm{ingot}}}\left[1 - \exp\left(-K^{\mathrm{d}}_{\mathrm{PbO}} S \rho_{\mathrm{slag}} \tau / m_{\mathrm{slag}}\right)\right]. \tag{8.33}$$

When lead was included in the list of elements in the overall mathematical model of the chemical processes that occur during ESR, the rate of reaction (8.20) was described by Eqs. 8.34 and 8.35 [8]:

$$J_{\mathrm{E}_i 2} = \frac{(K_{i2}x_2)^m [\mathrm{E}_i]_1 - (\mathrm{E}_i\mathrm{O}_m)}{(K_{i2}x_2)^m a_{i2} + b_{i2}}, \tag{8.34}$$

$$J_{\mathrm{Pb1}} = \frac{(K_{\mathrm{Pb1}}x_1)[\mathrm{Pb}]_0 - (\mathrm{PbO})}{(K_{\mathrm{Pb1}}x_1)^{0.8}/K^0_{\mathrm{Pb1}} + (K_{\mathrm{Pb1}}x_1)(a_{i2} + C_{i2}) + b_{\mathrm{Pb1}}}. \tag{8.35}$$

The concentration of lead in the steel was calculated in the electrode stage from the equation

$$[\mathrm{E}_i]_1 = [\mathrm{E}_i]_0 - J_{\mathrm{E}_i 1} S_1 M_{\mathrm{E}_i} 100/\dot{m} \tag{8.36}$$

and in the pool stage from the equations [6]

$$\begin{aligned} \Delta[\mathrm{E}_i]_2^{k+1,k} &= \left\{\dot{m}[\mathrm{E}_i]_j + v^n_M [\mathrm{E}_i]_n - J^k_{\mathrm{E}_i 2} S_2 M_{\mathrm{E}_i} 100 - \dot{m}_{\mathrm{cryst}} \kappa_{\mathrm{E}_i} [\mathrm{E}_i]_2^k\right\} \frac{\Delta\tau}{G_{M2}} \\ [\mathrm{E}_i]_2^{k+1} &= [\mathrm{E}_i]_2^k + \Delta[\mathrm{E}_i]_2^{k+1,k}. \end{aligned} \tag{8.37}$$

We took into account the variation of the slag mass using the equation

$$\begin{aligned} G^{k+1}_{\mathrm{slag}} = G^k_{\mathrm{slag}} &+ \sum_{i=1}^{i}\sum_{j=1}^{2} J^k_{\mathrm{E}_i j} S_j (M_{\mathrm{E}_i} - m M_{\mathrm{Fe}})\Delta\tau + V_{\mathrm{film}} S_{\mathrm{ox}} \rho_{\mathrm{ox}} \Delta\tau + V_{\mathrm{PSh}} \Delta\tau \\ &- V_{\mathrm{dr}} S_{\mathrm{sk}} \rho_{\mathrm{sk}} \Delta\tau - \sum_{i=1}^{i} V_{\mathrm{g}} \Delta\tau \end{aligned} \tag{8.38}$$

and the concentration of (PbO) in the slag for the $(k + 1)$-th time interval was calculated using expressions (8.34)–(8.39) [6]:

$$\begin{aligned} \Delta(\mathrm{E}_i\mathrm{O}_m)^{k+1,k} &= \left\{\sum_{j=1}^{2} J^k_{\mathrm{E}_i j} S_j \frac{M_{\mathrm{E}_i\mathrm{O}_m}}{M_{\mathrm{E}_i}} + V_{\mathrm{ox}} + V_{\mathrm{PSh}} - V_{\mathrm{g}} - V_{\mathrm{sk}}\right\} \frac{100\Delta\tau}{G_{\mathrm{slag}}} (\mathrm{E}_i\mathrm{O}_m)^{k+1} \\ &= (\mathrm{E}_i\mathrm{O}_m)^k + \Delta(\mathrm{E}_i\mathrm{O}_m)^{k+1,k} \end{aligned} \tag{8.39}$$

The general mathematical model of the chemical processes occurring during ESR under type ANF-6 flux was used to investigate the variation of the concentration of Pb in a grade 45 steel ingot and the concentration of PbO in the slag as a function of the feeding rate of the glass into the slag pool. The results of the calculations were compared with the experimental data. The following data were taken in the calculations for the ESR of electrodes with a diameter of 0.085 m in a mold with a diameter of 0.13 m: slag mass, 3 kg; $S_e = 0.0057$ m^2; $S_{po} = 0.0133$ m^2; $\beta_e = 7.4$ s$^{-1/2}$; $\beta_{po} = 0.9$ s$^{-1/2}$; thickness of the slag skin, 1.7×10^{-3} m; thickness of the oxide film on the electrode, 1×10^{-4} m; $\dot{m} = 94.3$kg/h; temperature of the metal on the surface of the liquid film on the electrode, 1,823 K, temperature on the surface of the metal pool, 1,863 K [9]. The chemical composition of the electrode metal (grade 45 steel) was as follows (wt.%): 98.65 Fe, 0.48 C, 0.56 Mn, 0.26 Si, 0.025 P, and 0.02 S. The chemical composition of type ANF-6 flux was as follows (wt.%): 0.5 FeO, 2.5 SiO_2, 30 Al_2O_3, with the remainder being CaF_2 and CaO.

The results of the calculations performed using the complete mathematical model of the chemical processes that occur during ESR and the experimental data are presented in Table 8.14 and in Figs. 8.5 and 8.6.

The calculated and experimental data reveal the nonuniform content of lead along the length of the ingot, especially in its lower part, regardless of the feeding rate of the glass into the slag (Fig. 8.5, curves 1–3). It is associated with the accumulation of PbO in the slag pool, which is most intense at the beginning of the melt (Fig. 8.6, curves 1–3). Therefore, the strongest heterogeneity in a transverse section of the ingot is also characteristic of its lower part (Table 8.14). On the first half of the length of the ingot, the concentration of (PbO) increases from 0 to 1.5 wt.% when $\dot{m}_{fl} = 1.7 \times 10^{-4}$ kg/s and from 0 to 2.2 wt.% when $\dot{m}_{fl} = 2.8 \times 10^{-4}$ kg/s, causing an appreciable increase in [Pb]: to 0.21 wt.% and 0.31 wt.%, respectively. In the upper part of the ingot the increase in the concentration of (PbO) is reduced. This results in slowing of the increase in [Pb]. Clearly, for each value of $\dot{m}_{fl}$, as a certain concentration of (PbO) is achieved, a corresponding constant value of [Pb] in the ingot should be expected. Therefore, it was postulated that to obtain ingots with an assigned uniform lead content, a certain required concentration of PbO in the slag should be ensured already in the initial melting period. This can be achieved, for example, already when the slag pool is created, and then the melting is carried out with an assigned feeding rate of the glass.

In order to test this hypothesis, calculations were performed (Figs. 8.5 and 8.6) for initial concentrations of PbO in the slag equal to 2 wt.% when $\dot{m}_{fl} = 1.7 \times 10^{-4}$ kg/s (curve 4) and to 3 wt.% when $\dot{m}_{fl} = 2.8 \times 10^{-4}$ kg/s (curve 5), respectively). It can be seen in Fig. 8.5 that in both cases some slight heterogeneity is observed only in the first 2–3 min, after which the lead content in the metal stabilizes and is equal to 0.28 wt.% when $\dot{m}_{fl} = 1.7 \times 10^{-4}$kg/s and to 0.43 wt.% when $\dot{m}_{fl} = 2.8 \times 10^{-4}$ kg/s. A trial melt conducted under factory conditions confirmed the correctness of the assumptions made and of the calculations performed.

Table 8.14 Chemical composition of the ingot (wt.%) during the ESR of grade 45 steel under type ANF-6 flux with the feeding of glass

Ingot height (m)	$\dot{m}_{fl} \cdot 10^4$ (kg/s)	Level	Sample collection site	Element					
				C	Si	Mn	P	S	Pb
0.36	1.7	Lower	Center	0.47 / 0.43	0.22 / 0.26	0.58 / 0.54	0.025 / 0.026	0.012 / 0.014	0.12 / 0.17
			Halfway	0.47 / 0.42	0.22 / 0.26	0.58 / 0.56	0.025 / 0.026	0.012 / 0.014	0.12 / 0.17
			Edge	0.42 / 0.47	0.23 / 0.26	0.58 / 0.55	0.024 / 0.026	0.012 / 0.011	0.06 / 0.10
		Middle	Center	0.48 / 0.43	0.21 / 0.17	0.55 / 0.57	0.025 / 0.025	0.012 / 0.020	0.21 / 0.27
			Halfway	0.48 / 0.42	0.21 / 0.18	0.55 / 0.57	0.025 / 0.024	0.012 / 0.020	0.21 / 0.27
			Edge	0.48 / 0.42	0.21 / 0.18	0.55 / 0.53	0.025 / 0.025	0.012 / 0.020	0.24 / 0.28
		Upper	Center	0.48 / 0.43	0.20 / 0.15	0.54 / 0.54	0.025 / 0.025	0.012 / 0.020	0.26 / 0.29
			Halfway	0.48 / 0.42	0.20 / 0.15	0.55 / 0.56	0.025 / 0.025	0.012 / 0.020	0.26 / 0.29
			Edge	0.48 / 0.43	0.20 / 0.15	0.55 / 0.55	0.025 / 0.025	0.012 / 0.020	0.25 / 0.29
0.60	1.7	Lower		0.47 / 0.51	0.23 / 0.26	0.57 / 0.51	0.025 / 0.023	0.012 / 0.010	0.09 / 0.10
		Upper		0.48 / 0.41	0.19 / 0.17	0.55 / 0.58	0.025 / 0.023	0.012 / 0.021	0.26 / 0.34
0.50	1.7	Lower		0.47 / 0.48	0.22 / 0.26	0.56 / 0.56	0.025 / 0.025	0.012 / 0.023	0.05 / trace
		Upper		0.43 / 0.45	0.19 / 0.14	0.54 / 0.59	0.025 / 0.023′	0.012 / 0.021	0.26 / 0.25
0.60	1.8	Lower		0.47 / 0.42	0.23 / 0.26	0.57 / 0.55	0.025 / 0.026	0.012 / 0.013	0.11 / 0.11
		Upper		0.48 / 0.42	0.19 / 0.10	0.54 / 0.51	0.025 / 0.023	0.012 / 0.021	0.27 / 0.26
0.60	2.8	Lower		0.47 / 0.50	0.22 / 0.24	0.57 / 0.55	0.025 / 0.023	0.012 / 0.014	0.16 / 0.19
		Upper		0.48 / 0.42	0.19 / 0.10	0.54 / 0.54	0.025 / 0.023	0.012 / 0.017	0.39 / 0.32

(The values above the bars are calculated, and the values below the bars are experimental.)

The calculated concentrations of carbon, silicon, manganese, and phosphorus are sufficiently close to the experimental values for the corresponding melting regimes (Table 8.14). According to the calculations, the sulfur concentration in the metal decreases from 0.02 to 0.012 wt.% over the entire length of the ingot. In the real process lowering of the extent of desulfurization of the metal is observed over the course of the melt, probably due to the accumulation of SiO_2 in the slag as it enters with the glass. This is reflected in the equilibrium constant of the desulfurization reaction of the metal, but was not taken into account in the model.

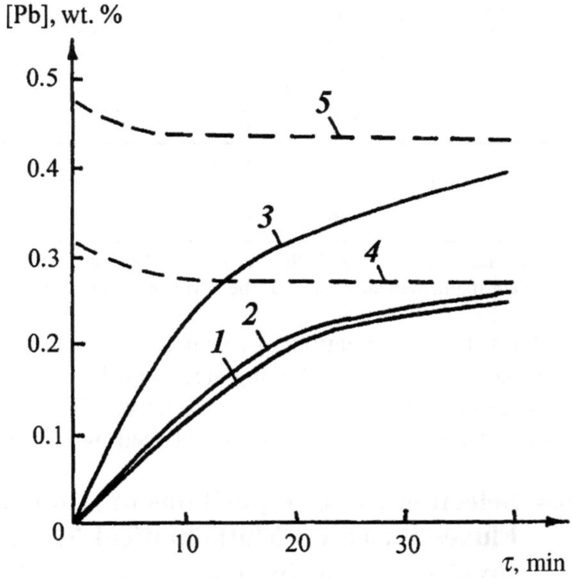

Fig. 8.5 Variation of the concentration of lead in the ingot during the ESR of grade 45 steel: *1*—$\dot{m}_{fl} = 1.7 \times 10^{-4}$ kg/s; *2*—$\dot{m}_{fl} = 1.8 \times 10^{-4}$ kg/s; *3*—$\dot{m}_{fl} = 2.8 \times 10^{-4}$ kg/s; *4*—$\dot{m}_{fl} = 1.7 \times 10^{-4}$ kg/s; $(PbO)_0 = 2$ wt.%; *5*—$\dot{m}_{fl} = 2.8 \times 10^{-4}$ kg/s; $(PbO)_0 = 3$ wt.%

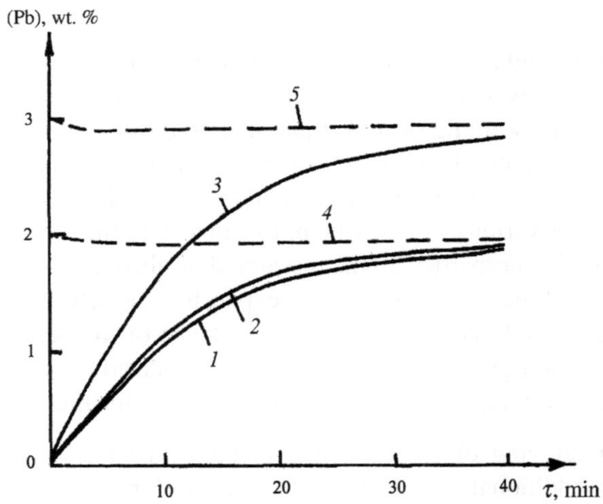

Fig. 8.6 Variation of the concentration of PbO in the slag pool during the ESR of grade 45 steel: *1*—$\dot{m}_{fl} = 1.7 \times 10^{-4}$ kg/s; *2*—$\dot{m}_{fl} = 1.8 \times 10^{-4}$ kg/s; *3*—$\dot{m}_{fl} = 2.8 \times 10^{-4}$ kg/s; *4*—$\dot{m}_{fl} = 1.7 \times 10^{-4}$ kg/s; $(PbO)_0 = 2$ wt.%; *5*—$\dot{m}_{fl} = 2.8 \times 10^{-4}$ kg/s; $(PbO)_0 = 3$ wt.%

In Table 8.15 the results of calculations performed with the aid of Eqs. 8.29, 8.30, and 8.33 and with the use of a general mathematical model of the chemical processes are compared with experimental data.

It can be seen that the best agreement with experiment was obtained when the complete mathematical model of the chemical processes was used. However, in some cases (when no computer is available etc.) it is not difficult to quickly obtain an estimate of the lead content in an ingot from Eq. 8.29 or 8.33, as well as from Eq. 8.30 when data for (PbO) are available.

Table 8.15 Calculated and experimental lead content in an ESR ingot

$\dot{m}_{fl} \cdot 10^4$ (kg/s)	[Pb] (wt.%)				
	Equation number			Complete model	Experiment
	(7.29)	(7.30)[a]	(7.33)		
1.7	0.31	0.30	0.29	0.26	0.25
1.8	0.33	0.32	0.31	0.27	0.26
2.8	0.49	0.50	0.48	0.39	0.32

[a] When Eq. 8.30 was used, the values of (PbO) were taken from calculations based on the complete mathematical model for the corresponding values of $\dot{m}_{fl}$and the melting time τ

Thus, the investigations performed enabled us to recommend introducing the method considered for alloying steel with lead during ESR by feeding lead-containing glass into the slag pool at the Izhstal Corporation. The flux additive for the ESR of lead-containing steels is protected by a patent [17].

8.4 Selecting the Compositions of Alloying and Ceramic Fluxes Based on Mathematical Modeling of the Processes Involved in the Interaction of the Metal and the Slag during Welding

The mathematical model developed (Sect. 5.2) has been used to select the compositions of ceramic fluxes for the wear-resistant surfacing of steels [18], as well as for the welding of nickel and nickel alloys [19].

A block diagram of the calculation algorithm is presented in Fig. 8.7.

Parts that operate under conditions of abrasive wear are welded and surfaced using various fluxes, which, together with the electrode wire, ensure the required service properties. In the research described in this section, the problem of creating a welding flux based on inexpensive, readily available materials for ensuring increased hardness and wear resistance of the deposited metal and good welding technology properties was solved. The stated purpose can be achieved by selecting a flux composition which would provide for the following:

- alloying of the deposited metal with vanadium and titanium, as well as the additional introduction of carbon to form carbides of these elements;
- better stirring of the weld pool;
- relatively low viscosity and good fluidity of the slag;
- good gas permeability of the slag and detachability of the slag crust.

The use of a high-carbon ferromanganese smelting slag, a vanadium converter slag, carbon black, and aluminum powder as the starting materials was proposed.

A high-carbon ferromanganese smelting slag (HCFSS) that is similar in composition (Table 8.16) to high-manganese high-silicon fluxes is added for the purpose of obtaining a flux with the required technological properties. It raises the arcing stability, improves the formation of the deposited metal, and promotes good detachability of the slag crust.

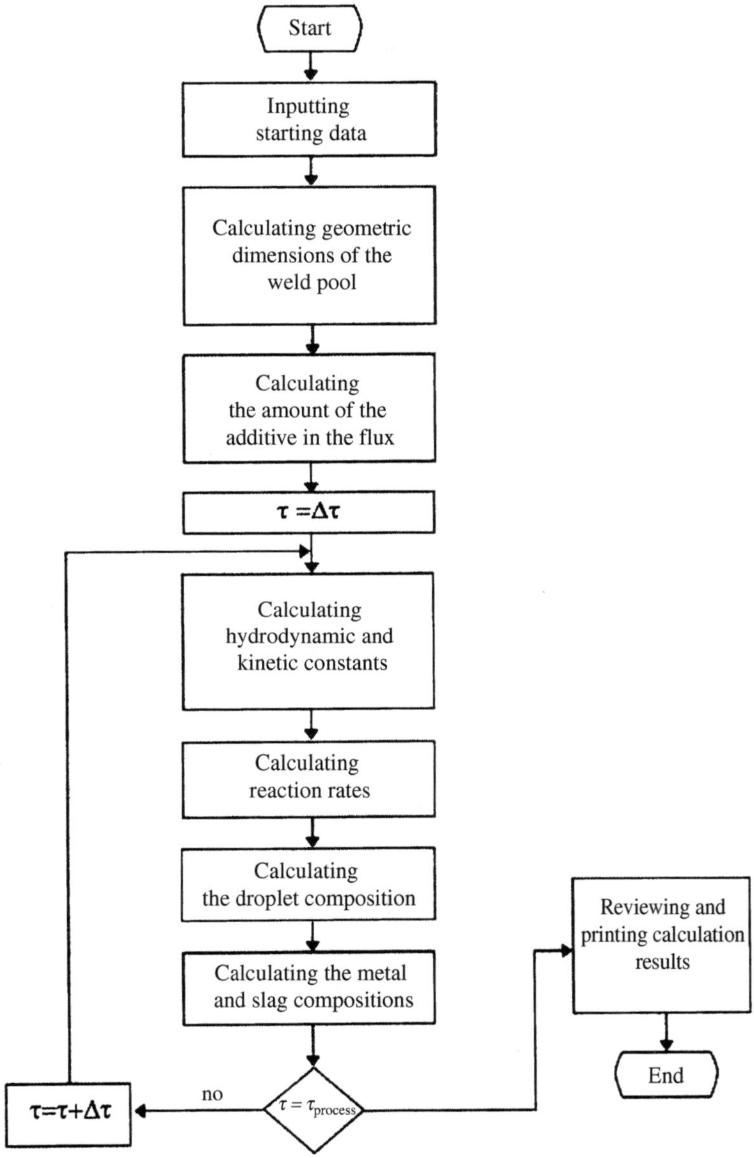

Fig. 8.7 Simplified diagram of the calculation algorithm

Vanadium converter slag (VCS) (Table 8.16) is added to the composition of the slag for the purpose of alloying the deposited metal with vanadium and titanium in order to increase the hardness and wear resistance.

Type GLS-1 carbon black (C) (State Standard 17022-81) is added to the composition of the flux as a reducing agent for obtaining vanadium and titanium from oxide compounds, as well as for forming the corresponding carbides.

Table 8.16 Chemical composition of slags (wt.%)

No.	Slag	SiO_2	CaO	Al_2O_3	MgO	FeO	MnO	V_2O_5	NiO	S	P
1	High-carbon ferrochromium smelting slag	29–36	9–38	4–5	1.5–3	0.2–0.8	17–39	–	–	0.1–1.0	0.05–0.1
2	Vanadium converter slag	15.4–21.0	1.3–3.0	0.4–1.0	1.2–2.0	34–40	5.9–9.0	14.0–18.5	6–11	≤ 0.05	≤ 0.05

Table 8.17 Flux compositions (wt.%)

Composition No.	Vanadium converter slag	High-carbon ferrochromium smelting slag	Carbon black	Aluminum powder
1	33.0	62.0	2.0	3.0
2	27.0	55.5	2.5	5.0
3	43.0	43.0	4.0	10.0
4	50.0	30.0	5.0	15.0
5	55.0	22.0	6.0	17.0

Table 8.18 Ratios between the flux components

Composition No. from Table 8.17	$\frac{HCFSS}{VCS}$	$\frac{VCS}{C + AP}$	$\frac{AP}{C}$
1	1.9	6.6	1.5
2	1.5	4.9	2.0
3	1.0	3.1	2.5
4	0.6	2.5	3.0
5	0.4	2.4	2.8

When oxides of alloying elements react with carbon black, carbon monoxide is generated, which imparts high porosity and gas permeability to the slag crust. In addition, the adsorption of gaseous carbon monoxide dissolved in the metal and its removal from the reaction zone can occur in such a system.

Type PA-4 aluminum powder (AP) (State Standard 6058-73) is added to the composition of the flux as a reducing agent. The aluminum oxide formed also imparts the necessary viscosity and gas permeability to the slag.

The objective of the calculation part of the work was to select ratios between the components which should ensure, on the one hand, the required concentrations of the alloying components in the deposited and, on the other hand, minimum losses of the reducing agents and the alloying components.

Calculations were performed for the following deposition regime: electrode wire, Sv-08A with a diameter equal to 3 mm; base metal, grade 3 steel; welding current 280–320 A; arc voltage, 24–28 V; surfacing speed, 16 m/h; wire feed rate, 50 m/h; reversed-polarity direct current.

The slag temperature $T_{slag} = 1{,}823$ K [20], the droplet temperature $T_{dr} = 2{,}500$ K [21], and the weld pool temperature $T_{po} = 2{,}070$ K [21] were taken for calculations of the reaction rates.

The results of the calculations show that from the standpoint of adding components to the composition of the deposited metal, the concentrations of the components can be varied over fairly broad ranges, wt.%:

- high-carbon ferromanganese smelting slag, 30.0–55.0;
- vanadium converter slag, 37.0–50.0;
- carbon black, 2.5–5.0;
- aluminum powder, 5.0–15.0.

Table 8.19 Results of investigations

Characteristic	Fluxes					
	ANK-19	Compositions from Table 8.17				
		1	2	3	4	5
Hardness, HRC	56	50	57	60	59	55
Wear resistance coefficient	1.0	1.1	1.2	1.4	1.3	1.1
Arcing stability	Good	Satisfactory	Good	Good	Good	Satisfactory
Slag crust detachability	Good	Satisfactory	Good	Good	Good	Satisfactory
Presence of defects in the metal	Does not exceed the permissible defect concentration range					

At the same time, the mechanical properties of the deposited metal and the technological characteristics of the flux can be determined only after experimental surfacing operations are performed.

Special experiments were performed to test the stated recommendations, as well as to determine the mechanical characteristics of the deposited metal and the technological properties of the flux.

Five flux compositions, including some with component concentrations that exceed the ranges determined by the calculation, were prepared (Table 8.17). The ratios between the individual components that influence the properties of the flux in these compositions are presented in Table 8.18.

As a result of a series of surfacing experiments under fluxes with compositions 1–5 (Table 8.17) under the regimes described above, it was established that the optimal alloying of the weld metal with vanadium and titanium, as well as the required technological properties of the flux and mechanical characteristics (Table 8.19), are achieved when the fluxes corresponding to variants 2–4 are used.

An additional advantage of the proposed flux is the fact that it contains inexpensive components and no scarce alloying additives are employed, making it economic and accessible for preparation and use. In addition, it may be noted that a deposited metal with high hardness and wear resistance is obtained, even when the surfacing is carried out with an ordinary low-carbon welding wire.

A patent was obtained for this flux [18].

When nickel and its alloys are welded, transition metal carbides, whose components can pass into the welded joint, are added to the welding materials by various methods to improve the mechanical and anticorrosion properties of the weld metal. At the same time, there is a definite range of concentrations of carbides, below which their influence is scarcely manifested, and when it is exceeded the carbides begin to separate as an independent phase in the metal and improve its properties. In particular, when electrodes were developed for welding nickel in [22], it was recommended that 10–15 wt.% TiC should be added to the composition of the coating to realize the objectives indicated. As a result, the concentration of carbon in the weld metal is about 0.05 wt.%, and the concentration of titanium is 0.11–0.13 wt.%.

Table 8.20 Composition of ceramic fluxes

Component	Content of components in the flux mixture (wt.%)				
	0	1	2	3	4
Marble	14	14	14	14	14
Alumina	17	17	17	17	17
Fluorspar	69	68	67	66	63
Titanium carbide	0	1	2	3	6

To develop the composition of a ceramic flux for welding NP-2 nickel, which contains titanium carbide, the composition of the weld metal was estimated using a mathematical model in [23].

The compositions of the ceramic fluxes used in the calculations are presented in Table 8.20.

The $CaO–CaF_2–Al_2O_3$ system, into which the appropriate amounts of titanium carbide were introduced, was taken as a basis.

When ceramic fluxes whose composition includes marble are used, the reactions that occur when the flux is heated and melted must be taken into account:

$$CaCO_3 = CaO + CO_2, \tag{8.40}$$

$$TiC + 3CO_2 = TiO_2 + 4CO, \tag{8.41}$$

$$TiC + 2O_2 = TiO_2 + CO_2. \tag{8.42}$$

Reactions (8.41) and (8.42) apparently develop fairly well, since, according to the data in [24], the extent of oxidation of the flux components when the $CaCO_3$ content is ≈ 14 wt.% is $\sim 30\%$ in the presence of atmospheric oxygen and $\sim 70\%$ in the presence of the atmosphere which is created upon the dissociation of marble.

We will make some assumptions:

- all the marble dissociates when the flux is heated and melted, and the corresponding quantity of CaO enters the slag weld pool;
- the titanium from the carbide is oxidized to TiO_2 due to reactions (5.75) and (5.76).

On this basis, it can be expected that the slag will contain the following components (without consideration of the binders): CaO, CaF_2, TiO_2, Al_2O_3, and TiC. Water glass in an amount equal to $\sim 15\%$ of the mass of the dry mixture is used to bind the flux components.

These data enabled us to obtain the composition of the slag at the time of formation of the slag pool and to use the mathematical model developed to predict the composition of the deposited metal.

Calculations were performed for the following surfacing regimes: welding wire diameter, 4 mm; welding current, 525 ± 5 A; arc voltage, 30 ± 2 V; welding speed, 22 ± 1 m/h; wire feed rate, 65 ± 2 m/h; current, reversed-polarity direct current.

In the calculations of the reaction rates, as in [25], the following process was regarded as the main process:

Table 8.21 Composition of the welding materials (wt.%)

Material	Ni	Cu	Fe	Ti	Cr	Mn	Si	C
NP-2 nickel	99.5	<0.1	<0.1	0.02	0.1	0.05	0.15	–
Welding wire	99.98	<0.1	<0.1	–	–	0.05	0.15	<0.1

Table 8.22 Chemical composition of the weld metal

Flux No. (Table 8.20)	Concentrations of components (wt.%)		
	C	Ti	Si
0	$\frac{0.02}{0.02}$	–	$\frac{0.05}{0.05}$
1	$\frac{0.02}{0.035 - -0.043}$	$\frac{0.02}{0.02}$	$\frac{0.07}{0.072 - -0.085}$
2	$\frac{0.06}{0.054 - -0.068}$	$\frac{0.02}{0.03 - -0.05}$	$\frac{0.07}{0.095 - -0.120}$
3	$\frac{0.09}{0.070 - -0.094}$	$\frac{0.04}{0.05 - -0.07}$	$\frac{0.11}{0.118 - -0.155}$
4	$\frac{0.16}{0.126 - -0.173}$	$\frac{0.13}{0.10 - -0.14}$	$\frac{0.26}{0.185 - -0.280}$

The values above the bars are calculated, and the values below the bars are experimental

$$1/m[E_i] + (NiO) = 1/m(E_iO_m) + [Ni]. \tag{8.43}$$

The original compositions of the welding materials are presented in Table 8.21.

Taking into account the composition of the welding materials, we can assume that mainly the concentrations of Ti, C, and Si will change when surfacing is carried out under ceramic fluxes containing titanium carbide.

Calculations were performed for TiC particles of two sizes, namely, 0.06 and 0.1 mm.

The results obtained are presented in Table 8.22.

In accordance with the data in [22], it may be assumed that the most promising results can be expected when compositions 3 and 4 are used (Table 8.20).

To test these assumptions, nickel plates were welded without the edge section on a copper base under fluxes that were similar in composition to the fluxes used in the calculations.

The results of the chemical analysis of the deposited metal are presented in Table 8.22.

A comparison of the calculated and experimental data reveals close agreement between them.

In addition, mechanical testing of the welded joints and a metallographic investigation of the joints were performed. The results are presented in Table 8.23.

An analysis of the testing results reveals that the addition of TiC to the flux composition has a positive effect on the properties of the welded joints. The optimal concentrations of TiC in the ceramic flux mixture are in the range from 3.0 to 6.0 wt.%.

A Russian Federation patent was obtained for the flux developed [19].

Table 8.23 Mechanical properties of the weld metal after welding and heat treatment

Property		TiC content in the flux mixture (wt.%)				
		0	1	2	3	6
Ultimate strength, σ_B, MPa						
After welding		300	345	375	390	390
After annealing:	473 K 12 h	300	360	390	425	450
	673 K 12 h	370	380	400	450	460
	873 K 12 h	360	370	380	450	460
	1,073 K 12 h	360	365	370	445	460
Impact viscosity, d_M, kg m/cm^2						
After welding		2.0	3.0	30.0	34.0	24.0
After annealing:	473 K 12 h	2.5	4.0	35.0	35.0	30.0
	673 K 12 h	3.0	5.0	36.0	36.0	32.5
	873 K 12 h	25.0	10.0	38.0	38.0	35.0
	1,073 K 12 h	25.0	5.0	36.0	38.0	30.0
Angle of bend, α, deg						
After welding		60	30	120	180	180
After annealing:	473 K 12 h	90	45	180	180	180
	673 K 12 h	180	60	180	180	180
	873 K 12 h	180	90	180	180	180
	1,073 K 12 h	180	90	180	180	180
Hardness, HRA						
After welding		26	35	29	30	33
After annealing:	473 K 12 h	25	32	27	30	33
	673 K 12 h	22	27	25	27	30
	873 K 12 h	23	22	27	29	31
	1,073 K 12 h	20	24	30	27	30
Grain size, S, μm^2						
After welding		36.0	35.5	33.0	32.0	26.5
After annealing:	473 K 12 h	38.0	40.0	35.5	33.0	31.0
	673 K 12 h	42.0	41.0	37.0	35.0	33.5
	873 K 12 h	42.5	42.3	41.5	38.0	34.5
	1,073 K 12 h	47.0	47.5	45.0	43.5	37.5

8.5 Practical Implementation of the Concept of Creating Advanced Welding Materials

8.5.1 Design, Preparation, and Testing of an Experimental Batch of Surfacing Materials

The new approach to designing welding materials was tested in the Materials Science Center of the Ariel University Center of Samaria (Israel).

8.5.1.1 Setting up the Experimental Studies

The problem was to ensure the reliable operation of the walls of the hopper of a loading device made from grade 20 K steel under the conditions of impact-abrasive wear caused by the action of a chemically inert material. One of the possible solution variants is to deposit a special material on the surface of the hopper walls.

The proposed approach included formulating a technical specification for designing a new welding material, in which the following were defined:

- the purpose and principal characteristics of the material being designed;
- the expected set of functions and properties of the design object;
- the area of application;
- the boundary values of the ranges of the parameters of the surfacing process;
- the characteristics of the surfacing equipment employed;
- the desirable relative wear resistance of the deposited metal compared to the material being protected, whose wear resistance was taken as unity;
- the design route;
- the set of output technical documentation.

The first stage of designing is analysis of the external environment and the nature of the loading of the working body. It is realized in an operator–expert-system interactive mode with the generation of a package of input data for the expert system as a result of each stage. In the second stage an analysis of the working body, i.e., its composition, set of properties, geometric characteristics, design features, and more, is performed in an interactive mode. A package of input data regarding the working body is created in parallel. The third stage determines the nature of the interaction of the working body and the external environment on the basis of the package of input data from the first two stages and the editing actions performed by the system operator. Knowledge regarding this interaction, which is recorded in a knowledge base and can be used by the system for subsequent designs, is created, and a model of the interaction with descriptions of the processes taking place is constructed [18, 26]. In the system developed this knowledge base contains a model of the knowledge that has been created on the basis of the results of the fundamental research of Russian and non-Russian scientists regarding the fitness of machine parts and designs under conditions of abrasive and impact wear [26–41] and establishment of a relationship between the structure and the wear resistance of the metal [28, 30, 35]. As a result of this stage it was determined that in order to protect the walls of a vessel made from grade 20 K steel from heavy wear under the conditions of abrasive wearing caused by the combined effects of dry friction, microcutting, and impact loading, the internal surface of the hopper walls must be protected by applying a protective layer of a metal that is capable of withstanding such combined effects. It is expedient to implement the application of a coating using electric-arc surfacing. A special welding material must ensure the formation of a deposited metal that has the base structure of finely acicular or acicular

Table 8.24 Predicted structure and properties of the deposited metal as a result of secondary crystallization

Code number of the deposited metal	Structural components (wt.%)			Hardness, HRC	Load hardening, ΔHRC
	Martensite	Austenite	Carbides		
RI 1/01	55.52	33.97	10.47	56.16	+ (3–4)
RI 1/11	44.94	45.29	9.74	46.19	+ (4–5)
RI 1/08	77.12	13.65	9.20	56.39	+ (2–3)
RI 1/10	81.16	9.76	9.05	57.23	+ (2–3)

martensite and metastable austenite (30–50 wt.%) with carbides uniformly distributed in it (5–15 wt.%).

Under heavy impact loading, metastable austenite absorbs part of the impact energy and is transformed into martensite, which strengthens the deposited metal and improves its wear resistance. An alloying system, in which the region of the martensitic transformation is displaced into the room-temperature range ($M_s < 120°C$, $M_f < 0°C$), was calculated [42–53]. It permits maintenance of a considerable amount of austenite supersaturated with alloying elements, including elements that form substitutional solid solutions with austenite (Ti, C, and N), until completion of the crystallization of the deposited metal, which makes the austenite unstable under plastic deformation [51]. Thus, the austenite decomposes under impact loading or microcutting into martensite and carbides, i.e., additional hardening of the deposited metal or an increase in its wear resistance occurs.

This representation of the required secondary structure of the deposited metal enables us to move on to the calculation of the primary structure and chemical composition of the weld metal. The calculation involves solving an inverse problem, in which the input parameters are the secondary structure of the weld metal and values of the parameters of the thermal-straining cycle of the welding process. The calculation is performed in an expert–expert-system interactive mode and is based on the phenomenological model developed for the nonequilibrium secondary crystallization of the weld metal, the accompanying processes that lead to the formation of secondary carbides, and a mathematical description of the new phase diagram. The input data regarding the secondary structure of the calculated welding materials are presented in Table 8.24.

The results of the calculation of the primary structure and chemical composition of the deposited metal for four welding materials designed are presented in Tables 8.25 and 8.26, respectively.

The information obtained regarding the required primary structure and chemical composition of the deposited metal (Tables 8.25 and 8.26) were used to determine the electrode formulas of the special welding materials that were designed.

The electrode formula contains complete information regarding the structure, properties, and functions of the welding material that were designed, the list of starting materials, the manufacturing technology (route), and the recommended

Table 8.25 Predicted structure of the deposited as a result of primary crystallization

Code number of the deposited metal	Structural components (wt.%)		
	Martensite	Austenite	Carbides
RI 1/01	–	Base	6.1
RI 1/11	–	Base	5.3
RI 1/08	–	Base	5.1
RI 1/10	–	Base	4.8

welding (surfacing) regimes. The calculation is performed in an operator–expert-system interactive mode. The basis of the calculation is a package of applications that was developed on the basis of a mathematical model of a technological welding process [54–64] and a phenomenological model of the primary non-equilibrium crystallization of a weld metal.

To test the results obtain, we performed a calculation that involved solving the direct problem, i.e., determining the chemical composition, primary structure, secondary structure, and service characteristics of the weld metal.

The design results, that is, the electrode formulas of the welding materials, serve as the input parameters for the calculation.

8.5.1.2 Technological Experiments

The next stage of the experimental work was to prepare an experimental batch of specialized welding materials in accordance with the technological requirements.

Flux-cored wires were prepared from a cold-rolled steel ribbon (08kp steel) with a 12 × 0.5 mm cross section on a special laboratory wire-drawing machine. The control system of the machine can be connected to a computer system in order to communicate the parameters of the manufacturing route for flux-cored wires with the possibility of on-line correction. The required mixture of powered materials (the flux mixture) for the core of the flux-cored wires was prepared in a laboratory mixer. The composition of the mixtures prepared for each flux-cored wire corresponded to the calculation. All the wires had a diameter of 2.0 mm with closing of the steel ribbon in a butt joint. Altogether, 4 kg of flux-cored wires (1 kg of each) were prepared.

To carry out the surfacing of a model workpiece, a special apparatus which provides for the performance of surfacing operations on a vertical wall (the actual conditions for surfacing the hopper of the loading device) was designed and fabricated. Thus, one of the requirements of the new approach for developing specialized materials, that is, the performance of trial and experimental studies under conditions that closely approximate the full-scale conditions, was satisfied. The surfacing was carried out in three layers on a model workpiece (a vertical sheet of grade 20 K steel with a thickness of 16 mm) in accordance with the recommended surfacing regimes. Samples measuring 80 × 35 mm [their thickness was equal to the thickness of the sheet (16 mm) plus the thickness of the deposited layer (5–6 mm)] were cut from the model workpiece. The cutting was

Table 8.26 Predicted chemical composition of the deposited metal

Code number of the deposited metal	Concentration of elements (wt.%)					Remarks
	C	Si	Mn	Cr	Ti	
RI 1/01	1.58	0.68	0.35	8.54	3.61	The remainder is Fe.
RI 1/11	1.62	0.70	0.38	6.89	3.60	The remainder is Fe.
RI 1/08	1.31	0.53	0.38	7.87	2.57	The remainder is Fe.
RI 1/10	1.32	0.70	1.12	7.93	2.62	The remainder is Fe.

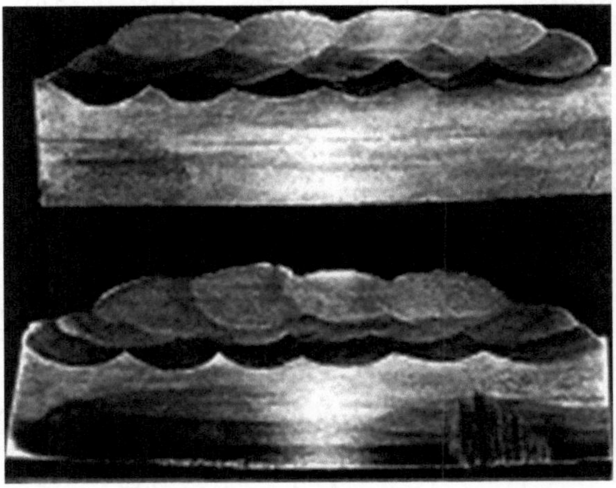

Fig. 8.8 Macroscopic transverse sections of samples RI 1/01.7 and RI 1/11.2

performed in such a way that the direction of the axes of the samples would coincide with the surfacing direction for some of the samples and would be perpendicular to it for others. The first samples were used for x-ray structural analysis, chemical analysis, and hardness measurements. The samples in the second group were used to perform metallographic investigations and full-scale testing. Templates for studying both transverse sections and the surface of the upper (third) layer of the deposited metal were cut from the samples obtained (Fig. 8.8).

8.5.1.3 Metallographic Investigations and Full-Scale Testing

The chemical composition of the deposited metal in the samples was determined in the upper deposited layer. Table 8.27 compares the expected (calculated) composition and the experimentally determined chemical composition of the deposited layers.

To determine the wear rate of the deposited layers, they were subjected to full-scale testing on a laboratory testing bench that simulates the real picture of the wear of hopper walls. Samples surfaced with the expensive high-alloy wear-resistant Vautid-100 alloy (from Vautid Ltd., Germany), as well as samples of

Table 8.27 Chemical composition of the deposited metal in the samples investigated

Code number of the sample	Concentration of elements (wt.%)					Remarks
	C	Si	Mn	Cr	Ti	
RI 1/01.7	1.58	0.68	0.347	8.54	3.61	The remainder is Fe
	~1.40	0.554	0.427	8.50	3.35	
RI 1/11.2	1.62	0.70	0.378	6.89	3.60	The remainder is Fe
	~1.80	0.29	0.447	6.63	2.63	
RI 1/08.3	1.306	0.53	0.385	7.87	2.57	The remainder is Fe
	~1.50	0.60	0.362	8.15	2.10	
RI 1/10.1	1.319	0.70	1.12	7.93	2.62	The remainder is Fe
	~1.40	0.54	0.60	5.62	2.28	

Note The values above the bars are calculated, and the values below the bars are experimental

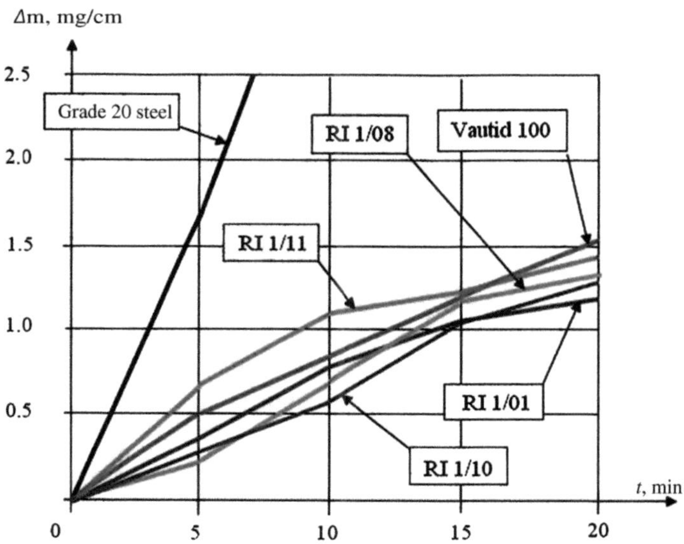

Fig. 8.9 Mass loss from samples as a function of the testing time

20 K steel, were tested simultaneously. According to the operating principle of the test bench, an abrasive material (quartz sand) is continuously supplied to the sample during testing, and a rotating steel pipe made from grade 20 steel, which periodically strikes the sample with a frequency of 25 Hz, serves as the counterbody.

Figure 8.9 presents plots of the dependence of the mass loss of the samples on the duration of the testing. It can be seen from plots presented that the surfacing materials developed provide for the formation of a deposited metal that is resistant to the effects of impact-abrasive wear. The wear rate of the deposited layers is 5–6 times lower than that of 20 K steel and is equal to more than 80% of the wear

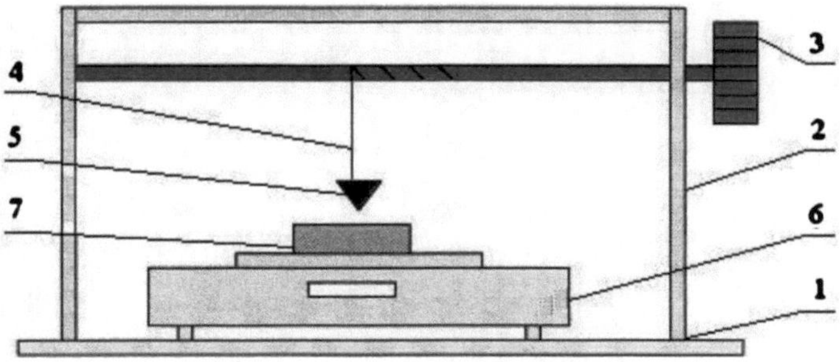

Fig. 8.10 Laboratory apparatus for determining the pull force of a magnetic sensor: *1*—base, *2*—frame, *3*—microstep motor, *4*—steel thread, *5*—magnetic sensor, *6*—digital dynamometer, *7*—sample under investigation

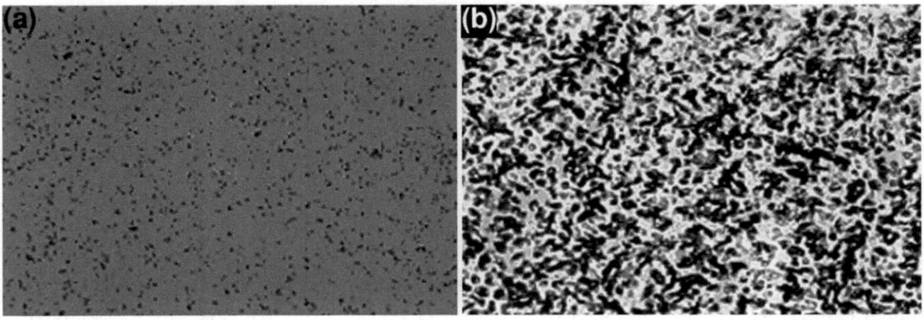

Fig. 8.11 Microstructure of the third deposited-metal surface layer of alloy RI 1/01 in its original state after surfacing: **a**—polished section not subjected to etching (×100), **b**—etching in 4% nital (×200)

rate of layers deposited by the expensive high-alloy (with more than 32 wt.% Cr) welding material Vautid-100.

To investigate the phase changes in an alloy after impact loading, samples were subjected to impact loading in a special laboratory device. The impact loading is performed by a striker with an energy equal to 13.5 J, which acts on the deposited surface of the sample through a metal sphere with a diameter of 4.7 mm made from 100Cr6 steel (DIN, Germany) with a hardness equal to 67 HRC.

The phase changes ($\gamma \rightarrow \alpha$ transformations) at the site of impact loading were evaluated from the change in the pull force of a magnetic needle sensor before and after impact loading. A diagram of the corresponding laboratory apparatus is presented in Fig. 8.10. The operating principle of the apparatus is clear from the figure presented and its description.

Figures 8.11 and 8.12 show the microstructure of a deposited surface layer of alloy RI 1/01. The polished section in Fig. 8.11a was not etched, but strengthening phases of the carbide type were manifested in relief on it. The results of computer

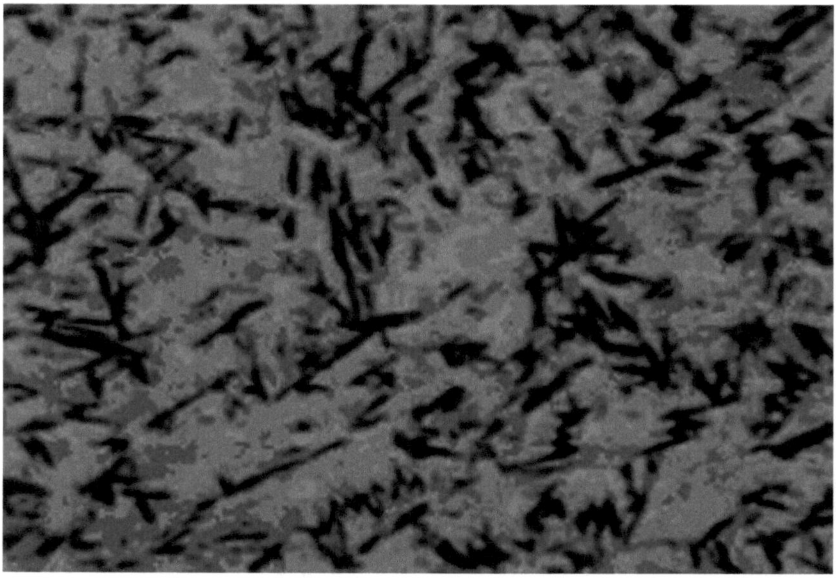

Fig. 8.12 Microstructure of the third deposited-metal surface layer of alloy RI 1/01 after impact loading (×200)

Table 8.28 Carbide content in alloys according to estimates of the SIAMS-600 software

Global parameters	RI 1/01	RI 1/11	RI 1/08	RI 1/10
Number of fields analyzed	10	10	10	10
Area analyzed (μm^2)	1.77×10^6	1.53×10^6	1.30×10^6	1.29×10^6
Number of particles	2.65×10^3	2.20×10^3	1.70×10^3	1.70×10^3
Volumefraction(vol.%) / Mass fraction (wt.%)	18.85 / 12.01	16.21 / 10.02	12.97 / 8.03	12.96 / 8.01

evaluation of the carbide content are presented in Table 8.28. Finely acicular martensite with a large amount of residual austenite is observed in the structure of the deposited metal (Fig. 8.13).

The metallographic investigation did not reveal the martensite structure with sufficient contrast, but the presence of the α phase was demonstrated by x-ray structural analysis.

In Fig. 8.12 a change in the structure of the metal after impact loading can be discerned. Acicular formations of martensite are clearly visible on slip planes within the austenite grains.

Summarized data on the change in the pull force of the magnetic sensor (an increase in the pull force of the magnetic sensor indicates a decrease in the content of the nonmagnetic γ phase in the deposited layer, i.e., occurrence of the $\gamma \rightarrow \alpha$ transformation under the action of impact loading) are presented in Table 8.29.

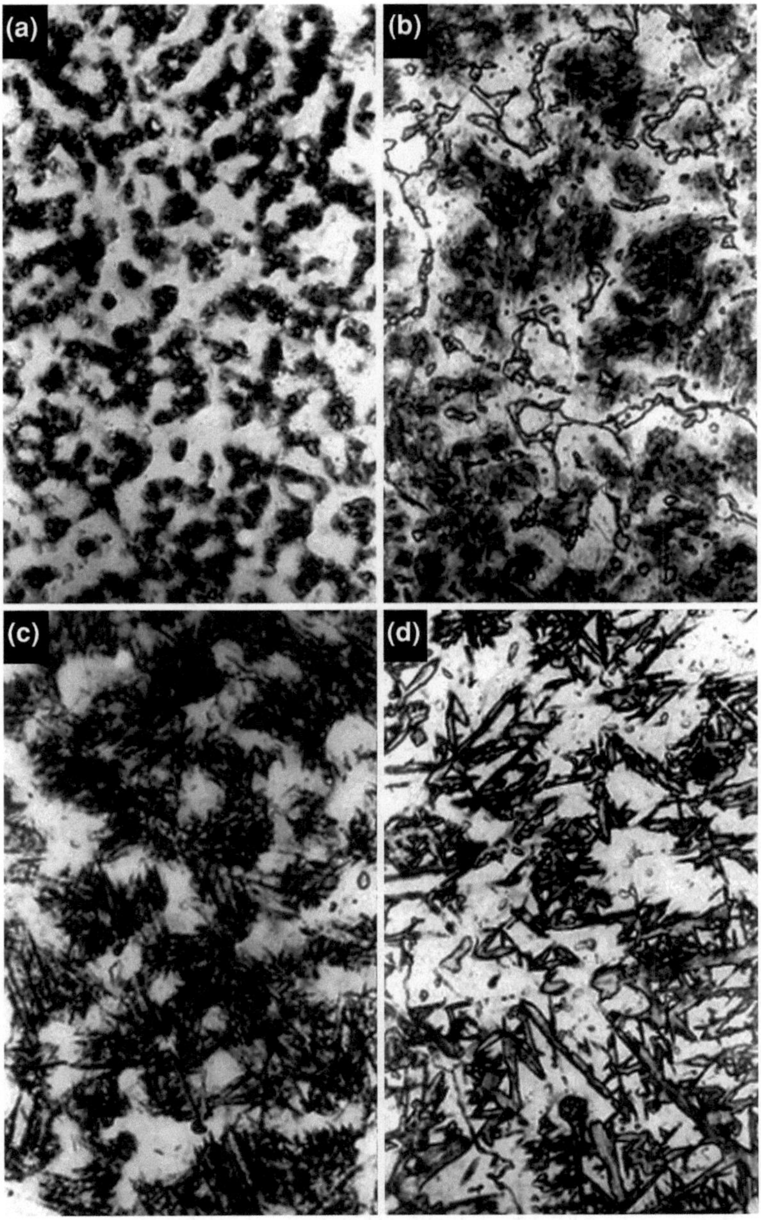

Fig. 8.13 Structure of surfaced samples (×800): **a**—RI 1/01.7, **b**—RI 1/11.2, **c**—RI 1/08.3, **d**—RI 1/10.1

It can be seen that alloy RI 1/11, which contains the largest amount of residual austenite (approximately 47%), provides for the most effective hardening. The hardness of the deposited metal as it was after surfacing and after impact

Table 8.29 Structure and properties of the deposited metal

Code number of the sample	Structural components (wt.%)			Hardness, HRC	Load hardening, ΔHRC	Change in the pull force of the magneto sensor, N	Integral width of the (110) x-ray line α	Microhardness HV (MPa)
	Martensite	Austenite	Carbide phase					
RI 1/01.7	55.52	33.97	10.47	56.16	+(3 – –4)	+0.050	53	7200–7400
	base	36.00	12.00	56 – –57	~(+4)			
RI 1/11.2	44.94	45.29	9.74	46.19	+(4 – –5)	+0.070	67	5500–6300
	43.00	47.00	10.00	46 – –50	~(+6)			
RI 1/08.3	77.12	13.65	9.20	56.39	+(2 – –3)	+0.020	47	6900–7200
	base	11.00	8.00	54 – –56	~(+3)			
RI 1/10.1	81.16	9.76	9.05	57.23	+(2 – –3)	+0.015	49	7100–8000
	base	8.00	8.00	55 – –58	~(+2)			

Note. The values above the bars are calculated, and the values below the bars are experimental

loading was studied using a Rockwell–Wilson hardness tester. Table 8.29 presents calculated and experimental values of the hardness of the deposited layers as they were after surfacing and the increases in hardness (ΔHRC) after impact loading.

The microstructure of the deposited metal was studied by standard metallographic methods on an Axiolab optical microscope after etching microsections in a 4% ethanolic solution of HNO_3. The microhardness on the surface of the microsections was determined in parallel under a load of 100 g by a Buehler microhardness tester. The phase composition of the deposited metal was investigated on a DRON-0.5 x-ray structural diffractometer in Fe Kα radiation. Profiles of the (110) line of the α phase and the (111) line of austenite were recorded. The ratio between the intensities of these lines was analyzed to determine the quantity of residual austenite.

After surfacing, all the samples had a structure that is typical of the cast deposited metal, namely, a cellular dendritic structure. The nature of the structures of the samples investigated is presented in Fig. 8.13. Because of the high carbon content and the presence of alloying elements (the total content of the alloying elements did not exceed 12%), which inhibit the decomposition of austenite in the 600–400°C temperature range, as well as the high cooling rate (50–70 K/s), the $\gamma \rightarrow \alpha$ transformation occurred in the metal with the resultant predominant formation of a martensitic matrix structure with a varying amount of residual austenite, depending on the chemical composition, and carbide phases, which were distributed relatively uniformly in the matrix. Finely acicular martensite and residual austenite were observed in all the samples (Fig. 8.13a, c, and d).

The presence of residual austenite in the structure is related to the position of the range for the martensitic transformation, which, in turn, depends on the chemical composition. The martensite point M_s for these alloys lies above the room-temperature range, and the temperature of the end of the martensite transformation lies below this range.

Sample RI 1/11.2 (Fig. 8.13b) is distinguished from the other three samples in this batch by having the maximum content of residual austenite (approximately 47%). This is attributed to the fact that it contains a considerable amount of carbon (approximately 1.8% according to the analysis) and a relatively small amount of carbide-forming elements compared with sample RI 1/01. Thus, the percentage of free carbon that is not bound in carbides is higher in it than in the other samples. This is reflected in the structure of martensite, which, as follows from the integral width of the (110) x-ray lines of the α phase (see Table 8.29), has a tetragonal structure. In addition, the morphology of martensite in this sample differs from the morphology of martensite in the other samples (tiny flat needles). In this sample the martensite needles have a lenticular shape, and the martensite crystals are joined in frameworks (see Fig. 8.13b), which are characteristic of alloys containing a large amount of nickel. If the fact that the austenite-forming ability of carbon is 30 times higher than that of nickel [49, 50, 65–68] is taken into account, such a picture is fully explicable. Summarized data from the metallographic investigations, x-ray structural analysis, and measurements of the hardness and microhardness of the samples are presented in Table 8.29.

The quantitative carbide content was determined using the SIAMS-600 image analysis software. The program provides for the sharp separation of any graphical image that is represented in a digital format. Scanning a field of vision provides for the integration of contrasting image elements of the same type that have common structural characteristics, coloring, and dimensions. As a result of the scanning, an integral estimate of the area analyzed over all the fields of vision is generated in the form of a total number of similar particles, and the area occupied by these particles is converted into the volume fraction of these particles in the overall pseudovolume analyzed. With the aid of such an analysis, it is possible to reveal the structural components of the carbide phase (to separate the primary and secondary carbides and obtain an integral estimate of their volume fraction). In all the samples analyzed the total content of the carbide phase determined by the SIAMS-600 software was close to the predicted content that was calculated using the proposed approach (Table 8.28). Image analysis shows that, despite the fact that the titanium content in the samples investigated is 2–3 times smaller than the chromium content, the percentage of primary TiC carbides amounts to 60–70% of the total fraction of the carbide phase, confirming the proposed conception of carbide formation in alloys. Titanium carbides are also clearly visible in a metallographic investigation according to their characteristic edged shape and pink coloring.

On the basis of an analysis of the results of the investigations and full-scale testing, it was determined that all the materials developed can be used to apply a protective coating, since they provide a resistance of the deposited layer to wear and impact loading that is significantly (5–6 times) greater than the analogous characteristics for 20 K steel. However, the optimal properties are exhibited by alloy RI 1/01, which provides the lowest wear rate, the highest hardness, a high extent of occurrence of the $\gamma \rightarrow \alpha$ transformation, and significant relative hardening after impact loading, as is indicated by the large change (increase) in the pull force of the magnetic sensor (see Table 8.29).

8.5.2 Creation of Advanced Welding Materials

The method developed was tested for creating real welding and surfacing materials.

The first and extremely important endeavor in this area was the creation of electrodes for applying layers with anticavitation properties based on Ni_3Al alloys. The development work was carried out under scientific grants from the Israeli Ministry of Science and the Commission for Science of the Peoples Republic of China (PRC). A special feature of this order was the fact that alloys of the $Ni_3Al(Ti)$ type have not only carbide strengthening, but also dispersion strengthening, which required making some adjustments to the model. As a result, five types of surfacing electrodes were created, and they were used successfully both for welding an $Ni_3Al(Ti)$ alloy and for cladding the hydroturbine rotors of hydroelectric power plants on the Yangtze and Huang He Rivers in China. First,

Ni_3Al alloy electrodes for welding and surfacing that eliminate the cracking of the alloy caused by the nonuniform precipitation of the γ' phase under the action of the thermal-straining cycle during welding were created [69, 70].

On the basis of an agreement between the General Motors Corporation (USA) and the Ariel University Center of Samaria (Israel), several types of flux-cored wires for welding galvanized AISI 1008/1010 sheet steels were designed and fabricated with strict requirements for ensuring high welding speeds, high plasticity and strength of the metal and for eliminating harmful gas and smoke emissions. It followed from the statement of the problem that the deposited metal must contain a minimum of strengthening phases and that an additional factor, namely, ensuring the retention of zinc in the zone near the joint, is imposed. The problem was successfully solved using mathematical modeling [60]. Flux-cored wire RI 6/139 provides for a welding speed for the steel indicated as high as 90 m/h (the welding speed usually used in General Motors plants is 20–40 m/h) without harmful gas and smoke emissions. The deposited metal has an ultimate strength equal to 5,560 kg/mm^2 and an angle of bend of a longitudinal sample with a welded joint equal to 180° (the requirement of the corresponding American standard is 120°).

On the basis of an agreement for scientific cooperation between the research department of the tool manufacturer Hanita Metal Works Ltd. (Israel) and the Ariel University Center of Samaria (Israel), flux-cored wires based on a low-cost alloy for restoring high-speed tools were designed and fabricated. The company's requirement was that a deposited metal with characteristics no poorer than those of heat-treated R6M5 tool steel must be obtained. The series of wires developed provide for the formation of a deposited metal (which does not require subsequent heat-treatment) with properties on the level of R6M5 steel. An additional property of the wires developed is that adhesion of the material being processed on the cutting edge of a tool coated by these wires has been eliminated.

All the materials developed have good welding-technological properties and exhibit high values for the service characteristics of the deposited metal.

8.6 Conclusion

The results of computational experiments based on mathematical models developed for optimizing existing technological processes and materials and for developing new ones have been presented in this chapter.

References

1. Zinigrad, M.I., et al.: Improving the technology for remelting tungsten-containing scrap based on mathematical modeling. Izv. Vyssh. Uchebn. Zaved. Chern. Metall. **12**, 59–64 (1991)
2. Bornatskii, I.I.: Theory of Welding Processes (in Russian), 288 pp. Kiev (1978)
3. Sokolov, G.A.: Ladle Refinement of Steel (in Russian), 208 pp. Metallurgiya, Moscow (1977)

4. Shalimov, M.P., et al.: Improving the technology of the ladle treatment of steel using mathematical modeling. In: Theory and Practice of the Ladle Treatment of Steel (in Russian), pp. 128–130. Moscow (1985)
5. Boronenkov, V.N., et al.: Mathematical modeling of the interaction of a metal and a slag in a ladle. Izv. Vyssh. Uchebn. Zaved. Chern. Metall. **1**, 36–41 (1983)
6. Boronenkov, V.N., Shanchurov, S.M.: Mathematical model of the kinetics of chemical processes that occur during the electroslag welding of steel. Avtom Svarka **6**, 22–27 (1985)
7. Shanchurov, S.M., Boronenkov, V.N., Yakovlev, V.V.: Predicting the composition of the weld metal during the electroslag welding of press columns. In: Theory and Practice of Industrial Welding (in Russian), pp. 9–14. Sverdlovsk (1986)
8. Boronenkov, V.N., Shanchurov, S.M.: Mathematical model of the interaction processes of a multicomponent metal and a multicomponent slag during electroslag remelting, Ural Polytechnical Institute, Article deposited in Chermetinformatsiya, vol. 700. Sverdlovsk (1979)
9. Medovar, B.I. (Ed.): Thermal Processes in Electroslag Remelting (in Russian), 304 pp. Naukova Dumka, Kiev (1978)
10. Boronenkov, V.N., Shanchurov, S.M.: Method for determining the parameters of the mass transfer between the metal and the slag during electroslag remelting. In: Physicochemical Investigations of Metallurgical Processes (in Russian), No. 11, pp. 89–95. Sverdlovsk (1983)
11. Mikhailov, G.G., Tanklevskaya, N.M.: Phase equilibria in the Fe–Si–C–O system during the crystallization of steel. In: Physicochemical Investigations of Metallurgical Processes, vol. 10, pp. 73–79. Sverdlovsk (1982)
12. Boronenkov, V.N., et al.: Mathematical modeling of the interaction processes of the metal and the slag during the electroslag remelting of an alloy steel. In: Physicochemical Investigations of Metallurgical Processes, (in Russian), No. 9, pp. 104–111. Sverdlovsk (1981)
13. Boronenkov, V.N., et al.: Application of the mathematical modeling of the interaction processes of the metal and the slag during the electroslag remelting of an alloy steel. In: Fourth All-Union Scientific Conference on Advanced Problems in the Electrometallurgy of Steel: Abstracts of Reports (in Russian), pp. 112–113. Chelyabinsk (1960)
14. Teacd, M.: Procese de oxidare in eursul relopirii electrice suo zgura. Posidilitatl de diminuare a piarderilor de titan la retopirla otelurilor inoiidabile. Gere Met **23**, 167–175 (1982)
15. Kulikov, I.S.: Thermodynamics of Oxides (in Russian), 344 pp. Metallurgiya, Moscow (1986)
16. Grigoryan, V.A., Belyanchikov, L.N., Stomakhin, A.Ya.: Theoretical Principles of Steelmaking Processes (in Russian), 272 pp. Metallurgiya, Moscow (1987)
17. Boronenkov, V.N., Brusko, I.E., Shanchurov, S.M., Afans'ev, N.D.: USSR Patent No. 1556111, Flux additive for remelting lead-containing steels, (not for publication in print) (1987)
18. Il'inskii, V.I., et al.: USSR Patent No. 1352828, Int Cl V23 K 35/36. Ceramic flux for wear-resistant surfacing, Application No. 4012423, 28 January 1986
19. Ignatov, M.N., et al.: Russian Patent No. 1808591 Int Cl B23 K 35/36. Material for welding nickel, Application No. 4886711, 30 November 1990; published 15 April 1993. Bulletin No. 14
20. Kirdo, I.V.: Thermal balance during submerged-arc welding. In: Collected Papers on Automatic Submerged-Arc Welding (in Russian), vol. 1, pp. 95–128. Kiev (1948)
21. Paton, B.E., (ed.): Technology of Electric Fusion Welding of Metals and Alloys (in Russian), 432 pp. Moscow (1977)
22. Ignatov, M.N.: Improving the quality of welded joints of nickel and alloys based on it by using transition metal carbides in electrode coatings, 145 pp. Candidate Dissertation, Perm' (1987)
23. Shalimov, M.P., Ignatov, M.N., Kazantsev, S.A.: Calculation of the composition of the weld metal during the welding of nickel under a ceramic flux containing titanium carbide. In: New Welding and Surfacing Materials and their Application in Industry (in Russian), pp. 70–73. St. Petersburg (1992)
24. Bagryanskii, K.V.: Submerged-Arc Welding and Surfacing with Ceramic Fluxes (in Russian), p. 184. Kiev (1976)

25. Davydov, A.K.: Physicochemical laws governing the formation of copper foil and optimization of the composition of the working surface of the cathode, 146 pp. Candidate Dissertation, Sverdlovsk (1987)
26. Kragel'skii, I.V., Dobychin, M.N., Kombalov, V.S.: Foundations of Friction and Wear Calculations (in Russian), 526 pp. Mashinostroenie, Moscow (1977)
27. Tkachev, V.N.: Fitness of Machine Parts during Abrasive Wear (in Russian), 336 pp. Moscow (1995)
28. Rybakova, L.M., Kuksenova, L.I.: Structure and Wear Resistance of Metal (in Russian), 212 pp. Mashinostroenie, Moscow (1982)
29. van Vlack, L.H.: Textbook of Materials Science. Addison-Wesley, Reading (1973)
30. Razikov, M.I., Tolstov, I.A.: Handbook for a Surfacing Unit Foreman (in Russian), 200 pp. Moscow (1966)
31. Meshkov, Yu.Ya.: Physical Foundations of the Failure of Steel Structures (in Russian), 240 pp. Naukova Dumka, Kiev (1981)
32. Bayer, R.G.: Prediction of wear in a sliding system. Wear **11**, 319–332 (1968)
33. Makhnenko, V.I.: Calculation Methods for Investigating the Kinetics of Welding Stresses and Strains (in Russian), 320 pp. Naukova Dumka, Kiev (1976)
34. Bitter, J.G.: A study of erosion phenomena. Wear **6**(1), 5–21 (1963)
35. Buckley, D.H., Johnson, R.L.: The influence of crystal structure and some properties of hexagonal metals on friction and adhesion. Wear **11**(6), 405–419 (1968)
36. Broek, D.: Elementary Engineering Fracture Mechanics, 437 pp. Sijthoff & Noordhoff, Alphen aan den Rijn (1978)
37. Gilmore, R.: Catastrophe Theory for Scientists and Engineers, p. 666. Wiley, New York (1981)
38. Goddard, G., Wilman, H.: A theory of friction and wear during the abrasion of metals. Wear **5**(2), 114–135 (1962)
39. Kragelsky, I.V., Bronovectz, M.A.: On the nature of friction during impact. In: First European Tribology Congress, Conference Publ. 18, pp. 109–118. Institute of Mechanical Engineers, London (1973)
40. Rabinowicz, E.: Friction and Wear of Materials, 244 pp. Wiley, New York (1965)
41. Neale, M.J. (ed.).: Tribology Handbook. Butterworths, London (1973)
42. Grachev, S.V., Baraz, V.R.: Heat-Resistant Corrosion-Resistant Spring Steels (in Russian), 144 pp. Metallurgiya, Moscow (1979)
43. Babich, V.K., Gul', Yu.P., Dolzhenkov, I.E.: Strain Aging of Steel (in Russian), p. 320. Metallurgiya, Moscow (1972)
44. Baraz, V.R.: Strain-aging austenitic steels. Metalloved Term Obrab Met **29**(10), 17–20 (1987)
45. Bogatov, A.A., Mezhiritskii, O.I., Smirnov, S.V.: Plasticity Margin of Metals during Pressure Shaping (in Russian), p. 144. Metallurgiya, Moscow (1984)
46. Beresnev, B.I., Ezerskii, K.I., Trushin, E.V., Kamenetskii, B.I.: High Pressures in Modern Technologies of Material Treatment (in Russian), p. 245. Nauka, Moscow (1988)
47. Gridnev, V.N., Gavrilyuk, V.G., Meshkov, Yu.Ya.: Strength and Plasticity of Cold-Strained Steel (in Russian), 232 pp. Naukova Dumka, Kiev (1974)
48. Lysak, L.I., Nikolin, B.I.: Physical Fundamentals of the Heat Treatment of Steel (in Russian), p. 304. Tehnika, Kiev (1975)
49. Lifshits, L.M., et al.: Physical Metallurgy of Welding (in Russian), 386 pp. Mashinostroenie, Moscow (1988)
50. Vinokurov, V.A., (ed.).: Welding in Mechanical Engineering: A Handbook (in Russian), vol. 3, 567 pp. Mashinostroenie, Moscow (1979)
51. Beres, L., Balogh, A., Irmer, W.: Welding of martensitic creep-resistant steels. Weld. J. **80**, 191–195 (2001)
52. Akulov, A.I. (ed.): Welding in Mechanical Engineering: A Handbook (in Russian), vol. 2, 462 pp. Mashinostroenie, Moscow (1978)
53. Baraz, V.R., Gladkovskii, S.V., Ishina, E.A.: Plastic Deformation of Steels with Metastable Austenite: Phase Transformations, Microstructure Evolution, and Mechanical Properties (in Russian), 282 pp. Moscow (1996)

54. Zinigrad, M., Mazurovsky, V., Gafri, O.: Development of the new welding materials made of waste products, In: Welding Conference, pp. 63–67, Israel (1994)
55. Zinigrad, M., Mazurovsky, V.: Modeling of metallurgical welding processes for the purpose of creating new welding materials. In: Proceedings of the International Conference "Modeling and Simulation in Metallurgical Engineering and Materials Science" (MSMM'96), pp. 669–673, Beijing, China (1996)
56. Zinigrad, M., Mazurovsky, V., Aksyutin, O.: Creating new welding materials on the basis of metallurgical processes modeling. In: Proceedings of the 7th International Conference on Computer Technology in Welding, pp. 291–297, USA (1997)
57. Zinigrad, M., Zalomov, N., Mazurovsky, V., Aksyutin, O.: Computer approach to the development of welding materials. In: Proceedings of the 8th Biennial International Conference: Computer Technology in Welding, pp. 185–192, Liverpool, UK (1998)
58. Zinigrad, M., Mazurovsky, V.: Computer modeling of metallurgical technologies. In: Proceedings of the International Conference on Computer Technology in Welding, pp. 164–171, Michigan, USA (1999)
59. Zinigrad, M., Mazurovsky, V., Zinigrad, A.: Mathematical modeling of phase interaction taking place during fusion welding processes. In: Proceedings of the Yazawa International Symposium, pp. 667–680, San Diego, USA **1** (2003)
60. Zinigrad, M., Mazurovsky, V., Shumyakov, V., Zinigrad, A: Optimization of the composition of a flux-cored wire for welding galvanized steel based on computer simulation of the welding process. In: Proceedings of the 10th International Conference "Computer Technology in Welding," pp. 340–47, Copenhagen, Denmark, Session (2000)
61. Zinigrad, M., Mazurovsky, V.: Development of new welding materials on the base of mathematical modeling of metallurgical processes. Part 1. Phase interaction analysis and development of the basic model. In: Proceedings of the Bi-National Russia–Israel Workshop, pp. 277–291, Ekaterinburg, Russia (2002)
62. Zinigrad, M., Mazurovsky, V.: Development of new welding materials on the base of mathematical modeling of metallurgical processes. Part 2. Development of solution algorithm and software. In: Proceedings of the Bi-National Russia–Israel Workshop, pp. 292–303, Ekaterinburg, Russia (2002)
63. Zinigrad, M., Mazurovsky, V., Zinigrad, A.: The development of electrode coating compositions basing on the mathematical modeling of physicochemical processes on metal–slag boundary. In: Proceedings of the Mills Symposium, pp. 545–553, London, UK (2002)
64. Mazurovsky, V., et al.: The phenomenological model of non-equilibrium crystallization and strengthening-phase-formation processes in the weld. In: Proceedings of the Bi-National Israeli–Russian Workshop, pp. 155–167, Jerusalem, Israel (2003)
65. Schaeffler, A.L.: Constitution diagram for stainless steel weld metal. Met Prog **56**(5), 680–680b (1949)
66. Zaks, I.A.: The Schaeffler constitution diagram and the chromium–nickel equivalent. Svarochn Proizv **8**, 44–46 (1957)
67. Hume-Rothery, W., Raynor, G.V.: The Structure of Metals and Alloys (The Institute of Metals Monograph and Report, Series 1), p. 391. Institute of Metals, London (1956)
68. Zhitnikov, N.P., Zaks, I.A.: Influence of nitrogen on the structure of an austenite weld metal. Svarochn Proizv **8**, 5–7 (1971)
69. Zinigrad, M., et al.: Modeling of the build up process for production of Ni-Al based high temperature wear resistant coatings. In: Proceedings of ASM Materials Solution'99: International Symposium on Steel for Fabricated Structures, pp. 230–233, Cincinnati, Ohio, USA (1999)
70. Feng, Di., et al.: Micro-features of the CE surface of a Ni3Al-based alloy. In: Proceedings of International ASTM Symposium on Hydraulic Failure Analysis: Fluids, Components, and System Effects, pp. 172–179, Reno, Nevada, USA (1999)

Index

M. Zinigrad et al., *Phase Interaction in the Metal–Oxide Melts–Gas System*, Engineering Materials, DOI: 10.1007/978-3-642-22377-8,

MIX
Papier aus verantwortungsvollen Quellen
Paper from responsible sources
FSC® C105338

If you have any concerns about our products,
you can contact us on
ProductSafety@springernature.com

In case Publisher is established outside the EU,
the EU authorized representative is:
Springer Nature Customer Service Center GmbH
Europaplatz 3, 69115 Heidelberg, Germany

Printed by Libri Plureos GmbH
in Hamburg, Germany